ICoSI 2014

T. Taufik · Indira Prabasari
Innaka Ageng Rineksane
Rizal Yaya · Retno Widowati
Sri Atmaja Putra Rosyidi
Slamet Riyadi · Puji Harsanto
Editors

ICoSI 2014

Proceedings of the 2nd International Conference on Sustainable Innovation

Springer

Editors
T. Taufik
Department of Electrical Engineering
California Polytechnic State University
San Luis Obispo, CA
USA

Retno Widowati
Faculty of Economics
Universitas Muhammadiyah
Yogyakarta
Indonesia

Indira Prabasari
Faculty of Agriculture
Universitas Muhammadiyah
Yogyakarta
Indonesia

Sri Atmaja Putra Rosyidi
Faculty of Engineering
Universitas Muhammadiyah
Yogyakarta
Indonesia

Innaka Ageng Rineksane
Faculty of Agriculture
Universitas Muhammadiyah
Yogyakarta
Indonesia

Slamet Riyadi
Faculty of Engineering
Universitas Muhammadiyah
Yogyakarta
Indonesia

Rizal Yaya
Faculty of Economics
Universitas Muhammadiyah
Yogyakarta
Indonesia

Puji Harsanto
Faculty of Engineering
Universitas Muhammadiyah
Yogyakarta
Indonesia

ISBN 978-981-13-5701-5 ISBN 978-981-287-661-4 (eBook)
DOI 10.1007/978-981-287-661-4

Printed on acid-free paper

This Springer imprint is published by Springer Nature
The registered company is Springer Nature Singapore Pte Ltd.
The registered company address is: 152 Beach Road, #21-01/04 Gateway East, Singapore 189721, Singapore

Advisory Committee

Dr. Idris F. Sulaiman, Australian National University, Australia

Contents

Part I
Agriculture

Application of Organic Fertilizer and *Plant Growth-Promoting Rhizobacteria* (PGPR) to Increase Rice Yield and Quality

Oktavia S. Padmini, Sri Wuryani and Ruly Aryani

Abstract Generally, the application of high doses of chemical fertilizer continuously causes leveling production and decreasing rice quality. The objective of this research was to examine the effect of organic fertilizer and PGPR on rice yield and quality. The experiment was conducted at Bener Village, Ngrampal subdistrict, Sragen district, Indonesia. The experiment was arranged into randomized completely block design with three replicates. The main plot was application dose of organic fertilizer consisting of two treatments, i.e., 1 and 2 tons/ha. The subplot was spraying frequency of *plant growth-promoting rhizobacteria* (PGPR), consisting of four treatments, i.e., without PGPR and spraying PGPR three times, four times, and five times. The data were subjected to analysis of variance followed by Duncan's multiple range test. Plants treated with organic fertilizer of 1 ton/ha and sprayed with PGPR four times resulted in the highest production and rice quality.

Keywords Organic fertilizer · PGPR · Rice yield and quality

1 Introduction

Increased rice production has become one of the Indonesian agricultural development priorities [1, 2]. Conventional agriculture, which is oriented only to increasing maximum yield using high doses of chemical fertilizers, has continuously resulted in environmental degradation. It shows that a decrease in soil fertility resulted in lower organic matter content and nutrient use efficiency. Farmers suffer the consequences of this environmental impact, among others that plants do not respond to fertilization, though the doses are increased, and the increase in plant productivity is not comparable with the addition of fertilizer intake [3–5]. Reliance on chemical fertilizers as a source of nutrients, besides tending to reduce land productivity,

O.S. Padmini (✉) · S. Wuryani · R. Aryani
Faculty of Agriculture, University of Pembangunan Nasional
"Veteran" Yogyakarta, Jl. SWK 104, Yogyakarta, Condongcatur, Indonesia
e-mail: oktaviasarhesti@yahoo.com

T. Taufik et al. (eds.), *ICoSI 2014*, DOI 10.1007/978-981-287-661-4_1

decreases rice quality, e.g., rice structure becomes hard due to high amylose content and low amylopectin [6].

Beneficial soil microorganisms are categorized as growth stimulants as well as natural fertilizers that play a mayor roles in (1) suppressing disease growth (bioprotectant); (2) producing phytohormone (biostimulant); (3) increasing the availability of nutrient for plants (biofertilizer); (4) reducing organic matter; (5) stabilizing soil aggregate; (6) remodeling agrochemical compounds. The reduction in chemical fertilizers by applying organic fertilizers available on site and has not been used optimally is expected to optimize the balance between the components of biotic and abiotic soil, give the nutrient availability in the soil, which in turn ensures the sustainability of land productivity and improves the quality of rice.

The basic principle of organic farming is the best maintenance of soil and how to give life in the soil [7]. Organic farming systems aim to improve the health of agroecosystems, biodiversity, and biological activity [8]. Organic and semi-organic farming are able to increase the protein content of seeds and the quality of rice taste compared with conventional farming. Organic farming systems over the past three seasons tend to increase the content of amylopectin, but not significantly different in amylase content [6]. Organic fertilizers for three planting seasons produce amylopectin content significantly higher, whereas the amylase content is lower than that of NPK fertilization [3]. The higher the amylopectin content, the more fluffy and sticky the rice gets, whereas the higher the amylase content, the drier or harder the rice gets [9].

Plant growth-promoting rhizobacteria (PGPR) are beneficial bacteria that are in the area of plant roots and enhance plant growth by various mechanisms. PGPR usage continues to increase in an effort to replace chemical fertilizers, pesticides, and supplements [10]. PGPR is a bacterium that can be found in rhizosphere, a thin layer of soil that surrounds the roots of plants, and provide a positive effect on the plant. PGPR contain bacteria such as Pseudomonas, Enterobacter, Azospirillum, Azotobacter, Burkholderia, Bacillus, and Serratia [10, 11]. The bacteria must be able to defend themselves (survive) from the inoculation into seeds, multiplication in spermosphere (the area around the seed) in response to seed exudates, attack on the root surface, and colonize/inhabit the root zone to strengthen the root system. Application of organic fertilizers combined with PGPR is expected to substitute for the use of NPK fertilizers as well as to improve the quality of rice. The objective of this research was to examine the effect of organic fertilizer and PGPR on rice yield and quality.

2 Materials and Methods

The experiment was conducted at Bener Village, Ngrampal, Sragen Regency. The research method used was Split Plot with a plot size of 6 m $\times$ 10 m.

The experiment tools and materials were Mekongga variety rice, fermented cow manure, decomposer, PGPR, analytical scales, and oven.

The main plot was the use of manure that consisted of two treatments:
Use of organic fertilizer of 1 ton/ha,
Use of organic fertilizer of 2 tons/ha.
The subplot was application of PGPR that consisted of four treatments:
Without PGPR,
Three times PGPR spraying,
Four times PGPR spraying,
Five times PGPR spraying.

The data were analyzed to determine the variability and the influence of each factor of both the main plot and the subplot with Duncan's multiple range test (DMRT) on the significance level of 5%.

3 Results

3.1 Growth Components

Table 1 shows that the number of leaves and plant dry weight were significantly higher in the treatment of 2 tons/ha of organic fertilizer application with four times PGPR spraying compared with the application of 1 and 2 tons/ha of organic fertilizers with three times PGPR spraying and without PGPR spraying. Both were not significantly different from the application of 1 ton/ha organic of fertilizer with four and five times PGPR spraying.

The number of tillers was better in the treatment of 2 tons/ha organic fertilizer application with four times PGPR spraying and 1 ton/ha of organic fertilizers with five times PGPR spraying compared to the application of 1 and 2 tons/ha organic fertilizer with three times PGPR spraying and without PGPR spraying, although not significantly different from the number of tillers with 2 tons/ha of organic fertilizer application with three and five times PGPR spraying and 1 ton/ha of organic fertilizer application with four times PGPR spraying.

3.2 Yield Component

Table 2 shows that the rice yield components were significantly higher in the treatment of 2 tons/ha of organic fertilizer application with four and five times PGPR spraying compared with the application of 1 and 2 tons/ha of organic fertilizer with three times PGPR spraying and without PGPR spraying. Both were not significantly different from the application of 1 ton/ha of organic fertilizer with four and five times PGPR spraying.

Table 1 The effect of organic fertilizer and PGPR application on number of leaves, plant dry weight, and number of tillers

Treatment		Number of leaves	Dry weight of plant (g)	Number of tillers
Organic fertilizers 1 ton/ha	Without PGPR	73.13 c	50.03 d	18.93 c
	Three times PGPR spraying	75.67 b	90.03 b	21.00 bc
	Four times PGPR spraying	77.87 ab	84.20 bcd	23.53 ab
	Five times PGPR spraying	76.73 ab	73.67 cd	24.67 a
Organic fertilizers 2 tons/ha	Without PGPR	76.33 ab	57.57 d	20.20 bc
	Three times PGPR spraying	74.67 bc	62.90 cd	22.80 abc
	Four times PGPR spraying	79.43 a	89.20 abc	25.87 a
	Five times PGPR spraying	78. 99 a	99.10 a	21.87 abc
Interactions		(+)	(+)	(+)

Note Numbers followed by similar alphabet in each column show not significantly different in Duncan's multiple range test analysis at 95% level of confidence. (+) Positive interaction

Table 2 The effect of organic fertilizer and PGPR application on component of rice yield

Treatment		Number of panicle/clump	Length of panicle (cm)	Weight of grains/clump (g)	Weight of grain/plot (kg)
Organic fertilizers 1 ton/ha	Without PGPR	13.83 d	22.850 b	18.80 c	1.316 b
	Three times PGPR spraying	18. 87 b	23.00 b	23.96 b	1.367 b
	Four times PGPR spraying	20.16 ab	23.800 ab	24.70 ab	1.467 ab
	Five times PGPR spraying	20.33 ab	23.53 ab	25.27 ab	1.456 ab
Organic fertilizers 2 tons/ha	Without PGPR	14.83 d	22.83 b	19.53 c	1.333 b
	Three times PGPR spraying	17.87 c	22.83 b	20.33 bc	1.350 b
	Four times PGPR spraying	21.17 a	24.58 a	28.33 a	1.533 a
	Five times PGPR spraying	22.67 a	25.33 a	29.67 a	1.587 a
Interactions		(+)	(+)	(+)	(+)

Note Numbers followed by similar alphabet in each column show not significantly different in Duncan's multiple range test analysis at 95% level of confidence. (+) Positive interaction

Table 3 The effect of organic fertilizer and PGPR application on content of fat, protein, carbohydrate, and energy of rice

		Fat (%)	Protein (%)	Carbohydrate (%)	Energy
Organic fertilizers 1 ton/ha	Without PGPR	0.58 ab	8.99 d	74.71 ab	340.01 bc
	Three times PGPR spraying	0.58 ab	9.27 bc	74.71 ab	341.12 ab
	Four times PGPR spraying	0.52 bc	9.45 a	74.58 b	340.82 abc
	Five times PGPR spraying	0.52 bc	9.18 c	74.68 ab	340.10 abc
Organic fertilizers 2 tons/ha	Without PGPR	0.58 ab	8.79 e	74.91 ab	340.03 abc
	Three times PGPR spraying	0.61 a	8.84 e	74.61 ab	339.26 bc
	Four times PGPR spraying	0.23 d	9.47 a	74.74 ab	338.91 c
	Five times PGPR spraying	0.46 bc	9.41 ab	75.15 a	342.39 a
Interactions			(+)	(+)	(+)

Note Numbers followed by similar alphabet in each column show not significantly different in Duncan's multiple range test analysis at 95% level of confidence. (+) Positive interaction

Table 3 shows that the increase in protein content in the application of 1 ton/ha of organic fertilizer with three times PGPR spraying was followed by a decrease in fat and carbohydrate content. Increase in fat occurred in the application of 2 tons/ha of organic fertilizers with three times PGPR spraying, but not different from application of 1 ton/ha organic fertilizer with three times PGPR spraying and without PGPR in 1 and 2 tons/ha organic fertilizer. Increase in fat and carbohydrates occurred in the application of 2 tons/ha organic fertilizers with five times PGPR spraying compared with application of 1 ton/ha organic fertilizers with four times PGPR spraying, but not different significantly from others. The amount of energy was determined by the content of carbohydrates in rice as carbohydrate is the highest compound in rice.

Table 4 shows an increase in amylose content in the application of 2 tons/ha of organic fertilizer with five times PGPR spraying. Increased amylopectin occurred in the 1 ton/ha of organic fertilizer application with four and five times PGPR spraying and 2 tons/ha of organic fertilizer application with five times PGPR spraying. Amylum digest is significantly higher in the treatment of 2 tons/ha of organic fertilizer application with four and five times PGPR spraying compared with other treatment. Whiteness is significantly highest in the treatment of 2 tons/ha of organic fertilizer application with five times PGPR spraying.

Table 4 The effect of organic fertilizer and PGPR application on amylose and amylopectin content, amylum digest, and whiteness of rice

Treatment		Amylose	Amylopectin	Amylum digest	Whiteness
Organic fertilizers 1 ton/ha	Without PGPR	19.66 d	58.96 bc	35.46 d	56.61 g
	Three times PGPR spraying	20.15 c	57.17 c	37.46 c	56.43 g
	Four times PGPR spraying	19.61de	60.38 a	39.62 b	59.64 d
	Five times PGPR spraying	19.57 e	60.11 a	40.33 b	62.63 b
Organic fertilizers 2 tons/ha	Without PGPR	20.38 b	58.05 bc	35.57 d	57.49 e
	Three times PGPR spraying	17.58 g	59.66 b	39.55 b	58.97 d
	Four times PGPR spraying	19.14 f	59.56 b	42.25 a	59.49 c
	Five times PGPR spraying	22.28 a	61.00 a	42.38 a	62.91 a
Interactions		(+)	(+)	(+)	(+)

Note Numbers followed by similar alphabet in each column show not significantly different in Duncan's multiple range test analysis at 95% level of confidence. (+) positive interaction

4 Discussion

Application of 1 and 2 tons/ha of organic fertilizer into rice plants with 4 and 5 times PGPR spraying was not significantly affected, and they provided growth and yield of rice rather than without and 3 times PGPR spraying. Based on the analysis results, the soil was dominated by low to very low, so that the soil in the research area was considered as infertile. Analytical results on Vertisol, the soil used in the experiment, had a pH (H_2O) of 6.64 and pH (KCl) of 5.99. Specific gravity and weight volume were in the amount of 2.65 and 1.82 g/cm^3. Organic matter content was low (1.45%), cation exchange capacity was 23.75 me %, levels of N total were 0.09%, and CN ratio was 9. Levels of P available were 7.23 ppm, and K exchanged was 0.41 me% [12]. Improved results were obtained mostly due to PGPR spraying. However, PGPR spraying was able to provide nutrition to increase growth and yield of rice.

Plant growth-promoting rhizobacteria (PGPR) are beneficial bacteria at the root zone of plants contain a variety microbes that provide nutrient and growth-stimulant compound as well [11]. Soil microbes are utilized to maintain and improve soil fertility that is environmentally friendly. Good soil fertility will create good planting conditions, especially root growth, availability of macro- and micronutrients, and increasing microorganism activity [10]. Padmini and Irawati [13] in their research reported that rice plants treated with Custom-Bio sprays into the soil and soaking rice seedlings with PGPR increased the total microbial population in the soil, resulting in a population of solvent P microbial and N_2 fixation microbial, as well as improved the content of total N and total P in the soil, compared with without Custom-Bio and PGPR. Soaking rice seedling with PGPR also improves the growth and yield [14].

Microorganism found in PGPR was active in helping the plants growth by providing nutrients (N fixation, solvent P). The function of nitrogen increases plant growth and yield by increasing the number of seedlings and leaf area growth. Phosphate serves as constituting a source of energy for the metabolic processes in plants. Potassium serves as a transport medium that brings nutrient including available N and P from the roots to the leaves and uptake assimilates from the leaves to the whole plant tissues [15]. Phosphorus also serves in the generative growth, namely the process of primordial initiation, rice grain filling, and synchronized grain ripening. PGPR spraying repeatedly is necessary for the need of nutrient in every phase of plant growth. Phosphorus given at early growth stages have partial productive efficiency (PPE) for higher grain, because P is required for the seedling formation and total P requirement is less than N, if absorbed P at the beginning of the growth can be redistributed to growing organs easily [15].

The highest fat content was obtained on the applications 1 and 2 tons/ha of organic fertilizer with three times PGPR spraying, but not significantly different from that without PGPR spraying. This implies that the fat content is affected only by organic fertilizer from manure. Fat is basically a decomposition of carbohydrate that stored as energy reserves [16]. However, because rice is not a source of fat, its content does not affect the quality of the rice. On the other hand, the highest protein was obtained from 1 and 2 tons/ha organic fertilizer application with 4 times PGPR spraying. Manure is needed for microbial proliferation medium contained in PGPR including N fixation microbes that produce nitrogen for plant growth. Nitrogen is an essential element in protein synthesis.

The application of 1 ton/ha organic fertilizer with four times PGPR spraying reached 9.7% brought optimum protein content. Based on the research carried out by Wuryani et al. [17], this value is relatively higher than protein content in rice grown using chemical fertilizers which is only 8.8%. Obviously, using organic fertilizer is better to produce protein than using chemical fertilizers. Meanwhile, four times PGPR spraying also produce high carbohydrate content and not significantly different between applying 1 and 2 tons/ha of organic fertilizer. The implications of this treatment were evident also in the energy results, that indicate no significant different from the highest yield that was treated with 1 ton/ha of organic fertilizer application with four times PGPR spraying.

The application of 1 ton/ha organic fertilizer with four times PGPR spraying produced the highest amylopectin. Amylopectin was correlated positively with the level of rice fluffiness [9], and synergies with protein will give a nice flavor of rice. Instead, amylase makes rice hard (*pera* and undesirable) because the bonds between the simple sugars are straight and strong so it is more difficult to disentangle to be the simple sugars' components.

Therefore, the rice with high amylase content was more filling. It also deals with the nature of amylase which absorbs more water so that the rice becomes fluffy. Related to digestibility, the starch that digests easily is the starch that is hydrolyzed easily by the enzyme into smaller units (simple sugars). Amylopectin, as a type of starch carbohydrate polymer −1.4—α glycoside bond that branched with −1.6—glycoside bond, is an open bond so it is easier to unravel. Therefore, rice with high amylopectin content tends to have high digestibility, and as a result eating fluffier rice makes one feel hungry more easier. The data shown by the treatment of 1 ton/ha of manure and four times PGPR spraying showed high digestibility. These results support the proposed idea by Ikameilaty [18]. While the level of rice whiteness basically deals with the level of polishing, the higher the level of polishing the whiter the rice will get [19]. Recently, the consumer prefers non-white rice, so the results of this research that the treatment of producing white rice with a not optimal whiteness level do not significantly affect the quality of the rice.

5 Conclusion

Plants treated with organic fertilizer 1 ton/ha and sprayed with PGPR four times resulted in the highest production and rice quality.

References

1. Sukristiyonubowo, R., Wiwik, H., Sofyan A, Benito, H.P., De Neve, S. 2011. Change from conventional to organic rice farming system: biophysical and socio economic reasons . International Research Journal of Agricultural Science and soil Science1(5):172–182.
2. Padmini, O.S. 2013. Productivity, Soil Fertility, And Economic Benefit In Changes From Conventional To Organic Rice Farming System At Sragen District. International Conference on Green Agro-Industry. Faperta UPNVY Yogyakarta 11–13 Nov 2013
3. Padmini, O. S. 2009. Peran Residu Legum - Padi Dan Campuran Pupuk Organik-Anorganik Terhadap Hasil Padi Dan Kandungan Amilosa–Amilopektin Beras. Jurnal Ilmu Pertanian Agrivita terakreditasi B.Vol.31(2):167–178
4. Padmini, O. S., R.R.R. Brotodjoyo, Sri Wuryani, and Dwi Aulia, P. 2013. Increasing Rice Yield By Utilizing Sugarcane Waste And Liquid Organic Fertilizer. International Conference on Sustainable Agriculture and Environment ICSAE. Univ Sebelas Maret UNS 27–29 June 2013.
5. Deore, G. B., Limaye A.S., Shinde B.M.,andS.L. Laware.2010. Effect of Novel Organic Liquid Fertilizer on Growth and Yield in Chilli (*Capsicum annum*,L.). Asian J. Exp. Biol. Sci. Spl.2010:15–19.
6. Jiann-Feng Lee, Yung-Wu Chen, Shih-Shiung chen, Yi-Fong Tsai. 2004. Effects of Long Term Application of Rape Seed Mealon The Grain Quality of Rice. Department of Agriculture National Chung Hsin University.
7. Kosit, P. 2011. Kyusei Nature Farming and the adaptation of farmers in the Isan Region of Thailand. European Journal of Social Science 21(3):471–482.
8. Dubey, K and D.K. Dubey. 2010. Evaluation Of Cost Effective Organic Fertilizers. Research & Development Centre, Kilpest India Ltd., Govindpura, Bhopal, 462023, (M.P), India.
9. Houston, D. F. 1972. RICE, Chemistry and Technology. American Association of Cereal Chemists, Inc. Minnesota.

10. Ashrafuzzaman,M.,Farid Akhtar Hossen1, M.Razi Ismail, Md. Anamul Hoque, M. ZahurulIslam,S.M. Shahidullah, and Sariah Meon. Efficiency of plant growth-promoting rhizobacteria (PGPR) for the enhancement of rice growth. African Journal of Biotechnology Vol.8(7),pp. 1247–1252.
11. Khalimi H., Suprapta, D. N.,and Nitta. 2012. Effect of *Pantoea agglomerans* on growth promotion and yield of rice. Agricultural Science Research Journals Vol.2 (5); pp. 240–249.
12. Padmini, O. S. 2010. Peran Legum dan Pupuk Organik dalam Rotasi Tanaman Berbasis Padi di Vertisol Kabupaten Sragen.
13. Padmini, O. S. dan Endah B. Irawati. 2012. Aplikasi Custom-bio dan PGPR untuk Meningkatkan Populasi MikrobaTanah, Ketersediaan Unsur Hara dan Pertumbuhan Tanaman padi.
14. Padmini, O. S. 2011. Environmentally Friendly Rice Production Increased By Plant Growth Promoting Rhizo-bacteria (PGPR) To Develop Agrotourism. Proceeding Seminar International Sustainable Agriculture and Development Seminar, Yogyakarta, INDONESIA/UPN Veteran Yogyakarta.
15. Yoshida, S. 1981. Fundamentals of Rice Crops Science. International Rice Research Institute. Los Baños. Philippines.
16. Conn,E.C.,P. K. Stumpf,,G. Bruening and Roy H. Doi. 1987. Outlines of Biochemistry. John Wiley and Sons, Inc. Canada.
17. SriWuryani., O. S. Padmini and D.A. Puspitaningrum. 2012. Postharvest Quality Enhancement of Rice Var. Mekongga Response of Liquid Organic Fertilizer Application. Paper presented on Asia Pacific Symposium,18–20 September 2012 at Yogyakarta-Indonesia.
18. Ikameilaty. 2011. Laporan Evaluasi Nilai Gizi: Daya Cerna Pati. http://ikameilaty.wordpress.com (diakses tanggal 15 Januari 2013).
19. Sutrisno. 2007. Penanganan Pascapanen Padi diIndonesia. Jurnal Keteknikan Pertanian. Vol.21(2):1–9.

Clustering of High-Yielding and Early-Maturing Soybean Genotypes

M. Muchlish Adie and Ayda Krisnawati

Abstract Global warming has potentially increased the intensity of drought and pest population. Soybean with early maturing (<80 days) tends to be more tolerant to those stresses through escape mechanism. The objective of this research was to map and classify the seed yield and days of maturity of soybean lines, as well as selecting soybean lines for both high yield and early maturing. The experiment was conducted at Genteng Research Station in Banyuwangi, East Java. A total of 666 F5 soybean lines and ten of check cultivars (Argomulyo, Kaba, Panderman, Anjasmoro, G 511 H, Grobogan, G100H, Rajabasa, Malabar, and Burangrang) were identified and grouped based on their yield and days of maturity. The performance of 666 lines and ten soybean cultivars showed optimum agronomic growth. Flowering date of 666 lines ranged from 25 to 34 days (average of 28 days), days of maturity varied from 70 to 79 days (average of 75 days), weight of 100 seeds ranged from 10.44 to 26.26 g/100 seeds (average of 16.22 g/100 seeds), and the seed yield was 0.16–4.06 t/ha (average of 1.47 t/ha). Seed yield from ten check cultivars ranged between 1.29 and 3.79 t/ha (average 1.99 t/ha), and days of maturity has a diverse range from 73 to 78 days (average of 76 days). The result indicated an opportunity to obtain both high-yielding and early-maturing soybean lines. Most of the lines (39.79% or 265 lines) had days of maturity between 75 and 76 days and as many as 261 lines (39.18%) categorized as super-early maturing (<75 days). Based on cluster analysis, ten lines with seed yield between 3.2 and 4.06 t/ha were chosen. Two check cultivars, Anjasmoro and Malabar produced yield of 3.79 and 2.48 t/ha, respectively. Line of Arg//Ljtg/Sbg-12 had the highest yields of 4.06 t/ha with days of maturity 78 days, followed by G511/Mlgg 1029 (84)//M 1029 (84)///1029 (3.75 t/ha). There was a positive correlation between soybean seed yield and days of maturity. For Indonesia's tropical regions, it was suggested the maturing day of soybean to produce high yield is 74 days.

M. Muchlish Adie (✉) · A. Krisnawati
Indonesian Legumes and Tuber Crops Research Institutes,
PO Box 66, Malang, Indonesia
e-mail: mm_adie@yahoo.com

T. Taufik et al. (eds.), *ICoSI 2014*, DOI 10.1007/978-981-287-661-4_2

Keywords Seed yield · Days to maturity · *Glycine max* · Selection

1 Introduction

Soybean is known as the third most important food commodities in Indonesia after rice and maize. The widest area of soybean cultivation during dry season is at second planting time (dry season II) between June/July and September/October. The main obstacles of soybean cultivation in the dry season II are scarcity of water, increasing pest, and disease infestations. However, some of soybean varieties have tolerance on drought stress and pest attack [1] through the escape resistance mechanism.

Soybean is considered as a short-day plant which only flowering when the day length is shorter than the maximum critical value. Based on the maturing days, soybean is classified into two categories, i.e., based on days of flowering and days of maturity. Both categories are controlled by a single dominant gene. Various studies revealed that there were at least eight genes that controlled time of flowering and maturity in soybean: *E1* and *E2* [2], *E3* [3], *E4* [4], *E5* [5], *E7* [6], *E6* [7], and J [8] for long juvenility. Of these loci, E1, E3, E4, and E7 are involved in the response of flowering to artificially induced long day length [3, 4, 6, 9]. The E3 locus controls the insensitivity to fluorescent long day length obtained by extending natural day length to 20 h using cool white fluorescent lamps with a high red to far-red quantum (R:FR) ratio [3]. The dominant alleles of all eight genes determine late flowering and maturity. Flowering represents the transition from the vegetative to the reproductive phase in plants. Various external cues, such as photoperiod and temperature, are known to initiate plant flowering under the appropriate seasonal conditions. Of these cues, light is the most important, being received by several photoreceptors, including the red light (R)-absorbing and the far-red light (FR)-absorbing phytochromes and the blue-/UV-A-absorbing cryptochromes and phototropins [10]. On cowpea, photoperiod sensitivity was found to be partially dominant to insensitivity [11].

The grouping of soybean maturity is different for different countries. In Indonesia, based on the days of maturity, soybean is grouped into late maturity (>90 days), medium maturity (80–90 days), and early maturity (<80 days). In Thailand, early-maturing soybean varieties require less than 85 days to maturity. Until now, it has been released as many as 86 varieties of soybean with a various flowering day between 28 and 50 days and days of maturity ranged between 66 and 105 days. Early maturity soybean provides many benefits that are to minimize the yield loss due to drought stress through escape mechanism, and to increase the cropping intensity within a year [12].

The objective of this research was to map and classify the seed yield and days of maturity of soybean lines, as well as selecting soybean lines for both high yield and early maturing.

2 Research Method

The study was conducted in paddy fields during dry season II (June–September 2013) in Genteng Research Station, Banyuwangi, East Java. The research station has a brown latosol soil type, 168 m elevation above sea level, and D2-type climate. The research material consists of 666 F5 soybean lines and ten check cultivars (Argomulyo, Kaba, Panderman, Anjasmoro, G511H, Grobogan, G100H, Rajabasa, Malabar, and Burangrang). Each line was planted on a 0.8 m × 4.5 m plot size, with a spacing of 40 cm × 15 cm. Pests and diseases were controlled optimally. Drainage was applied to maintain optimum soil moisture. Fertilization with 50 kg urea, 100 kg SP36, and 75 kg KCl per hectare was given spread evenly before planting.

The data were collected on flowering day (when 50% of the plant population have been flower), maturity day (calculated if 95% of the leaves have turned yellow), weight of 100 seeds (randomly taken from the seed yield per plot), and the seed yield (weighed at 12% of moisture content which is subsequently converted to yield per hectare). The data were summarized by descriptive statistics. The grouping of the genotypes based on seed yield and days to maturity were conducted by using cluster analysis. These analysis were using MINITAB ver. 16.

3 Result and Discussion

The performance of 666 soybean lines and ten check cultivars showed the optimum agronomic growth. Flowering date of 666 lines ranged between 25 and 34 days (average of 28 days), days to maturity varied from 70 to 79 days (average of 75 days), weight of 100 seeds ranged from 10.44 to 26.26 g/100 seeds (average of 16.22 g/100 seeds), and seed yield ranged from 0.16 to 4.06 t/ha (average of 1.47 t/ha) (Table 1).

The results from Table 1 showed that the seed yield had a very wide range and all of the lines were categorized as early maturing (<80 days). Seed yield of check cultivars ranged from 1.29 to 3.79 t/ha (average of 1.99 t/ha), and the days to maturity ranged from 73 to 78 days (average of 76 days). The average seed yield of the lines was lower than the general average of check cultivars, but the upper limit yield of lines (4.06 t/ha) was higher than the best check cultivar (3.79 t/ha). This result indicated an opportunity to obtain higher-yielding soybean lines than check cultivars. Most of the lines (265 lines or 39.79%) had days to maturity between 75 and 76 days. In Indonesia, soybean with days to maturity less than 75 days is classified as super-early maturity. The result also showed that a total of 261 lines (39.18%) belong to the super-early maturity.

Seed yield of 666 soybean genotypes were classified into three groups, i.e., average, highest, and lowest yield. Data on Table 2 showed that there was a positive correlation between early-maturing days and seed yield of soybean. The

Table 1 Agronomic character of early-maturing soybean genotypes

Genotypes	Characters			
	Flowering (days)	Maturity (days)	Seed size (g/100 seeds)	Seed yield (t/ha)
Six hundred and sixty-six lines				
Average	28	75	16.22	1.47
Standard deviation	1.74	2.29	2.43	0.66
Maximum	34	79	26.28	4.06
Minimum	25	70	10.44	0.16
Ten check cultivars				
Argomulyo	27	73	16.08	1.31
Kaba	30	78	12.41	2.04
Panderman	30	78	14.22	1.29
G 511 H	28	76	14.67	0.85
Grobogan	29	74	22.75	2.10
Anjasmoro	30	78	16.17	3.79
G 100 H	32	78	13.13	1.83
Rajabasa	31	75	14.68	1.96
Malabar	30	73	13.75	2.48
Burangrang	32	77	19.22	2.21
Average	30	76	15.71	1.99
Standard deviation	1.51	2.00	2.97	0.76
Maximum	32	78	22.75	3.79
Minimum	27	73	12.41	0.85

super-early-maturing soybean lines had an average seed yield between 1.01 and 1.29 t/has, whereas soybean with days to maturity between 75 and 79 days had an average seed yield from 1.49 to 2.54 t/has. However, based on the highest yield from each maturity group, it appeared that the highest yield as 4.06 t/ha was obtained by soybean line with days to maturity 78 days, then followed by seed yield as 3.71 t/ha with days to maturity 77 days, and seed yield as 3.69 t/ha with days to maturity 74 days. This fact showed the possibility to obtain both high yield and early maturing of soybean lines.

Various studies have stated that the soybean seed yield was associated with days to maturity [13, 14] although [15] claimed that correlation between the yield and days to maturity was not important, except with days to flowering which showed significant correlation. Therefore, the extension of days to flowering is important to increase the potential yield of early-maturing soybean. Late flowering has been reported to be associated with higher soybean yield [16]. Many studies have shown that days to flowering [17] and days to maturity have relatively high heritability [18, 19].

Table 2 Seed yield of 666 soybean genotypes

Maturity (days)	Genotypes	Yield (t/ha)		
		Average	Highest	Lowest
70	79	1.01	2.19	0.25
72	36	1.03	1.89	0.31
73	86	1.14	2.39	0.42
74	60	1.29	3.69	0.56
75	154	1.49	3.64	0.67
76	111	1.66	3.27	0.16
77	89	1.84	3.71	0.54
78	50	2.12	4.06	0.75
79	1	2.54	2.54	2.54
Total	666	1.47		

Table 3 Soybean grouping based on yield and days to maturity

Cluster	Number of genotype	Maturity (days)	Yield (t/ha)
I	78	70	0.25–1.56
II	104	77–78	0.54–2.48
III	74	76	0.75–1.77
IV	190	74–75	0.56–2.10
V	122	72–73	0.31–1.94
VI	58	75–76	1.83–3.06
VII	40	77–78	2.37–4.06
VIII	4	75	3.27–3.65
IX	2	73	2.40–2.48
X	1	74	3.69
XI	1	76	0.16
XII	1	70	2.19
XIII	1	79	2.54

Cluster analysis was used to classify high-yielding soybean lines at various days to maturity (Table 3). A dendogram cutting on 80% (based on 80% similarity) has successfully clustering 676 genotypes (lines and cultivars) into 13 groups. Groups I to V and groups IX, XI, XII, and XIII were characterized by low yield, while the high-yield category was in groups VI, VII, VIII, and X.

Data on Table 3 showed that it was difficult to obtain a high-yield soybean line with the days to maturity under 74 days. On the other side, it could be stated that the average days to maturity for the tropical region such as Indonesia was 74 days. Research by [20] showed that soybean with early maturing had lower agronomic traits than the late maturing one. Early-maturing soybeans had low yield due to their small plant structure and short vegetative period [15].

The selection of high-yielding soybean lines is shown in Table 4. Ten soybean lines with seed yield of 3.42–4.06 t/ha were selected. The best two of check cultivars (Anjasmoro and Malabar) have seed yield 3.79 and 2.48 t/ha, respectively.

Table 4 Soybean lines selected for high yield

Genotype	Maturing (days)	Yield (t/ha)
Lines		
Grobogan/Kaba-19	77	3.42
L Jtg/Sbg//Burangrang-10	78	3.42
Grobogan/Wilis-42	78	3.46
G 511 H/Argomulyo-3	77	3.56
Arg//Ljtg/Sbg-11	75	3.56
Rajabasa/Grobogan-21	75	3.65
Rajabasa/Grobogan-20	74	3.69
Grobogan/Kaba-10	77	3.71
G511/Mlgg 1029 (84)//M 1029(84)/// 1029	78	3.75
Arg//Ljtg/Sbg-12	78	4.06
The best check cultivars		
Anjasmoro	78	3.79
Malabar	73	2.48

Line of Arg//Ljtg/Sbg-12 had the highest yield (4.06 t/ha) with days to maturity 78 days. It was followed by line G511/Mlgg 1029 (84)//M1029 (84)///1029 (3.75 t/ha) with days to maturity 78 days. The super-early-maturing soybean lines (74 days) with high yield was Rajabasa/Grobogan-20, which reached 3.69 t/ha (Table 4). The best line Arg//Ljtg/Sbg-12 had days to maturity that is equivalent to Anjasmoro (78 days). Line Rajabasa/Grobogan-20 had 48% higher yield than super-early-maturing cultivar of Malabar (2.48 t/ha).

The climate change has increased the El Nino intensity (drought) and accelerates the development of pests and diseases. The availability of early-maturing cultivars and high yield of soybean is needed as an adaptation effort to climate change. Ten selected lines of soybean with both high yield and early maturing are needed to be proceeded to the next breeding step to be released as improved varieties in Indonesia.

4 Conclusion

1. There was a positive correlation between soybean seed yield and days to maturity. For Indonesia's tropical regions, it was suggested the maturing day of soybean to produce high yield is 74 days.
2. The identification of high yielding and early maturing of soybean lines is important to be tested in various soybean production centers in Indonesia and being proposed to be released as national improved soybean varieties.

References

1. Heatherly, L. G., and Elmore, R. W. 2004. Managing inputs for peak production, in: Soybeans: Improvement, Production, and Uses, Agronomy Monographs 3rd ed. No. 16, H. R. Boerma, and J.E. Specht, eds., ASA-CSSA-SSSA, Madison, WI, USA, pp. 451–536.
2. Bernard, R.L. 1971. Two genes for time of flowering in soybeans. Crop Sci. 11: 242–244.
3. Buzzell, R.I. 1971. Inheritance of a soybean flowering response to fluorescent-day length conditions. Can J Cytol. 13: 703–707.
4. Buzzell, R.I., and H.D. Voldeng. 1980. Inheritance of insensitivity to long day length. Soybean Genet. Newsl. 7: 26–29.
5. McBlain, B.A., and R.L. Bernard RL. 1987. A new gene affecting the time of flowering and maturity in soybean. J. of Hered. 78: 160–162.
6. Cober ER, and H.D. Voldeng. 2001. A new soybean maturity and photoperiod-sensitivity locus linked to E1 and T. Crop Sci. 41: 698–701.
7. Bonato, E.R., and N.A. Vello. 1999. E_6, a dominant gene conditioning early flowering and maturity in soybeans. Genet. Mol. Biol. 22:.229–232.
8. Ray, J.D, K. Hinson, J.E.B. Manjono, and M.F. Malo. 1995. Genetic control of long-juvenile trait in soybean. Crop Sci. 35: 1001–1006.
9. Cober, E.R., J.M. Tanner, H.D. Voldeng. 1996. Genetic control of photoperiod response in early-maturing, near-isogenic soybean lines. Crop Sci. 36: 601–605.
10. Chen, M., J. Chory and C. Fankhauser, 2004. Light signal transduction in higher plants. Annu. Rev. Genet. 38 87–117.
11. Ishiyaku, M. F., B.B. Singh, and P.Q. Craufurd, P. Q. 2005. Inheritance of time to flowering in cowpea *(Vigna unguiculata (L.) Walp.)*. Euphytica 142: 291–300.
12. Krisnawati, A and M.M. Adie. 2008. Variability of days to maturity on soybean and contribution to seed yield. Agritek 17 (1): 1–7.
13. Ishibashi, T., C.H. Sneller, and G. Shannon, 2003. Soybean yield potential and phenology in the ultra-short-season production system. Agron. J. 95: 1082–1087.
14. Malik, M.F.A., M. Ashraf, A.S. Qureshi, and A. Ghafoor. 2006. Utilization of diverse germplasm for soybean yield improvement. Asian J. Plant Sci. 5 (4): 663–667.
15. Machikowa, T. and P. Laosuwan. 2009. Effects of extension of days to flowering on yield and other characters of early maturing soybean. Suranaree J. Sci. Technol. 16: 169–174.
16. Dunphy, E.J., J.J. Hanway, and D.E. Green. 1979. Soybean yields in relation to days between specific developmental stages. Agron. J. 71: 917–920.
17. Niessl, H., J. Vollmann, and P. Ruckenbauer. 1995. Selection for time to maturity in soybean (*Glycine max* (L.) Merrill) and seed protein content of early maturing genotypes. Die Bodenultur 46: 311–320.
18. Xu, H. and J.R. Wilcox. 1992. Recurrent selection for maturity and percent seed protein in *Glycine max* based on S0 plant evaluations. Euphytica 62: 51–57.
19. Hasan, E., M. Arshad, M. Ahsan, and M. Saleem. 2008. Genetic variability and interrelationship for grain yield and its various components in chickpea (*Cicer arietinum* L.). J. Agric. Res. 46 (2): 109–115.
20. Abayomi, Y.A. 2008. Comparative growth and grain yield response of early and late soybean maturity groups to induced soil moisture stress at different growth stages. World J. of Agric. Sci. 4: 71–78.

Cyperus Rotundus L. Extracts as Botanical Fungicides to Control Soybean Rust Disease

Eriyanto Yusnawan and Alfi Inayati

Abstract Soybean rust disease caused by *Phakopsora pachyrhizi* Syd. is one of the most important diseases on soybean. Severe infection on soybean crops causes early defoliation. The yield loss caused by this disease has been reported up to 80% on susceptible cultivars. Applications of chemical fungicides that reduce the infection rate, however, are not considered environmental friendly. The use of botanical fungicides is one of the alternative controls to reduce the rust disease. This study aimed to obtain effective concentrations of methanolic and *n*-hexane extracts of roots, leaves, and flowers of *Cyperus rotundus* as well as to investigate secondary metabolites in each extract. To evaluate the effectiveness of each extract, a spore germination test was conducted in vitro. Methanolic extracts of the roots, leaves, and flowers inhibited more germinated spores compared to the applications of *n*-hexane extracts at the same concentrations. Applications of 2.5 and 5% of methanolic extracts of the roots resulted 75 and 78% of ungerminated spores. The numbers of lysed spores as a result of these two concentration treatments were 5 and 6%, respectively. A similar inhibition was observed when 5% of polar extract of the flowers was applied, and the numbers of ungerminated and lysed spores were 71 and 10%. Phytochemical screenings showed that the polar extracts contained more diverse in secondary metabolites compared to those in the non-polar extracts. The methanolic extract of the roots which was the most effective to inhibit the spore germination contained alkaloids, flavonoids, tannins, saponins, and terpenoids. A further test to confirm the presence and to separate these compounds with thin-layer chromatography showed that alkaloids, flavonoids, saponins, and terpenoids were separated into 2, 4, 3, and 2 spots, respectively. These compounds in the methanolic extract of the roots may be responsible to suppress the spore germination of *P. pachyrhizi* when these extracts at concentrations of 2.5 and 5% were applied.

Keywords *Cyperus rotundus* · Botanical fungicide · *Phakopsora pachyrhizi* · Secondary metabolite

---

E. Yusnawan (✉) · A. Inayati
Indonesian Legumes and Tuber Crops Research Institute,
Jalan Raya Kendalpayak Km 8, PO Box 66, Malang, East Java, Indonesia
e-mail: yusnawan@yahoo.com

T. Taufik et al. (eds.), *ICoSI 2014*, DOI 10.1007/978-981-287-661-4_3

1 Introduction

Soybean rust disease caused by *Phakopsora pachyrhizi* Syd. is one of the important diseases on soybeans. Yield loss caused by this disease is estimated up to 80% depending on the susceptibility of the genotypes [1]. The use of natural fungicides to control this disease is considered more environmental friendly and expected to reduce the yield loss. Natural fungicides extracted from plants have antimicrobial agents and anti-insects because of its secondary metabolite contents, including flavonoids, alkaloids, terpenoids, saponins, and tannins [2–4].

Cyperus rotundus L. is an invasive weed and able to dominate land because of the ability to produce allelochemical compounds. *C. rotundus* contained secondary metabolites which benefits as antimalaria, anticancer, and antimicrobial agents [5–7]. As antimicrobial agents, the active compounds inhibited the growth of *Bacillus subtilis, Escherichia coli, Staphylococcus aureus,* and *Pseudomonas aeruginosa*. In addition, the active compounds also inhibited fungal growth such as *Aspergillus fumigatus* and *Candida parapsilosis.* The spore formation of *A. flavus* and *Fusarium oxysporum* was also disturbed [5, 7].

The application of Cyperus crude extract to control soybean rust disease has not been reported yet. Therefore, this study aimed to obtain effective concentrations of methanolic and *n*-hexane extracts of roots, leaves, and flowers of *C. rotundus* to inhibit spore germination of *P. pachyrhizi* as well as to investigate secondary metabolites in each extract.

2 Materials and Methods

Sample preparation and extraction. Roots, leaves, and flowers of Cyperus were separated, air dried, and ground to obtain fine particles. The particles were macerated in methanol as a polar solvent and *n*-hexane as a non-polar solvent for 18 h separately after shaking the suspension using an orbital shaker at 100 rpm for 4 h. Supernatant was collected and evaporated using a vacuum rotary evaporator. The crude extracts were stored at 4 °C in the dark before used [8].

Spore germination test. Experiments were arranged in a completely randomized design with three factors and three replicates. The first factor was polarity of the solvents, the second factor was three parts of Cyperus (roots, leaves, and flowers), and the third factor was four concentration levels (0.1, 1, 2.5, and 5%). Uredospores of *P. pachyrhizi* were collected from infected crops cultivated in a greenhouse. The infected leaves were harvested and incubated in petridishes for two days. The humidity inside the petridishes was maintained at around 95% by placing wet cotton layers. Mature uredospores were harvested and suspended in sterile water. Spore germination test was conducted in the polar and non-polar extracts. As a control was used phosphate buffer. The numbers of germinated and ungerminated spores were recorded after 24 h of incubation [8].

Phytochemical screening. To screen plant secondary metabolites (i.e. alkaloids, flavonoids, tannins, saponins, and terpenoids) in the polar and non-polar extracts, a phytochemical test was conducted [2, 9]. Briefly, alkaloids were detected using Mayer dan Wagner reagents. Flavonoids were determined using Mg and HCl. Tannins were detected using $FeCl_3$ and gelatin. Saponins were detected by the ability of the extracts to form foam for at least 30 s after shaking. Terpenoids were detected using Liebermann–Burchard reagent and H_2SO_4 [2, 9].

Separation of active compounds using thin-layer chromatography (TLC). Silica gel plates (F_{254}) were used to separate the active compounds. Crude extracts (5 μl) were spotted on the plates and separated using suitable mobile phases. Alkaloids were separated with methanol:chloroform (0.5:9.5 v/v) [10], flavonoids were separated with chloroform:methanol (9:1 v/v) [2], saponins were separated with *n*-hexane:acetone (4:1 v/v) [11], and terpenoids were separated with *n*-hexane: ethyl acetate (2:8 v/v) [10].

3 Results and Discussion

Spore germination test. To determine the effectiveness of the root, leaf, and flower extracts to inhibit spore germination of *P. pachyrhizi*, four concentration levels of the polar and non-polar extracts were tested. In general, methanolic extracts of the roots, leaves, and flowers inhibited more germinated spores than application of non-polar extracts at all concentration levels (Fig. 1). Applications of 2.5 and 5% of polar extract of the roots resulted 75 and 78% of ungerminated spores. The numbers of lysed spores as a result of these two concentration treatments were 5% and 6%, respectively. A similar inhibition was observed when 5% of methanolic extract of the flowers was applied, and the numbers of ungerminated and lysed spores were 71 and 10%. Only non-polar extract of the roots at 5% showed similar effectiveness to the application of 2.5% of methanolic extract of the roots. This application inhibited 73% of ungerminated spore and resulted 4% of lysed spore.

Plant secondary metabolites extracted with methanol or other alcoholic group solvents resulted more effective to inhibit fungal growth than the use of non-polar solvents [12, 13]. Compositions and groups of chemical compounds of the secondary metabolites were influenced by extraction solvents [14]. Polar solvents were commonly used to extract organic compounds from natural products. Endocellular secondary metabolites were extracted more using polar solvents than non-polar solvents since the polar solvents were able to enhance cell permeability and to penetrate inside the cells. The polar solvents dissolved both polar and less polar compounds, whereas non-polar solvents only dissolved non-polar compounds [15, 16]. The active compounds obtained from the extraction of the roots, leaves, and flowers of Cyperus with methanol may enhance the effectiveness to inhibit *P. pachyrhizi* spore germination as observed in this study.

Phytochemical screening. Extraction using the polar solvent resulted more diverse in the active compound groups (Table 1). The polar extract of the roots

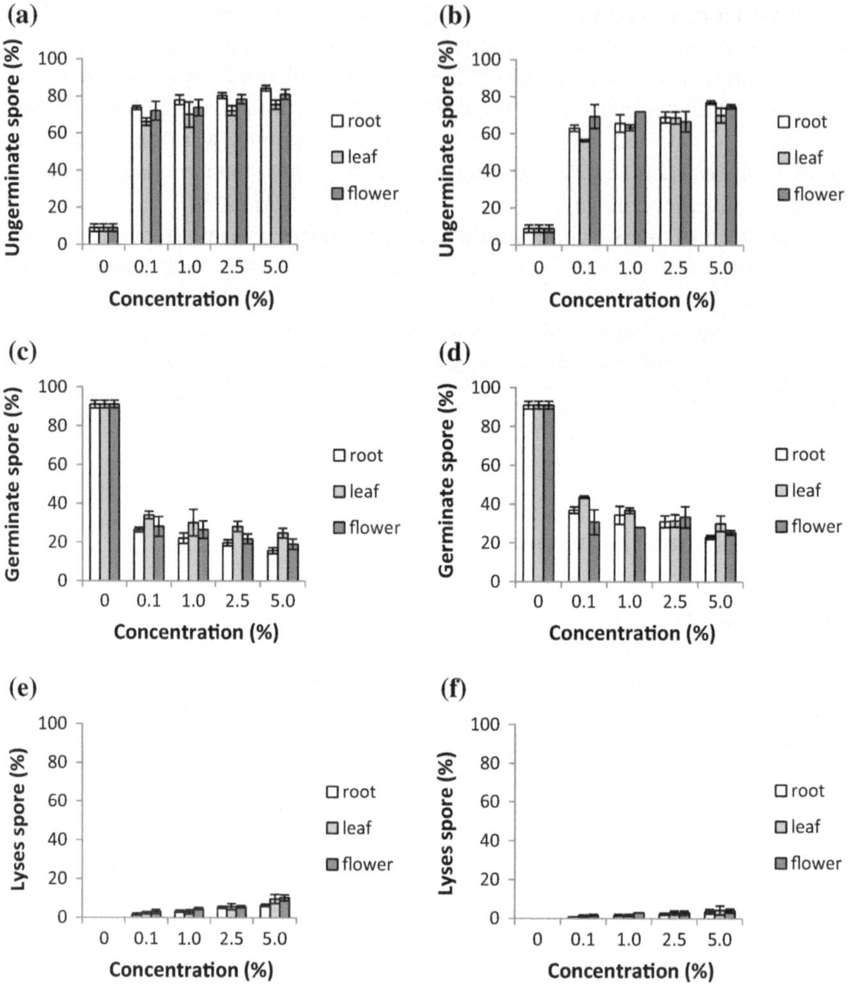

Fig. 1 Percentages of ungerminated spores of *P. pachyrhizi* treated with methanolic extracts (**a**), ungerminated spores treated with *n*-hexane extracts (**b**), germinated spores treated with methanolic extracts (**c**), germinated spores treated with *n*-hexane extracts (**d**), lysed spores treated with polar extracts (**e**), and lysed spores treated with non-polar extracts of *C. rotundus*. *Bars* show deviation standard

qualitatively showed more intense in colour changes after being reacted with the suitable reagents compared to those of the leaf and flower extracts. All polar extracts of the roots, leaves, and flowers contained alkaloids, flavonoids, tannins, saponins, and terpenoids. The non-polar extract of the roots at 5% concentration which had similar effectiveness to inhibit spore germination to 2.5% polar extract of the roots contained alkaloids, flavonoids, saponins, and terpenoids.

Table 1 Phytochemical screenings of the methanolic and *n*-hexane extracts of the roots, leaves, and flowers of *C. rotundus*

Active	Methanolic extract			*n*-hexane extract		
Compound	Root	Leaf	Flower	Root	Leaf	Flower
Alkaloid	+	++	++	+	+	−
Flavonoid	++	+	+	+	−	−
Tannin	++	+	+	−	−	−
Saponin	++	++	+	++	−	−
Terpenoid	++	++	++	+	−	+
Steroid	−	−	−	−	+	−

++ showed more compound/dark colour, + showed less compound/light colour, − showed no compound detected

A high polarity index of methanol was able to extract the active compounds which had polar properties such as flavonoid glycosides, tannins, and several alkaloids [16]. Furthermore, this solvent was effective to extract phenolic compounds with low molecular weights and medium levels of polarity [17], aglycons of flavonoids [14], anthocyanins, terpenoids, saponins, flavons, and polyphenolic compounds [18]. A low polarity index of the solvents such as *n*-hexane only dissolved liphophilic compounds, including alcanas, waxes, colour pigments, sterols, several terpenoids and alkaloids, therefore, extracted less secondary metabolites [16].

Separation of active compounds using TLC. The methanolic extract of the roots which significantly inhibited spore germination was separated using a TLC method. This experiment aimed to confirm the presence of the active compounds as a further study following phytochemical screenings. Also, the purpose of this study was to determine the number of active compounds visualized as spots on the TLC plates. More spots were detected in flavonoids compared to spots of alkaloids, saponins, and terpenoids (Table 2). Tannins were not detected using this technique; therefore, other techniques with more sensitive detection and high resolution may be required to confirm the presence of tannins in the extract.

Secondary metabolites in plants have several functions, one of which is to combat pathogen infections [19]. Alkaloids inhibit cell respiration, esterases, and DNA as well as RNA polymerases [20]. Flavonoids act to form complex compounds on the extracellular protein which disturb cell membrane integrity and as protein coagulators. Flavonoids also function to protect plants from pathogen infections [21, 22]. Saponins with have bitter properties are effective chemicals to combat pathogens and pest insects [23]. Terpenoids are one of the phytoalexins effective to suppress the spread of pathogen infection [2, 18]. Tannins may disturb protein metabolism of pathogens; therefore, the pathogens are unable to continue their growth normally [18, 24]. Alkaloids, flavonoids, saponins, terpenoids, and tannins in *C. rotundus* may play important roles to inhibit spore germination of *P. pachyrhizi*, particularly when the applications of 2.5 and 5% of the methanolic extracts of the roots were conducted.

Table 2 Number of spots and secondary metabolite characteristics of the methanolic extract of the roots after separation using thin-layer chromatography

Secondary metabolite	Number of spot	Rf value	Spot colour		
			Visible colour	UV light (λ = 254 nm)	UV light (λ = 366 nm)
Alkaloid	2	0.60	No colour	Brown	Orange
		0.70	No colour	Brown	Orange
Flavonoid	4	0.07	No colour	Brown	Purple
		0.12	Yellow	Brown	Green
		0.22	No colour	Brown	Orange
		0.35	No colour	Brown	Purple
Saponin	3	0.15	No colour	No colour	Orange
		0.25	No colour	Brown	Purple
		0.31	No colour	Brown	No colour
Terpenoid	2	0.76	Light green	Brown	Dark orange
		0.82	Dark green	Brown	Dark orange
Tannin	–	–	–	–	–

Rf = retardation factor

4 Conclusion

The methanolic extract of the roots at a concentration of 2.5% may be potential to be utilized as a botanical fungicide to suppress the growth of *P. pachyrhizi*. Further studies including greenhouse and field experiments should be conducted to determine the effectiveness of this concentration. Purification of the active compounds can be considered to determine the effectiveness of each purified compound.

Acknowledgements Author would like to thank to Tri Kustono Adi, M.Sc, who gave suggestion during the experiment and Rina Rosita S., S.Si who helped this research.

References

1. Twizeyimana, M, PS Ojiambo, T Ikotun, JL Ladipo, GL Hartman, and R Bandyopahyay. 2008. Evaluation of soybean germplasm for resistance to soybean rust (*Phakopsora pachyrhizi*) in Nigeria. *Plant Dis.* 92: 947–952.
2. Harborne, JB. 1998. *Phytochemical Methods: a Guide to Modern Techniques of Plant Analysis*. 2nd ed. Chapman and Hall, London.
3. Kamboj, A & AK Saluuja. 2008. *Ageratum conyzoides* L.: a review on its phytochemical and pharmacological profile. *Internat. J Green Pharmacy*. 2(2): 59–68.
4. Patil, RP, MS Nimbalkar, UU Jadhav, VV Dawkar, & SP Govindwar. 2009. Antiaflatoxigenic and antioxidant activity of an essential oil from *Ageratum conyzoides* L. *J. Sci. Food Agric.* 90(4): 608–614.

5. Bisht, A, GRS Bisht, M Singht, R gupta, & V Singht. 2011. Chemical composition and antimicrobial activity of essential oil of tubers of *Cyperus rotundus* Linn. collected from Dehradun (Uttarakhand). *Internat. J. Res. Pharma. and Biomed. Sci.* 2229–3701.
6. Khamsan, S, B Liawruangrath, S Liawruangrath, S Teerawutkulrag, GS Pyne, & JM Garson. 2011. Antimalarial, anticancer, antimicrobial activities and chemical constituents of essential oil from the aerial parts of *Cyperus kyllingia* Endl. *Rec. nat. product.* 324–327.
7. Parekh, J. & S Chandra. 2007. *In vitro* antimcrobial activities of extracts of Launaea Roxb. (Labiateae), *Vitis vinifera* L. (Vitaceae) and *Cyperus rotundus* L. (Cyperaceae). *African J. Biomed. Res.* 9: 89–93.
8. Yusnawan, E. 2013. Skrining golongan senyawa aktif fraksi polar *Amaranthus spinosus* dan efektivitasnya terhadap penyakit karat daun kacang tanah *in vitro*. Dalam Purnomo, J, M Harisudin, D Praseptiangga, A Magna, Rahayu, Widiyanto, R Indreswari, Y Yanti, BS Hertanto (Eds). *Prosiding Sem Nas Akselerasi Pembangunan Pertanian Berkelanjutan Menuju Kemandirian Pangan dan Enargi*. Pp. 377–382. UNS, Surakarta Oktober, 2013.
9. Trease, GE & WC Evans. 1983. *Pharmacognosy*. 12th ed. Bailliere Tindall Press, London.
10. Wagner, H & S Bladt. 1996. Plant Drug Analysis: a Thin Layer Chromatography Atlas. Springer, Germany.
11. Marliana, SS, V Suryanti & Suyono. 2005. Skrining fitokimia dan analisis kromatografi lapis tipis komponen kimia buah labu siam (*Sechium edule* Jacq. Swartz.) dalam ekstrak etanol. *Biofarmasi*. 3: 26–31.
12. Cimanga, RK, K Kambu, L Tona,T de Bruyne, A Sandra, J Totte, L Pieters, & AJ Vlietinck. 2004. Antibacterial and antifungal activities of some extracts and fractions of *Mitracarpus scaber* Zucc. (Rubiaceae). *J. Nat. Remedies*. 4(1): 17–25.
13. Sharma, M, S Puri & PD Sharma. 2007. Antifungal activity of *Millingtonia hortensis*. *Indian J. Pharmaceutical Sci.* 69: 599–601.
14. Dehkharghanian, M, H Adenier & MA Vijayalakshmi. 2010. Analytical methods study of flavonoids in aqueous spinach extract using positive electrospray ionisation tandem quadrupole mass spectrometry. *J. Food Chem.* 121: 863–870.
15. Cannell, RJP. 1998. *Natural Products Isolation*. Humana Press, Totowa.
16. Seidel, V. 2012. Initial and bulk extraction of natural product isolation. In: Sarker, SD & L Nahar (Eds). *Natural Product Isolation, Methods in Molecular Biology*. 864: 27–41.
17. Yu Lin, H, YH Kuo, YL Lin, & W Chiang. 2009. Antioxidative effect and active component from leaves of lotus (*Nelumbo nucifera*). *J Agric. Food Chem.* 57: 6623–6629.
18. Cowan, MM. 1999. Plant products as anti-microbial agents. *Clin. Microbiol. Rev*. 12: 564–582.
19. Vickery, ML & B Vickery. 1981. *Secondary Plant Metabolism*. The Macmillan Press, London.
20. Aniszewski, T. 2007. *Alkaloid-Secrets of Life: Alkaloid Chemistry, Biological Significance, Applications and Ecological Role*. Elsevier, Amsterdam.
21. Boue, SM, TE Cleveland, C Carter-Wientjes, BY Shih, D Bhatnagar, JM McLachlan, & ME Burow. 2009. Phytoalexin-enriched functional foods. *J. Agric. Food Chem.* 57: 2614–2622.
22. Samantha, A, G Das, & KD Sanjoy. 2011. Roles of flavonoid in plants. *Int. J. Pharm. Sci. Tech*. 6(1): 12–35.
23. Robinson, T. 1995. *Kandungan Senyawa Organik Tumbuhan Tinggi*. Diterjemahkan oleh Kosasih Padmawinata. ITB, Bandung.
24. Das, K, RKS Tiwari, & DK Shrivastava. 2010. Techniques for evaluation of medicinal plant products as antimicrobial agent: Current methods and future trends. *J. Med. Plants Res.* 4(2): 104–111.

Development of Breakfast Meals from Local Taro Using Extrusion for Food Security

Ermi Sukasih and Setyadjit

Abstract Extrusion technology is growing fast and been demanded by food processor. The selection of material for formulation (composite) is important to produce high quality and performance product. Taro, mung bean and banana are good combination to make breakfast meals. To produce breakfast meals which are expanded, light, porous and crunchy, it is required that starch has special characteristics, i.e., cassava flour. Improvement of the formulation can be done by adding protein, developer material and emulsifier. The aim of the research was to develop formula based on taro flour with the addition of egg flour, baking soda, GMS and arabic gum so it can produce best taro crunch. Formulation was done in sequence. First was addition of cassava flour to obtain the best expansion. Second was improvement of the dough of first formulation by addition of egg flour, baking soda, arabic gom and GMS. The results showed that the best formula is taro crunch with the percentage of taro composite flour 43.75%, egg flour 5%, GMS 1.5%, vegetable oil 5%, salt 1%. The produced taro crunch has characteristics of protein 1.62%, moisture content 5.5%, fat 5.29%, crude fiber 4.07%, carbohydrate 81.36%, energy 379.57 Kal/100 g with the expansion degree of 0.54%.

Keywords Breakfast meal · Taro · Extrusion · Taro composite flour · Food security

1 Introduction

Taro, bananas and mung bean are potential food sources, when they mixed made of composite has many advantages because it can be an complement each other. Taro is a local product of Indonesia, is widely cultivated by farmers, has advantages compared with other tubers, in addition such as containing a high carbohydrate, i.e.,

E. Sukasih (✉) · Setyadjit
Indonesian Centre for Agriculture Postharvest Research and Development, Cimanggu Agriculture Campus, JL. Tentara Pelajar no. 12, Bogor 16114, Indonesia
e-mail: ermi_sukasih@yahoo.co.uk

T. Taufik et al. (eds.), *ICoSI 2014*, DOI 10.1007/978-981-287-661-4_4

84% in the form of flour, and also has a oligosaccharide content as a prebiotic compounds and dietary fiber. Composite flour of taro–banana–mung bean have been used for breakfast meal high-energy with oven method [1]. Taro is an underutilized root crop with promising economic value. It is rich in energy, readily available and quite affordable. The utilization of Taro could be increased by developing suitable processing technology and securing consumer acceptance [2].

Along with the rapid development of science and technology in the field of food processing as well as the response of the general consumers, the requirement of food products which has high demand is easy to prepare, safe, nutritious, and has an interesting organoleptic characteristics as well as affordable. Thus, to fulfill the demand of the consumers, the extrusion technology is even thrive and has growing demand from the food processor.

Extrusion technology well known can provide various shapes, in expensive and efficient. Raw material in the form of starch, oil, other gradients and food additive is fed into the extruder without going to frying processes [3]. However, this process sometimes requires an iteration to have a better understanding and control for excellent final product. The variables of extrusion processes determine the final product quality [4].

Some of the advantages of extrusion process is a high productivity, very distinctive product shape and many variations, and so high-quality products because the cooking is done at high temperature in a short period of time, similar to the effect of UHT (ultra-high temperature) for killing microbes, but little nutritional damage as well as low costs and energy consumption per unit of production process. Process with a high temperature for a short time produces a high-quality end product that can be easily digested and has high nutritional value. Thus, by reducing time of using process at high temperatures, it is reducing the destruction reaction which cause loss of nutrient content [5]. One of the most important technologies which have shown great potential for the development of new snack products is extrusion cooking. Extrusion is a continuous food processing technique classified as a high-temperature-short-time operation in which raw food materials are thermo-mechanically cooked in a screw barrel assembly by a combination of moisture, pressure and temperature in order to be mechanically sheared and shaped [6].

Extrusion process is done similarly to UHT for killing microbes, and they are only required low energy consumption per unit process production and only little nutritional damage. The high temperature is short-time processes cause the product easier to be digested but still contain high nutrient [5]. Thus, the extrusion technique is included in the modern processing which has great potential for the development of new snack, and it is continuous. The basic principle of the processes is a thermo-mechanically cooking of row material, in screw barrel assembly using a combination of moisture, pressure and temperature in order to be mechanically sheared and shaped [6].

Raw materials of extrudate are usually made from grains and will affect the final product. Still a few research works about the processing of taro flour into extruded products have been conducted. Maga, Bang-Liu and Rey produced extruded snacks using taro flour as starting raw material [9]. During the extrusion process, the starch granules burst and gelatinization occurs. The addition of water during the process resulted in starch particles swell and lose its bond cohesiveness and part of the amylose diffuses out due to the heat effect [10, 11]. The aim of the research was to develop formula based on taro flour with the addition of egg flour, baking soda, glycerol monostearate (GMS) and arabic gum so it can produce best taro crunch.

2 Materials and Methods

The raw material used for the manufacture of composite flour was taro Cv.Mentega, bananas Cv.Kepok Kuning, peeled mung beans from Anyar Market, Bogor. Additional materials used were egg flour, baking powder, cooking oil, gum arabic, GMS (glycerol monostearate), salt obtained from the Yuks Store, Bogor. The tools used are glassware, double-screw extruder, ovens, scales, baking pan. Extrusion manufacturing process is performed in the laboratory, Seafast Center, Bogor Agricultural University. The analysis was conducted in the laboratory of Indonesian Centre for Agriculture Postharvest Research and Development from April to September 2012.

3 Preparation of Composite Flour-Based Talas

Each of these ingredients (taro, bananas and mung beans) were ground to made flour; then, they were mixed with comparison taro flour:banana flour:mung bean flour was 50:30:20. For taro flour was needed special treatment to reduce oxalate contents by soaking and steaming [12]. Taro has been shredded and then soaked with 3% salt solution for 5 min and steamed for 20 min before being dried in an oven temperature of 60°C.

4 Development of Taro Crunch Formulation

The formulation consisted of two phases: the first with the addition of tapioca at 10%, 30%, 50%, 70% and 90% levels. As the control was 100% tapioca and 100% taro composite flour. These formulations to obtain the ratio between taro flour composite and the tapioca flour taro that produce a good degree of expansion. Observations were on starch contents, amylose and amylopectin, fat, protein, moisture content, ash content, color, degree of expansion and organoleptic test. The second stage was the formulation for improving products with the addition of egg powder, baking soda, GMS, arabic gum, cooking oil and salt. The composition of cooking oil, salt and GMS made the same, while others were made formulations as given in Table 1.

Table 1 Formula of improvements of taro crunch

Formula	Percentage of materials (part)							
	Composite flour	Tapioca flour	Egg flour	Baking soda	Glycerol monostearate	Cooking oil	Gum	Salt
F1	45.25	45.25	0	1	1.5	5	1	1
F2	45.75	45.75	0	0	1.5	5	1	1
F3	45.75	45.75	0	1	1.5	5	0	1
F4	46.25	46.25	0	0	1.5	5	0	1
F5	42.75	42.75	5	1	1.5	5	1	1
F6	43.25	43.25	5	0	1.5	5	1	1
F7	43.25	43.25	5	1	1.5	5	0	1
F8	43.75	43.75	5	0	1.5	5	0	1
F9	40.25	40.25	10	1	1.5	5	1	1
F10	40.75	40.75	10	0	1.5	5	1	1
F11	40.75	40.75	10	1	1.5	5	0	1
F12	41.25	41.25	10	0	1.5	5	0	1

Manufacturing of these products using a double-screw extruder-type Bex 2256, extruder screw consists of two equally long and attached to each other, with length of thread 100 cm and speed of 50 Hz. The materials were weighed according to the formula and added as much as 13–19% of boiled water while stirring until evenly distributed. The dough was then placed in the entrance hopper and then fallen into the auger feeder that serves to mix and solidify the dough before going to screw. The thread was covered by barrel made of stainless steel, in which the material was heated at a temperature of 135 °C for 10–15 s [13]. The next expanded dough went into die and cut by a knife with a rotating speed of 25 Hz. The materials that came out was placed into the pan and oven-dried and then packaged. Observation parameters were protein content, moisture content, ash content, fat, crude fiber, carbohydrates, color, calories and degree of expansion. To determine the best formula, organoleptic test was done with a score of 1: bad, 2: moderate and 3: good, involving 20 semi-trained panelists.

5 Results and Discussion

5.1 Early Stage of Taro Crunch Formulation

Extrusion process can be used to produce several products such as pasta, breakfast meals and snacks. Correct component of the formulation will produce a product that expands. Puffing would happen if the gelatinization perfectly happened, and need appropriate materials, the most important ingredient is starches. Some starch-based materials are preferred as raw materials to enhance the puffing of extruded snacks [14]. Extrusion process caused changes in the chemical and nutrients on the resulting extrudates. Chemical characteristics of extruded product from taro flour composites are presented in Table 2.

Table 2 shows that the starch content decreased from 82.85% in the original composite flour taro to be 73.38–78.46% in the final product. This resulted in levels of amylose and amylopectin tended to decrease although it was irregular. Amylose plays an important role in gelling and retrogradation processes of starch during extrusion, which influences the texture and nutritional characteristics of final product [15]. The relative quantities of amylose and amylopectin also affect the direction in which product expands during extrusion; for example, amylose contributes with a greater longitudinal expansion, while amylopectin favors the radial expansion [14]. The whole formulation showed a greater amount of amylopectin than amylose in produced taro crunch. Starch which has a high amylopectin will produce extrusion products that are fluffy, light, porous, crisp and crunchy. This happens because the branched structure of amylopectin underwent greater damage and unable to effectively structured inside the barrel and die, thus helping the expansion of the product. Amylose is more resistant to mechanical damage during the process rather than amylopectin [16]. Levels of fat and protein have considerable variation, while the moisture content and ash content were relatively uniform. It was probably caused by a less homogeneous mixing process as it was done manually.

The lightness value (L value) showed no significant differences except for the control which has the highest L value. To get the ideal crispness using the sorghum extrusion products as a benchmark, it was required approximately 24% amylose with a protein content of 9.9%. From the experimental results it was known that the closest amylose obtained in the mixture formulation tapioca 30 and 60%, but no different from the degree of development of the formulations on tapioca up to 50% which is equal to 1.1% so that this formulation would be applied in the next formulation. The protein content in the first formulation reached 0.39 to 5.6% so that the treatment required the addition of protein to about 9.9% in the subsequent formulation.

6 Development of Taro Crunch Formulation

From the results of the first formulation for the ratio between tapioca and taro flour were each 50%. In accordance with the results of the previous experiment and sorgum extrudate as the benchmark, then in the formulation next experiment were added protein from egg powder, baking soda and gum arabic as developer ingredient. Improved formula with sorgum extrudates as a benchmark is presented in Table 1. Chemical characteristics of extrudate made from taro flour composites are presented in Table 3.

Table 2 Chemical characteristics of extruded product from taro flour composites

No	Treatments	Starch	Amylose	Amylopectin	Fat	Protein	Water content	Ash content	L (Lightness)	Expansion degree (%)
1	Tapioca 10%	78.46	18.81	59.65	6.94	5.60	6.34	1.81	58.62	1.1
2	Tapioca 30%	74.94	21.23	53.71	7.42	0.98	6.52	1.09	50.37	1.1
3	Tapioca 50%	74.87	19.86	55.01	6.20	1.59	4.37	1.29	52.24	1.1
4	Tapioca 70%	74.80	18.06	56.81	7.72	3.19	5.32	1.10	48.13	0
5	Tapioca 90%	73.38	17.98	55.40	6.79	2.17	6.30	0.09	46.11	0
6	Tapioca 100%	71.90	21.47	50.43	3.67	0.39	5.45	1.03	60.91	1.4
7	Composite flour 100%	82.85	18.49	64.36	1.55	8.36	7.20	1.33	46.05	0

Table 3 Chemical characteristics of extrudate made from taro flour composites

Code	Protein	Water content	Ash content	Fat	Crude fiber	Carbohydrate	Expansion index	Color			Calories
	(%)							L	a	b	Kal/100 g
F1	1.08	5.86	2.01	4.71	4.83	81.52	0.44	55.41	8.81	22.64	372.81
F2	1.30	6.25	2.11	1.89	4.20	84.26	0.38	49.47	9.40	21.97	359.27
F3	0.96	5.79	1.88	1.19	4.42	85.77	0.40	42.15	11.45	22.46	357.60
F4	1.30	5.78	2.00	1.47	1.73	87.71	0.16	45.54	11.62	22.98	369.28
F5	1.64	5.68	2.25	3.42	2.47	84.55	0.22	44.09	9.53	23.28	375.51
F6	1.93	5.98	2.15	4.79	2.96	82.19	0.27	48.72	10.02	25.01	379.59
F7	1.62	5.50	2.15	5.29	4.07	81.36	0.37	53.85	9.38	24.21	379.57
F8	1.41	6.62	2.03	5.72	3.73	80.50	0.34	50.90	8.80	23.04	379.10
F9	1.45	5.93	2.03	803	6.47	76.10	0.59	49.89	8.96	26.51	382.43
F10	1.49	6.06	2.21	8.64	6.89	74.71	0.63	54.48	8.37	27.49	382.57
F11	1.49	5.03	1.71	9.41	6.35	76.01	0.58	48.14	8.23	25.96	394.70
F12	0.87	5.33	1.80	9.38	1.57	81.05	0.14	56.95	5.98	25.47	412.09

The addition of egg flour resulted in fat content and calories of the extrudates varied, however, not to the content of moisture content, ash, protein, crude fiber, carbohydrates, color and degree of expansion. Extrusion product moisture content was slightly higher than that required by the Indonesian National Standards maximum of 4% and US Patent 2 to 2.5% [17] and 3% [18]. Water content plays an important role in taro extrusion, and the feeding was the most important factor [4]. Lower moisture content will result in the smaller expansion index. Puffing was caused by heat and pressure release when extrudate came out from die, causing water trapped in the extrudate to evaporate. Water vapor is trapped by a layer of film that formed due to gelatinization and protein denaturation and expands to form air cavities. The ash content ranged from 1.88 to 2.15%. This value is lower than the extrudate taro mixed with maize, which is 3.78%. Taro belongs to the family Araceae recognized to be rich in calcium oxalate crystals which are the components responsible for the acridity of uncooked taro [19]. The protein content of extruded products ranged from 0.97 to 1.93%, and there is no requirement on the protein content of standard snacks. During extrusion, protein will be denatured; thus, its conformation was modified particularly at tertiary and quaternary structure. High pressure in the extruder will straighten the structure, and this provides a right product with the desired texture. Flow conditions in the die are very important in forming good straight alignment and structure [10]. Levels of fat extrusion products vary from 1.19 to 9.41%, and there is a tendency that the higher the added flour egg the fat is increasing, though is not consistently. The fat content is much lower than that required by Indonesian National Standards that a maximum of 30%. Fats and oils in the product extruder will change the texture, taste and flavor [20].

The formation of fatty acids and starch during the process can be increased by increasing the amount of amylose in starch [21]. The new structure actually limits the process of expansion of the product. Fat will form a layer on the outside of the starch granules and also will inhibit the penetration of water into the granules. Lower water penetration produces only lower gelatinization. Carbohydrates in extruded taro products are relatively uniform and ranged between 74.71 s/d 87.71%. There is no standard requirement of carbohydrate content in Indonesian National Standard. High pressure will increase the efficiency of gelatinization but still often the end product was not formed. Maillard reaction between sugars and aminoacids causes brown color on the product [22]. GMS will form complexes with amylose thus prevent the starch granules swell and cause a reduction in expansion strength and solubility. Degrees of expansion of extruded product of taro are ranged from 0.14 to 0.63%. Some researchers [23, 24] have reported that the maximum expansion degree of extruded snacks is not only related to starch content but also to the amylose content and degree of starch gelatinization during extrusion [25, 26]. The higher starch and amylose contents and lower fiber levels could be responsible for the higher expansion of extrudates. The taro extruded products have high calorie value because every 100 g contains more than 100 calories. Calorie value varies between 357.60 and 412.09 Kal. The higher percentage of composite flour added

Table 4 Response of panelist on taro crunch after improvements of formulations

Formulation	Response of panelist			
	Appearance	Color	Texture	Total
F1	1	2	1	4
F2	3	2	1	6
F3	3	2	1	6
F4	3	2	2	7
F5	3	2	3	8
F6	3	2	3	8
F7	3	3	2	9
F8	3	3	3	8
F9	3	2	2	7
F10	2	2	2	6
F11	1	2	2	5
F12	1	2	3	6

Remarks 1 bad, *2* moderate, *3* good

did not lower the brightness value of the products. In general, the results of this formulation are still not able to expand to extrudate to the maximum. According to the panelists with organoleptic scoring (Table 4), best products in terms both of shape, color and texture are of formula 7, which is the mixture of composite flour 42.25–43.75 and 5% egg flour without the addition of baking soda; however, the crispness is still required for an improvement.

7 Conclusion

The optimal formula for making taro crunch is on the ratio between the composite flour taro and tapioca is 1:1. From this formula, it can be developed to produce taro crunch better. The best formulas are taro crunch with the percentage of taro composite flour 43.75%, egg flour 5%, GMS 1.5%, vegetable oil 5%, salt 1%. The produced taro crunch has characteristics of protein 1.62%, moisture content 5.5%, fat 5.29%, crude fiber 4.07%, carbohydrate 81.36%, energy 379.57 Kal/100 g with the expansion degree of 0.54%.

8 Acknowledgement

We would like to thank to Heru Dwi Janarko for his contribution in some analyses and some literature cited.

References

1. Sukasih, E and Setyadjit. 2012. Taro-based formulations for the manufacture of flake breakfast meal high energy with oven method. J.Pascapanen.
2. Ikpeme-Emmanuel, C.A, Okoi, J and Osuchukwu, N.C. 2009. Functional, anti-nutritional and sensory acceptability of *taro* and soybean based weaning food. African Journal of Food Science Vol 3. (11) pp. 372–377.
3. Badan Standarisasi Nasional Indonesia. 2000.SNI01-2886-2000.Makanan ringan ekstrudat. Jakarta.
4. Nurtama,B.and JenShinn.2009. Effects of process variables on the physical properties of taro extrudate. Journal World Journal of Dairy & Food Sciences 2009 Vol. 4 No. 2 pp. 154–159.
5. Pilli, T.D.,Carbone,B.F.,Fiore, A.G and Severini, C.2007.Effect of some emulsifiers on the structure of extrudates with high content of fat.http:www.aisncommons.js.
6. Riaz, M. N. (2001). Selecting the right extruder. In R. Guy (Ed.), Extrusion cooking Technologies and applications (pp. 29e49). USA: Woohead Publishing in Food Science and Technology Ltd. CRC Press LLC.
7. Singh, S., Gamlath, S., and Wakeling, L. 2007. Nutritional aspects of food extrusion: a review. International Journal of Food Science and Technology, 2007; 42(8); 916–929.
8. [8] Chaiyakul, S., Jangchud, K., Jangchud, A., Wuttijumnong, P., and Winger, R. Effect of extrusion conditions on physical and chemical properties of high protein glutinous rice-based snack. Food Science and Technology, 2008; 42(3); 781–787.
9. Maga, J. A., Bang-Liu, M., & Rey, T. (1993). Taro (Colocasia esculenta) extrusion. Carbohydrate Polymers, 21, 177e178.
10. Janssen, L.P.B.M. 1993.Influence of process on raw material properties. In Extrusion Cooking. Encyclopedia of Food Science, Food Technology and Nutrition. Edited by Macrae, R., Robinson, R.K. and Sadler, M.J. Academic Press Ltd.London.
11. [11] Wang, w.M.,Klopfenstein, C.F and Ponte, J.G. 1993. Effects of tein screw extrusion on the physical properties of dietary fiber and other components of whole wheat and wheat bran and on the baking quality of the wheat bran. Cereal Chemistry. 70(6):707–711.
12. Hang, D.T., Preston, T.R.2010. Effect of processing taro leaves on oxalate concentrations and using the ensiled leaves as a protein source in pig diets in central Vietnam. Livestock Research for Rural Development 22 Article.
13. [13] Lazou, A and Krokida, M.2010. Structural and textural characterization of corn-lentil extruded snacks. Original Research Article. Journal of Food Engineering, 100 (3):392–408.
14. Guy, R. (2001). Raw materials for extrusion cooking. In R. Guy (Ed.), Extrusion cooking: Technologies and applications (pp. 5e28). USA: Woohead Publishing in Food Science and Technology Ltd. CRC Press LLC.
15. Agama, A. E., Ottenhof, M. A., Farhat, I. A., Paredes, O., Ortíz, D. J., & Bello, L. A. (2004). Efecto de la nixtamalizaci6n sobre las características moleculares delalmid6n de variedades pigmentadas de maíz. Interciencia, 29(11), 643e649.
16. [16] Muchtadi, T.R.,Purwiyatno.,Basuki, A.1988. Teknologi Pemasakan Ekstrusi.Institut Pertanian Bogor. Bogor.
17. [17] Mesu, G.J and Boot, Jacobus. 2007. Manufacture of Granola and Snack Food Product. US Patens No 7, 169,422.
18. Malfait, J.L.2007.Puffed Starch Snack Product. US Patent No 7, 141,257.
19. Sefa-Dedeh, S., & Agyr-Sackey, K. (2004). Chemical composition and the effect of processing on oxalate content of cocoyam Xanthosoma sagittifolium and Colocasia esculenta cormels. Food Chemistry, 85, 479e487.
20. [20] Harper, J. M. (1979). Food extrusion. CRC Critical Review ofFood Science and Nutrition, 11, 155–215.
21. Artz,W.E.,Rao, S.K. and Sauer, R.M. Jr.1991. Lipid Oxidation in Extruded Products during storage as affected by extrusion temperature and selected antioxidants. In Food extrusion science and technology edited by Kokini, J.L., Ho. Chi Tong, Karwe,M.V. New York.

22. Johnson, I. 1993. Chemical and Nutritional Changes. In Extrusion Cooking.Encyclopaedia of Food Science, Food Technology and Nutrition. Edited by Macrae,R.,Robinson, R.K andSadler, M.J. Academic Press Ltd.London.
23. Chang, Y. K., Silva, M. R., Gutkoski, L. C., Sebio, L., & Da Silva, A. P. (1998). Development of extruded snacks using jatobá (Hymenaea stigonocarpa Mart) flour and cassava starch blends. Journal of the Science of Food and Agriculture, 78(1), 59e66.
24. Rampersad, R., Badrie, N., & Comissiong, E. (2003). Physicochemical and sensory characteristics of flavoured snacks from extruded cassava/pigeonpea flour. Journal of Food Science, 68(1), 363e367.
25. Case, S. E., Hanna, M. A., & Scwartz, S. J. (1992). Effect of starch gelatinization on physical properties of extruded wheat and corn based products. Cereal Chemistry, 69, 401e404.
26. Parker, M. L., Lopez-Andreu, J., Maina, G., Esteban, R. M., Smith, A. C., & Waldron, K. W. (1999). Modification to physicochemical and nutritional properties of hard to cook beans (Phaseolus vulgaris L.) by extrusion cooking. Journal of Agricultural and Food Chemistry, 47, 1174e1182.

Factors Determining the Practice of Sustainable Cassava Farming System in Tanjungsiang—West Java

Carolina Carolina

Abstract For almost 6 decades, farmers in Tanjungsiang of West Java Province cultivate cassava (*Manihot esculenta*) as main crop without any sign of decrease in area nor productivity. Locally known as *sampeu manggu* or manggu cassava, the plant variety is Adira-2. With the average land productivity of 20–30 tonnes per hectare, the steady output is considered special due to the crop's character that tends to cause soil nutrition losses. Investigation on factors determining the implementation of sustainable cassava farming strategy was conducted, utilizing socio-ecological approach to perform a case study. Data and information are obtained through group discussion and participatory observation and analyzed descriptively, and the data indicate that farmers' preference to plant cassava as their main crop all year round is manifestation of their determination to pursue long-term land productivity. Market demand is accommodated accordingly without extreme conduct of boosting land productivity. Despite the relatively low economic value, market certainty provides assurance to the farmers to keep cultivating cassava and maintain productivity. Moreover, its special quality of the crop serves as important symbol to the community's pride which leads to a collective decision to apply ecological friendly land management system leading to sustainability.

Keywords Agriculture ecosystem · Cassava · Environmental management · Sustainable innovation · Tanjungsiang—West Java

1 Introduction

Indonesia has long been known as an agricultural-based country in the world producing many important commodities. Being a tropical country with two seasons, wet and dry, agricultural activities are mainly dependent on the seasons. Farmers

C. Carolina (✉)
Development Center for Appropriate Technology Indonesian Institute of Sciences,
Jalan KS Tubun no. 5, Subang, West Java, Indonesia
e-mail: carolina.4q@gmail.com

T. Taufik et al. (eds.), *ICoSI 2014*, DOI 10.1007/978-981-287-661-4_5

adapt themselves to the climatic conditions, and it influences crop choice and time of planting. Although number of traditional farmers in Indonesia decreases overtime due to industrialization, their presence cannot be neglected because their important role as staple food provides for the nation. Besides rice—the staple food for Indonesians—food diversity policy direction has been inclined to root crops promotion since it is still widely consumed locally by rural communities. One of root crops species being promoted is cassava.

Indonesia grows cassava in an area of 1,066,254 ha producing 23,824,008 tonnes with a productivity of 22.4 tonnes per hectare. Of those, West Java Province placed as the 4th highest producers of cassava after Lampung, East Java and Middle Java Provinces. For many purposes, 95,505 ha of land is cultivated with cassava. Nevertheless, West Java lost around 77,731 ha of cassava land during the last decade [1]. Widely known as poor smallholder tuber crops, cassava was cultivated to fulfill family need, but lately the use of cassava as raw material for many processed products was introduced. The latter induced extensification and intensification, not only as a support to food security but also to provide raw materials for variety of industries [2].

The trend to extend and intense should be cautiously observed, for cassava is also known as a crop that cause a significant depletion of soil nutrition; environmental degradation due to long period of cultivation evidently occurs in many areas. This phenomenon occurs basically due to the whole root removal from the soil and generally short turn around time for soil recovery [3, 4]. At the end, it leads to productivity decrease, an unfavorable circumstances for farmers.

Sustainable cassava cultivation is not impossible to achieve. A small region in Subang area of West Java Province called Tanjungsiang Sub-district is known for long history of cassava planting. No sign of productivity depletion is exceptional, thus understanding the phenomenon is utmost important in order to acquire relevant knowledge to be implemented elsewhere since sustainable cassava cultivation is not commonly found. A thorough exploration of cassava farming system will reveal not only the agricultural innovation practiced by farmers' community, but more importantly uncovering the factors promoting their decision to innovate underlining their decision to achieve sustainability.

2 Methodology

This paper is constructed based on a field study that has been conducted during the period of October 2013 to March 2014 in order to gain understanding on cassava farming system practiced by the farmers' community in Tanjungsiang Sub-district of West Java Province. Socio-ecological approach is utilized to allow holistic perception to take place. The principle of human interaction with the environment is used as main reference to be explored reciprocality exemplified as phenomenon in environmental management [5, 6]. This framework of thinking is selected because focus of my study is the fact that farmers' community has been successfully survived a long period of agricultural land exploitation to grow cassava, a crop species known as soil quality degrader. The farmers' perspective and their practice are integratedly perceived in the agroecosystem [7]. All those were enfolded in a case study [8], the most appropriate strategy to pose how and why questions and the data collected through group discussion and participatory observation to be analyzed qualitatively and specifically on factors that motivate farmers to practice environmentally friendly farming system.

3 Results and Discussion

3.1 *The Study Area*

Tanjungsiang in West Java Province is known for its cassava production. Its long popularity has made the people proudly call themselves cassava community. Cassava has been always present in their agricultural land to fulfill their need as complement to daily meal. Since early 1990s, cassava production increased significantly triggered by the development of a cassava chips production which required as much as 5 tonnes per day. Due to the demand of that one producer, farmers were motivated to plant cassava extensively. As small industry grew, more people were inspired to produce various kinds of snacks based on cassava. Its prime quality has made the demand for cassava from Tanjungsiang increased and eventually more plots are cultivated. Starting from an intercrop species, cassava presently is the main crop for most of the rainfed areas in Tanjungsiang reaching approximately 1500 Ha. Cultivated as the main cash crop for almost 6 decades, there are almost no indications of shifting strategy despite introduction of new cash crops. The number of cassava growers seems not to decrease overtime nor the land productivity. In fact, the area is quite well known for its cassava production, mainly because the prime quality in terms of taste and texture is most favorable for food [9]. Cassava from Tanjungsiang is locally known as *sampeu manggu* or *manggu* cassava. It is a generic plant variety of generic Adira-2 that can be found elsewhere in many regions in Java and Sumatera. The average land productivity of *sampeu manggu* is within the range of 20–30 tonnes per hectare.

The above-mentioned facts are undoubtedly valuable to be utilized as reference for those who plan to expand the crop cultivation. Although Taufiqurrahman et al. [10] pointed that 19.01% land in Tanjungsiang Sub-district is categorized as degraded, the level is still considered light. Tanjungsiang farmers believe that cassava does not always have to be a soil nutrient miner, and sustainable farming system of cassava is not an impossible mission to achieve. However, the challenge is how should we support the practice of ecologically balanced strategy and at the same time it ensures prosperity of the farmers. It is, therefore, utmost important to gain thorough understanding on what are the determinant factors to achieve sustainability of cassava farming practices, more than the technology itself.

4 The Practice of Sustainable Cassava Farming System

Rarely found in a vast area of agricultural land, cassava is averagely cultivated in relatively small plots of approximately 3500 sqm. Most of the cassava farmers are indeed smallholders. On their relatively limited arable agricultural land, they plant Adira-2, the variety of cassava which specifically utilized for food. Most plots are cultivated with cassava all year round, and its dependency on rainwater has compelled the farmers to learn to wisely manage the agricultural ecosystem for their utmost benefit. Although cassava is known as a crop that can withstand dry seasons, many farmers will not do planting in the month of July to September. Their long experience teaches them not to plant cassava during those months when water availability is the least in a year. They will, sometimes, shorten the fallow period if they consider the soil moisture is sufficient to support cassava stem to grow.

Theoretically, cassava reaches maturity at about 9–12 months after planting. However, farmers harvest their cassava tubers when buyers are available and the price is just right. There are cases, although rare, of cassava being harvested when the crop is 8 months of age. Those who can afford sometimes withhold harvest and wait until the selling price is agreeable.

In general, agricultural activities in cassava farming system in a year round can be depicted as follows in Table 1.

As indicated from the table, agricultural activities of cassava (*Manihot esculenta*) farming can be found in almost every month of the year. Extricated from series of group discussions, and in-depth interviews with key informants, it can be concluded that cassava's popularity among the farmers of Tanjungsiang is attributed to 4 main factors of technology and economy, i.e., attainable agricultural technology, manageable pest problem, constant market demand of cassava and easy access to market the product. However, another important factor is that the practice of cassava farming provides opportunity to nurture social relations in the community.

Those are the factors that promote longevity of cassava farming implementation. Farmers do not have to obtain new farming knowledge since the cropping system management has long been known in the community. They do not have to endure the hassle of getting new technology as experienced by Jamaican farmers who were introduced Yam Minisetts that requires intensive approach for them to adopt [11]. If there is a need of change or development of new technology, it does not differ much from the traditional technology. Less complexity in handling and managing the crops, and relatively no significant pest problems ever occur beyond their capacity to tackle, has helped in developing strong confident to reckon the crop as their main source of family income. In summary, the characters of cassava farming system in the study area are depicted in Table 2.

Cassava agricultural system is not an independent entity in rural area of Tanjungsiang. Its existence is associated ecologically with rice fields through production of chicken farms used as main fertilizer for cassava to grow. Rice fields provide rice husks used by the poultry as media to catch chicken droppings. It

Table 1 Agricultural activities in cassava farming system in Tanjungsiang, Subang

Type of Activities	J	F	M	A	M	J	J	A	S	O	N	D
Land Preparation												
ploughed, harrowed, ridged									■			
Application of chicken manure									■			
Planting												
Cassava										■		
Peanuts										■		
First Chem-Fertilizer Application										■		
First Weeding											■	
Second Weeding	■											
Peanuts Harvesting	■											
Second Chem-Fertilizer Application		■										
Cassava Harvesting						■						
Fallow Land							■	■				

Table 2 Characters of cassava farming system in Tanjungsiang Sub-district of West Java Province

Seeds	Adira-2 or locally known as sampeu manggu Planting stocks are available locally (among the neighborhood) for free
Labors	Family and neighbors;
Technology	Utilization of simple agricultural tools, i.e., hoe, machete, golok
Cropping system	Polyculture; mixed with kacang suuk or groundnuts (*Arachis hypogaea* L.), jagong or *Zea mays* L, cikur or galanga or aromatic ginger (*Kaempferia galanga*), *Jahe or ginger root* (*Zingiber officinale)*
External inputs	Low inputs of chemical fertilizers and mechanization; but high quantity of chicken manure obtained in affordable price
Harvest	Manually
Markets	Through middleman; sold locally to supply small industries; and also to other town within the province

absorbs moist and serves as the manure particle holder to make handling easier. Carolina and Novianti [12] concluded that the utilization of chicken manure is an expression of how well farmers adapt themselves to the dynamic change of agroecosystem. Chicken farming activity is relatively new to the locals. Prior to their existence, cassava cultivation obtained its organic fertilizer from cow manure as cattle is quite common in the area. However, when poultry activities become popular, they leave the habit of using cow manure and shift to the use of chicken manure in the knowledge that the later has a better quality of nutrition content. Chicken manure is, indeed, a good soil amendment; it adds organic matter and increases the water-holding capacity and beneficial biota in soil. The content of nitrogen, phosphorus and potassium that can be absorbed by the plants is higher than those of cow manure. The traits are well understood by the farmers.

Besides utilization of organic fertilizer available locally, cassava farmers implement multiple cropping strategy not only to gain additional income from the system, but also to provide cover crop which is important to reduce nutrient losses occurrence from soil particle dislodges. This situation is commonly happened in cassava monocultures with no cover crops to limit runoffs from occurring as found in cassava plantation in Vietnam [13]. Furthermore, with or without fertilizer, serious soil loss due to erosion in cassava monoculture system is evident as the case found in Lampung Indonesia [14].

Coupled with simple technology implementation with almost no mechanization, cassava farming system is indeed environmentally friendly. The benefit of preparing land heavy mechanized machineries is that they keep the top soil in place. Aside from the land preparation process, weeding is never carried out aggressively, so that nitrogen and phosphorus from the weed plants will be recycled in the system.

The agricultural technology practice seems to become a standard procedure for farmers to implement. Overall, the steps conducted could be called as an ecologically sound agricultural innovation. What is interesting is this develops to be a collective action rather than individuals. This could signify the fact that cassava farming system has been successfully maintained without any sign of environmental degradation nor productivity decrease. Marten [5] and Altieri [15] noted the phenomenon as agroecosystem that comprises interacting factors within each system and integration of many, in which, at the end, capability of human to adapt to the biogeophysical system is important determinant factors to the attainment of survivorships. By and large, pursuance of sustainability is inseparable from appreciation of cassava value chain. Understanding and appreciation of the value chain will guide us to gain exposure of many plausible strategy to be implemented as clearly achieved by Nigeria that promotes variety of programs directed at the development of the root crops for the sake of millions people who engage on the activities [16].

5 Factors Promoting the Practice of Sustainable Cassava Farming System

Aside from the less complexity of agriculture technology in cassava farming system management, the study indicates that Tanjungsiang's farmers' preference on cassava as their main crop is basically motivated by the constant market demand and the easy access to market the product. Despite the relatively low economic value, market certainty provides assurance to the farmers to keep growing cassava and strive to maintain the productivity. Moreover, its special quality of the community's pride leads to a collective decision to attain sustainability. And this is to some extent related to an expression of risk adjustment strategy as indicated by Gupta [17], an ability to bounce back from any adversity. As noted by Altieri [18] and

Berardi et al. [19], adaptiveness to the agriculture ecosystem to include biotic and abiotic factors, as well as the social system, will eventually lead to achievement of sustainability. And this phenomenon found in the study area deserves to be explored further.

The reasons why should they strive to devote themselves to cassava, lead to important realities, a combination of economic and social value of cassava. Cassava production remains as the main source of family income, although we found that from 3500 sqm of land farmers' average income is in the area of poverty level ($ 1.25 ppp[1]). Despite the low revenue, cassava planting system provides opportunity for them to have certainty. Local demand, as well as regional, has never been depleted. Furthermore, the price tends to stay stable and never fluctuates significantly. Thus, economically, cassava production is a potential source of income.

Socially, cassava farming system activity provides compliant span for land holders to support landless farmers to work in the field. Farm works such as land preparation, weeding, harvesting are usually made available. This is expressed by all farmers' respondents, indicating that socially cassava farming system plays an important role to nurture silaturahmi and attain a sense of good social cohesion in local neighborhood.

Both underlying factors, i.e., economy and social, cannot be perceived as separate to the farming system. Reciprocally, they interact, the interaction of which shapes socio-ecological resilience of rural area. As the key to balance, it is important to always evaluate the resilience of each system so that sustainability state will always be achieved. This finding is important for those who need to promote cassava-based agricultural development without leaving farmers' well-being behind. Referring to elaboration by Awoyinka [20] and consideration of the interacting determinant factors of cassava cultivation in Tanjungsiang agroecosystem, developing the area as a center of cassava production to fulfill different market niche is quite plausible. However, the prime factor to be considered should always be the farmers' sustainable livelihood. In other words, farmers as integral part of the agroecosystem should be the focus of any plan of strategic agricultural development. The case of Tanjungsiang cassava farmers shows us that through minimal intervention from government institution, they are able to maintain the agroecosystem productivity. However, this situation should be improved because as cassava producer their role is utmost important. The government obliges to provide necessary services, especially to ensure farmers' sustainable livelihoods which also mean securing the sustainable provision of cassava. Many schemes can be promoted and supported by the government, to include pro-farmers' business networking [21], and also incorporating grassroots innovation as an important agent of development [22]. All those strategies are implemented to promote farmers prosperity that has become one of most important issue for Indonesia to achieve sustainable development goals.

---

[1]ppp: purchasing power parity.

6 Conclusions and Recommendations

Indicated from the farming system practices of farmers' community in Tanjungsiang Sub-district of West Java Province in Indonesia, cassava cultivation does not always have to cost a significant environmental loss. Their strategy to manage the agricultural land is an indication of a co-adaptation of the farmers with their environment.

To some extent, the economic and social role of cassava farming system has become farmers' main motivation to always pursue ecological balance. Although it is proven that for the last 6 decades, there is no case of significant decrease in productivity, we must always keep an open mind for any possible intervention due to socio-economic change in the region that will induce a significant distortion to the resilience of the ecosystem.

Ecologically sound cassava farming practices are evidently achievable and, however, should not be considered as granted. Farmers deserve to have higher income—than what is presently gained—from their cassava production. That is attainable only by creating economic activities based on higher value of cassava processing and organized by the community itself. It should be the long-term strategy to ensure sustainable sources of equitable returns for cassava farmers, the main actor and supporter to achieve ecological balance.

To maintain the sustainability, it is therefore important to foster a strategic plan of cassava development perceived from the crops value chain which actively includes participation of government, businesses, research and development centers and the community.

Acknowledgements Gratitude goes to the Development Center for Appropriate Technology Indonesian Institute of Sciences for providing the opportunity to conduct the research. The data collection is made possible with the support of Ms. Fithria Novianti and kind cooperation of local farmers in any occasion.

References

1. Anonymous, 2015. Tanaman Pangan Ubikayu. http://data.go.id/dataset/tanaman-ubi-kayu-per-provinsi/resource/6f64a9f0-b3b4-45b4-98e5-1f18c3493a63. Accessed on 8 February, 2015.
2. Burns, A., Gleadow, R., Cliff, J., Zacarias, A. and Cavagnaro, T. 2010. Cassava: The Drought, War and Famine Crop in a Changing World. *Sustainability,* 2010, 2, 3572–3607; doi:10.3390/su2113572. Accessed on 5 May 2015.
3. Daellenbach, G.C., Kerridge, P.C., Wolfe, M.S., Frossard, E., Finckh, M.R. 2005. Plant productivity in cassava-based mixed cropping system in Columbian hillside farms. *Agriculture, Ecosystems and Environment*, vol.105. 595–614.
4. Ruysschaert, G., Poesen, J., Verstraeten, G., and Govers, G. 2007. Soil loss due to harvesting of various crop types in contrasting agro-ecological environments. *Agriculture, Ecosystems and Environment*, vol.120. 153–165.

5. Marten, Gerald G. 2001. *Human Ecology – Basic Concepts for Sustainable Development.* Earthscan Publication.
6. Vayda, A.P. 1983. Progressive Contextualization: Methods for Research in Human Ecology. *Human Ecology*, vol 11, no.3.
7. Wezel, A., Bellon, S., Dore, T., Francis, C., Vallod, D. and David, C. 2009. Agroecology as a Science, a Movement and a Practice. A Review. *Agronomy for Sustainable Development,* vol. 29. 503–515.
8. Yin, Robert K. 2003. *Case Study Research Design and Methods.* Thousand Oak.Sage Publications.
9. Carolina, 2009. Appraising Opportunity of Manggu Cassava Base Economic Activity. (In Indonesian). *Proceedings of Grassroots Innovation Workshop.* Center for Appropriate Technology Development. Indonesian Institute of Sciences. Bandung.
10. Taufikqurrahman, B.P., Suriadikusuma, A. and Haryanto, R. 2014. Analysis of Potential Land Degradation of Cassava (*Manihot utilissima*) in Dryland Ecosystem of Tanjungsiang Subdistricts – Subang Regency (In Indonesian). *Jurnal Agro,* vol 1, no. 1.
11. Beckford, C.L. 2009. Sustainable Agriculture and Innovation Adoption in a Tropical Small-Scale Food Production System: The Case of Yam Minisetts in Jamaica. *Sustainability,* 2009/81-96; doi:10.3390/su1010081. Accessed on 5 May 2015.
12. Carolina and Novianti, F. 2013. Ecological Innovation of Cassava (*Manihot esculenta*) Cultivation in Rancamanggung – Subang District: A Local Initiative (In Indonesian). *Proceeding of National Seminar and Workshop. Enhancement of Innovation in Poverty Alleviation.* Indonesian Institute of Sciences.
13. Phien, T and Vinh, N.C. 1998. Soil Organic Matter Management for Sustainable Cassava Production in Vietnam. *In Proceeding of the 7th Regional Workshop of Cassava Research and Development in Asia : Exploring New Opportunity for an Ancient Crop (Ed. RH Howeler). Page 234–250.*
14. Wargiono and Ispandi, 1998. Cassava Agronomy Research and Its Contribution to a Secure Food System in Indonesia. *In Proceeding of the 7th Regional Workshop of Cassava Research and Development in Asia : Exploring New Opportunity for an Ancient Crop (Ed. RH Howeler). Page 174–182.*
15. Altieri, M. 2002. *Agroecology: The Science of Natural Resource Management for Poor Farmers in Marginal Environment.* Elsevier Science.
16. Odemero, A.F. 2015. Impact of Cassava Value Chain Intensification Intervention on Nigerian Economy: Evidence from Delta State, Nigeria. *Turk Tarim ve Doga Bilimleri Dergisi*, vol 2 (1). 18–25.
17. Gupta, Anil K. 1988. Survival Under Stress: Socio Ecological Perspective on Farmers Innovation and Risk Adjustments. *Proceedings of the International Congress of Plant Physiology,* New Delhi. India. February 15–20.
18. Altieri, M. 2004. Linking Ecologist and Traditional Farmers in the Search for Sustainable Agriculture. *Frontiers in Ecology and the Environment* 2: 35–42. http://dx.doi.org/10.1890/1540-9295(2004)002.0035.LEATFI.2.0.CO.2 Accessed on 8 February, 2015.
19. Berardi, G., Green, R., and Hammond, B. 2011. Stability, Sustainability, and Catastrophe: Applying Resilience Thinking to U.S. Agriculture. *Human Ecology Review*, vol 18, No. 2. 2011.
20. Awoyinka, Y.A. 2009. Cassava Marketing: Option for Sustainable Agricultural Development in Nigeria. *Ozean Journal of Applied Science* 2(2). 175–183.
21. Saptana and Ashari. 2007. Sustainable Agricultural Development through Business Partnerships (In Indonesian). *Jurnal Litbang Pertanian*, 26 (4).
22. Seyfang, G. and Smith, A. 2007. Grassroots Innovations for Sustainable Development: Towards a New Research and Policy Agenda. *Environmental Politics,*vol 16, No. 4. 584–603.

In Vitro Sterilization and Shoot Induction of Fig (*Ficus carica* L.) Using MS Containing GA_3 Medium Supplemented with BAP and NAA

Innaka Ageng Rineksane, Rahman Budiawan and Gunawan Budiyanto

Abstract The purpose of this research was to obtain the proper sterilization method for fig (*Ficus carica* L.) shoot and determine the best combination of BAP and NAA for in vitro shoot induction of fig. The sterilization research was arranged in completely randomized design (CRD), a single factor with five treatments. The treatments tested were dipping in NaClO (0.1, 5% for 7 min), fungicide and bactericide 4 g/L (5 min, 30 min, 1 h, 2 h, 3 h). Each treatment was repeated three times. Parameters measured were the percentage of browning explants, the percentage of contaminated explants and the percentage of survive explants. Shoots induction research was arranged in completely randomized design (CRD), a single factor with seven treatments. The treatments tested were the addition of BAP (0, 2, 4, 6 mg/L) and NAA (0, 0.5, 1 mg/L) to the MS medium containing GA_3. Each treatment was repeated six times. Parameters measured were the percentage of contaminated explants, the percentage of survive explants, percentage of shoot formation, percentage of browning explants, number of shoots, shoot height, number of leaves and percentage of rooted explants. The best method for fig shoot sterilization was dipping explants in NaClO10%, 5′ + NaClO 5%, 7′ + fungicide and bactericide 4 g/L for 3 h with the percentage of survival explants 100% during 30 days of observation. The addition of 2 mg/L BAP and 0.5 mg/L NAA into MS containing GA_3 medium was favorable to produce among the highest percentage of fig shoot formation (33.33%) and the browning explants (0.00%).

Keywords Fig (*Ficus carica* L.) · Sterilization · Shoot induction · BAP (6-benzylaminopurine) · NAA (a-naphthalene acetic acid) · GA_3 (gibberellin acid)

I.A. Rineksane (✉) · R. Budiawan · G. Budiyanto
Department of Agrotechnology, Faculty of Agriculture,
University Muhammadiyah Yogyakarta, Yogyakarta, Indonesia
e-mail: rineksane@gmail.com

R. Budiawan
e-mail: rahmanmu_415@yahoo.com

G. Budiyanto
e-mail: goenb@yahoo.com

T. Taufik et al. (eds.), *ICoSI 2014*, DOI 10.1007/978-981-287-661-4_6

1 Introduction

Fig (*Ficus carica* L.) is known as relatives of banyan tree. This plant is widely grown in coastal areas Balkans to Afghanistan, and then developed in Australia, Chile, Argentina, the USA, Middle Eastern countries and Mediterranean region. The Global Diabetes Community [1] stated that the fig leaf is known as diabetes drugs in Spain and southwestern Europe. Fig fruit contains many nutrients such as carbohydrates, proteins, vitamins, minerals, fiber and others. Fig fruit also contains high dietary fiber. Each 100 g of dried fig contains 10.95 g fiber in which this concentration is higher than those in apple and orange fruit, i.e., 3.33 and 3.4 g. Fig also contains unsaturated fatty acids such as omega-3 23.87 + 6.27%, omega-6 23.04 + 0.48% and omega-9 19.72 + 1.07% that function to prevent coronary heart disease [2]. Each 100 g or dried figs contain 180.75 mg calcium, 75.75 mg magnesium, 759.75 mg potassium and 2.26 mg iron [3].

Based on its benefit and it is still a rare fruit in Indonesia, the tin has a great opportunity to be cultivated. Fig trees are still planted in some areas in Indonesia, especially in Java [4]. Therefore, a technique such as in vitro culture to multiply large quantities of shoots in a short amount of time is needed. In vitro culture is influenced by the use of plant growth regulator as a driver of growth and development of explants, such as naphthalene acetic acid and Benzylaminopurine. Those NAA and BAP can be added alone or together into medium, and the explants morphogenesis will be depend on interaction of both growth regulators [5]. The objective of this study was to obtain proper sterilization method and to determine the best concentration of BAP and NAA in fig shoots induction in the MS medium containing GA_3.

2 Materials and Methods

The experiment was conducted in In Vitro Culture Laboratory, Faculty of Agriculture, University Muhammadiyah Yogyakarta. The materials used in this research were fig shoot explants, plant growth regulators such as BAP, NAA and GA_3, sterilization compounds, i.e., bactericide, fungicide, NaClO 5%, ethanol 70%, detergent, iodine and vitamin C. The basic culture medium used in this study was Murashige and Skoog (MS) medium.

This study consisted of two experiments, first was the optimization of sterilization, and second was the effect of plant growth regulators on the induction of fig shoots in MS medium. The treatments in this research are: types, concentration and dipping time of sterilization compounds at first experiment, whereas the concentration of BAP (0, 2, 4 and 6 mg/L) and NAA (0, 5 and 1 mg/L) was used at second experiment.

2.1 *Optimization of Sterilization*

In this study, the variables were the concentration of sterilization compounds and dipping time in the sterilization liquid as shown in Table 1. Experimental units were arranged in completely randomized design (CRD) one factor. Each treatment combination was replicated five times.

2.2 *Shoot Induction*

Shoots induction research was arranged in completely randomized design (CRD), a single factor with seven treatments. The treatments tested were the addition of BAP (0, 2, 4, 6 mg/L) and NAA (0, 0.5, 1 mg/L) to the MS medium containing GA_3. Each treatment was repeated six times. Parameters measured were the percentage of contaminated explants, the percentage of survives explants, percentage of shoot formation, percentage of browning explants, number of shoots, shoot height, number of leaves and percentage of rooted explants.

The treatments were as follows

A = Medium MS + BAP 0 mg/L + NAA 0 mg/L + GA_3 0.5 mg/L
B = Medium MS + BAP 2 mg/L + NAA 0.5 mg/L + GA_3 0.5 mg/L
C = Medium MS + BAP 4 mg/L + NAA 0.5 mg/L + GA_3 0.5 mg/L
D = Medium MS + BAP 6 mg/L + NAA 0.5 mg/L + GA_3 0.5 mg/L
E = Medium MS + BAP 2 mg/L + NAA 1 mg/L + GA_3 0.5 mg/L
F = Medium MS + BAP 4 mg/L + NAA 1 mg/L + GA_3 0.5 mg/L
G = Medium MS + BAP 6 mg/L + NAA 1 mg/L + GA_3 0.5 mg/L

3 Result and Discussion

3.1 *Optimization of Sterilization*

Explants sterility is one of the factors that influence the success of in vitro culture; therefore, the appropriate method of sterilization is very important to provide sterile explants. The result of sterilization observation was provided in Table 2.

Percentage of Browning Explant

The highest percentage of browning (100%) was obtained on treatment C which used NaClO 10 and 5%, respectively, then followed by the addition of fungicide 4 g/L and bactericide 4 g/L for 1 h (Table 2). Otherwise, there is no explants browning which achieved on treatment NaClO 1% 5′ + NaClO 0.1% 7′ + fungicide and bactericide 4 g/L for 30′ (B), NaClO 10%, 5′ + NaClO 5% 7′ + fungicide

Table 1 Type of sterilization compounds and their concentration, dipping time in sterilization liquid used for optimization of shoot fig sterilization

Treatment	Step 1		Step 2		Step 3		Step 4		Step 5		Step 6	
	Compound	Time	Compound	Time	Compound	Time	Compound	Time	Compound	Time	Compound	Amount
A	NaClO 1% + liquid detergent 2 drops	5′	Fungicide 4 g/L + bactericide 4 g/L	5′	Ethanol 70%	30″	NaClO 0.1%	7′	Vitamin C 0.1%	4–5 x	Iodine	5 drops
B	NaClO 1% + liquid detergent 2 drops	5′	Fungicide 4 g/L + bactericide 4 g/L	30′	Ethanol 70%	30″	NaClO 0.1%	7′	Vitamin C 0.1%	4–5 x	Iodine	5 drops
C	NaClO 10% + liquid detergent 2 drops	5′	Fungicide 4 g/L + bactericide 4 g/L	1 h	Ethanol 70%	30″	NaClO 5%	7′	Vitamin C 0.1%	4–5 x	Iodine	5 drops
D	NaClO 10% + liquid detergent 2 drops	5′	Fungicide 4 g/L + bactericide 4 g/L	2 h	Ethanol 70%	30″	NaClO 5%	7′	Vitamin C 0.1%	4–5 x	Iodine	5 drops
E	NaClO 10% + liquid detergent 2 drops	5′	Fungicide 4 g/L + bactericide 4 g/L	3 h	Ethanol 70%	30″	NaClO 5%	7′	Vitamin C 0.1%	4–5 x	Iodine	5 drops

Table 2 Effect of sterilization method on explants sterility after 30 days of planting

Treatment	Percentage of browning explants(%)	Percentage of contaminated explants (%)	Contaminants	Day of contamination	Percentage of survive explants (%)
A	66.67	100	Fungi	14.33	0
B	0	100	Fungi, bacteria	7.33	0
C	100	100	Bacteria	12	0
D	0	66.67	Bacteria	3	33.33
E	0	0	–	–	100

and bactericide 4 g/L for 2 h (D) and the treatment NaClO 10%, 5′ + NaClO 5% 7′ + fungicide and bactericide 4 g/L for 3 h (E). Browning occurs due to the high content of phenolic compounds which are oxidized when cells are injured, consequently the isolated tissue turn into brown and failure to thrive [5].

Percentage of Contaminated Explant

The result of sterilization showed that among five treatments tested, four of them caused explants contamination in different time (Table 2). It can be explained that dipping time in sterilization compound was too short and microorganism-infected explants may be used. The lowest contaminated explants (0%) was observed on treatment E which explants sterilized in fungicide 4 g/L and bactericide 4 g/L for 3 h. This result showed that dipping explants in fungicide and bactericide for 3 h were able to eliminate fungi and bacteria.

Percentage of Survive Explant

The effect of sterilization compound and dipping time showed that the longest dipping (3 h) in fungicide and bactericide resulted the highest percentage of survive explants (100%). Fungicide and bactericide used in this study is systemic which their active ingredients penetrate into plant tissue as stated by Wudianto [6]. So that, either fungicide or bactericide eliminates the fungi and the bacteria inside the plant tissue. This study showed that the longest dipping in fungicide and bactericide (treatment E) gave the lowest percentage of browning explants (0%), the lowest percentage of contaminated explants (0%) and the highest percentage of survive explants (100%), therefore, treatment E was chosen as sterilization method in shoot induction of Fig.

3.2 Shoot Induction of Fig

Shoot induction of Fig was strongly influenced by the survive explants and growth rate. The effect of BAP and NAA on shoot induction of Fig was shown in Table 3.

Table 3 Effect of BAP and NAA on the percentage of contaminated explants, the percentage of survive explants, the percentage of callus formation, the percentage of shoot formation and the percentage of browning explants after 8 weeks of culture

Treatment	Percentage of contaminated explants (%)	Percentage of survive explants (%)	Percentage of browning explants (%)	Percentage of shoot formation (%)
0 mg/L BAP + 0 mg/L NAA	40.48	50.00	16.67	0.00
2 mg/L BAP + 0.5 mg/L NAA		50.00	0.00	33.33
4 mg/L BAP + 0.5 mg/L NAA		66.67	33.33	16.67
6 mg/L BAP + 0.5 mg/L NAA		66.67	16.67	0.00
2 mg/L BAP + 1 mg/L NAA		66.67	16.67	16.67
4 mg/L BAP + 1 mg/L NAA		50.00	16.67	16.67
6 mg/L BAP + 1 mg/L NAA		66.67	16.67	16.67

Percentage of Contaminated Explants

To survive and grow properly, in vitro plant cultures need to be largely free from pest, fungi and bacterial infections. Results showed that the percentage of contaminated explants was 40.48% (Table 3) which caused by fungi and bacteria. The contamination was occurred in week 1 until 6. This contamination may be caused by endogenous microbial inside plant tissues. This was in agreement with Fahmadi [7] who stated that endogenous contamination derived from microbial contamination in plant tissues. The non-sterile explants which shipped from outstation may cause contamination. Microorganisms penetrate into explants via the water and the air in the shipping.

Percentage of Browning Explants

The results in Table 3 showed that 4 mg/L BAP + 0.5 mg/L NAA produced the highest percentage of browning explants (33.33%, Fig. 1a), otherwise 2 mg/L BAP + 0.5 mg/L NAA produced the lowest percentage of browning explants (0%, Fig. 1b). The browning explants may caused by the sap released cut explants, which continuously released hydroxyphenol. This was in agreement with Tabiyeh et al. [12] whereby in vitro browning caused by increasing of phenolic exudate from explants which then followed by their oxidation. Fig is perennial plant which containing sap; therefore, cutting explants induce the oxidized sap released from the shoots.

Percentage of Shoot Formation

Shoots began to grow at day 25 in the nodes. The results showed that treatment 2 mg/L BAP + 0.5 mg/L NAA was among the concentration which induced the highest percentage of shoot formation (33.3%, Fig. 2b) which was not significantly different with the percentage of shoot formation obtained on 4 mg/L

Fig. 1 Browning explants cultured on MS + 4 mg/L BAP + 0.5 mg/L NAA (**a**) and non-browning explants cultured on MS + 2 mg/L BAP + 0.5 mg/L NAA (**b**)

Fig. 2 Explants cultured on MS containing GA_3 + 2 mg/L BAP + 0.5 mg/L NAA after 7 weeks of culture

Table 4 Effect of BAP dan NAA on number of Fig shoots, shoot height, number of leaves and percentage of rooted explants after 8 weeks of culture

Treatment	Number of shoot	Shoot height (cm)	Number of leaves	Percentage of rooted explants (%)
0 mg/L BAP + 0 mg/L NAA	0.00	0.27	1.67	0.00
2 mg/L BAP + 0.5 mg/L NAA	0.67	1.27	1.33	0.00
4 mg/L BAP + 0.5 mg/L NAA	0.25	1.08	2.50	0.00
6 mg/L BAP + 0.5 mg/L NAA	0.00	0.55	1.75	0.00
2 mg/L BAP + 1 mg/L NAA	0.25	1.43	1.50	0.00
4 mg/L BAP + 1 mg/L NAA	0.33	1.33	2.67	0.00
6 mg/L BAP + 1 mg/L NAA	0.25	0.63	0.63	0.00

BAP + 0.5 mg/L NAA (Fig. 2c), 2 mg/L BAP + 1 mg/L NAA (Fig. 2e), 4 mg/L BAP + 1 mg/L NAA (Fig. 2f) and 6 mg/L BAP + 1 mg/L NAA (Fig. 2g) (16.67%). The results also showed that all treatment-induced shoots except those explants cultured on MS containing GA_3 medium without BAP + NAA (control, Fig. 2a) and MS containing GA_3 + 6 mg/L BAP + 0.5 mg/L NAA (Fig. 2d). This could be explained that medium without BAP did not able to induce the adventitious shoots. This result was in agreement with the study reported by Julianti [8], whereby 0.1 mg/L NAA and 2.5 mg/L BAP induced the highest shoots formation (12 shoots) in *Aquilaria malaccensis* Lamk. However, treatment 6 mg/L BAP + 0.5 mg/L NAA was not able to induce shoot formation, because explants may had not responded to BAP and NAA supplemented into MS medium.

Number of Shoot

The results showed that there was no significant difference on the number of shoots produced by the various concentration of BAP and NAA supplemented into MS containing GA_3 medium (Table 4). The treatment 2 mg/L BAP + 0.5 mg/L NAA produced 0.67 shoots. This result contrasted with results reported by Kumar et al. [9] which stated that 2 mg/L BAP induced 4.8 shoots/explants and the average of shoot height achieved was 5.7 cm. The growth of shoots on the different medium may be caused by the interaction between BAP and NAA in stimulating growth and differentiation of explants. Wareing and Phillips [10] similarly demonstrated cytokinin induce plant cells division and interact with auxin to lead the differentiation of cells. The higher concentration of cytokinin than auxin may stimulate shoots and leaves formation.

Shoot Height and Number of Leaves

The results showed that there was no significant difference on the shoot height and number of leaves produced by the various concentration of BAP and NAA supplemented into MS containing GA_3 medium (Table 4). The increasing concentration of BAP showed the decreasing of shoot height as shown in Table 4. The results above were supported by George and Sherrington [5] who stated that the effect of cytokinin in tissue cultures is to inhibit apical dominance and encourage the growth of axillary buds.

Percentage of Rooted Explants

The results showed that all treatments did not produce roots after 8 weeks of culture. This study contrasted with results reported by Kumar et al. [9] which stated that medium MS + 2 mg/L NAA was able to induce 45.2 ± 1.1% fig roots. These different results may be explained that the NAA concentration supplemented into MS medium was too low; therefore, the explants did not have enough auxin to produce the adventitious roots. This result was in agreement with those reported by Armini et al. [11] who stated that in vitro cultured explants require auxin for inducing roots. On the other hand, the growth of adventitious roots required longer observation than 8 weeks to induce explants in producing roots.

4 Conclusion

The best method for fig shoot sterilization was dipping explants in NaClO10%, 5′ + NaClO 5%, 7′ + fungicide and bactericide 4 g/L for 3 h with the percentage of survival explants 100% during 30 days of observation. The addition of 2 mg/L BAP and 0.5 mg/L NAA into MS containing GA_3 medium was favorable to produce among the highest percentage of fig shoot formation (33.33%) and the browning explants (0.00%).

References

1. The Global Diabetes Community. 2011. *Diabetes and Herbal Remedies*. http://www.diabetes.co.uk/Diabetes-herbal.html.accesses on November 27, 2011
2. Guvenc, M., Tuzcu, M. and O. Yilmaz. 2009. *Analysis of fatty acid and some lipophilic vitamins found in the fruits of the Ficus carica variety picked from the adiyaman district*. Elazig, Turkey J. Bio Scie 4 (3): 320–323. http://docsdrive.com/pdfs/medwelljournals/rjbsci/2009/320-323.pdf.accesses on November 29, 2011.

3. Wind, D. 2009. *Uncommon Nutrition from the Common Fig ~ Ficus carica.* http://davesgarden.com/guides/articles/view/1787/#b.accesses on November 27, 2011.
4. Haris, M. 2010. *Buah Surga. Balai Besar Pelatihan Pertanian (BBPP) Ketindan-Malang Jawa Timur.* http://www.deptan.go.id/bpsdm/bbppketindan/. accesses on March 26, 2011.
5. George. E. F and P. D. Sherrington. 1984. *Plant propagation by Tissue Culture. Handbook and Directory of Commercial Laboratories.* Exegetics Limited. England.
6. Wudianto, R. 2002. Petunjuk Penggunan Pestisida. Jakarta: Penebar Swadaya.
7. Fahmadi, *A,. 2006. Induksi tunas aksiler secara in vitro, optimasi sterilisasi dan induksi tunas jarak pagar (Jatropha curcas LINN) secara in vitro. Skripsi Fakultas Pertanian Universitas Muhammadiyah Yogyakarta. Tidak dipublikasikan.*
8. Julianti. 2013. *Penambahan NAA Dan BAP Terhadap Multiplikasi Subkultur Tunas Gaharu* (Aquilaria malaccensis Lamk). jurnal.untan.ac.id/.accesses on November 22, 2013.
9. Kumar V, Ridha A, Chitta SK. 1998. *In vitro plant regeneration of fig (Ficus carica L. cv. gular) using apical buds from mature trees.* Plant Cell Report (1998) 17:717–720. Sri Krishnadevaraya University, Anantapur. India.
10. Wareing, P.F. and I.D.J. Phillips. 1970. *The Control og Growth and Differentiations in Plants.* Pergamon. Press. Oxford.
11. Armini, G, G. A. Watimena, dan L.W. Winata. 1992. Perbanyakan Tanaman. *Dalam Bioteknologi Tanaman I.* Wattimena, G.A. *et al* (ed). PAU. Bioteknologi. IPB. Dirjen Dikti Dept. P&K. Hlm. 12–48
12. Tabiyeh, D.T., F. Bernard, and H. Shacker. 2006. Investigation of glutathione, salicylic acid and GA_3 effects on browning in Pistacia vera shoot tips culture. ISHS Acta Hort. 726.

Induction of Callus Cultures from the Leaves of *Syzygium cumini* (Linn.) Skeels in Woody Plant Medium with Variations of Growth Hormones

Tjie Kok, Xavier Daniel and Stevanus Soegiono

Abstract *Syzygium cumini* (Linn.) Skeels, known as *jamblang*, belongs to the family of guava (Myrtaceae). The plant can be found in India, Southeast Asia and Eastern Africa. This plant has a potential use for treatment of type 2 diabetes mellitus. Although still visible, especially in rural areas, it is becoming very rare in big cities. Nowadays, not many people use the fruit of this plant because of its scarcity. To overcome the problem, in this research, *jamblang* plant was propagated in vitro using the leaves as explants to produce callus cultures instead of the regeneration of whole plants. The sterilization was conducted by immersing them in the Mushi Guard solution (5 mL/L), followed by the immersion in the solution of NaOCl. The callus cultures were here induced using WPM (Woody Plant Medium) with different concentrations of growth hormones, i.e., naphthalene acetic acid (NAA) and benzylaminopurine (BAP), in order to determine their optimum concentrations for the induction. The results showed the optimum media of calluses induction was WPM with NAA 5 ppm without any addition of BAP with an induction time of 30 days and percentage of callus cultures induction up to 90%. The optimum media for calluses growth was also WPM with the addition of NAA 5 ppm. Their average growth index was 2.212 after the 30 days of subculturing.

Keywords *Syzygium cumini* (Linn.) Skeels · Callus cultures induction · NAA · BAP

1 Introduction

Syzygium cumini (Linn.) Skeels belongs to the family of guava (Myrtaceae), which can be found in India, Southeast Asia and Eastern Africa. Indonesian people call this plant as *jamblang* or *jambulan* (West Java), *juwet* or *duwet* (East Java), and *jambu kaliang* (West Sumatra) [1].

T. Kok (✉) · X. Daniel · S. Soegiono
Faculty of Biotechnology, University of Surabaya, Surabaya, Indonesia
e-mail: tjie_kok@staff.ubaya.ac.id

T. Taufik et al. (eds.), *ICoSI 2014*, DOI 10.1007/978-981-287-661-4_7

Fig. 1 *Syzygium cumini* [4]

Some parts of the plant, which are fruits, seeds and barks, have a potential to be used to treat type 2 diabetes mellitus. In addition, the fruits can also be utilized to combat diarrhea, whooping cough, Scurvy [2]. The bark of this plant is also useful for the treatment of diarrhea, constipation, stomach ache, flatus, fever and as antibacterial agent. Moreover, the leaves are beneficial to strengthen teeth and gums [3] (Fig. 1).

The information about the growth of the plant is very limited. The cultivation of this plant is still not too many. It is no wonder that the plant is becoming very rare in big cities, even though it is still visible in several rural areas. That is why nowadays not many people use the fruit of this plant because of its scarcity [5].

Syzygium cumini (Linn.) Skeels contains several useful secondary metabolites. In its fruits, there are tannins, essential oils, resins, gallic acids and triterpenoids, whereas in its seeds, there are tannins, gallic acid, phytomelins, alpha-phytosterols, essential oils, alkaloids and triterpenoids. The ethanolic extract of its leaves and aqueous extracts of it seeds contain antimicrobial agents useful to prevent the growth of bacteria, both gram-positive and gram-negative [6].

Based on the matters, this research is conducted to obtain the beneficial secondary metabolites of *Syzygium cumini* (Linn.) Skeels in future via in vitro propagation using the leaves as explants to form callus cultures instead of the regeneration of whole plants. In in vitro technique, callus cultures to produce secondary metabolites have several advantages compared to the common/classic techniques that are efficient in terms of time and energy, can yield products in large amounts in controllable conditions in accordance with what are required, and reduce the need of areas for cultivation [7].

Sterilization of the explants is conducted by cleansing the explants using detergent and removing the dirty parts of them under flowing water. After being cleansed, the explants are immersed in the solution of fungicides or antibiotics [7]. The induction of the cultures was conducted in artificial media with the addition of several concentrations of growth hormones. Auxines and cytokinins are among the most widely used growth hormones in plant tissue cultures and should be added in appropriate the concentration as needed [8]. As the callus cultures have been succeeded to grow for several periods of time, callus subculturing was carried out in media with several concentrations of growth hormones, too in order to determine the optimum media for callus growth. The average callus growth index is determined afterward to know how fast the calluses grow. The concentration of nutrients and growth hormones are the determining factors in the growth of plant tissue cultures [7].

2 Materials and Methods

Sterilization of explants in this research was initiated by first washing the leaves of *Syzygium cumini* (Linn.) Skeels, followed by immersing them in the Mushi Guard solution (5 mL/L), containing bactericides and fungicides. The mixtures were shaken for 2 h and then flushed with flowing water afterward.

The next sterilization step was conducted in the laminar air flow cabinet. The explants were immersed in the several solutions of NaOCl/sterile distilled water with the ratio of 1/2:1 for 5 min, followed by the ratio of 1/3: 1 for 10 min. The explants were then flushed with distilled water. The damaged parts of explants were removed. The clean parts of explants were cut with the approximate size of 1 cm^2 and then cultivated on cultures media, i.e., WPM media containing several concentrations of NAA and BAP, as growth hormones. The name of media, representing the types of media used in this research, is:

Media 1: WPM with 0 ppm NAA and 0 ppm BAP
Media 2: WPM with 2.5 ppm NAA and 2.5 ppm BAP
Media 3: WPM with 5 ppm NAA and 0 ppm BAP
Media 4: WPM with 0 ppm NAA and 5 ppm BAP
Media 5: WPM with 5 ppm NAA and 5 ppm BAP
Media 6: WPM with 0 ppm NAA and 2.5 ppm BAP
Media 7: WPM with 2 ppm NAA and 1 ppm BAP
Media 8: WPM with 3 ppm NAA and 2 ppm BAP
Media 9: WPM with 3.5 ppm NAA and 2.5 ppm BAP
Media 10: WPM with 4 ppm NAA and 3 ppm BAP
Media 11: WPM with 5 ppm NAA and 4 ppm BAP
Media 12: WPM with 1 ppm NAA and 1 ppm BAP
Media 13: WPM with 2 ppm NAA and 2 ppm BAP
Media 14: WPM with 3 ppm NAA and 3 ppm BAP
Media 15: WPM with 4 ppm NAA and 4 ppm BAP

The cultures were then incubated for ±60 days in dark condition at the temperature of 25 °C. The growing calluses weighing ±0.5 g were subcultured every 30 days. After 30 days, the cultures were then harvested, washed with water and weighed. The determination of growth index was done by weighing the initial weight of calluses in the beginning of cultivation and their final weight at time of harvesting. The growth index was calculated by dividing the latter with the former.

$$\text{Growth index} = \frac{\text{Initial weight of callus}}{\text{Final weight of callus}}$$

3 Results and Discussions

In the calluses induction step, there were three types of data, i.e., the rate of calluses induction, percentage of calluses induced and the visual appearance of calluses formed on several WPM media with different concentrations of NAA and BAP.

It can be seen from Fig. 2 that media 1, 4 and 6 could not induce calluses; meanwhile, media 3 was able to induce them, yet it needed much longer time to do it. Other media just required 12–17 days to produce calluses from explants. The optimum concentration for calluses induction of each nutrient varies with the species of plant and type of culture. In addition, it is also related with the age and size of explants.

Figure 3 shows the percentage of calluses induced on the media aside from media 1, 4 and 6 at day-21 was ranged 70–90%, whereas up to day-30 the percentage was between 90 and 100%. Media 1, 4 and 6 cannot induce calluses because there were no auxins in the media. In general, the addition of auxins at low concentration will spur the formation of calluses; meanwhile, the higher ratio of auxins/cytokinins will promote the regeneration of calluses to form organs [9]. Media 3, i.e., WPM with NAA 5 ppm without any addition of BAP, is the best media to induce calluses (Fig. 4).

The observation was carried out at the time of calluses formed. Media 1, 4 and 6 could not induce calluses at all. Calluses formed on media 3 were greenish white and compact, those on media 7 were brownish white and compact, while on the other media the calluses produced were brown and compact. The white or green color is the characteristics of embryonic calluses [10]. Due to its compact and good

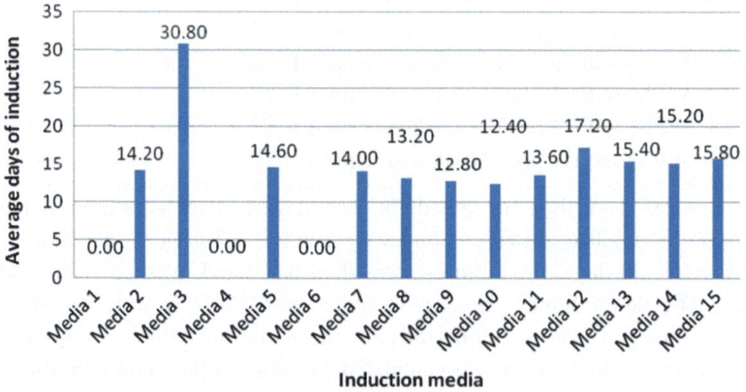

Fig. 2 Rate of calluses induction

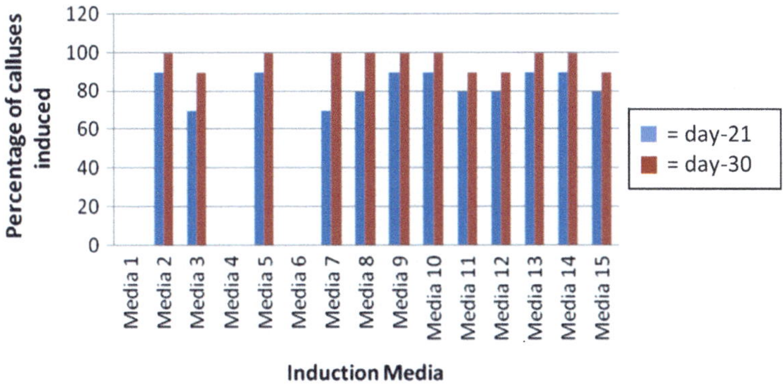

Fig. 3 Percentage of calluses induced

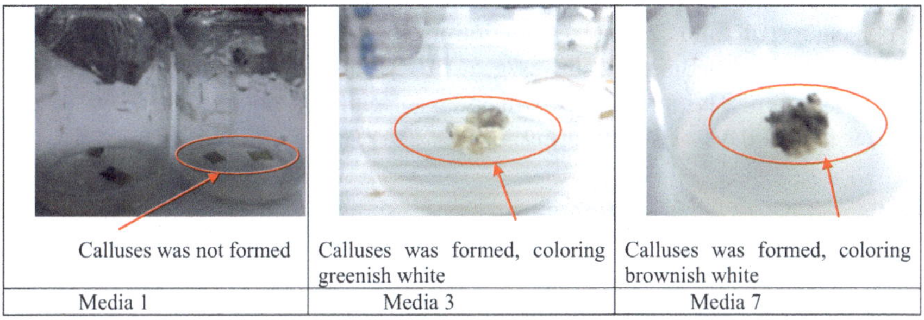

Fig. 4 Some visual appearances of calluses formed

Tabel 1 Growth index of callus cultures on Media 3

Type of media	Replication	Initial weight of calluses (g)	Final weight of calluses (g)	Growth index
Media 3 (WPM with 5 ppm NAA and 0 ppm BAP)	1	0.505	1.233	2.442
	2	0.480	1.057	2.202
	3	0.503	1.355	2.694
	4	0.459	0.910	1.983
	5	0.493	0.876	1.777
Average growth index				2.212

appearances, calluses cultured on media 3 and 7 were selected for subsequent work, i.e., determining the optimum media for their growth, representing by their growth index.

Tables 1 and 2 demonstrate the growth index of calluses. The average growth index of calluses on media 3 and media 7 was 2.212 and 1.911, respectively. The high concentration of cytokinins can inhibit cell proliferation [11] and also the brown color of calluses is undesirable, and then media 3, i.e., WPM with 5 ppm NAA and 0 ppm BAP, is there best media by far for calluses growth.

Tabel 2 Growth index of callus cultures on Media 7

Type of media	Replication	Initial weight of calluses (g)	Final weight of calluses (g)	Growth index
Media 7 (WPM with 2 ppm NAA and 1 ppm BAP)	1	0.629	1.384	2.200
	2	0.505	0.959	1.899
	3	0.495	0.852	1.721
	4	0.559	1.072	1.918
	5	0.503	0.907	1.814
Average growth index				1.911

4 Conclusion

1. The optimum media for calluses induction was WPM with NAA 5 ppm without any addition of BAP as shown by the parameter induction time of 30 days and percentage of callus cultures induction up to 90%.
2. The optimum media for calluses growth was also WPM with the addition of NAA 5 ppm. The average growth index of calluses obtained here was 2.212 after the 30 days of subculturing.

References

1. Anonim. 2013. Tropilap® Inc. "Exporter and Wholesaler of Medicinal Plants, Herb and Tropical Seed", *Syzyngium cumini* L.- Jamun." Available on http://www.tropilab.com/syzigium-jambolanum, accessed on December 19, 2013
2. Mudiana, D. 2007. Perkecambahan *Syzigium cumini* (L.) Skeels. Boidiversitas 8(1): 39–42
3. Saravan, G. and Leelavinothan, P. 2006. Effects of *Syzygium cumini* bark on blood glucose, plasma insulin and C-peptida in streptozotocin induced diabetic rats. Int J Endocorinol Metab 4: 96–105.
4. Anonim. 2014. http://www.stuartxchange.org/Duhat.html, accessed on March 5, 2014.
5. Tjitrosoepomo Gembong. 2007. Morfologi Tumbuhan, UGM Press, Yogyakarta.
6. Remashree A.B., Varghese Thomas, E. Nabeesa, N. Neelakandan and S. Nandakumar. 2003. Plant regeneration through kalus culture of *Syzyngium cumini* L. Plant Cell Biotechnol. Mol. Biol., 4: 197–200.
7. Kumar U. 2008. Methods in Plant Tissue Culture, Second edition, Agrobios, India.
8. Hidayat. 2007. Induksi Pertumbuhan Eksplan Endosperm Ulin dengan IAA dan Kinetin. Fakultas Pertanian Udayana. Agritop 26 (4): 147–152.
9. Gustian. 2009. Upaya Perbanyakan Tanaman Penghasil Gaharu (*Aquilaria malaccensis* Lamk) secara In Vitro, http://repository.unand.ac.id/562/1, accesses on Juni 19, 2012.
10. Arijanti S., Ribkahwati and Retna, S. 2009. *Teknik Kultur Jaringan Tanaman*: *Implementasi beserta Aplikasi, dan Hasil Penelitian*. Surabaya: Dian Prima Lestari.
11. Carimi F., Zottini M., and Formentin E. 2003. Cytokinins: new apoptotic inducers in plants, Planta (2003) 216: 413–421.

Nutritional and Physical Characteristic of Sweet Potato and Taro FlourModified by Amylolytic Enzyme

Badrut Tamam

Abstract Nutrition improvement and security could be implemented by utilizing local commodities, such as sweet potato (*Ipomoea batatas* L) and taro (*Colocasia esculenta* (L) *Schoot*). An effort on food security using both commodities is by producing them to become modified flour that is fermented using amylolytic enzyme. Due to similar characteristic to wheat flour, the modified sweet potato and taro flour could substitute the role of wheat flour, could be beneficial for people with autism. The objective of this research is to know the nutritional and physical characteristic of modified sweet potato and taro flour using amylolytic enzyme fermentation. The type of this study was experimental designed in group random. There were six treatments of each commodity (sweet potato and taro) with three duplications. In nutritional aspect, the administration of amylolytic enzyme with different concentration resulted in significant difference ($P < 0.05$) on starch content of the modified sweet potato flour as well as on protein content of the modified sweet potato and taro flour and very significant difference ($P < 0.01$) for the modified taro flour. In contrast, the addition of amylolytic enzyme with different concentration resulted in insignificant difference ($P > 0.05$) on fiber content of modified taro flour. Meanwhile, in physical aspect the addition of amylolytic enzyme with different concentration resulted in significant difference ($P < 0.05$) on luminosity score (whiteness degree) of the modified sweet potato and taro flour. Both modified flour could have characteristics of wheat flour.

Keywords Food security · Modified flour · Sweet potato · Taro

B. Tamam (✉)
Polytechnic of Health Denpasar, Denpasar, Indonesia
e-mail: badruttamam_70@yahoo.com

T. Taufik et al. (eds.), *ICoSI 2014*, DOI 10.1007/978-981-287-661-4_8

1 Background

Nutrition improvement and food security could be implemented by utilizing local commodities. Several local commodities that are easily afforded and developed as food products are sweet potato (*Ipomoea batatas* L) and taro (*Colocasia esculenta* (L) *Schoot*). Both commodities are mostly utilized and consumed in simple forms, such as boiled and fried products. In addition, they are less produced in paste and flour [1].

Sweet potato is rich in carbohydrate, vitamin, and mineral. Besides that, sweet potato contains bioactive compound such as anthocyanin and scopoletin. In addition, taro is also rich in carbohydrate and fiber. Moreover, taro also contains carotenoids and anthocyanin as antioxidant [2].

An effort on food diversification and security using both commodities (sweet potato and taro) is by producing them to become modified flour. The modified flour can be obtained by fermenting both commodities using amylolytic enzyme, such as α-amylase or gluco-amylase. The benefits of the modified flour may have highly digested ability, higher amount of soluble fiber, low oligosaccharide, no gluten, and similarity on physical characteristics with wheat flour.

Due to similar characteristic to wheat flour, the modified sweet potato and taro flour could substitute the role of wheat flour in cake production. The modified flour could be used for wide range of cake and bakery production, instead of wheat flour. Especially, for people with autism who should take diet with no gluten, the modified sweet potato and taro flour could be a good alternative to change wheat flour in their meals. The main objective of this research is to know the nutritional and physical characteristic of modified sweet potato and taro flour using amylolytic enzyme fermentation.

2 Methods

The type of this study was experimental designed in group random. There were six treatments of each commodity (sweet potato and taro) with three duplications. Treatments conducted on sweet potato consisted of: sweet potato without amylolytic enzyme (UJ0), sweet potato with 0.1% amylolytic enzyme solution (UJ1), sweet potato with 0.2% amylolytic enzyme solution (UJ2), sweet potato with 0.3% amylolytic enzyme solution (UJ3), sweet potato with 0.4% amylolytic enzyme solution (UJ4), sweet potato with 0.5% amylolytic enzyme solution (UJ5). Moreover, treatments conducted on taro consisted of: taro without amylolytic enzyme (TL0), taro with 0.1% amylolytic enzyme solution (TL1), taro with 0.2% amylolytic enzyme solution (TL2), taro with 0.3% amylolytic enzyme solution (TL3), taro with 0.4% amylolytic enzyme solution (TL4), taro with 0.5% amylolytic enzyme solution (TL5).

The stock solution of amylolytic enzyme (from a MOCAF industry in Mojokerto) was made by fermenting cassava (100 g), milk powder (7 g), and MOCAF enzyme (2 g) in 1 L of fresh water for 24 h.

The modified flour was made following these steps: selection of materials, cleaning, chipping 2–3 mm in width, fermentation with amylolytic enzyme (as treatments given) and citric acid (0.05%) for 24 h, washing with adding salt (0.01%), drying, and powdering.

The analysis of this research consisted of nutritional analysis (starch, protein, and fiber) and physical analysis (white intensity or luminosity). The data then were analyzed using analysis of variance (ANOVA) and continued to *Tukey's post hoc* test for significant different results ($P < 0.05$).

3 Results and Discussion

The administration of amylolytic enzyme with different concentration resulted in significant difference ($P < 0.05$) on starch content of the modified sweet potato flour and very significant difference ($P < 0.01$) for the modified taro flour, as shown in Fig. 1. The lowest number of starch content of the modified sweet potato flour and the modified taro flour was on the addition of the highest concentration of the enzyme (44.50 and 45.83%, respectively).

Starch is homopolymer of glucose that is linked by α-glycosidic bond. Fresh sweet potato and taro contain starch, 27.9 and 23.7%, respectively. This study showed that drying process increased the amount of starch on both commodities without the addition of amylolytic enzyme (67.51 and 64.13%, respectively). This is due to the removal of big amount of water during drying process. The administration of amylolytic enzyme could partially break down starch to become smaller

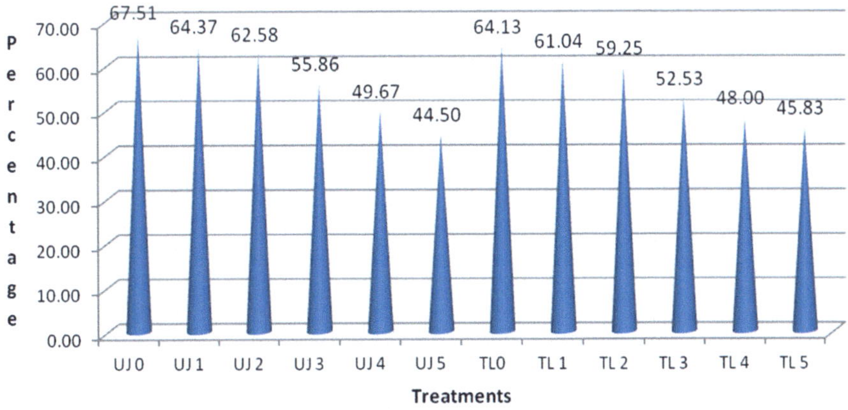

Fig. 1 Starch contents of the modified sweet potato and taro flour

compounds, such as glucose, maltose, or dextrin [3]. The higher the amount of amylolytic enzyme was added, the higher simpler saccharide was produced.

The addition of amylolytic enzyme with different concentration resulted in significant difference ($P < 0.05$) on protein content of the modified sweet potato and taro flour, as shown in Table 1.

In this study, the number of protein on the modified flour fluctuated with the addition of amylolytic enzyme. The highest number of protein on the modified sweet potato flour was 4.54% (treatment of UJ4). Meanwhile, the highest number of modified taro flour was 4.86% (treatment of TL3). It means that the addition of amylolytic enzyme doubled the protein content of both commodities flour. It may be caused by the decrease in starch content, and then, protein total would be higher. Degradation of starch would increase the accumulation of protein in the modified flour.

The addition of amylolytic enzyme with different concentration resulted in insignificant difference ($P > 0.05$) on fiber content of the modified sweet potato and taro flour, as shown in Fig. 2.

Dietary fibers are components of plan tissues that cannot be hydrolyzed by digestive enzymes in gaster and intestines [4]. The fibers are much located in plant cell walls containing cellulose, hemicellulose, and pectin. Due to their stability against enzymatic reaction, the fiber content on the modified sweet potato and taro flour remains stable in their numbers ($P > 0.05$).

The addition of amylolytic enzyme with different concentration resulted in significant difference ($P < 0.05$) on luminosity score (whiteness degree) of the modified sweet potato and taro flour, as shown in Fig. 3.

Table 1 Protein content of the modified sweet potato and taro flour in various treatments

No.	Treatments	Percentage
1	UJ0	3.23 a
2	UJ1	3.36 a
3	UJ2	3.29 a
4	UJ3	3.35 a
5	UJ4	4.54 b
6	UJ5	3.07 a
7	TL0	2.23 a
8	TL1	2.37 a
9	TL2	3.43 ab
10	TL3	4.86 b
11	TL4	3.71 ab
12	TL5	3.06 ab

Note Different letters behind the scores identify significant differences ($P < 0.05$)

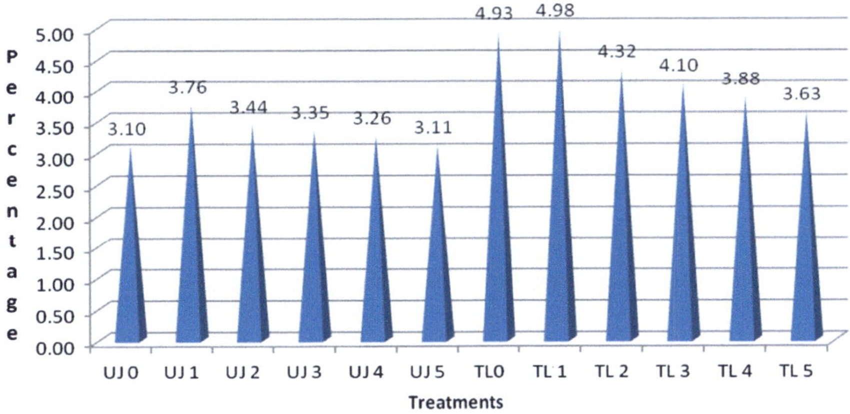

Fig. 2 Fiber content of the modified sweet potato and taro flour

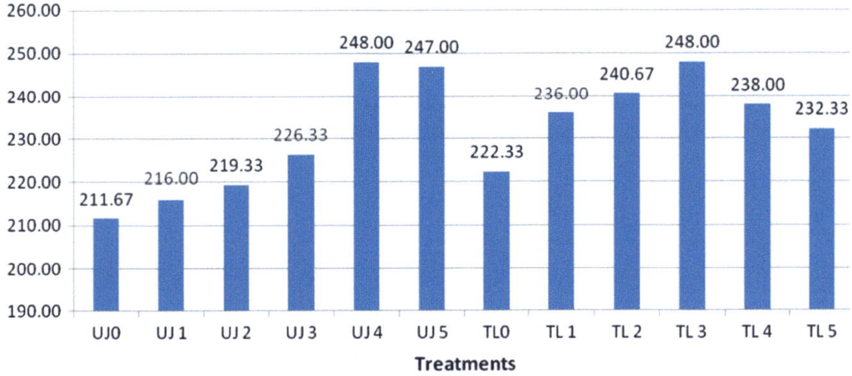

Fig. 3 Whiteness degree (luminosity) of the modified sweet potato and taro flour

The administration of amylolytic enzyme stimulated the decrease in pH during the fermentation of sweet potato and taro chips. Moreover, the addition of citric acid also whitened the modified flour produced. This is because citric acid can retard the action of enzymatic browning and degrade color to become pale [5]. The higher enzyme concentration was added, the higher acid was generated. The whitest flour on the modified sweet flour was on UJ4 treatment (248), while the whitest flour on the modified taro flour was on TL3 treatment (248). Wheat flour has the score of luminosity of 250. The measurement of whiteness degree referred to identification of luminosity score on computer. The higher the luminosity score is, the whiter the food is [6].

4 Conclusion

The administration of amylolytic enzyme with different concentration resulted in significant difference on starch content of the modified sweet potato flour and on protein content and luminosity score (whiteness degree) of the modified sweet potato and taro flour. Additionally, the addition of amylolytic enzyme with different concentration resulted in very significant difference on the modified taro flour, whereas the addition of amylolytic enzyme with different concentration resulted in insignificant difference on fiber content of the modified sweet potato and taro flour.

References

1. Direktorat Jenderal Industri Kecil dan Menengah (IKM). Teknologi Diversifikasi Produk Olahan Pangan Berbasis Tepung Ubi Jalar. Jakarta: Departemen Perindustrian; 2009.
2. Bradbury, J.H. and W.D. Holloway. Chemistry of Tropical Roots: Significance for Nutrition and Agriculture in the Pasific. Canbera: ACIAR; 1988.
3. Viera Horváthová, ©tefan Janeèek & Ernest ©turdík, Amylolytic enzymes: their specificities, origins and properties; Biologia, Bratislava, 2000.
4. Winarno FG. Enzim Pangan. Jakarta: PT Gramedia; 1986.
5. Winarno FG. Kimia Pangan dan Gizi. Jakarta: PT Gramedia; 1988.
6. Yuwono, Sudarmanto dan Tri Susanto. Pemeriksaan Fisik Bahan Makanan. Malang: Penerbit Universitas Brawijaya; 1998.

Optimation of Breakfast Meals from Local Taro Using Response Surface Methodology

Setyadjit and Ermi Sukasih

Abstract The simplest method in producing plain breakfast meals from taro by preparing a dough, making the dough flat, making a flake, then drying the flake in a cabinet dryer. In this experiment optimation of cooking factors such as baking temperature (°C), time (mins), and thickness of flakes (mm), have been done by using an RSM approach with 20 experimental units in central composite design. The products then were analysed on they hardness (g), total work (MJ) and fracturability (g). The selected formula was validated. The best processing conditions recommended by RSM was 120.8 °C baking temperature, 12.02 min baking time and 0.50 mm product thickness. The product has the hardness of ca. 139.0 g, total work ca. 0.37 (MJ) and fracturability ca. 139.0(g). Predicted and actual product is considered as the same since they are fall in range of 95% confidence. Thus, the above recommended condition of the treatment is validated as the best treatments in producing plain breakfast meals.

Keywords Breakfast meal · Response surface methodology · Flake · Taro composite flour · Food security

1 Introduction

Ready to eat breakfast shaped as flakes is one of the food products which is currently popular in Indonesia. The breakfast product is also popular in developed countries [1]. Currently, most of the breakfast food are made from cereals such as wheat, maize, and rice. Actually, breakfast meals can also be prepared from the tubers as a source of carbohydrates mixed with bean as a source of protein and also mixed with fruit as a source of fiber and vitamins. Thus the selection of materials for

Setyadjit (✉) · E. Sukasih
Indonesian Centre for Agriculture Postharvest Research and Development,
Indonesian Agency for Agriculture Postharvest Research and Development,
Cimanggu Agriculture Campus, JL. Tentara Pelajar no. 12, Bogor 16114, Indonesia
e-mail: pascapanen@yahoo.com

T. Taufik et al. (eds.), *ICoSI 2014*, DOI 10.1007/978-981-287-661-4_9

the formulation (composite) is important to be able to produce a good quality product.

Tubers and corms are an important source of carbohydrates for food energy in the tropics and subtropics and various products has been prepared from taro viz. extrudate, chips, bread [2–5]. Cooking of taro was also reported reducing the degree in tripsin inhibitor [6]. One of the local food resources that can be used as an alternative in the development of breakfast meals are corms of taro (*Colocasia esculenta*). Taro corms is potential as a high source of carbohydrate, i.e. 23.79 g per 100 g of fresh taro [7]. In addition, taro corms also contain fat, vitamins, and minerals although in small amounts. Vitamins contained in taro corms are vitamin A, B1, and a small amount of vitamin C [8]. Taro corms also contain minerals such as Ca at 28 mg and P at 61 mg per 100 g of fresh taro. These minerals are essential for the growth of bones and teeth.

Breakfast meals are categorized into several types: traditional cereal breakfast meals that have not been processed, ready to cook cereal (flour), ready to eat cereals (flakes, tortillas, shreded), and ready to eat cereal mix [9]. The flakes is as a form of fast food products that is suitable for breakfast. The way of preparation is quite easy such as by adding hot water or milk into the products.

The purpose of this research was to find an optimal formula for producing breakfast flakes made from taro-based composite flour, green beans and bananas. The optimal process parameters optimized were baking temperature, baking time, and the thickness of the flakes in the formulation optimal flakes by using Response Surface Methodology (RSM). Then, it was comparing the value derived from prediction and from actual processing.

2 Materials and Methods

2.1 Materials

Composite flour were made from taro flour (50%) of local Bogor, banana flour (30%), and mung bean flour (20%) was the best composite formula developed by Indonesian Centre Agriculture Postharvest Research and Development.

2.2 Experiment's Variable and Parameters

Three controlled experimental variables were baking temperature (°C), baking time (min), and the thickness of the flakes (mm). As input to the RSM tools, there were set maximum and minimum parameters values based on the research that has been done before, such as sweet potato flakes by [10]; taro flakes [11]; flakes sorghum [12]; and red sweet potato flakes by [13]. The RSM method referred to the study by [14].

Experiment's controlled variables used in the design process is the baking temperature, baking time, and the thickness of the flakes. While the response parameters used were from the texture profile analysis compression test. The design of RSM used was the central composite design. The results of the process design was based on the value of the variable which has been entered to the program resulted in 20 unit experiment were applied into production of flakes. The value used were temperature (°C) range between 110 and 150 °C, time was 10–20 min and for the thickness was from 0.25 to 1.5 mm. The design process output of the program can be seen in Table 1.

Twenty formula designed by the RSM which consisted of 6 tested formula which was the axial point, 8 point formula which was factorial and 6 formula that was the center point. The design formulas obtained then evaluated for the suitability of the design formulas to desired effect. This was done by an analyzing whether there is alias and analyzing Ri^2 value. The existence of an alias indicated there is lack of unique point in the models used. If in the model used there was not any alias, it means the design model is sufficient to predict the desired results.

Table 1 The combination of treatment with RSM

Formula	Treatment codes			Value		
	Temp. (°C)	Time (min)	Thickness (mm)	Temp. (°C)	Time (min)	Thickness (mm)
1	−1	−1	−1	118.11	12.02	0.50
2	−1	−1	1	118.11	12.02	1.25
3	−1	1	−1	118.11	17.97	0.50
4	−1	1	1	118.11	17.97	1.25
5	1	−1	−1	141.89	12.02	0.50
6	1	−1	1	141.89	12.02	1.25
7	1	1	−1	141.89	17.97	0.50
8	1	1	1	141.89	17.97	1.25
9	−1.682	0	0	110	15	0.88
10	1.682	0	0	150	15	0.88
11	0	−1.682	0	130	10	0.88
12	0	1.682	0	130	20	0.88
13	0	0	−1.682	130	15	0.25
14	0	0	1.682	130	15	1.50
15	0	0	0	130	15	0.88
16	0	0	0	130	15	0.88
17	0	0	0	130	15	0.88
18	0	0	0	130	15	0.88
19	0	0	0	130	15	0.88
20	0	0	0	130	15	0.88

2.3 Flakes Processing

There were two main ingredients first was composite flour which is the mixture of taro flour, banana flour and mung bean flour. Second was the additive materials such as sugar and salt. This would make base material of plain oven flake. First ingredient were mixed by hand and the second ingredient were diluted in water. The first and second ingredient were placed into the mixer and operated for 5 min to make a dough. The dough then were placed into the roller to make a sheet. The sheet were cut into the 1×2 cm rectangle. The cuts were placed at a pan and then placed in an oven for baking in various treatments. The plain breakfast meals flakes were ready for analyses.

2.4 Optimisation Processes

In the optimization process, first it was determining the order of polynomial equations corresponding to each of the response variables. Response variable used were the texture parameters from the texture profile analysis include hardness, total work, and fracturability. Determination of polynomial equations and the corresponding order of each response variable were done by selecting the highest order that the results of the analysis of variance which still had a significant difference. The model equations and order selected by the model has to have the smallest standard deviation with R^2 and adjusted R^2 values close to 1.

After the determination of the order and polynomial equations, then ANOVA analysis was performed to see if there was a significant difference in the response of each variable at the 95% confidence interval. The results of variance analysis indicated the response variable has a value that is significantly different if the 95% confidence interval, and at P value less than $\alpha = 0.05$. The results of the analysis of the response variable which was significant difference can be used as a predictive model in the optimisation. If response variables was not significantly different, it is not used as a major variable in the optimisation.

After ANOVA analysis and got the response variables which was significance at the 95% confidence interval, then the optimisation formula was performed. Optimisation was done by setting the value of the formula desired characteristics of the response variable that has a significant difference in the 95% confidence interval. Determination of the value of the variable characteristics was also done by comparing the value of the variable response of similar products available in the market, namely the product of Nestle breakfast flakes.

Optimisation was performed to obtain the optimal formula with variables predictive value of the response that best suits the desired value of the response variable. Correspondence between the predicted value of the response variable to the desired value of the response variable was described by the value of desirability.

2.5 Validation of the Model by Using Texture Analyser

Optimisation results obtained or predicted formula of Response Surface Method (RSM) was then validated with a texture analyzer to generate the actual response variable. The results were then compared with the predicted value of the response variable by Response Surface Method (RSM).

3 Results and Discussion

3.1 Optimisation Process by Response Surface Method

RSM is a method of regression analysis used to predict the value of the response variable based on controlled experimental variables [15, 16]. RSM was used in optimizing the quality of process or product [17, 18]. Response variables value, namely texture flakes of 20 formula RSM design data was not presented, but optimisation of response parameters were presented.

Hardness values obtained from the measurements ranged from 73.25 to 337.13 g. Hardness value is the amount of force required to reach the peak at the time of compression. Hardness value was obtained from the highest peak in the graph of texture analyser. Hardness depicts resistance of the products. According [19], a product hardness measurement can be performed using standard compression force in a certain time to determine the deformation that occurs in the product. Moreover, it can also use a specific compression distance and measure the force required to achieve the desired percent deformation [19].

The total value of work ranging from 0.14 to 1.4 mJ products. These value were obtained from the area under the peak of the first compression. Total work describes the amount of energy used to compress the product when analyzed the texture characteristics. Fracturability values obtained ranged from 73.25 to 337.13 g. This value was the first peak in the graph of texture analysis.

In optimisation done by setting targets according to the texture characteristics value of the benchmark product value. Product standard was a brand flakes (Nestle) that have a hardness value of 136.5 g, a total work of 0.295 mJ, and fracturability of 136.5 g. These value became the reference of texture characteristics which will be used to optimise the flakes.

Optimisation of the design process for 20 formulas using Response Surface Method (RSM) produce a value of the response variable texture characteristics that best fit the desired variable value response. Table 2 shows the optimal design process recommended by Response Surface Method (RSM) and Table 3 shows the predicted value of the response variable.

Recommended optimal process design by Response Surface Method (RSM) using temperatures of 120.82 (°C) with a time of 12.02 min baking and thickness of 0.50 mm flakes. The formula predicted to produce flakes which have a variable response to the hardness value of 136.50 g, the total work value of 0.33 mJ and fracturability value of 13.49 g (Table 3). The formula has a desirability value of 0.974. Response variables used in the form of texture parameters can be analyzed by texture profile analysis [20].

Desirability value indicates the correspondence between the desired response variable value with the predicted value of the response variable of the optimal formula recommended Response Surface Method (RSM). Figure 1 shows the desirability obtained from the optimisation.

Based on the results of the optimization was performed from the target value, the value of the average desirability of the overall parameter is 0.974. Desirability value in the optimization ranges from 0 to 1. Value close to 1 means that the formula obtained in accordance with the characteristics most desired response parameters. However, the goal of optimization is not just looking for a desirability value of 1, but also look for the optimal conditions of all the functions of the response variable.

Desirability value of 0.974 was obtained from the average desirability of each treatment variable and the response variable. Value desirability variable treatment include temperature, time, and thickness is 1. Value targeted for the treatment variable was the average value so that the desirability value obtained is 1. While for parameter values desirability response obtained hardness value was 0.999 and for

Table 2 Parameters of optimisation using RSM

Optimum parameters	Value
Temperature (°C)	120.82
Time (min)	12.02
Thickness (mm)	0.50

Table 3 Predicted value of response using optimum formula

Response variable	Predicted value
Hardness (g)	136.50
Total work (mJ)	0.33
Fracturabilty (g)	136.49

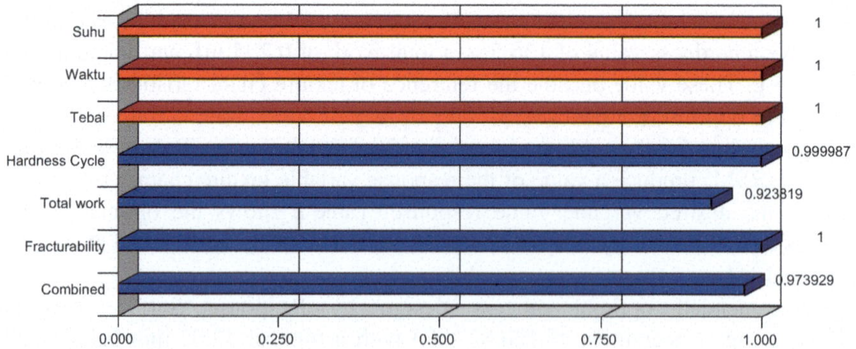

Fig. 1 Desirability optimisation of formula

the total work desirability value obtained was 0.923. Fracturability desirability value was 1.

Optimisation was done by setting targets according to the texture characteristics value of the desired value. This is done by using texture characteristics value of similar products available in the market as a comparison. A brand flakes product (N) has a hardness value of 136.5 g, a total work of 0.295 mJ, and fracturability of 136.5 g. This value becomes the reference characteristics of desired texture flakes.

With a time of 12.02 min and the product thickness of 0.5 mm. The formula predicted to produce flakes with a hardness value of 136.50 g, the total work value of 0.33 mJ, and fracturability of 136.49 g. Below is the graph of the results of optimisation of the model response variable.

(1) **Hardness**. Hardness values range is a combination of red to blue. The red color indicates the higher value while the blue color indicates that the lower the value. The green color indicates the average value. Hardness value describe the firmness of product. Graph the results of optimisation hardness response variables presented in Fig. 2.

The desired criteria on hardness is the target values according to the characteristics of the comparator product brand Nestle. The greater the value of the product produced hardness higher. Optimization of hardness values are on the vertical axis with a value of *Y* 136.50 g. The prediction values obtained from the treatment variable on the horizontal axis are composed of a time axis *X*1, *X*2 axis as temperature and thickness of 0.50 mm as the actual factor.

The relationship between baking temperature and time on the hardness of flakes is a negative correlation [14]. When baking temperature higher the level of firmness will decrease. The baking time is similar. However, the effect of baking temperature

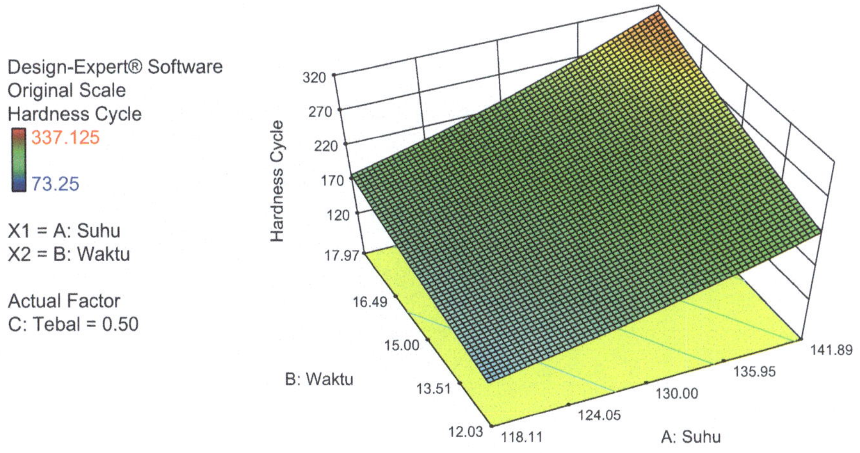

Fig. 2 Graph of optimation of variable response hardness

is more significant in lowering the level of firmness rather than baking time [14]. In contrast, the thickness is positively correlated to the hardness of the product.

(2) **Total Work**. The value of total work range is from 0.1375 to 1.4025 mJ with a combination of red to blue. The total value of the lowest work is shown by the blue area, the highest value is shown by the red area, while the average value is shown in green. Graph the results of the total response variable optimisation work presented in Fig. 3.

Optimisation of value of total work contained in the vertical axis with a value of *Y* 0.33 mJ. The prediction values obtained from the treatment variable on the horizontal axis are composed of a time baking axis *X*1, *X*2 axis as baking temperature and thickness of 0.50 mm as the actual factor.

(3) **Fracturability**. The lowest value of fracturability is indicated by the blue area, the highest value is shown by the red area, while the average value is shown in green. The graph is not presented.

The value of the response area meets the criteria. The desired criteria on fracturability at the target value comparison was done by comparing to the fracturability of standard product. Value optimisation of fracturability found on the *Y* axis with a value of 136.50 g. The prediction values obtained from the treatment variable on the horizontal axis are composed of a baking time axis *X*1, *X*2 axis as baking temperature and thickness of 0.50 mm as the actual factor.

Relationship of fracturability inverse to baking temperature. With increasing baking temperature, the lower the fracturability. The baking time also had a negative correlation with the level of fracturability. The longer the baking time, the product will be easier to be broken. In contrast to the temperature and time,

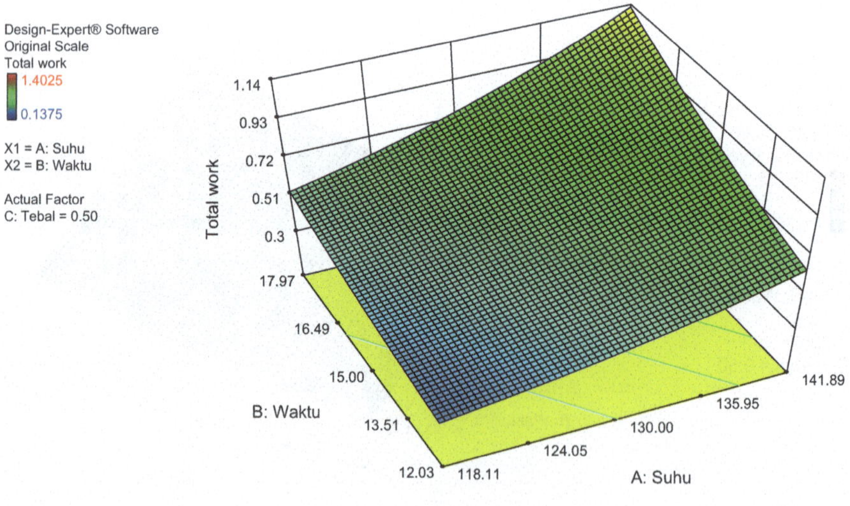

Fig. 3 Graph of optimisation of variable response total work

thickness is directly proportional to fracturability. If the product is getting thicker, then fracturability will also be greater.

3.2 Validation to Response Method Surface

Value of the response variable was derived from the formula with the variable baking temperature using 120.82 °C with a baking time of 12.02 min and the product thickness of 0.5 mm. The product of the formula was then analyzed by using TPA for hardness, total work, and fracturability.

Analyses was performed on the selected optimisation formula to look at the suitability of texture characteristics value predicted by the product obtained. Textural characteristics of the actual product compared to predicted value of the texture. The actual value should be between the minimum and the maximum value. Minimum and maximum predicted value was determined based on 95% confidence interval. The validated value obtained are in Table 4.

Validated product had hardness value of 139.44 ± 14.39 g. This value is slightly higher than the predicted value which is 136.50 g. The value of the validated total work is 0.37 ± 0.125 mJ This value is also higher than the predicted value which is 0.33 mJ. The difference between the predicted value and the validated value, however, does not exceed the minimum and maximum value, so it is still fall within the 95% confidence interval.

Predicted fracturability value is 136.49 g while validated fracturability value is 139.44 ± 14.39. This value is still in the region between the minimum and maximum predicted values i.e. within 95% confidence interval.

Thus, the optimal value of the response variable to produce the plain breakfast meals should be using the baking temperature of 120.82 °C, baking time of 12.02 min and the product thickness of 0.5 mm. The texture product between prediction by using RSM and its validation is similar since it is fall in 95% confidence level. The texture is not inferior compared to commercial flake products since the desirability value are closed to one. The breakfast meals is also nutritious since mung bean flour as a source of protein and banana flour as source of dietary fiber were added during preparation of taro composite flour, the main raw material of breakfast meals.

Acknowledgements We would like to thank to Tito Tegar for doing analyses and Dr. Sri Yuliani for consultation on RSM

Table 4 Value of prediction and validation

Response variable	Prediction (95% of confidence interval)			Validated value
	Average	Min	Max	
Hardness (g)	136.50	119.59	155.80	139.44±14.39
Total work (mJ)	0.33	0.27	0.40	0.37±0.125
Fracturabilty (g)	136.49	119.59	155.79	139.44±14.39

References

1. Morales, Francisco J., Cristina Delgado-Andrade, Jose ′ A. Rufia ′n-Henares,. 2005. Lysine availability is diminished in commercial fibre-enriched breakfast cereals. Consejo Superior de Investigaciones Cientı ′ficas, Instituto del Frı ′o, Jose ′ Antonio Novais 10, 28040 Madrid, Spain
2. James Hollyer J, Paul R, Alvin Huang A. 2000. Processing Taro chips. Food Manufacturing Technology. Feb. 2000:1–2.
3. Liu, Q, E Donner, Y Yin, RL Huang dan MZ Fan. 2006. The physicochemical properties and in vitro digestibility of selected cereals, tubers, and legumes grown in China. Food Chemistry 99: 470–477.
4. Ammar MS, Hegazy AE and Bedeir SH. 2009. Using of Taro Flour as Partial Substitute of Wheat Flour in Bread Making. World Journal of Dairy & Food Sciences 4 (2): 94–99.
5. Rodríguez-Miranda J, Ruiz-López II, Herman-Lara E, Martínez-Sánchez CE, Delgado-Licon E, Vivar-Vera MA, Rodríguez-Miranda J, Ruiz-López II, Herman-Lara E, Martínez-Sánchez CE, Delgado-Licon E, Vivar-Vera MA. 2011. Development of extruded snacks using taro (*Colocasia esculenta*) and nixtamalized maize (*Zea mays*) flour blends. LWT - Food Science and Technology 44: 673–680.
6. Kiran KS and Padmaja G. 2003. Inactivation of trypsin inhibitors in sweet potato and taro tubers during processing. Plant Foods for Human Nutrition 58: 153–163.
7. Direktorat Gizi Departemen Kesehatan. 1972. Daftar Komposisi Bahan Makanan. Bharata. Jakarta. 57pp.
8. Muchtadi, TR, dan Sugiyono. 1992. Ilmu Pengetahuan Bahan Pangan. PAU. Bogor.
9. Tribelhorn, R.E. 1991. Breakfast Cereal. *In*. K.J. Lorenz dan K. Pulp (ed). New York, Basel, Hongkong. pp:259–285.
10. Purwanto, Anto. 2009. Kajian *Scale Up* Industri Pembuatan *Sweet Potato Flakes*. Skripsi. Fateta. IPB. Bogor.
11. Fauzan, F. 2005. Formulasi *Flakes* Komposit dari Tepung Talas, Tepung Tempe dan Tapioka. Skripsi. Fateta. IPB. Bogor.
12. Felicia, A. 2006. Pengembangan Produk Pangan sarapan Siap Santap Berbasis Sorghum. Skripsi. Feteta. IPB. Bogor.
13. Hutami, Sri. 2004. Formulasi *Sweet Potato Flakes* dari Ubi Jalar Merah dan Ungu. Skripsi. Fateta. IPB. Bogor.
14. Bhattacharya S, and Sumithra B. 2008. Toasting of corn flakes: Product characteristics as a function of processing conditions. Journal of Food Engineering 88:419–428
15. Meilgaard, MC, GV Civille dan BT Carr, 2007. Sensory Evaluation Techniques, 4th edition. CRC Press, Boca Raton, FL, USA.
16. Lenth, Rusell V. 2009. Response-Surface Methods in R, Using rsm Updated to version 1.30. The University of Iowa Journal of Statistical Software.
17. Myers, RH and Montgomery DC. 2002. Response Surface Methodology: Process and Product Optimization Using Designed Experiment. A Wiley-Interscience Publication.
18. Raissi, S. 2009. Developing New Processes and Optimizing Performance Using Response Surface Methodology. World Academy of Science, Engineering and Technology.
19. Cauvain, SP dan Young LS. 2006. Baked products: Science, Technology and Practice. Blackwell Publishing Ltd. Garsington Road.
20. Sipos, Peter, Agnes Elek dan Zoltan Gyori. 2003. Application of Texture Analyzer in the Quality Analysis of Bakery Product. Institute of Food Science, Quality Assurance and Microbiology, Faculty of Agronomy, University of Debrecen, 138 Böszörményi. Hungary.

Selection of Soybean Lines (*Glycine max*) Tolerant to Drought

Apri Sulistyo, Suhartina, Novita Nugrahaeni and Purwantoro

Abstract Soybean cropping pattern in Indonesia, which follows the pattern of rice-rice-soybean, causing soybean cultivation often encounter the problem of water deficiency due to soybean growing season which falls in the dry season. Soybean varieties tolerant to drought can prevent yield loss due to water deficit. The aim of this research was to identify soybean lines tolerant to drought. A total of 85 soybean lines from F5 generation population were tested in Muneng Experiment Station, Probolinggo district, East Java province from July to September 2013. Six soybean genotypes namely: Dering 1, Tidar, Grobogan, ARG/GCP-335, Tanggamus and SU-17-1014 were used as check. All genetic material planted into five different blocks in the field and arranged as augmented design. Each block contains of 17 different lines and 6 soybean genotypes were planted randomly. Fertilization was given at planting and conducted according to the dosage recommendation, i.e. 100 kg ha^{-1} of Urea, 75 kg ha^{-1} of SP36, and 75 kg ha^{-1} of KCl. Irrigation was only given twice, first at planting time and the latter at the time of flowering with the purpose of conditioning the plant exposed to drought stress during the reproductive phase. The results showed that there were soybean lines with better performance than those of check. Line number 76, 81, and 84 (Tanggamus/Shr-3B-342-207, Tanggamus/Shr-3B-343-208, and Tanggamus/Shr-3B-351-64, respectively) had the number of fertile nodes and the number of pods higher when compared with Dering 1. All soybean lines were classified as ultra early maturing soybean (ranging from 60 to 69 days after planting) and early maturing soybean (ranging from 70 to 73 days after planting). A total of 78 lines (92%) more early maturing when compared with Dering 1, but only 20% (17 lines) more early maturing than those of Tidar. In general, almost all of lines had seed size larger compared to Tidar, but there were only 14 soybean lines with seed size greater than those of Dering 1. However, none of soybean lines are capable of producing seeds exceeds Dering 1 and Tidar.

A. Sulistyo (✉) · Suhartina · N. Nugrahaeni · Purwantoro
Indonesian Legumes and Tuber Crops Research Institute, Indonesian Agency for Agricultural Research and Development, Jl Raya Kendalpayak km 8, Po Box 66, Malang 65101, Indonesia
e-mail: apri.sulistyo@gmail.com

T. Taufik et al. (eds.), *ICoSI 2014*, DOI 10.1007/978-981-287-661-4_10

Keywords Drought tolerant · Early maturity · Medium seed size · Soybean line · F5 generation

1 Introduction

Soybean cultivation in irrigated lands in Indonesia is usually planted in the dry season following the cropping pattern of rice-rice-soybean. In East Java Province (especially in the Brantas river basin), the cropping pattern like that reached 20% of 219 ha of irrigation area [1]. Because it is grown in the dry season, resulting soybeans often face the problem of water deficiency. The direct impact on soybean that occurs due to drought stress are the decrease in yield and its component such as number of pods, number of seeds and seed weight [2].

Yield losses due to drought stress in soybean reach up to more than 30% [3], depending on variety [4], location and season [3]. The previous research showed that the lack of water during the generative phase led to a decrease in soybean yields as much as 25% [5]. Another study added that the water stress on the phase R3, R5 and R6 of soybean development cause a reduction in seed yield by 33%, 31%, and 50%, respectively [6].

Drought-tolerant varieties can be used to prevent yield loss on soybean due to water deficit. In Indonesia, there is only one variety that was developed specifically for drought stress, namely Dering 1 which released in 2013. In addition, two old high yielding varieties, namely Wilis and Tidar, indicated drought tolerance. Therefore, improvement of high yielding varieties tolerant to drought is still wide open. The purpose of this study was to identify soybean lines tolerant to drought.

2 Materials and Methods

The research was conducted at the Muneng Experiment Station, Probolinggo district, East Java province, from July to September 2013. A total of 85 soybean lines from F5 generation population tested in this study. These populations derived from crosses between drought tolerant genotypes (MLG 2805, MLG 3474, ARG/GCP-335, DV/MLG 2984, Tanggamus, and Willis) with high yielding varieties (Grobogan, Malabar, Baluran, and Gema). Six soybean genotypes namely: Dering 1, Tidar, Grobogan, ARG/GCP-335, Tanggamus and SU-17-1014 were used as check.

All the genetic materials planted in field and arranged in augmented design [7]. Each soybean lines were planted in a single row. Planting spacing used was 40 cm between rows and 15 cm within rows, two plants per hole. According to the augmented design that applied, all of genetic materials planted into five different blocks. Each block contains of 17 different lines and 6 soybean genotypes were planted randomly.

Fertilization was conducted according to the dosage recommendation, i.e. 100 kg ha^{-1} of Urea, 75 kg ha^{-1} of SP36, and 75 kg ha^{-1} of KCl. All fertilizers were given at planting. Irrigation was only given twice, first at planting time and the latter at the time of flowering with the purpose of conditioning the plant exposed to drought stress during the reproductive phase [6]. Intensive pest control was done by spraying insecticide every two weeks. Data was collected on the following characters: days to maturity, number of fertile nodes $plant^{-1}$, number of pods $plant^{-1}$, weight of 100 seeds, and yield.

Data analysis of the check genotypes using PKBT Stat 1.0 software followed a randomized completely block design. Meanwhile, the analysis data for the soybean lines using Microsoft Office Excel 2007 followed the augmented design. Furthermore, all of soybean lines were compared to the check genotypes, especially with Dering 1 and Tidar.

3 Results and Discussion

Statistical analysis of the character of days to maturity, number of fertile nodes, number of pods, weight of 100 seeds, and yield of six check varieties are shown in Table 1. Analysis of variance showed that there were significant differences among check genotypes for all characters. This suggests that the six varieties can be used as a comparison to identify the best soybean lines.

There were differences of days to maturity between soybeans line tested, varied between 60 and 73 days after planting (dap) (Table 2). In Indonesia, based on the age ripe of pods, then soybean can be grouped into five group, namely ultra early maturing (<70 dap), early maturing (70–80 dap), medium maturing (80–85 dap), late maturing (86–90 dap), and very late maturing (>90 dap) [8]. Based on these grouping, then all of soybean lines were classified as ultra early maturing soybean (68 lines) and early maturing soybean (17 lines). A total of 78 lines (92%) were more early maturing when compared to Dering 1, but only 20% (17 lines) were more early maturing than those of Tidar. This indicates that there were improvements in harvesting characters.

Legumes plants use some mechanism to overcome the deficit of water, among others, drought escape, drought avoidance and drought tolerance [9]. Early maturity is one of agronomic characters associated with drought resistance through a

Table 1 Analysis of variance of the six check varieties

Characters	Mean square	*F* value
Days to maturity	61.87	45.61**
Number of fertile nodes	126.27	29.08**
Number of pods	1310.75	26.04**
Weight of 100 seeds	84.28	79.56**
Yield	0.24	6.12**

Note ** = highly significant by *F*-test at the level of 1%

Table 2 The agronomic characteristic of 85 soybean lines

No.	Lines	Days to maturity (dap)	Number of fertile nodes	Number of pods	Weight of 100 seeds (g)	Yield (t ha^{-1})
1	MLG 2805/Mal-(3B)-235-170	64.87[a]	20.04	55.11	7.76[b]	1.16
2	MLG 3474/Mal-(3B)-241-171	63.87[a,b]	14.84	33.31	7.80[b]	0.93
3	MLG 2805/SHR-(3B)-317-58	70.87[a]	21.29	50.26	7.78[b]	0.98
4	MLG 2805/SHR-(3B)-297-51	65.87[a]	18.64	48.91	6.47	0.82
5	MLG 2805/SHR-(3B)-301-54	69.87[a]	22.04	53.31	7.35	1.03
6	MLG 2805/SHR-(3B)-334-62	63.87[a,b]	19.84	54.91	7.69[b]	1.01
7	MLG 2805/SHR-(3B)-328-60	68.87[a]	19.64	50.51	8.56[b]	0.98
8	MLG 2805/SHR-(3B)-322-59	68.87[a]	20.64	49.71	7.60[b]	1.10
9	MLG 2805/SHR-(3B)-303-56	70.87[a]	22.84[a]	53.11	7.47	0.73
10	MLG 2805/SHR-(3B)-302-55	67.87[a]	17.64	44.91	6.69	0.89
11	MLG 2805/SHR-(3B)-316-57	68.87[a]	20.84	45.51	8.13[b]	0.95
12	MLG 2805/SHR-(3B)-300-53	64.87[a]	24.44[a]	52.31	7.34	0.88
13	MLG 2805/SHR-(3B)-332-61	68.87[a]	23.64[a]	52.31	9.58[b]	0.76
14	ARG/GCP-335/Bal-(3B)-670-217	63.87[a,b]	20.64	43.51	7.36	0.95
15	ARG/GCP-335/Bal-(3B)-685-125	65.87[a]	15.64	33.11	13.87[a,b]	1.04
16	ARG/GCP-335/Bal-(3B)-610-190	64.87[a]	17.64	41.51	10.62	1.28
17	ARG/GCP-335/Bal-(3B)-581-183	67.87[a]	21.44	60.11	9.96[b]	1.07
18	ARG/GCP-335/Bal-(3B)-598-187	65.03[a]	20.27	47.18	10.16[b]	1.12
19	ARG/GCP-333/Bal-(3B)-191-165	63.03[a,b]	20.47	50.18	10.37[b]	1.15
20	ARG/GCP-335/Bal-(3B)-585-185	63.03[a,b]	20.07	45.58	8.63[b]	0.99
21	ARG/GCP-335/Bal-(3B)-604-188	62.03[a,b]	15.87	37.18	9.96[b]	0.89
22	ARG/GCP-333/Bal-(3B)-194-167	62.03[a,b]	22.67	51.18	7.76[b]	1.06
23	ARG/GCP-333/Bal-(3B)-197-168	66.03[a]	22.47	63.18	11.86[a,b]	0.85
24	ARG/GCP-335/Bal-(3B)-605-189	65.03[a]	19.47	44.78	9.70[b]	1.16

(continued)

Table 2 (continued)

No.	Lines	Days to maturity (dap)	Number of fertile nodes	Number of pods	Weight of 100 seeds (g)	Yield ($t\ ha^{-1}$)
25	ARG/GCP-335/Bal-(3B)-562-181	67.03[a]	18.87	36.98	10.04[b]	1.17
26	ARG/GCP-333/Bal-(3B)-192-166	64.03[a,b]	17.27	47.78	9.78[b]	1.32
27	ARG/GCP-335/SHR-(3B)-448-77	66.03[a]	24.47[a]	64.38	8.68[b]	1.21
28	ARG/GCP-335/SHR-(3B)-524-216	69.03[a]	24.27[a]	56.58	9.72[b]	0.87
29	ARG/GCP-335/SHR-(3B)-727-139	68.03[a]	24.87[a]	50.98	9.10[b]	0.87
30	ARG/GCP-335/SHR-(3B)-466-79	67.03[a]	21.87	43.98	8.92[b]	0.94
31	ARG/GCP-335/SHR-(3B)-467-80	60.03[a,b]	11.27	24.18	8.22[b]	1.10
32	ARG/GCP-335/SHR-(3B)-269-172	62.03[a,b]	18.67	43.18	8.34[b]	1.07
33	ARG/GCP-335/SHR-(3B)-521-214	66.03[a]	21.87	57.18	9.43[b]	0.83
34	ARG/GCP-335/SHR-(3B)-267-45	66.03[a]	17.67	45.78	11.82[a,b]	1.33
35	ARG/GCP-335/SHR-(3B)-728-140	68.20[a]	19.57	43.08	9.48[b]	1.05
36	ARG/GCP-335/SHR-(3B)-285-175	67.20[a]	20.77	48.28	9.6[b]	1.03
37	ARG/GCP-335/SHR-(3B)-286-202	69.20[a]	21.77	47.28	8.42[b]	0.97
38	ARG/GCP-335/SHR-(3B)-279-49	67.20[a]	23.97[a]	53.68	10.28[b]	0.98
39	ARG/GCP-335/SHR-(3B)-270-46	64.20[a,b]	21.77	46.68	7.21	1.27
40	ARG/GCP-335/SHR-(3B)-289-50	67.20[a]	19.77	42.88	8.24[b]	0.86
41	ARG/GCP-335/SHR-(3B)-463-179	70.20[a]	21.37	50.08	10.43[b]	0.88
42	ARG/GCP-335/SHR-(3B)-262-44	68.20[a]	20.97	54.28	11.10[a,b]	0.86
43	ARG/GCP-335/SHR-(3B)-284-174	71.20	22.97	51.88	8.97[b]	0.89
44	ARG/GCP-335/SHR-(3B)-522-215	69.20[a]	15.57	34.08	9.5[b]	0.98
45	ARG/GCP-335/SHR-(3B)-292-176	70.20[a]	22.97	60.28	8.86[b]	1.13
46	ARG/GCP-335/SHR-(3B)-725-138	64.20[a,b]	13.77	33.08	12.97[a,b]	1.32
47	ARG/GCP-335/SHR-(3B)-459-178	71.20	22.37	54.28	9.60[b]	0.96
48	ARG/GCP-335/SHR-(3B)-275-173	71.20[a]	22.77	57.68	10.83[a,b]	1.15

(continued)

Table 2 (continued)

No.	Lines	Days to maturity (dap)	Number of fertile nodes	Number of pods	Weight of 100 seeds (g)	Yield ($t\ ha^{-1}$)
49	DV/2984-330/Grob-(3B)-105-156	67.20[a]	22.77	56.08	10.47[b]	1.18
50	DV/2984-330/Grob-(3B)-148-164	69.20[a]	16.57	52.28	11.13[a,b]	1.20
51	DV/2984-330/Grob-(3B)-123-160	68.20[a]	24.17[a]	57.28	10.22[b]	1.08
52	DV/2984-330/Grob-(3B)-132-197	68.87[a]	15.97	40.61	10.10[b]	0.58
53	DV/2984-330/Grob-(3B)-100-154	68.87[a]	17.17	40.61	9.85[b]	0.73
54	DV/2984-330/Grob-(3B)-141-162	68.87[a]	20.57	52.81	11.45[a,b]	0.85
55	DV/2984-330/Grob-(3B)-124-161	66.87[a]	12.57	28.21	11.08[a,b]	1.02
56	DV/2984-330/Grob-(3B)-109-158	66.87[a]	17.77	53.21	11.23[a,b]	1.02
57	DV/2984-330/Grob-(3B)-104-155	68.87[a]	16.17	39.41	9.90[b]	1.08
58	DV/2984-330/Grob-(3B)-134-199	67.87[a]	24.17[a]	57.21	10.83[a,b]	0.88
59	DV/2984-330/Grob-(3B)-96-152	66.87[a]	18.17	39.41	11.34[a,b]	0.97
60	DV/2984-330/Grob-(3B)-108-157	67.87[a]	13.77	24.61	7.91[b]	0.48
61	DV/2984-330/Grob-(3B)-118-159	64.87[a]	12.77	20.21	9.53[b]	0.51
62	DV/2984-330/Grob-(3B)-103-196	66.87[a]	19.57	48.01	8.21[b]	0.48
63	DV/2984-330/Grob-(3B)-133-198	68.87[a]	21.37	45.21	10.56[b]	0.84
64	DV/2984-413/Mal-(3B)-238-200	69.87[a]	20.37	54.21	11.72[a,b]	0.46
65	DV/2984-330/Blr-(3B)-21-7	61.87[a,b]	10.77	16.61	8.85[b]	0.53
66	DV/2984-330/Blr-(3B)-34-18	61.87[a,b]	17.37	41.41	8.31[b]	0.60
67	DV/2984-330/Blr-(3B)-6-1	61.87[a,b]	17.97	36.61	8.13[b]	0.55
68	DV/2984-330/Blr-(3B)-3-151	60.87[a,b]	20.77	39.41	9.00[b]	0.68
69	DV/2984-330/Blr-(3B)-32-16	63.03[a,b]	13.54	35.41	7.72[b]	1.00
70	DV/2984-413/SHR-(3B)-187-209-1	69.03[a]	23.81[a]	57.88	7.22	0.82
71	DV/2984-413/SHR-(3B)-187-209-2	70.03[a]	20.74	44.81	7.80[b]	0.71

(continued)

Table 2 (continued)

No.	Lines	Days to maturity (dap)	Number of fertile nodes	Number of pods	Weight of 100 seeds (g)	Yield (t ha^{-1})
72	Tanggamus/SHR-(3B)-340-205	72.03	22.74	58.41	8.48^{b}	0.63
73	Tanggamus/SHR-(3B)-361-213	67.03^{a}	21.39	57.96	8.63^{b}	1.06
74	Tanggamus/SHR-(3B)-345-210	70.03^{a}	19.74	62.81	9.62^{b}	1.00
75	Tanggamus/SHR-(3B)-356-66	68.03^{a}	21.94	64.81	10.00^{b}	1.17
76	Tanggamus/SHR-(3B)-342-207	73.03	26.14^{a}	$83.01^{a,b}$	8.85^{b}	0.87
77	Tanggamus/SHR-(3B)-371-67	68.03^{a}	21.34	72.01^{a}	9.72^{b}	1.29
78	Tanggamus/SHR-(3B)-347-212	72.03	22.14	$78.61^{a,b}$	7.97^{b}	0.83
79	Tanggamus/SHR-(3B)-341-206	72.03	16.47	48.88	7.26	0.66
80	Tanggamus/SHR-(3B)-339-204	70.03^{a}	24.34^{a}	67.41	7.39	1.08
81	Tanggamus/SHR-(3B)-343-208	73.03	$28.34^{a,b}$	$97.81^{a,b}$	7.63^{b}	0.84
82	Tanggamus/SHR-(3B)-346-211	72.03	23.14^{a}	70.61^{a}	7.58^{b}	1.12
83	Tanggamus/SHR-(3B)-355-65	71.03	20.94	57.21	7.47	0.95
84	Tanggamus/SHR-(3B)-351-64	69.03^{a}	25.14^{a}	$80.01^{a,b}$	7.92^{b}	1.16
85	Wilis/Malabar-(3B)-817-224	68.03^{a}	18.47	49.54	8.03^{b}	1.41
	Dering 1	70.92	22.98	65.27	10.76	1.62
	Tidar	64.72	26.62	76.83	7.57	1.61
	Grobogan	63.32	12.22	31.35	18.76	1.12
	ARG/GCP-335	70.72	20.50	52.11	10.26	1.62
	Tanggamus	71.32	24.74	70.71	8.70	1.62
	SU-17-1014	69.92	20.82	56.87	8.39	1.29
	LSI 5%	2.88	5.14	17.51	2.54	0.49
	LSI 1%	3.93	7.01	23.89	3.47	0.67

a = significantly different compared to Dering 1 variety
b = significantly different compared to Tidar variety

mechanism of drought escape. Soybean with early maturity allows itself utilize existing water supply to complete its life cycle before the onset of drought [10]. Soybean early maturity gives a solution to the problem of drought which is frequently encountered in soybean cultivation [11].

In general, there are few soybean lines that have a number of pods better than Dering 1 and Tidar. In this genetic material, six soybean lines (7%) had the number of pods that significantly higher when compare to Dering 1, four of them even higher than those of Tidar (Table 2). A better result was obtained for the character of the number of fertile nodes. Approximately 17.6% lines tested were better than Dering 1. The number of fertile nodes and the number of pods are known as agronomic traits affecting seed yield of soybean under drought stress conditions [12].

There were 15 lines which have the number of fertile nodes that significantly higher than Dering 1, one of them even higher than those of Tidar (Table 2). Line number 81 (Tanggamus/SHR-(3B)-343-208) has 28.34 fertile nodes, higher than Dering 1 (22.98 fertile nodes) and Tidar (26.62 fertile nodes). In addition, line number 81 also has the highest number of pods and significantly higher than both of Dering 1 and Tidar. In this research, both check genotypes have 65.27 and 76.83 pods, respectively, while line number 81 has 97.81 pods (Table 2). Two other lines, i.e. line number 76 and 84 (Tanggamus/Shr-3B-342-207 and Tanggamus/Shr-3B-351-64) have a number of fertile nodes and number of pods more than those of Dering 1.

There are two strategies that may be used by soybean plants in effort to face drought during the stage of seed filling, i.e. reducing the number of pods and or the size of seeds [13]. In this study, both the strategy seems to be used by soybean lines in dealing with drought stress. It can be seen from the progeny of crosses involving Tanggamus and Dering 1 (formerly known as DV/2984-330). In the crosses between Tanggamus and SHR-(3B), all the offspring reduce the size of the seed as a response to drought stress. Based on description of soybean varieties, Tanggamus was classified as medium seed size (11.0 g per 100 seeds). But in present research, all its progeny (line number 72–84) were classified as small seed size (less than 10.0 g per 100 seeds). Meanwhile, another strategy used by crosses involving Dering 1 as a parent. According to description of soybean varieties, Dering 1 classified as medium seed size (11.0 g per 100 seeds). Almost all of its progeny had the same seed size, but with reduced number of pods (Table 2).

In general, almost all of lines have seed size larger than those of Tidar, but there were only 14 soybean lines with seed size greater compared to Dering 1. Seed size has been known to be genetically inherited [14, 15]. Soybean seed size determines the preferences of farmers and the food industry in Indonesia. Large seed size more preferably than the small or medium ones. Seed size is also an important factor in the effort to increase the production of soybean [16]. In present study, although there were improvements in the character of seed size, but none of soybean lines are capable of seed yielding more than Dering 1 and Tidar. Among 85 soybean lines tested, only 7 lines that capable of seed producing equivalent to the average of the national production (1.3 t/ha).

4 Conclusion

1. All of soybean lines were classified as ultra early maturing soybean (68 lines) and early maturing soybean (17 lines).
2. Line numbers 76, 81, and 84 (Tanggamus/Shr-3B-342-207, Tanggamus/Shr-3B-343-208, and Tanggamus/Shr-3B-351-64, respectively) had a number of fertile nodes and number of pods more than those of Dering 1.
3. Almost all of lines had seed size larger than those of Tidar, but there were only 14 soybean lines with seed size greater than those of Dering 1.
4. None of soybean lines that capable of producing seeds exceeds Dering 1 and Tidar.

References

1. M. Siregar and Sumaryanto. 2003. Estimating soybean production efficiency in irrigated area of Brantas River basin. Indonesian Journal of Agricultural Science 4(2): 33–39
2. O. Sadeghipour amd S. Abbasi. 2012. Soybean response to drought and seed inoculation. World Appl. Sci. J. 17(1): 55–60
3. B. Wiebold. 2012. Comparison of the 2012 drought to other drought for yield reductions (Part 2: soybean). Integrated Pest Crop Management 2(11): 113–114 http://ipm.missouri.edu/IPCM/archive/2012/v22n11.pdf [19 Oct 2015]
4. T. Oya, A.L. Nepomuceno, N. Neumaier, J.R.B. Farias, S. Tobita and O. Ito. 2004. Drought tolerance characteristics of Brazilian soybean cultivars – Evaluation and characterization of drought tolerance of various Brazilian soybean cultivars in the field –. Plant Prod. Sci. 7(2): 129–137
5. Suhartina and D.M. Arsyad. 2005. Tolerance of varieties and soybean lines to drought stress. National Workshop: Increased Production of Legumes and Tuber Crops to Support Self-Reliance of Food. (in bahasa Indonesia)
6. E. Dogan, H. Kirnak and O. Copur. 2007. Deficit irrigations during reproductive stages and CROPGRO-soybean simulations under semi-arid climatic conditions. Field Crops Res. 103: 154–159
7. W.T. Federer. 1961. Augmented design with one-way elimination of heterogeneity. Biometrics 17: 447–473
8. M.M. Adie. 2007. Individual testing guide of novelty, distinctness, uniformity and stability of soybean. The Center of Plant Variety Protection. Ministry of Agriculture the Republic of Indonesia. 12p
9. N.C. Turner, G.C. Wright, and K.H.M. Siddique. 2001. Adaptation of grain legumes (pulses) to water limited environments. Adv. Agron. 71: 193–231
10. L.P. Manavalan, S.K. Guttikonda, L.S.P. Tran, and H.T. Nguyen. 2009. Physiological and molecular approaches to improve drought resistance in soybean. Plant Cell Physiol. 50(7): 1260–1276 Doi:10.1093/pcp/pcp082
11. L.G. Heatherly and R.W. Elmore. 2004. Managing inputs for peak production. *In* Soybeans: Improvement, Production, and Uses. Agronomy Monographs, 3rd edn. No. 16. Edited by Specht, J.E. and Boerma, H.R. pp. 451–536. ASA-CSSA-SSSA, Madison, WI
12. J.R. Frederick, C.R. Camp, and P.J. Bauer. 2001. Drought-stress effects on branch and mainstem seed yield and yield components of determinate soybean. Crop Sci. 41(3): 759–763

13. D. Desclaux, T.T.Huynh, and P. Roumet. 2000. Identification of soybean plant characteristics that indicate the timing of drought stress. Crop Sci. 40: 716–722
14. J.A. Brian, W.R. Fehr, and G.A. Welke. 2002. Selection for large seed and high protein in two and three parent soybean population. Crop Sci. 42: 1876–1881
15. Z. Hu, H. Zhang, G. Kan, D. Ma, D. Zhang, G. Shi, D. Hong, G. Zhang, and D. Yu. 2013. Determination of the genetic architecture of seed size and shape via linkage and association analysis in soybean (*Glycine max* L. Merr.). Genetica 141(4): 247–254 Doi:10.1007/s10709-013-9723-8
16. S.L. Johnson, W.R. Fehr, G.A. Welke, and S.R. Cianzio. 2001. Genetic variability for seed size of two- and three-parent soybean populations. Crop Sci. 41(4): 1029–1033 Doi:10.2135/cropsci2001.4141029x

Synergism Between Sago Starch and Chitosan in Enhancing Biodegradable Film Properties

Indira Prabasari, Nafi Ananda Utama and Chandra Kurnia Setiawan

Abstract The research studied properties of biodegradable film from the mixture of sago starch and chitosan. The biodegradable film was obtained by solution casting method with addition of glycerol 1.5% as plasticizer. Mechanical properties of biodegradable film from sago starch and chitosan were determined including tensile strength, elongation, color and biodegradability. To study shape surface morphology of biodegradable film, the microfracture of biodegradable film was observed by scanning electron microscopy (SEM). Biodegradable film with the composition of sago starch and chitosan 1:1 significantly had the lowest number of tensile strength. The study revealed that addition of chitosan improved biopolymer elongation and brightness, but it weakened tensile strength of biopolymer and reduced its biodegradability.

Keywords Sago starch · Chitosan · Biodegradable film · SEM

1 Introduction

Biodegradable film from the mixture of polysaccharides and proteins has been studied intensively as a substitute for synthetic polymer. Researchers have their own definitions of biodegradable film/biodegradable plastic. Biodegradable film is a plastic material that its chemical structure is change under certin condition therefore affects its properties. The chemical structure changes can occur from the attack of microorganisms such as fungi, bacteria and algae. Another definition of the biodegradable plastic is a polymer in which the molecular weight is being lower because of the degradation wherein at least one step in the degradation process occurs through the natural metabolism of the organism.

I. Prabasari (✉) · N.A. Utama · C.K. Setiawan
Faculty of Agriculture, Universitas Muhammadiyah Yogyakarta,
Yogyakarta, Indonesia
e-mail: i.prabasari@umy.ac.id

T. Taufik et al. (eds.), *ICoSI 2014*, DOI 10.1007/978-981-287-661-4_11

One of the important properties of biodegradable film is its biodegradability that is an ability of materials to be degraded. The term of biodegradation sometimes confused with the term of deterioration. Deterioration can be defined as a loss in physical integrity of a material, whereas biodegradation is a biochemical transformation of compounds by microorganisms and results in mineralization or incorporation into microbial biomass. Biodegradation process produces CO_2 and H_2O under aerobic conditions or CH_4 and CO_2 under anaerobic conditions. From the definition, a distinction can be drawn that the deterioration will be fragmented plastic, but the plastic particles will still remain in the environment, while biodegradation will decompose the plastic particles through the process of mineralization.

Starch including from mung bean, potato, sweet potato and its derivatives has been widely studied in the manufacture of biodegradable films and capsules. Starch is used as a substitute for gelatin in the manufacture of these biodegradable films for its simplicity to be molded into films and its properties as high oxygen barrier, and it has a good mechanical strength [1]. Among other starches, sago starch has some advantages as the base material of biodegradable films because it has low temperature on gelatinization, low degree on sineresis, high degree on viscosity and it is easy to be attacked by mold. Another advantage is that sago starch has fairly high amylose content approx. 27% which is suitable for making a biodegradable film with solid gel [2].

Basically, polysaccharide-based biodegradable films have brittle properties and their mechanical properties are unfavorable. Therefore, in the manufacture of biodegradable film gelling agent, e.g., chitosan, and plasticizer, e.g., glycerol, sorbitol and polyethylene glycol are added to overcome fragility and to improve elasticity [3]. Chitosan is a natural cationic polysaccharide obtained from deacetylation of chitin which is widely available in nature. According to the chemical structure, chitosan consists of monomer 2-amino-2-deoxy-D-glucose (glucosamine) which shows the properties of biomedical polymers such as non-toxic, biocompatible and biodegradable. Chitosan structure is similar to cellulose, and its ability to form a gel in the acid is due to its properties to build a matrix in the polymer system. In the manufacture of biodegradable films, the addition of chitosan and the presence of polyacrylic acid (PAA) produce $CaCO_3$ crystals that will affect the morphology of film. There was areport that the addition of chitosan to maize starch biodegradable film produces a film with more strength.

The aim of the study was to produce biodegradable film which has similar properties with synthetic plastic. The synergism between sago starch and chitosan was then studied including the character of biodegradable film produced, the effect of sago starch and glycerol concentration on the physical and mechanical properties, i.e., tensile strength and elongation, and the rate of biodegradable film degradation.

2 Research Method

2.1 *Sample Preparation*

Biodegradable film was prepared from sago starch and mixture between sago starch and chitosan as follows: (1) sago starch without chitosan (BS); (2) sago starch blended with chitosan and the ratio was 1:1 (BKS I); and (3) sago starch blended with chitosan and the ratio was 2:1 (BKS II). The film was prepared by mixing polysaccharides with glycerol and distilled water at 50 °C. The solution was stirred using an agitator at 100 rpm for 30 min. A commercial sago starch and chitosan were then added to the glycerol solution, while it was stirred and distilled water was added until the solution reached 200 mL. The stirring process was continued until it became a clear solution. The solution then poured onto five pieces of 20 cm $\times$ 30 cm acrylic glass plate and stood for 15 min. After the frame was released, biodegradable films on acrylic glass plate were then dried in oven at 50 °C. After drying, the film was released from acrylic glass plate and wrapped by aluminum foil.

2.2 *Tensile Strength and Elongation Measurement*

Biodegradable film was cut and linked horizontally on Instron Universal Testing Machine tools (Zwick Z.05 texture analyzer) connected to a computer for data analysis.

2.3 *Brightness Test*

Measurement of brightness used a color index Chroma Meter Minolta CR-600 with the attribute "L" indicating the level of brightness, "$+a$" degree of redness, "$-a$" degree of greenness, "$+b$" yellowish level and "$-b$" the bluish level.

2.4 *Biodegradability Test*

Biodegradability test was conducted using soil burial test of Behjat et al. [4] with slight modifications. Testing was done by cutting the film with the size of 5 $\times$ 10 cm. It was then buried approx. 20 cm below the ground for one week. The film was observed and the film remained was measured.

2.5 *Scanning Electron Microscopy (SEM)*

Observations using electron microscopy were conducted to determine the microstructure of biodegradable films.

3 Results and Discussion

3.1 *Film Properties*

Physical characteristic of biodegradable film was shown in tensile strength and elongation (Table 1). Chitosan affected characteristics of biodegradable film. The cross-link between sago starch and chitosan (BKS I and BKS II) lowered tensile strength of film produced due to decrease in starch concentration. The decrease in starch concentration was in line with the decrease in amylose content in biodegradable film. The result was similar to the previous research that mung bean starch with the highest amylose content (30%) had the highest tensile strength [1]. The increase in amylose content also increased aggregation from the formation of hydrogen bonds between polymers that formed amylose microcrystal that resulted in film with high tensile strength [5].

The result showed that BKS I film had a higher tensile strength than BKS II film. Studies showed that increase in the ratio of starch and chitosan composite decreased tensile strength of the film. The decrease in tensile strength that was in line with the increase in starch ratio was caused by the formation of intramolecular hydrogen bonds' starch that was larger than the intermolecular bonding, resulted in separation phase between the two main components [6].

Starch contains macromolecules, such as amylose and amylopectin, which form a solution when heated with water and become a gel after cooling. During the process of forming a gel, inter- and intramolecular cross-linking is established to form microcrystalline regions [7]. Starch components without plasticizer resulted in increase in crystallinity that occurred during the formation of the film which increased tensile strength. This was more obvious in more rigid starch films [8]. It also occurred in elevated concentrations of chitosan that result in decreased crystallinity of the films [6].

Table 1 Mechanical properties of sago starch–chitosan biodegradable film

Sago starch–chitosan concentration	Tensile strength (MPa)	Elongation (%)
BKS I	29,277^{b}	493,239^{a}
BKS II	24,502^{b}	447,872^{a}
BS	69,004^{a}	318,846^{a}

Description The value is the mean of three replicates of analysis ± standard deviation; same superscript letters indicate no significant difference ($P > 0.05$)

For elongation, BSI film had the highest elongation values though not significantly different from other treatment. It showed that the addition of chitosan can increase the percentage elongation value of the film. Elongation value was inversely proportional to the film tensile strength values. Previous study showed that the combination of starch–chitosan film had significantly higher elongation value than the film made from starch or chitosan itself [9]. Higher amylose content in the film will result in a stronger starch intermolecular force, therefore the film with less flexibility and stronger in tensile strength has lower number in elongation [6].

3.2 Film Brightness

Table 2 shows that addition of chitosan affected starch color. L value stated brightness parameters with the range value 0–100 (black to white), while the b value stated chromatic color mixture of blue–yellow with +b values of 0–70 for yellow and 0 to −70 for blue.

Increasing the *L* value indicated the level of brightness of the starch. The results of the study showed that the reduced levels of sago improved the brightness of biodegradable films. In *b* value, decreasing the levels of sago in biodegradable films was in line with the decrease in b value. Previous study from Chillo et al. [10] indicated the same phenomenon where the L and b values increased with increasing chitosan concentrations.

3.3 Biodegradability Test

Biodegradation is the result of an enzymatic process and involves both microorganism and macroorganism. Microorganism accelerates degradation by enzymes and may occur in aerobic or anaerobic conditions. Linear polymer is generally more easily degraded than the branched polymers [11]. Table 3 shows the result of film biodegradation.

Sago starch–chitosan film (BKS I and BKS II) had lower degradation compared to sago–starch film (BS). It indicated that addition of chitosan inhibited the

Table 2 Effect of chitosan on degree of white sago starch films

No	Film	Color		
		L	*A*	*B*
1	BKS I	31.49^{b}	-0.67^{a}	3.85^{c}
2	BKS II	25.47^{a}	-0.39^{a}	2.21^{b}
3	BS	25.98^{a}	-0.73^{a}	1.86^{a}

Description The value was the mean of three replicates of analysis ± standard deviation; same superscript letters indicate no significant difference ($P > 0.05$)

Table 3 Biodegradation of sago starch–chitosan film

Sago starch–chitosan concentration	Soil burial test (cm^2/week)
BKS I	62
BKS II	24
BS	104

degradation by microbes because chitosan was antimicrobial and antifungal, the same phenomenon which was found in Fernandez-Saiz et al. [12] and Ziani et al. [13]. In other work, hydrogen bond interactions between tapioca starch and chitosan reduced the availability of hydrophilic groups, diminishing their interactions with water molecules. It was also observed that when chitosan coating was applied to salmon fillet pieces, the antimicrobial effectiveness was greater than when chitosan–tapioca starch blends were applied [14]. However, observations of the rate of degradation showed that degradation of BKS II film was smaller than BKS I. It might be due to the presence of other organisms in the treatment media that were accelerating the degradation process.

3.4 Microstructure Observation Using SEM

Observations using electron microscopy were conducted to determine microstructure of the film. The SEM analysis was only conducted for BKS I that had better mechanical properties compared to BKS II and BS. The results of SEM analysis of BKS I are presented in Fig. 1.

Figure 1 shows the outer structure of BKS I that was compact, continuous, prevalent and not porous. The result was similar to the chitosan film SEM from Valenzuela et al. [15] and sago starch film Afiq and Azura [16]. The structure was not homogeneous that might be caused by the imperfection of intramolecular bond between chitosan and sago starch. The imperfection was due to the starch melting point that did not reach maximum temperature. It prevented starch granule from fully shedding so that it became soluble and lost its intermolecular bond [14].

4 Conclusion

The study shows that chitosan improved mechanical properties of sago starch film in elongation value and also improved film brightness. However, addition of chitosan decreased tensile strength value and degradation rate. The investigation of permeability characteristics and mechanical properties with addition other resources was needed to obtain comprehensive knowledge of biodegradable film.

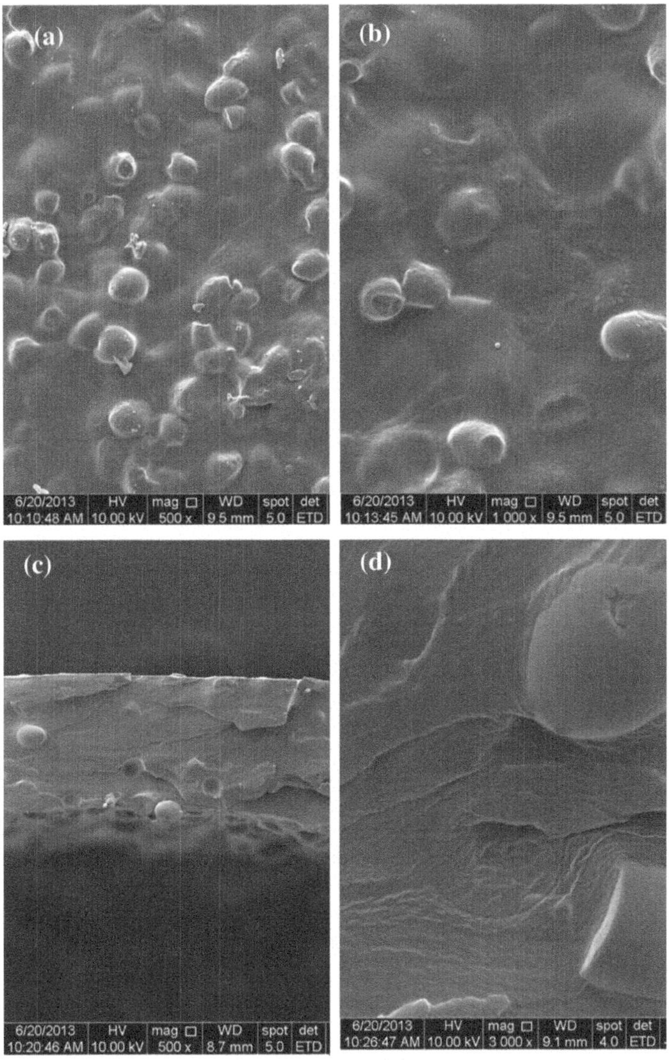

Fig. 1 Scanning electron microscopy observations of BKS I with various magnifications after soil burial test for 1 week. Biodegradable film looks over 500× magnification (**a**) and 1000× (**b**), 500× magnification side view (**c**) and 3000× (**d**)

References

1. Bae, H.J., D.S. Cha, W.S. Whiteside, H.J. Park. 2008. Film and pharmaceutical formation properties of mungbean, waterchestnut, and sweet potato starches. Food Chem. 106: 96 – 105
2. Krochta, J. M., E. A. Baldwin, dan M. O. Nisperos-Carriedo. 1994. *Edible Coatings and Film to Improve Food Quality*. Technomic Publishing Company, New York

3. Gilleland, G.M., J.L. Turner, P.A. Patton, and M.D. Harrison. 2001. *Modified starch as a replacement for gelatin in soft gel films and capsules*. World intellectual Property Organization. International Patent Number, WO 01/91721 A2
4. Behjat, T., Russly, A. R., Luqman, C. A., Yus, A. Y. and Nor Azowa, I. 2009. Effect of PEG on the biodegradability studies of Kenaf cellulose -polyethylene composites. International Food Research Journal 16: 243–247
5. Liu, Z. dan J.H. Han. 2005. Film-forming Characteristics of Starches. J. Food Science. 70 (1): E31-E36
6. Xu, Y.X., K.M. Kim, M.A. Hanna dan D. Nag. 2005. Chitosan-starch Composite Film: Preparation and Characterization. Industrial Crops and products. 21:185–192
7. Embuscado, M. E. and K.C. Huber. 2009. Edible films and Coating for Food Applications. Springer
8. Talja, R.A., H. Helen, Y.H. Roos, dan K. Jouppila. 2007. Effect of various polyols and polyol contents on physical and mechanical properties of potato starch based films. Carbohydrate Polymers 67: 288–295
9. Mathew, S., M. Brahmakumar, T.E. Abraham, 2006. Biodegradable films 82:76–187.
10. Chillo, S., S. Flores, M. Mastromatteo, A. Conte, Lia Gerschenson, M.A. Del Nobile. 2008. Influence of glycerol and chitosan on tapioca starch-based edible film properties. Journal of Food Engineering 88: 159–168
11. Briassoulis, D. 2004. An Overview on the Mechanical Behaviour of Biodegradable, Agricultural Films, Journal of Polymers and the Environment. Vol. 12, No. 2
12. Fernandez-Saiz, P., Lagaron, J. M., & Ocio, M. J. 2009. Optimization of the biocide properties of chitosan for its application in the design of active films of interest in the food area. Food Hydrocolloids, 23, 913–921.
13. Ziani, K., Fernandez Pan, I., Royo, M., & Maté, J. 2009. Antifungal activity of films and solutions based on chitosan. Food Hydrocolloids, 23:2309–2314.
14. Jiménez A., M. José Fabra, P. Talens, A. Chiralt, 2012, Edible and Biodegradable Starch Films: A Review, Food Bioprocess Technology 5:2058–2076.
15. Valenzuela, C., L. Abugoch, C. Tapia, 2013, Quinoa protein chitosan sunflower oil biodegradable film: Mechanical, barrier and structural properties. Food Science and Technology 50: 531–537
16. Afiq. M.M., A.R. Azura, 2013, Effect of sago starch loadings on soil decomposition of Natural Rubber Latex (NRL) composite films mechanical properties, International Biodeterioration & Biodegradation, 85: 139–149

Variability of Soybean Genotypes Based on High Yield and Seed Size Supporting Industrial Raw Material

Ayda Krisnawati and M. Muchlish Adie

Abstract The main utilization of soybean with large seed size (>14 g/100 seeds) in Indonesia is for tempeh raw material. The aim of this research was to identify soybean genotypes based on high yield and seed size. Experiment was conducted at Jambegede Research Station (Malang), from February to May 2013 using 429 soybean genotypes. 'Rajabasa,' 'Grobogan,' and 'Mutiara' were used as check varieties. Each soybean line was planted on 4.5 m row, 40 cm × 15 cm plant distance, 2 plants/hill. Plant was fertilized with 50 kg urea, SP36 100 kg, and KCl 75 kg/ha. The seed size and seed yield varied among 429 genotypes. A total of 167 genotypes (38.93%) have large seed size (14–16 g/100 seeds). Most of genotypes have yield at range 2.5–3.0 t/ha, counted for 212 genotypes (49.42%), and only three genotypes have seed yield above 2.94 t/ha. Range of seed yield was 0.17–3.11 t/ha (average 2.09 t/ha), and 100 seeds weight ranged from 8.20 to 21.34 g (average 14.31 g). The average seed yield and seed size of check variety of 'Rajabasa' was 2.32 t/ha and 11.56 g/100 seeds, respectively. Check variety of Grobogan had average seed yield and seed size of 2.30 t/ha and 21.34 g/100 seeds, respectively, while the seed yield of 'Mutiara' was 1.67 t/ha with seed size up to 17.80 g/100 seeds. Simultaneously, two genotypes were successfully selected (Grobogan × Wilis-52 and Grobogan × Wilis-17) with yield above 2.94 t/ha and seed size above 14 g/100 seeds. Genotype of G511H/Kaba//Kaba///Kaba//// Kaba-19 had highest yield, but the seed size was categorized as medium size (11.35 g/100 seeds). All the three selected genotypes had higher yield than the check varieties and have a chance to proceed to the next breeding step, so it could be released as new improved varieties.

Keywords Soybean · Selection · High yield · Large seed size

A. Krisnawati (✉) · M. Muchlish Adie
Indonesian Legumes and Tuber Crops Research Institutes,
PO Box 66, Malang, Indonesia
e-mail: my_ayda@yahoo.com

T. Taufik et al. (eds.), *ICoSI 2014*, DOI 10.1007/978-981-287-661-4_12

1 Introduction

The main soybean function in Indonesia is as raw material for industry, both domestic scale and large industries, such as tofu and tempeh. Both products require different soybean seed size. Generally, tofu industry requires soybean with medium seed size. On the contrary, soybean with large seed size is preferred for tempeh production to produce tempeh with bright colors and large volume [1]. During 2002–2012, the average consumption of tempeh reached 7.61 kg/capita/year [2]. It is not comparable with the demand, both in terms of the number of existing varieties and production. Therefore, it must be anticipated by providing highly yielding soybean varieties with large seed size in accordance with industry preference.

As part of complex quantitative trait, yield is influenced by various yield components. The yield components such as seed number and seed size are primary components which determined the soybean yield and its products [3]. In USA, soybean cultivars with small seed size (≤ 80 mg seed^{-1}) are developed to fulfill the needs for natto production in Japan [4]. The effect of seed size on the seed yield has attracted of research workers over the years, since seed size has become one of essential part of seed quality which play important role in supporting the performance of a crop [5, 6]. The differences in the seed yield and seed size were determined by assimilates supply during the reproductive stage. The key factor which determines the pod and seed number, and also seed weight per plant, is assimilate availability [7, 8].

A research by [9] on the relationship between seed size, seed quality, and seed yield on soybean, had found that small seed size had maximum seed germination and emergence while large seed size produced highest seed per plant, pod per plant, and seed yield per plant. Another study by [10] showed that high heritability for yield-related and some morphological characters were observed. Significant positive correlation existed between grain yield and seeds/pod, 100-seeds weight, pods/plant, pod length, days to flowering and plant height, nodes/plant; seeds/pod and plant height, LAI, flowering, pod length; 100-seeds weight and flowering, maturity, pods/node, and seeds/pod. Both high heritability and high genetic advance as well as positive associations of the characters obtained can be explored in the selection of soybean for high yield improvement.

A study related to determinate and indeterminate soybeans in four regions in China showed that a higher yield performance was produced by differential increases in pods per plant, seed per pod, and seed size [11]. On the other hand, more recent research comparing between old and new Midwestern soybean cultivars has clearly shown that seed m^{-2} was more strongly related to yield improvement than to its seed size [12]. Yield improvement within cultivar development has reported can be obtained through different yield component mechanisms. Improvement in soybean varieties in Indonesia performed through conventional method, i.e., by crossbreeding and followed by the selection process.

The aim of this research was to identify soybean genotypes based on high yield and seed size.

2 Research Method

The research was conducted in Jambegede Research Station from February to May 2013 (dry season I). The research materials consist of 426 soybean lines and three check of large seed size cultivars ('Rajabasa', 'Grobogan', and 'Mutiara'). The study was conducted in lowlands after rice planting and without any soil tillage. Soybean was planted 8 days after the rice harvest, and irrigation channels were made to maintain optimum soil moisture for soybean growth. Each soybean line was planted on 4.5 m row, 40 cm × 15 cm plant distance, 2 plants/hill. Plant was fertilized with 50 kg urea, SP36 100 kg, and 75 kg/ha KCl. Pests, diseases, and weeds were controlled optimally.

The data were collected on 100 seeds weight and seed yield. The 100 seed weights were observed by randomly taking and weighing at 11% moisture content. The seed yield was weighed at 11% of moisture content which is subsequently converted to yield per ha.

3 Results and Discussion

Descriptive statistic from 426 soybean lines and three check varieties is presented in Table 1. Seed weight character showed variability among the genotypes with a range of 8.20–21.34 g/100 seeds with average of 14.31 g/100 seeds. The results were quite similar to those of [13, 14] who reported ranges of 13.9–20 g and 14.67–25.67 g for 100 seeds weight, respectively.

The seed yield ranges from 0.17 to 3.11 t/ha with an average 2.09 t/ha. The average of seed size from three check varieties 'Rajabasa', 'Grobogan', and 'Mutiara' was 11.56, 21.34, and 17.89 g/100 seeds, respectively, while the yield of each were 2.32, 2.30, and 1.67 t/ha, respectively. The yield of 'Grobogan' was

Table 1 Descriptive statistic of 429 soybean genotypes. 2013

Parameter	Seed weight (g/100 seeds)	Yield (t/ha)
Average	14.31	2.09
Minimal	8.20	0.17
Maximal	21.34	3.11
Standard Deviation	1.9160	0.4267
Skewness	0.4409	−0.9302
Kurtosis	0.7844	2.3328
Check varieties		
'Rajabasa'	11.56	2.32
'Grobogan'	21.34	2.30
'Mutiara'	17.80	1.67

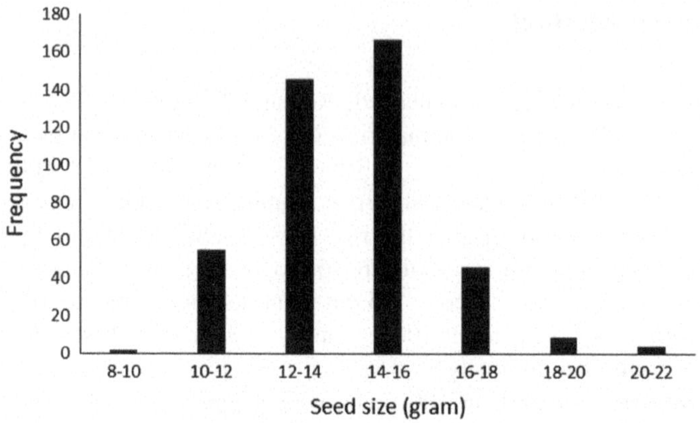

Fig. 1 Histogram of seed size of 426 soybean lines. 2013

comparable to the ‘Rajabasa’, whereas the ‘Mutiara’ had a low yield. Among the three check varieties, ‘Grobogan’ had the highest yield and largest seed size (21.34 g/100 seeds).

Seed size is one of the important factors of soybean yield. Grouping of seed size was varied among soybean production countries. In Indonesia, soybean seed categorized as large if the weight of 100 seeds >14 g. From 429 genotypes, a total of 167 lines (38.93%) have large seed size (14–16 g/100 seeds), 46 genotypes have seed size from 16 to 18 g/100 seeds, and 13 genotypes have seed size 18–22 g/100 seeds (Fig. 1). Neither of 426 lines have seed size larger than ‘Grobogan’ which reached 21.34 g/100 seeds.

The economic value of seed size in soybean is related to its suitability for tempeh raw materials and contributes to the soybean productivity. Soybean varieties with large seed size produced maximum seed per plant, pod per plant, and high yield per plant [9]. This excellence due to the high seedling vigor resulting from extra food reserves stored in large seed [15, 16]. Soybean seed size is controlled by multiple genes which may affects the performance of seed yield, quality, and appearance [4, 17]. Numerous researches have shown that seed size may be controlled by different genetic factors [18–20].

Apart from seed size, soybean seed yield is also influenced by genetic and environmental (G $\times$ E) factors; several studies stated that it was determined by the combination of genetic, agronomic, and environmental factors [21–23]. In this study, most of lines (212 lines or 49.42%) have seed yield in the range of 2.0–2.5 t/ha, and only three lines have seed yield above 2.94 t/ha. The classification of the yield was based on the general mean ($\bar{x}$) and the standard deviation (SD) [24]. By adopting these criteria, then the selection limit for very high yielding genotype if the seed yield more than $\bar{x} + 2\text{SD}$. Thus, the limit selection for high yielding line was more than 2.94 t/ha. Based on that, three lines with yield 2.95, 3.00, and 3.11 t/ha were obtained (Fig. 2).

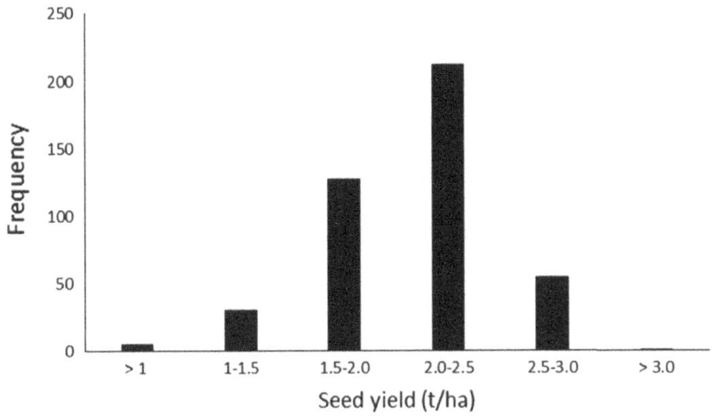

Fig. 2 Histogram of seed yield of 426 soybean lines. 2013

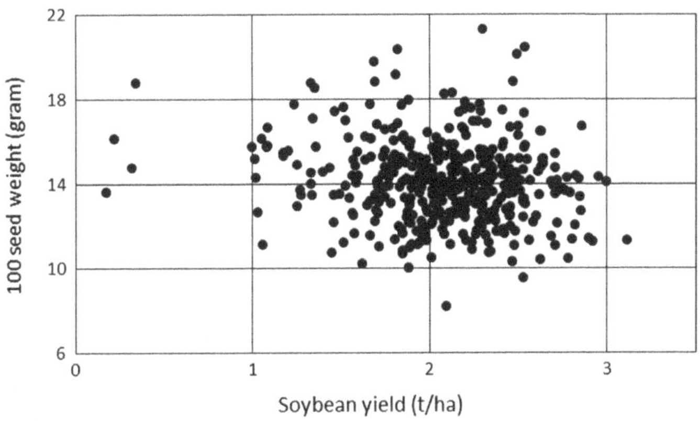

Fig. 3 Mapping of soybean yield vs seed size. 2013

Nowadays, the user preference is not limited to soybean varieties with high yield but also to large seed size, which is suitable for industrial raw materials. Simultaneous selection for high yield and large seeds size is mapped in Fig. 3.

Simultaneous selection to obtain both of high yield and large seed size resulted in two lines (Grobogan × Wilis-52 and Grobogan × Wilis-17) with yield above 2.94 t/ha and the seed size more than 14 g/100 seeds. These lines were prospected to be released as high yielding and large seed size varieties for industrial raw materials. Line of G511H/Kaba//Kaba///Kaba////Kaba-19 had the highest yield, but the seed size was categorized as medium size (11.35 g/100 seeds). All the three selected genotypes had higher yield than the check varieties and could be proceed to the next stage of breeding to be released as new improved varieties.

4 Conclusion

1. Genotypes Grobogan × Wilis-52 and Grobogan × Wilis-17 with both of high yield and large seed size were suitable for industrial raw materials.
2. Genotype G511H/Kaba//Kaba///Kaba////Kaba-19 had the highest yield and medium seed size.
3. Genotypes Grobogan × Wilis-52, Grobogan × Wilis-17, and G511H/Kaba//Kaba///Kaba////Kaba-19 could be released as new improved varieties.

References

1. Krisdiana, R. 2005. Preferensi Industri Tahu dan Tempe dalam menggunakan bahan baku Kedelai di Jawa Timur (Preference of tofu and tempe industry in using soybean raw materials). Research Performance in Supports on Legumes and Tuber Agribusiness. Indonesian Center for Food Crop Research and Development. Bogor. [in Bahasa].
2. Pusdatin. 2013. Pusat Data dan Sistim Informasi Pertanian (Data Center and Agricultural System Information). Buletin Konsumsi Pangan 4 (3), 57p. [in Bahasa].
3. Quijano, A., and E.N. Morandi. 2011. Post-flowering leaflet removals increase pod initiation in soybean canopies. Field Crops Res. 120: 151–160.
4. Jonson, S.L., W. R. Fehr, G.A. Welke, and S.R. Cianzio. 2001. Genetic variability for seed size of two and three parent soybean population. Crop Sci. 41:1029–1033.
5. Adebisi, M.A. 2004. Variation, stability, and correlation studies in seed quality and yield components in sesame (*Sesame indicum* L.). PhD Thesis, University of Agriculture, Abeukota, Nigeria.
6. Adebisi, M.A., T.O. Kehinde, M.O. Ajala, E.F. Olowu, and S. Rasaki. 2011. Assesment of seed quality and potential longevity in elite tropical soybean (*Glycine max* L.) Merrill grown in southwestern Nigeria. Niger. Agric. J. 42: 94–103.
7. Egli, D. B. 1998. Seed biology and the yield of grain crops. CAB International, Oxford. pp. 178.
8. Egli, D. B. and W.P. Bruening. 2006. Temporal profiles of pod production and pod set in soybean. European J.of Agron. 24: 11–18.
9. Adebisi, M.A., T.O. Kehinde, A.W. Salau, L.A. Okesola, J.B.O. Porbeni, A.O. Esuruoso, K. O. Oyekale. 2013. Influence of different seed size fractions on seed germination, seedling emergence and seed yield characters in tropical soybean (*Glycine max* L. Merrill). Int. J. of Agric. Res. 8 (1): 26–33.
10. Okonkwo, F.A. and D. O. Idahosa. Heritability and correlation studies of yield characters in some soybean (Glycine Max) varieties in Ekpoma. 2013. American J. of Res. Communication. Available on: www.usa-journals.com, ISSN: 2325–4076. (Access 20 March 2014).
11. Cui, Y.S. and D.Y. Yu. 2005. Estimates of relative contribution of biomass, harvest index, and yield components to soybean yield improvements in China. Plant Breed. 5: 473–476.
12. De Bruin, J.L. and P. Pedersen. 2009. Growth, yield, and yield component changes among old and new soybean cultivars. Agron. J. 101: 123–130.
13. Maestri, D.M; Labuckas, D.O; Guzman, C.A and Giorda, L.M. 1998. Correlation between seed size, protein and oil contents and fatty acid composition in soybean genotypes. Grasses Aceites 49: 450–453.
14. Arshad, M. Ali, N. and Ghafoor, A. 2006. Character correlation and path coefficient in soybean (*Glycine max.* (L) Merrill). Pakistan J. of Botany 38 (1): 121–130.

15. Sexton, P. J., J.W. White and K.J. Boote. 1994. Yield-determining processes in relation to cultivar seed size of common bean. Crop Sci. 34: 84–91.
16. Gholami, A., S. Sharafi, A. Sharafi and S. Ghasemi, 2009. Germination of different seed size of pinto bean cultivars as affected by salinity and drought stress. Food, Agric. Environ., 7: 555–558.
17. Harnowo, D. 2004. Effect of time of harvest and seed size on seed quality of soybean. Thesis. Univ. Putra Malaysia. 204p.
18. Brian J. A., W. R. Fehr, and G.A. Welke. 2002. Selection for large seed and high protein in two and three parent soybean population. Crop Sci. 42:1876–1881.
19. Li, W., H. L. Ning, W. B. Li and W.H. Lu. 2006. Developmental Genetic Analysis of Seed Size in Soybean (*Glycine max*). Acta Genetica Sinica 33 (8): 746–756.
20. Hu, Z., H. Zhang, G. Kan, D. Ma, D. Zhang, G. Shi, D. Hong, G. Zhang, and D. Yu. 2013. Determination of the genetic architecture of seed size and shape via linkage and association analysis in soybean (*Glycine max* L. Merr.). Genetica 141: 247–254.
21. Specht, J.E., D.J. Hume, and S.V. Kumudini.1999. Soybean yield potential: A genetic and physiological perspective. Crop Sci. 39: 1560–1570.
22. Wilson, W., B.J. Havercamp, R.G. Laurenz, D.A. Marbuger, J. Orlowski, S. Casteel, S. P. Conley, P.D. Esker, C. Lee, E.D. Nafziger, K.L. Roozeboom, W.J. Ross, K.D. Thelen, and S.L. Naeve. 2012. Agronomic maximization of soybean yield and quality: population x management interactions. Water, Food, Energy & Innovation for a Sustainable World. ASA, CSSA and SSSA International Annual Meetings, November 3–6, 2013. Tampa, Florida, USA.
23. Suhre, J.J. 2012. Genetic gain × management interaction in soybean: seeding rate effect. Thesis. University of Illinois. USA. 54p.
24. Chiang, H.S. and N.S. Talekar. 1980. Identification of sources of resistance to the bean fly and two other agromyzid flies in soybean and mungbean J. Econ. Entom. 73:197–199.

Vegetables Mapping Using Production and Socioeconomic Indicators Approach

Weksi Budiaji, Juwarin Pancawati and Suherna

Abstract Founded in 2007, Serang was considered as one of the new cities in Indonesia. That was why vegetables commodities mapping was not available yet. This mapping was essential for a new city to create an accurate policy in order to develop and maintain sustainable agriculture practices. Many agriculture commodities mappings were constructed by using location quotient (LQ) method. Unfortunately, it was based on productivity only. In this paper, we presented a vegetables commodities mapping by using not only production factor, but also farmer's socioeconomic indicators, namely: (1) land width (preservation), (2) future land width (enhancement), and (3) household expenditure. The production factors and the farmer's socioeconomic indicators were mapped and drawn in a biplot as the first and the second layer. These two layers were overlayed in order to obtain a priority commodity in each district. Finally, this study resulted in four commodity priority categories, and they are: first, Cipocok Jaya district as the main priority with cucumber as its potential commodity; second, Curug district with chili as its commodity; third, Walantaka district with chili and Taktakan district with pea bean as their potential commodities; and finally, Kasemen and Serang districts that have no identified commodity yet.

Keywords LQ (location quotient) · Biplot · Mapping · Agriculture commodity

W. Budiaji (✉) · J. Pancawati · Suherna
Agribusiness Department, Sultan Ageng Tirtayasa University, Serang, Indonesia
e-mail: budiaji@untirta.ac.id

J. Pancawati
e-mail: juwarinpancawati@yahoo.com

Suherna
e-mail: shsuherna@yahoo.com

Suherna
Department of Management, Diponegoro University, Semarang, Indonesia

T. Taufik et al. (eds.), *ICoSI 2014*, DOI 10.1007/978-981-287-661-4_13

1 Introduction

National food security can be considered as an essential matter to build a national stability because food security is identical to the national security. Food security is related to self-sufficiency concept, a nation capacity to provide food needs without any imported food [1]. The degree of self-sufficiency is measured by a proportion of the utilized food in the internal and external systems, i.e the more self-sufficiency, the less utilized food from the external system and vice versa.

The self-sufficiency can be supported by the optimal agriculture potential exploration. Banten Province as a new province, for example, has a potential in the agriculture labor force that is 20% from its total [2]. This manpower potency must be optimally and sustainably utilized in order to have a greater income from the agriculture sector. Serang city as a new city in Banten Province must also identify its agriculture potential. A map of agriculture commodities is important in Serang city because it has about 30% of the total land use [3].

Decentralization regulation in Indonesia plays an important role in the optimal agriculture potential exploration [4] by managing and determining the superior commodity. Agriculture land and water resource mapping supported regulatory policy and public investment [5]. Amaliah and Julia [6], moreover, emphasized the importance of agriculture commodities mapping as an agriculture commodity exploration activity to improve their competitiveness when there is abundance of the imported agriculture commodities. This activity intended for superior commodity determination, the leading commodity that has a strategic position to be expanded, depends on the socioeconomic condition of its local community [7, 8]. Preservation and enhancement willingness, on the other hand, are the other important factors because these factors determine the future superior agriculture commodity.

One of the analyses for commodities mapping is location quotient (LQ) using production factor perspective [9–11]. LQ is usually employed as a prior analysis before applying another analysis [12, 13]. Shift share analysis, for example, is often used as a further analysis completing LQ [12, 14]. Yulianti [7], furthermore, utilized attractive–competitive potency in which the commodities were categorized into the first, second, third, and fourth superior commodity in a four quadrant. Land appropriateness analysis can also be applied to evaluate the LQ results to the land use [5], land structure, and composition [15].

Those further analyses are performed after the LQ analysis because it does not allow any additional variable. This constrain also encourages the multivariate analysis approach such as biplot to be applied in the agriculture commodities mapping. The biplot gives a chance for not only the productivity variable but also the other variables to be included in the agriculture commodities mapping simultaneously.

The vegetables commodities mapping which is still unavailable in Serang city becomes problem. There are at least two reasons why it is essential to set up this mapping, which are: (1) to create an accurate policy in order to support the sustainability of beneficiary agribusiness practices and (2) the lowest growth rate of vegetables commodity when compared to the other [16].

Since socioeconomic of the farmers affects the sustainability commodity production, it is insufficient if the vegetables mapping in Serang city uses productivity perspective (LQ). The intention of enclosing the socioeconomic factors in the mapping construction is to produce a map that is responsive to the policy changing. That is why the vegetables mapping using production and socioeconomic factors becomes the main objectives of this paper.

2 Methods

Descriptive quantitative research, in which there was no statistical test [17], was applied in this study. The primary and secondary data were collected from farmers in Serang city and some institutions, respectively. A set of questionnaire was administered to the farmers chosen by stratified random sampling. Stratification level, which consisted of six districts, was applied to the districts in Serang city. Each district was sampled in order to obtain the farmers using land width producing vegetables approach. The following formula was applied when the number of farmers was taken as samples:

$$n = \frac{\sum_{i=1}^{6} N_i^2 \sigma_i^2 / w_i}{N^2 D + \sum_{i=1}^{6} N_i^2 \sigma_i^2} \quad (1)$$

where N_i is the number of the land width in district i, σ_i is the standard deviation of the land width in district i, w_i is the proportion between the land width in district i and the total land width in Serang city, N is the total land width in Serang city, and D is computed from $\frac{B^2}{4}$ in which B is the bound of the error [18]. Using secondary data of the vegetables land width [19], the land width sampled was 0.7 ha for each district. If each farmer was assumed having 1000–2500 m^2, the number of farmers that should be taken for each district was 3–7.

Primary data were gathered to obtain the farmer socioeconomic characteristics that were household expenditure, farming system expenditure, accessed land width (preservation variable), and future accessed land width (enhancement willingness). These characteristics were intended to construct the second layer of the commodity mapping. On the other hand, the secondary data contained vegetables production formed the first layer. The overlay layers were purposed to take into account the all variables simultaneously.

Both layers were formed using biplot analysis by which there is a link between biplot and location quotient (LQ), an analysis used to agriculture commodities mapping. LQ is based on the proportion between the ratio of a commodity production to the total commodity's production and the ratio of the total commodity production in its district to the total commodity production of all districts in Serang city. It was computed from a contingency table of the commodity production. The relation between district and commodity produced in a table contingency, nevertheless, can also be measured using chi-squared statistic/χ^2 formula

$$\chi^2 = \sum_{i=1}^{n}\sum_{j=1}^{p}\frac{\left(x_{ij} - E_{ij}\right)^2}{E_{ij}}, \tag{2}$$

where x_{ij} is the value in the cell (i, j) and E_{ij} is the expected value of the cell (i, j) that is computed using $E_{ij} = \frac{x_{i\cdot}x_{\cdot j}}{x_{\cdot\cdot}}$, with $x_{\cdot\cdot} = \sum_{i=1}^{n} x_{i\cdot}$ [20].

The advantage of the chi-squared statistic in a two-way contingency table of district and commodity is the possibility of chi-squared decomposition yielding in a two-dimensional projection of district and commodity simultaneously. The projection has a Euclidean distance interpretation by which a district that closes to another is interpreted as they have similar profile and vice versa. Moreover, a district that closes to a commodity means that the commodity has an important value in this district. The other advantage is that adding variables in the analysis can be applied. It is important because the commodity mapping can not only using the production factor, like in LQ, but also including the socioeconomic of the farmers to create a responsive map to the policy changing.

The first and second layers result in a four-quadrant figure of district and commodity. The first quadrant (the main quadrant) is a district with a high in both the commodity production and the enhancement willingness of the farmers, while the combination of a low commodity production and high enhancement willingness lays in the second priority (second quadrant). The next priority is the third quadrant determined by a high commodity production yet low enhancement willingness. Finally, the last quadrant as the least priority has a low score in both the commodity production and the enhancement willingness of the farmers.

3 Results and Discussion

3.1 Respondent Characteristics

There were 33 respondents consisted of vegetable farmers where 94% of them are male. Their age range was 25–65 years and dominantly (79%) above 40 years. The majority respondents had low educational background. About 95% of them were elementary school and junior high school. Their farming system expenditure had a very large variation from Rp 500,000 to Rp 45,000,000 corresponding to their farming system land width (positively correlated) with 70% of them having less than Rp 6,000,000 for their farming system expenditure. On the other hand, their household expenditure is from Rp 600,000 to Rp 7,500,000 monthly, with 73% of them having the household expenditure less than Rp 2,400,000. Meanwhile, as much as 91% of respondents had enhancement willingness for their farming system and 9% exception was applied for the farmers in Taktakan district, because of having 0.2–0.5 ha of the accessed land width.

In this study, the following are the important socioeconomic indicators : (1) household expenditure, (2) the farming system expenditure, (3) the number of family members, (4) the accessed land width (preservation factor), and (5) the land width that will be accessed (enhancement willingness factor). The two latter factors were retained to form a commodity mapping as the second layer. The preservation factor is important because it determines the commodity production, whereas the enhancement willingness is closely related to both the future commodity production and the policy development that can be implemented.

The household expenditure, the farming system expenditure, and the number of family members, in contrast, were related to the age and the education level so that they were further explored. Because of the positive correlation between the farming system expenditure and the preservation factor, the variables retained to draw the second layer, the farming system expenditure was excluded in the second layer mapping. Meanwhile, the household expenditure and the number of family members were correlated with the age and the education level. Table 1 shows the correlation between variables.

Table 1 shows that the correlation between variables is not significant (ns). For further analysis, a simple linear regression to detect the variables affecting the enhancement willingness can be applied without any problem. The simple linear regression of the respondent characteristics to the enhancement willingness as the dependent variable yielded in an analysis of variance table (Table 2). Table 2 shows that the household expenditure has an effect to the enhancement willingness, indicated by *. To sum up, the second layer of the vegetables mapping was drawn using three variables, namely the preservation factor, the enhancement willingness, and the household expenditure.

3.2 Commodity Mapping

The contingency table produced is the production and the socioeconomic indicator tables in which the row–column definition is the commodity table and the

Table 1 Correlation variables

	Age	Education level	Household expenditure	The number of family members
Age	1.00			
Education level	0.22^{ns}	1.00		
Household expenditure	0.00^{ns}	0.40^{ns}	1.00	
The number of family members	0.09^{ns}	0.10^{ns}	-0.03^{ns}	1.00

Table 2 Analysis of variance of the regression model

Source	df	SS	MS	F	P value
Age	1	0.7861	0.7861	1.1091	0.3020
Education level	3	2.8479	0.9493	1.3394	0.2832
Household expenditure	1	3.1828	3.1828	4.4906	0.0438*
Number of family members	1	1.8077	1.8077	2.5505	0.1223
Residual	26	18.4281	0.7088		
Total	32	27.0526			

* $p < 0.05$

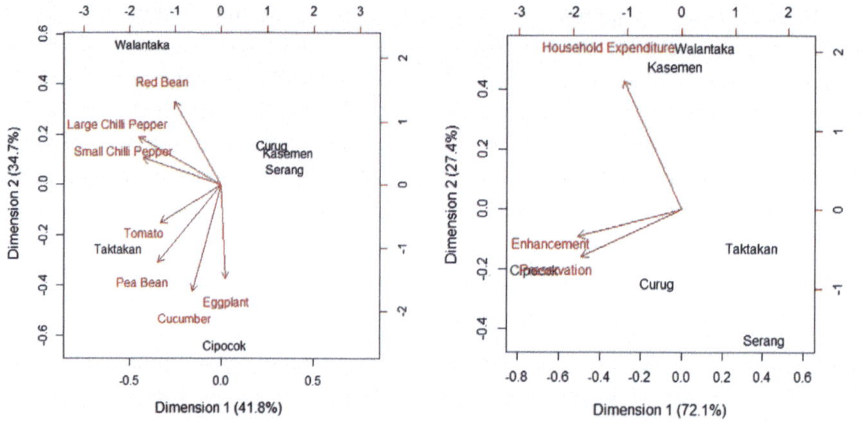

1a Biplot of Production Indicators

1b Biplot of Socio-economic Indicators

Fig. 1 Vegetables commodity biplot based on production and socioeconomic indicators. **a** Biplot of production indicators. **b** Biplot of socioeconomic indicators

socioeconomic variables in each district, respectively. The former table is a secondary data table obtained from Statistics Bureau of Serang city [19]. It has six rows filled by the districts in Serang city that are Cipocok, Curug, Taktakan, Walantaka, Serang, and Kasemen and seven columns filled by the commodity productions in Serang city that are cucumber, eggplant, pea bean, tomato, large and small chili pepper, and red bean. The other table has six rows of the districts in Serang city and three columns of the socioeconomic indicators obtained in the prior analysis that are the preservation factor, the enhancement willingness, and the household expenditure. These two original table data can be provided by the author upon request.

Chi-squared decomposition of the two contingency tables produced two correspondence analyses that can be represented in two-dimensional spaces in a biplot [20]. The first biplot (Fig. 1a) shows the six districts position and seven vegetables commodity. To interpret this biplot, one must consider the commodity arrow direction and the district positions orthogonally from the commodity lines. For example, the most of the eggplant, production is possessed by Cipocok district, followed by Taktakan, Serang, Curug, Kasemen, and Walantaka districts, respectively.

Figure 1a shows that the commodities that have a potency to become a superior commodity in each district based on its production are eggplant and cucumber in Cipocok district, pea bean and tomato in Taktakan district, and large and small chili pepper in Walantaka district. Curug, Kasemen, and Serang districts do not have any potency in all commodities. The total variance that can be explained by the biplot 1a, moreover, is 76.5% from the both dimensions. It indicates that the biplot well describes the relation between commodities and districts in a two dimensions based on their production.

On the other hand, Fig. 1b shows the district positions and socioeconomic farmers' indicators. Similar to Fig. 1a, socioeconomic indicators arrow direction must be considered because its line can be drawn orthogonally to determine the district position. Cipocok district, for example, is a district that possesses a very high preservation and enhancement willingness indicator compared to the other districts. Curug district is in the second place. Then it is followed by Kasemen, Serang, Taktakan, and Walantaka, respectively. If they are projected on the preservation and enhancement line orthogonally, these four districts are below the average.

The preservation and enhancement willingness indicator has a high correlation indicated by the small angle between these two lines. It means that the farmers who have a wider land width tend to expand their land width in the future. Household expenditure, in contrast, does not have any correlation with the other two indicators representing by the almost 90° angle between the household expenditure and the other two. Furthermore, there is 99.5% of total variance explained by the both two dimensions in the biplot. It means the biplot very well depicts the relation between socioeconomic indicator and districts.

3.3 Priority Commodity per District

The two-layer biplot combination results in a four-quadrant figure of district and priority commodity (Fig. 2). The main priority is Cipocok district possessing cucumber because it has a very high production and high enhancement willingness. The second priority is Curug district due to its low production and high enhancement willingness. The small chili pepper commodity is preferred in this district based on the primary data that the farmers often plant small than large chili pepper. The policy in this district should be focused on the small chili pepper production so that it can be shifted to the first quadrant. The third priority is Taktakan and Walantaka districts with pea bean and small chili pepper commodity, respectively. In these two districts, the farmers should be attracted using empowerment programs that can increase their enhancement willingness so that these districts can become a district with a main priority.

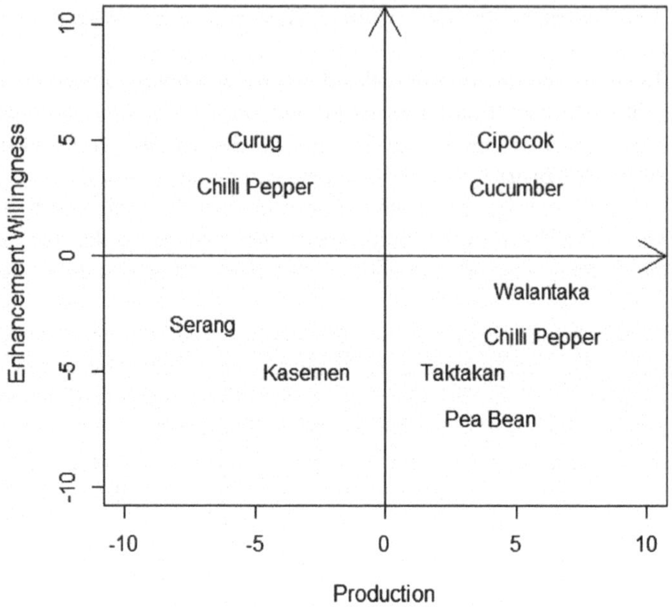

Fig. 2 A four quadrant of district and priority commodity

The last two districts lying in the fourth quadrant are Serang and Kasemen districts. These districts have low in both commodity production and enhancement willingness. Because of low commodity production, specific commodity that can be developed in these districts is unidentified. Geographically, the priority commodities are mapped in Fig. 3.

4 Conclusion

Socioeconomic analysis applied in the data resulted in household expenditure indicator affected to the enhancement farmers' willingness. This indicator, the preservation and enhancement willingness variables simultaneously form the second layer that is combined with the first layer based on commodity production yielded in a map of district and commodity. The district and commodity map produced four priority areas. The first priority was Cipocok district with cucumber commodity. The second priority was Curug district having small chili pepper. Meanwhile, Walantaka and Taktakan districts with small chili pepper and pea bean commodity, respectively, were the third priority. Last, the fourth priority was Serang and Kasemen districts that had unidentified commodity. These two districts need more exploration to determine their priority commodities.

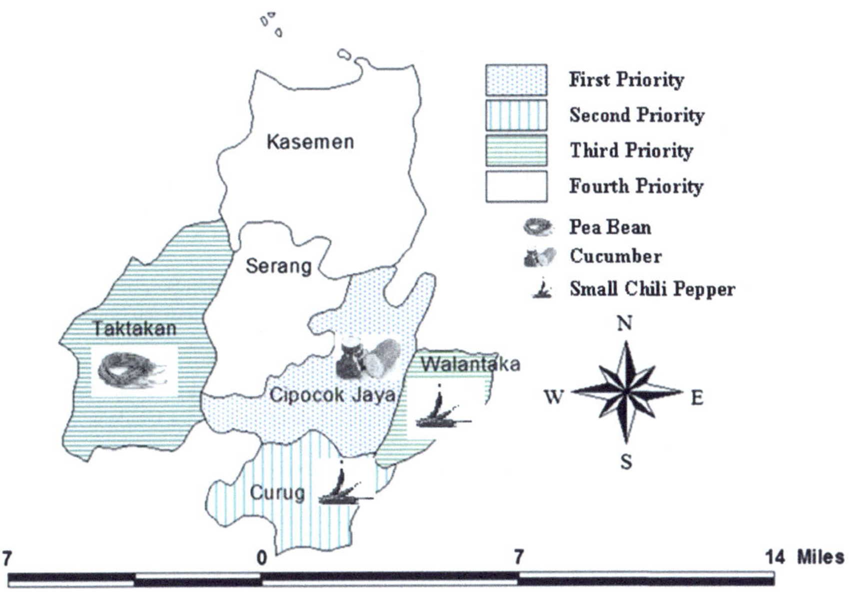

Fig. 3 District and priority commodity map

References

1. Peljor, N, and N Minot. 2010. Food security and food self-sufficiency in Bhutan. International Food Policy and Research Institute: Washington.
2. RPJM Banten. 2010. Banten Province: the first draft.
3. Dinas Pertanian Kota Serang. 2013. Pemanfaatan Lahan Pertanian bagi Kebijakan Pemerintah Daerah. Presented in BPS 2013 Agriculture Census Presentation: 26 March 2013 in Serang.
4. Zakaria, Amin. 2003. Decentralization Extensions to Local Government: Indonesia Experience. Regional Workshop on Operational zing Reform in Agricultural Extension in South Asia, New Delhi May 6 – 8 2003.
5. Bryan, A A, S Hajkowicz, S Marvanek, and M D Young. Mapping Economic Returns to Agriculture for Informing Environment Policy in the Murray – Darling Basin Australia. Environmental Modelling and Assessment Vol. 14 Issue 3, pp 375 – 390.
6. Amaliah I and Julia A. 2012. Pemetaan Faktor Penentu Daya Saing Komoditas Hortikultura Unggulan di Jawa Barat. *Prosiding Seminar Nasional Penelitian dan PKM: Sosial, Ekonomi, dan Humaniora* pp: 225 – 231.
7. Yulianti M. 2011. Penentuan Prioritas Komoditas Unggulan Buah-buahan di Kabupaten Minahasa Utara Provinsi Sulawesi Utara: Aplikasi Analsisi LQ dan Daya Tarik – Daya Saing. *Jurnal Agribisnis Perdesaan* Vol. 01 No. 03 pp: 206 – 221.
8. Isard, W, K Basset, C Choguill, J Furtado, R Izumita, J Kissin, E Romanoff, R Seyfarth, and R Tatlock. 1968. On the Linkage of Socio-Economic and Ecological System. Journal of the Regional Science Association International Vol 21. Issue 1, pp 79–99.
9. Chiang, Shu-hen. 2009. Location Quotient and Trade. Ann Reg Sci Vol. 43, pp 399 – 414.
10. Mack, RS, and DS Jacobson. 1996. Core Periphery Analysis of the European Union: a Location Quotient Approach. The Journal of Regional Analysis and Policy Vol. 26 No. 1, pp 3 – 21.

11. Ismayani. 2013. Development Strategy of Prime Commodities of Plantation in the District of Aceh Besar. Developing Countries Studies Vol. 3, No. 7, pp 1 – 11.
12. Suliyanto, -, and D Purnomojati. 2012. Analysis of Determining Basis Commodity and Pricing at Basis Commodity in Banyumas Indonesia. Interdisciplinary Journal of Research in Business Vol. 2 Issue. 2 pp, 27–36.
13. Bakhtiari, S, and M Dehghani zadeh. 2012. Proposing a New Version of Location Quotients for Estimating Regional Input-Output Coefficients: A Case Study of Iran's Yazd Province. African Journal of Business Management Vol. 6 Issue. 23, pp 6903 – 6909.
14. Kiser, Don. 1992. A Location Quotient and Shift Share Analysis of Regional Economies in Texas [Thesis]. Southwest Texas State University: Texas.
15. Syafruddin, Kairupan AN, Negara A, and Limbongan J. 2004. Penataan Sistem Pertanian dan Penetapan Komoditas Unggulan Berdasarkan Zona Agroekologi di Sulawesi Tengah. *Jurnal Litbang Pertanian* Vol. 23 No. 2 pp: 61 – 67.
16. BPS (Badan Pusat Statistik). 2013. Kota Serang dalam Angka 2012.
17. Hutchinson, SR, and CD Lovell. 2004. A Review of Methodological Characteristics of Research Published in Key Journals in Higher Education: Implications for Graduate Research Training. Research in Higher Education Vol. 45 No. 4, pp 383 – 403.
18. Scheafer, Richard L, Mendenhall W, and Ott L. 1990. *Elementary Survey Sampling.* PWS – Kent Publising: Boston.
19. BPS (Badan Pusat Statistik). 2012. Kota Serang dalam Angka 2011.
20. Hardle W and Simar L. 2007. *Applied Multivariate Statistical Analysis.* Springer-Verlag: Berlin.

Willingness to Pay and Willingness to Work to Avoid Deforestation and Forest Degradation

A. Solikin

Abstract Contingent valuation method (CVM) is widely applied to value environmental goods and services. Initially developed in developed countries, CVM now is also widely applied in developing countries. However, in many cases, value estimates from CVM studies carried out in developing countries are smaller than expectations. One possible explanation for this phenomenon is that people in developing countries may have different preferences toward environmental goods and services. Another explanation argues that it is effect of poverty on the valuation, i.e., effect of limited income instead of deficiencies in preferences. A suggestion to resolve the problem according to the latter argument is by using nonmonetary contribution as payment vehicle. Following this argument, this paper reports result of CVM surveys using working time without pay (or called willingness to work, WTW) in addition to standard monetary contribution (or called willingness to pay, WTP) implemented for a contingent scenario to avoid deforestation and forest degradation in Segah watershed in Berau District, East Kalimantan, Indonesia. Both WTP and WTW are elicited in open-ended format, i.e., directly asking respondents about their maximum contributions for the proposed program. Results show that in terms of WTP, respondents' average benefit from the proposed project is Rp124,088 for onetime payment, whereas average WTW is 20.25 days per year. When WTW is converted to WTP using casual daily wage, it turns out that WTW is 8.15 times higher than WTP. The results confirm findings of previous studies that value estimated in terms of WTW is substantially higher than WTP. In addition, among similar sets of socioeconomic and forest use variables, variables of income, gender, family having baby, living in mixed tribe villages, and extent of using river services are significant determinants for WTP, whereas migration, living in mixed tribe villages, extent of using forests, practicing ritual related to forests, having gardens located in forests, and belief in the scenario applicability are significant

A. Solikin (✉)
Institute of Development Research and Development Policy,
Ruhr University Bochum, Bochum, Germany
e-mail: akhmad.solikin@rub.de; akhsol@kemenkeu.go.id

A. Solikin
State Finance Polytechnic of STAN, Ministry of Finance, Jakarta, Indonesia

T. Taufik et al. (eds.), *ICoSI 2014*, DOI 10.1007/978-981-287-661-4_14

determinants for WTW. Although determinants of the WTP and WTW are found to be different, in general the variables exhibit expected signs, which shows validity of the model. Therefore, the finding supports for using nonmonetary contribution for CVM study implemented in developing countries, especially in community with limited cash income.

Keywords Willingness to pay · Willingness to work · Deforestation · Forest degradation · Contingent valuation method

1 Introduction

CVM is a stated preference method which directly asks respondents about values of any goods or services. It is widely used to value environmental goods or services due to its ability to capture nonmarket values, but basically, it can be used to value any goods and services due to flexibility in its contingent scenario. Initially applied in developed countries, CVM now is also widely applied in developing countries.

While general guidelines to conduct CVM are also successfully applied in developing countries, some specific problems emerged. One of the problems is about asking willingness to pay in monetary terms for people with limited money income. Respondents with limited cash income may find it difficult to cite WTP, and instead, they answer smaller, zero, or near-zero WTP although genuinely they value the good or service highly. In addition to eliciting negligible WTP, the respondents may refuse to answer altogether. One way to curb the problem is by using nonmonetary unit of measurement.

This article presents results of an empirical study in Indonesia by applying both monetary and nonmonetary unit of measurements to value a program aimed at avoiding deforestation and forest degradation. By doing so, the article aims at comparing values elicited in terms of monetary and labor and WTW, and attesting validity of the unit of measurement in Indonesian context. Based on this objective, the rest of this paper is organized as follows: Sect. 2 discusses lessons from previous CVM studies which use WTW. Section 3 describes theoretical framework and data gathering method. Section 4 shows results, and then Sect. 5 offers brief conclusion.

2 Previous Studies

Nonmonetary contributions as unit of measurement are gaining importance in the CVM studies in developing countries. Several alternatives unit measurements are found in the literature, i.e., in terms of (1) labor or working time, e.g., study by [1–17], (2) commodity, e.g., rice [18], and (3) borrowing, e.g., [1]. It is clear that working time is more popularly used as alternative to monetary contribution in part due to its abundance for cash-strapped respondents in developing countries.

Among the previous studies, WTW is successfully used to value contribution to rural water service [1, 8], large dam [7], irrigation project [17], urban river ecosystem restoration [2], sport facility [10], animal disease control [13], human disease control [3], biodiversity conservation [11], and an exotic species [14]. Score of articles applied WTW in forestry sector, e.g., to value community forest management [4–6], forest fire prevention [9], forest restoration [15], forest environmental services [12], and an exotic tree species conservation [16]. However, specific study to value avoidance of deforestation and forest degradation is still missing. This article thus enriches literature in CVM using WTW by adding variables related to extent of forest use and contingent scenarios.

Although nonmonetary unit of measurement is gaining wider acceptance, disagreements can also be found. Several disagreements include the following: (1) Using different units of measurement will result in different values [19], (2) nonmonetary unit of measurement is difficult to convert to utility which may make aggregation difficult [20], and (3) different numeraires favor to different groups of people [19]. In response to the critiques, the proponents of using nonmonetary units argue that it is not about choice of numeraire that matters. When people are familiar with the unit of measurement, then the value elicited should be similar [21, 17]. In addition, the difference in value may also be attributed to different marginal social utilities of money or other units of measurement across different groups of people rather than caused by choice of standard unit [21].

Besides the theoretical debates, several general conclusions can be drawn from the previous literature. First, the percentage of respondents who agree to contribute is higher in WTW than in WTP. Second, the value of WTW is substantially higher than that of WTP. The first and second general conclusions hold except results of [10] in a case study in Germany. Third, determinants of WTP and WTW are sometimes different. For WTP, income is found to be crucial determinant of WTP and thus becomes an important validity check for WTP model. On the other hand, determinants for WTW are less clear. The most notable determinant for WTW is gender, i.e., male respondents usually prefer to contribute labor more than female respondents.

3 Methods

3.1 Theoretical Foundation

CVM is the stated preference method in which valuation is contingent upon scenario presented to respondents. The method is capable of eliciting consumers' WTP and willingness to accept (WTA) for quantity or quality changes of goods and services. More importantly, it is the single method able to capture nonmarket value, i.e., values of goods and services which are not traded in markets. In addition, it is also capable of capturing values of nonexistence goods and services, which is very suitable to problem at hand.

Theoretical background for valuation of a contingent scenario starts from individual preference function. The following is exposition based on [22]. The preference function is determined by combination of bundles of private (***x***) and public (***q***) goods and services, where the private goods and services are available at price vector ***p***. An individual tries to maximize preference subject to budget constraint or income (***y***) or to achieve highest level of utility which requires minimum expenditure. For the former, indirect utility function is given by

$$V(p, q, y) = \max(u(x, q) | p.x \leq y), \quad (1)$$

and for the latter is given by

$$m(p, q, u) = \min(p.x | u(p, q) \geq u). \quad (2)$$

The WTP is defined as maximum money willing to sacrifice to ensure quantity and quality improvement for ***q***. Assume that ***q**** is public goods after improvement, then

$$WTP = m(p, q, u) - m(p, q*, u), \quad (3)$$

where $u = V(p, q, y)$. To allow for using WTW as standard of measurement, Eq. (3) should be adapted accordingly, i.e., WTW substitutes WTP and time budget substitutes income (y) [3].

3.2 Contingent Valuation Method: Questionnaire Designs and Survey Method

The scenario is a program to avoid deforestation and forest degradation. The scope of the project is a watershed known to the respondents. Specifically, the valuation question is asking about maximum WTP and WTW to ensure implementation of the program. In the scenario, it is disclosed that contributions of families in the surrounding area are required to make the program implemented. Contributions from respondents are required, and without the contribution, it is impossible to implement the program. This additional information is presented to respondents in order to increase consequentiality of the scenario.

Face-to-face surveys were administered in August to September 2011 to local people living in or around forests in ten villages in Berau District, East Kalimantan Province, Indonesia. The villages can be categorized into Dayak villages (4 villages), transmigration villages (3 villages), and mixed villages (3 villages). The ten villages are located in upper, middle, and lower stream of Segah watershed.

Forest condition in the survey areas varies, from preserved in upper stream area to totally altered to plantations and mining in lower stream area. In total, 410 interviews were successfully administered. The maximum WTP and WTW questions reported in this article are additional to double-bounded dichotomous choice question for identical contingent scenario which is reported elsewhere [23].

Among the total interviews, 159 respondents were asked about their maximum WTP, while 343 respondents were asked about their willingness to work. For the maximum WTP and WTW, question format is open-ended, i.e., directly asking respondents to elicit maximum value they put on the proposed program. Although open-ended elicitation method is widely applied due to its straightforward design and analysis, less starting point bias, and needs only small samples, nevertheless it has some problems, i.e., adds cognitive burden to respondents, high nonresponses, and possibility of not achieving maximum amount [24, 25]. On the other hand, open-ended elicitation method is a standard method to elicit WTW as practiced by [3, 8, 14, 15, 20], except, e.g., [5, 10] who used dichotomous choice method. In order to facilitate direct comparison between WTP and WTW, open-ended method for both WTP and WTW is used.

3.3 Empirical Model

Since the question format is open-ended, thus ordinary least square (OLS) is applied. Empirical model used WTP or WTW as independent variable and some socioeconomic, forest use, and scenario designs as independent variables.

Some socioeconomic variables are included in model to capture effects of the factors to WTP and WTW. Working hypotheses regarding the independent variables are as follows. Income is expected to have positive sign for WTP, but to have negative sign for WTW. It has positive sign to reflect that higher respondents' income is associated with higher WTP, but higher respondents' income is associated with reluctance to participate in working without pay. There is no a priori expectation of effect of age on WTP, but age is expected to have negative sign for WTW since elder age may prohibit respondents from participating in field work. Furthermore, male respondents are expected to have higher WTW, but no a priori expectation for effects of gender on WTP. Household size is expected to have negative signs for both WTP and WTW. Both negative signs are related to more resources, in terms of income and time needed to take care of the family member. Families who have baby members is expected to have smaller WTP and WTW due to more resources to take care of the family. With similar logic, families who have senior members are also expected to have less WTP and WTW. Variable of tribe, i.e., Javanese, is expected to have negative sign, meaning that Javanese are less willing to contribute than Dayak tribe. This is due to assumption that Javanese tribe is more independent compared to Dayak people in terms of forest extraction. Respondents' length of stay is expected to have positive sign, i.e., the longer the respondents stay in the area, the higher their WTP and WTW due to more benefits they receive from the forests. Similar to variable of length of stay, migration variable (i.e., nonmigrant = 1) is expected to have positive sign. Respondents living in mixed villages are expected to have less WTP and WTW compared to respondents living in Dayak villages which is consistent with assumption that indigenous people are more dependent to nature and thus have more WTP and WTW.

Several variables related to extent of resource uses and scenario designs are also included in the model. Respondents use more forest resources, receive more river services, practice ritual related to forests, have gardens located in forests, and perceive that benefits they receive form forests are higher than the corresponding costs are expected to have higher WTP and WTW. Similarly, respondents who perceive that the contingent scenario is implementable are expected to cite higher WTP and WTW. This variable is related to trusting the scenario. It has been found in the literature, e.g., [26], that perception of scenario effectiveness positively determines WTP. Lastly, respondents who perceive that higher share of costs should be borne by community are expected to have higher WTP and WTW. The maximum share of community's contribution reflects respondents' perception about community's participation in the proposed program.

4 Results

4.1 Descriptive Statistics

Descriptive statistics of the respondents are presented in Table 1. Average monthly income of sample is about 3 million rupiah (about US$350 with exchange rate 1 $ = Rp8,571 at the time of survey). Average respondents age is slightly more than 41 years. Male respondents represent 51.7% of sample, which is smaller than average District's and Province's percentage which is 53.94 and 52.68%, respectively. In addition, on average respondents have household size of 5 and have been staying for long time in the respective villages, i.e., 20 years, although substantial percentage of respondents have migration roots. Almost all (92.3%) respondents believe that CVM scenario is implementable which add validity to this study. Regarding forest extractions, respondents on average exploit about one-third of forest resources and river services. Respondents also perceive that small amount of costs (20%) should be shouldered by community, set aside larger parts to government and international actors.

Maximum WTP elicited by respondents ranges from Rp5,000 to Rp800,000. Average WTP is Rp124,088 which represents 4.15% of respondents' monthly income. On the other hand, maximum WTW ranges from 0 to 365 days with median and modus of 12 days. When three outliers of 365 contribution days are deleted from the sample, mean number of days the respondents are willing to contribute to the program is 20.25 days per year. Assuming casual wage is

Table 1 Descriptive statistics of samples

Variable	Description	Obs.	Mean	Std. dev.	Min	Max
Income	Monthly, categorical, 1: $\leq$Rp0.5 mil.; 2: 0.5 mil.-Rp1 mil.; 3: Rp1 mil.-Rp2 mil.; 4: Rp2 mil.-Rp3 mil.; 5: Rp3 mil.-Rp4 mil.; 6:Rp4 mil.-Rp 5 mil.; 7: Rp > 5 mil.	320	2.991	1.370	1	7
Age	Continuous, in years	349	41.266	14.396	17	100
Male	Gender, binary, 1: male, 0: female	352	0.517	0.500	0	1
Hhsize	Household size, continuous data	352	4.787	2.005	1	14
Havebaby	Binary, 1: have baby, 0: do not have	352	0.443	0.497	0	1
Havesenior	Binary, 1: have senior age, 0: do not have	352	0.230	0.422	0	1
Tribejava	Category, 1: Java, 0: others	351	0.353	0.479	0	1
Stay	Length of stay, continuous, in years	347	20.027	13.321	0.083	100
Migration	Binary, 1: indigenous, 0: migrants	346	0.202	0.402	0	1
Mixvillage	Category, 1: stay in mix village	352	0.366	0.483	0	1
Forestuse	Extent of forest products extraction, aggregation score	346	3.847	3.331	0	13
Riverservice	Extent of using river services, aggregation score	352	2.656	2.011	0	6
Ritual	Perform ritual related to forests, binary	344	0.84	0.278	0	1
Gardenloc	Whether garden located in forests, binary	287	0.314	0.465	0	1
Bvsc	Comparison benefits vs. costs, category, 1: costs much bigger, 2: costs bigger, 3: equal, 4: benefits bigger, 5: benefits much bigger	322	3.627	0.956	1	5
Scimplement	Whether scenario implementable, binary	285	0.923	0.267	0	1
Maxshare	Maximum costs share be borne by community, percentage	237	0.290	0.226	0	1

Rp50,000 per day, the average WTW turns to be Rp1,012,500 per year. Thus, WTW is substantially higher than WTP, as expected.

4.2 Determinants of WTP

Results of OLS regressions are provided in Table 2. In Model 1, dependent variable is WTP, while in Model 2, dependent variable is WTW, both with different sets of independent variables. Initially, both models have similar independent variables. As backward elimination procedure is applied [27], the final models only include combination of independent variables with best goodness of fit, F statistics, and R^2 [28]. Both Model 1 and Model 2 have acceptable explanatory power, i.e., more than 16% and almost 24%, which are higher than standard 10% [29].

As shown in Table 2, the determinants for WTP and WTW are different. Different determinants of WTP and WTW are also observed in [13]. For WTP,

Table 2 Determinants of WTP and WTW

Variable	Model 1: WTP		Model 2: WTW	
	Coefficient	SE	Coefficient	SE
Income	18.806***	6.996	–	–
Age	–	–	0.161	0.132
Male	28.887*	16.882	4.542	3.311
Hhsize	−7.152	4.507	–	–
Havebaby	−30.570*	18.075	–	–
Havesenior	–	–	−6.800	4.177
Tribejava	−19.882	17.778	–	–
Stay	−0.964	0.661	−0.176	0.151
Migration	15.062	9.728	4.027**	1.978
Mixvillage	−36.721*	20.332	−10.672***	3.452
Forestuse	–	–	1.316*	0.719
Riverservice	12.193**	5.180	1.546	1.096
Ritual	–	–	10.906*	5.975
Gardenloc	–	–	8.536*	4.487
Bvsc	–	–	−2.645	1.781
Scimplement	–	–	12.165*	6.326
Maxshare	–	–	8.620	7.280
Constant	75.220	50.781	−6.219	10.946
N	151		157	
Prob. > F	0.0027		0.0001	
R-squared	0.1604		0.2389	
Adj. R-squared	0.1068		0.1697	

Note ***$p < 0.01$, **$p < 0.05$, *$p < 0.1$

variables of income, gender, having baby, living in mixed village, and using river services significantly determine WTP. On the other hand, for WTW, variables of migration, living in mixed village, using forests, practicing ritual related to forests, having garden located in forests, and belief in implementability of the contingent scenario are significant determinants of WTW.

Income, gender (male), and extent of using river services have positive impact on WTP, as expected. Positive sign for income variable means respondents with higher income have higher WTP, an important basic economic theory. Respondents who receive higher environmental services from rivers also have higher WTP, as expected. On the other hand, respondents who have baby and reside in mixed village have lower WTP. As mentioned previously, respondents who live in mixed villages are considered as pragmatic and have fewer cultural ties with forests and thus logically have lower WTP.

For WTW, respondents residing in mixed villages have lower WTW, which is similar to the WTP case. Respondents who are not migrating from other districts have higher WTW. Similarly, respondents who are extracting more forest resources, practicing rituals related to forests, having gardens located in forests, and believing the contingent scenario have higher WTW. Thus, respondents who get bigger benefits from forests are willing to contribute labor more, in conformity with economic theory.

5 Conclusions

As mentioned previously, average and modus WTP are 4.5 and 3.34% of monthly income, respectively. These numbers are larger than average WTP of CVM studies in developing countries, which are usually low, i.e., represent 0.2–0.4% of household income [30]. Mean WTW found this study, i.e., 20.25 days per year, is also bigger compared to other CVM forest studies in developing countries. For example, Ref. [14] in an Indian case study to value habitat function of planted forests found that WTW varies from zero to 120 days per year, with average 16.77 man days per year. In addition, Das and Mahanta [31] found that local people were willing to contribute 2–4 h per week (or equivalent to 14–30 days per year) for biodiversity conservation in Assam, India. On the other hand, Casey [5] found bigger WTW, i.e., 2.67 days per month (or 32 days per year) for community forest conservation in Mexico.

It is shown that determinants of WTP and WTW are different. Nevertheless, they all show conformity with economic theory. Most notably for WTP, income is significant and positive determinant of WTP. On the other hand, benefits from forests are significant and positive determinants of WTW, while gender variable is not a significant determinant.

As also found in previous literature, when compared, WTW is substantially higher than WTP, i.e., about 8.15 times. This phenomenon is frequently found in the literature and usually used as argument for supporting WTW. However, one might interpret the result with care. One possible counter-argument is that respondents interpreted WTP as installments rather than onetime lump-sum payment, whereas in WTW is clearly defined as labor contribution per year. Set aside the caveat, this article empirically shows that WTW can be successfully applied in developing country context, especially when contributing labor is a social norm in the community as observed in the survey areas.

Acknowledgements The author would like to express special thanks to Government Financial Management and Revenue Administration Project (GFMRAP) for providing financial supports to undertake this research. Financial and administrative assistances from Fiscal Policy Agency Secretariat which enable the author to present this research in the 2nd ICOSI conference are also appreciated.

References

1. Abramson, A., Becker, N., Garb, Y., & Lazarovitch, N. (2011). Willingness to pay, borrow, and work for rural water service improvements in developing countries. *Water Resources Research*, 47, W11512.
2. Alam, K. (2013). Factors affecting public participation in river ecosystem restoration: Using the contingent valuation method. *The Journal of Developing Areas*, 47(1), 223–240.
3. Arbiol, J., Borja, M., Yabe, M., Nomura, H., Gloriani, N., & Yoshida, S. (2013). Valuing human leptospirosis prevention using the opportunity cost of labor. *International Journal of Environmental Research and Public Health*, 10, 1845–1860.
4. Arifin, B., Swallow, B., Sutanto, & Coe, R. (2008). *A conjoint analysis of farmer preferences for community forestry contracts in the Sumber Jaya Watershed, Indonesia*. ICRAF Working Paper 63. Nairobi, Kenya: World Agroforestry Centre.
5. Casey, J. F. (2003). Partners in forest conservation: "Willingness to work" to protect local forest resources in Calakmul, Campeche, Mexico. *Problemas del Desarrollo*, 34(135), 125–142.
6. Girma, W. & Beyene, F. (2012). Willingness to contribute to collective forest management: Evidence from Godere in the Gambela Region of Ethiopia. *The Journal of Socio-Economics*, 41, 79–86.
7. Gorkhali, S. P. (2009). *Economic Analysis of Infrastructure Investments: National Benefits and Local Costs- Case Study of a Hydropower Project in Nepal*. Göttingen, Germany: Cuvillier Verlag.
8. Hardner, J. J. (1996). Measuring the value of potable water in partially monetized rural economies. *Water Resources Bulletin*, 32(6), 1361–1366.
9. Hung, L. T., Loomis, J. B., & Thinh, V. T. (2007). Comparing money and labour payment in contingent valuation: The case of forest fire prevention in Vietnamese context. *Journal of International Development*, 19, 173–185.
10. Kiefer, S. (2014). *Are Riding Club Members Willing to Pay or Work for an Overall Quality Improvement*. Discussion Paper DP-IO 1/2014, Institute for Organizational Economics, Muenster University, Germany.
11. Ninan, K. N. & Sathyapalan, J. (2005). The economics of biodiversity conservation: A study of coffee growing region in the Western Ghats of India. *Ecological Economics*, 55, 61–72.
12. Notaro, S. & Paletto, A. (2011). Links between mountain communities and environmental services in the Italian Alps. *Sociologia Ruralis*, 51(2), 137–157.
13. Pokou, K., Kamuanga, M. J-B., & N'Gbo, A. G. M. (2010). Farmers' willingness to contribute to tsetse and trypanosomosis control in West Africa: The case of northern Cote d'Ivoire. *Biotechnology, Agronomy, Society, and Environment*, 14(3), 441–450.

14. Saxena, A. K., Bisht, N. S., & Singh, C. J. (2008). The value of the Indian gazelle (Gazella gazelle): A case study in Haryana, India. *The Indian Forester*, 134(10), 1289–1295.
15. Schiappacasse, I., Vasquez, F., Nahuelhual, L., & Echeverria, C. (2013). Labor as a welfare measure in contingent valuation: The value of a forest restoration project. *Cience e Investigacion Agraria*, 40(1), 69–84.
16. Tilahun, M., Vranken, L., Muys, B., Deckers, J., Gebregziabher, K., Gebrehiwot, K., Bauer, H., & Mathijs, E. (2012). *Rural Households' Demand for Frankincense Forest Conservation in Tigray: A Contingent Valuation Analysis*. Bioeconomics Working Paper 2012/2. Leuven, Belgium: Department of Earth and Environmental Sciences, University of Leuven.
17. Vondolia, G. K., Eggert, H., Navrud, S., & Stage, J. (2011). *What Do Respondents Bring to Contingent Valuation? A Comparison of Monetary and Labor Payment Vehicles*. Environment for Development Discussion Paper Series No. DP 11–13. Gothenburg, Sweden: University of Gothenburg & Washington, USA: Resource for the Future.
18. Shyamsundar, P. & Kramer, R. A. (1996). Tropical forest protection: An empirical analysis of the costs borne by local people. *Journal of Environmental Economics and Management*, 31, 129–144.
19. Brekke, K. A. (1997). The numèraire matters in cost-benefit analysis. *Journal of Public Economics*, 64, 117–123.
20. Ahlheim, M., Frör, O., Heinke, A., Duc, N. M., & Dinh, P. V. (2010). *Labour as a utility measure in contingent valuation studies: How good is it really?* FZID Discussion Paper No. 13–2010. Stuttgart, Germany: Center for Research on Innovation and Services, University of Hohenheim.
21. Johansson, P.-O. (1998). Does the choice of numeraire matter in cost-benefit analysis? *Journal of Public Economics*, 70, 489–493.
22. Haab, T. C. & McConnell, K. E. (2002). *Valuing Environmental and Natural Resources: The Econometrics of Non-Market Valuation*. Cheltenham, UK & Northampton, MA, USA: Edward Elgar.
23. Solikin, A. (2014). *Willingness to pay to avoid deforestation and forest degradation: Evidence from Berau District, Indonesia*. Proceeding the 5th International Conference on Global Resource Conservation (ICGRC) 2014, pp. 53–56. Batu, East Java, Indonesia: Faculty of Mathematics and Natural Science, University of Brawijaya.
24. Ahmed, S. U. & Gotoh, K. (2007). The choice of elicitation methods in CVM and their impact on willingness to pay in environmental assessment. *Reports of the Faculty of Engineering Nagasaki University*, 37(68), 47–52.
25. Bennett, R. & Tranter, R. (1998). The dilemma concerning choice of contingent valuation willingness-to-pay elicitation format. *Journal of Environmental Planning and Management*, 41(2), 253–257.
26. Brouwer, R., Brander, L., & van Beukering, P. (2008). "A convenient truth": Air travel passengers' willingness to pay to offset their CO2 emissions. *Climatic Change*, 90, 299–313.
27. Field, A. (2009). *Discovering Statistics Using SPSS* (3rd ed.). London, UK: Sage.
28. Bandara, R. & Tisdell, C. (2003). *The Net benefits of Saving the Asian Elephant: A Policy and Contingent Valuation Study*. Economics, Ecology and the Environment Working Paper No. 87. School of Economics, The University of Queensland, Australia.
29. Bateman, I. J., Carson, R. T., Day, B., Hanemann, M., Hanley, N., Hett, T., … Swanson, J. (2002). *Economic Valuation with Stated Preference Techniques: A Manual*. Cheltenham, UK/Northampton, MA, USA: Edward Elgar.
30. Whittington, D. (2010). What have we learned from 20 years of stated preference research in less-developed countries? *Annual Review of Resource Economics*, 2, 209–236.
31. Das, D. & Mahanta, R. (2013). Willingness to spend labour hour for biodiversity conservation: A case study with special reference to village forest dwellers and encroachers in Assam. *Journal of Biodiversity*, 4(1), 45–51.

Part II
Economics

A Cluster Model for Increasing Performance of Small and Medium-Scale Enterprises (A Case Study in Bogor, Indonesia)

Resista Vikaliana, Desi Harsanti, Dewi Sri Wulandari and Asti Andayani

Abstract Small Medium Enterprises (SME) is one of economic fundament in Indonesia. Their endurance has already known since Monetary Crisis in 1998. But, most of them have not managed well. Many problems occurs on them, such as restricted resources in human resources, assets, capital and others. For increasing their performance, they need supports from others stakeholders. The better performance of them, the higher profit they will get. For larger, it will awake the economic based on community in Indonesia and give more progress. Based on those reasons, the purpose of this research is to increase the performance of SME's in Kecamatan Cibinong, Kabupaten Bogor by Cluster Model. This model built a place as the center of industiesy (Kecamatan Cibinong). The center (Kecamatan Cibinong) consists of three area productions: SME of "Kerupuk Mie", SME of Bambu Hitam Meubeul, and SME Traditional Cake (Dodol, Wajik, Bugis). Ishikawa diagram was used to identify problems occurs on each of SME. This research's result shows that marketing is the most difficult problem of them. The solution of that problem is building a cluster with Digital Marketing Center (DMC) in the center of cluster (Kecamatan Cibinong).

Keywords SME · Kecamatan cibinong · Ishikawa diagram · Cluster model

R. Vikaliana (✉) · D. Harsanti · D.S. Wulandari · A. Andayani
Stie Dewantara, Bogor, Indonesia
e-mail: resistav31@gmail.com

D. Harsanti
e-mail: desipinuji@gmail.com

D.S. Wulandari
e-mail: dewi_wulandaripg@yahoo.com

A. Andayani
e-mail: astiandayani83@yahoo.com

T. Taufik et al. (eds.), *ICoSI 2014*, DOI 10.1007/978-981-287-661-4_15

1 Introduction

Small Medium Enterprises (SME) is one of economic fundament in Indonesia. Their endurance has already known since Monetary Crisis in 1998. But, most of them have not managed well. Many problems occurs on them, such as restricted resources in human resources, assets, capital and others. For increasing their performance, they need supports from others stakeholders. The better performance of them, the higher profit they will get. For larger, it will awake the economic based on community in Indonesia and give more progress.

According to Bank Indonesia [1], these are the inherent characteristics that also cause some problems of SME:

1. Low quality of human resources in SME's
2. Low productivity so that the wages of man is low
3. Low quality of product (no standardization)
4. Less innovations and new technologies
5. Low of capital structure and low access from financial institution for funding
6. Low access of potential marketing

Based on those, the purpose of this research is to increase the performance of SME's in Kecamatan Cibinong, Kabupaten Bogor by Cluster Model. This model is built a place as the center of industries (Kecamatan Cibinong). The center (Kecamatan Cibinong) consists of three area productions: SME of "Kerupuk Mie", SME of Bambu Hitam Meubeul, and SME Traditional Cake (Dodol, Wajik, Bugis).

2 Literature Review

Decision of President of Indonesia (Keputusan Presiden RI No. 99 tahun 1998) says that SME is an activity based on economic community the small scale which needs to be protected, to prevent unfair competition of business.

SME contributes 60% of GDP (Gross Domestic Product) and absorbes about 97% human resources as worker. From the census of civilization in 2008, SME is the solution for demography problems, especially in worker absorption.

Competitive advantages of SMEs is from the product that they produce, their marketing share, consumer's satisfaction, and human resources. To raise SMEs competitive advantage, there are two factors: internal and external factors. The internal factors are productivity and innovation. The external factors are volume of production, market share, market oriented.

3 Methodology

Ishikawa diagram was used to identify problems occurs on each of SME. *Ishikawa Diagram* or cause effect diagram is one of analysis tools that can help identification of quality problems. The way to start is using five categories: material, machine, man, method [2].

Ishikawa Diagram is founded by Dr. Kaoru Ishikawa, an expert of quality control from Japan as one of 7 basic quality tools. Ishikawa Diagram is used to identify the possibility of problem causes, especially in routine activities [3].

The procedure or steps to make Ishikawa Diagram:

1. Identify Effect that will be analyzed
2. Identify the categories (man, material, method, machine, measurement)
3. Identify potential causes with brainstorming way
4. Determine and analyze causes defined (Fig. 1)

Cluster in this research based on the definition of cluster from Agriculture Department of Republik Indonesia (Departemen Pertanian). The characters of cluster are:

1. Cluster in business
2. Has supply chain
3. Coordination to improve added value and value chain
4. Running business by using economic infrastructure reserved
5. Business in a group

In other word, Cluster Area is an area designed in systematic in order to make a unity of business.

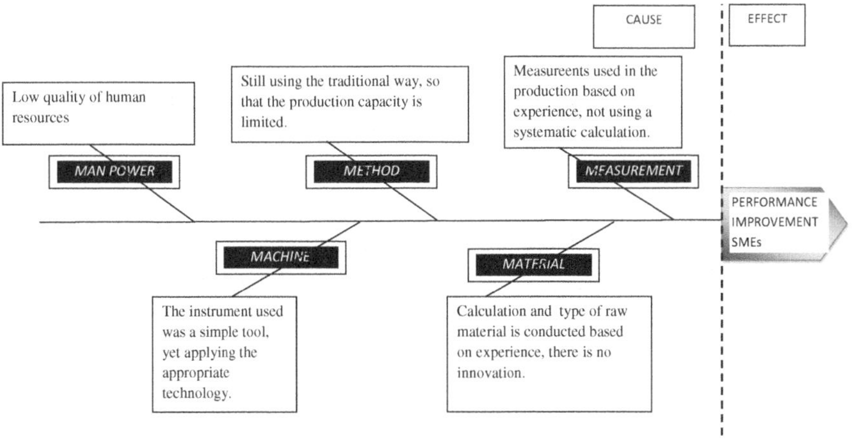

Fig. 1 The example of Ishikawa diagram

4 Analysis of SME Bambu Hitam Meubeul

See Fig. 2.

5 Analysis of SME Traditional Cake (Dodol, Wajik, Bugis)

See Fig. 3.

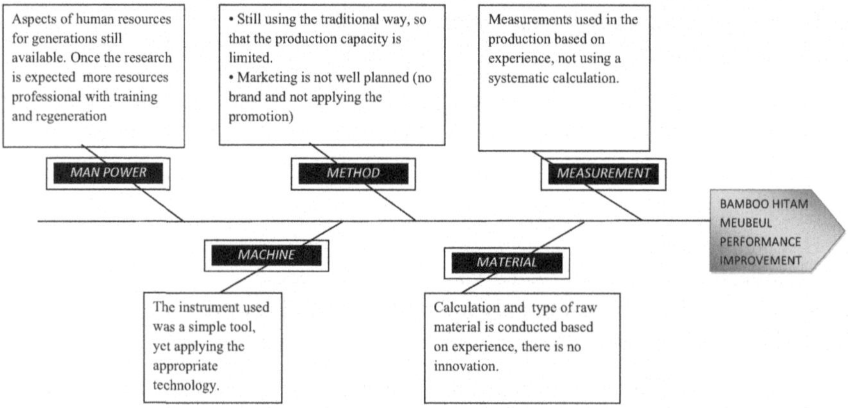

Fig. 2 Ishikawa diagram of SME Bambu Hitam Meubeul

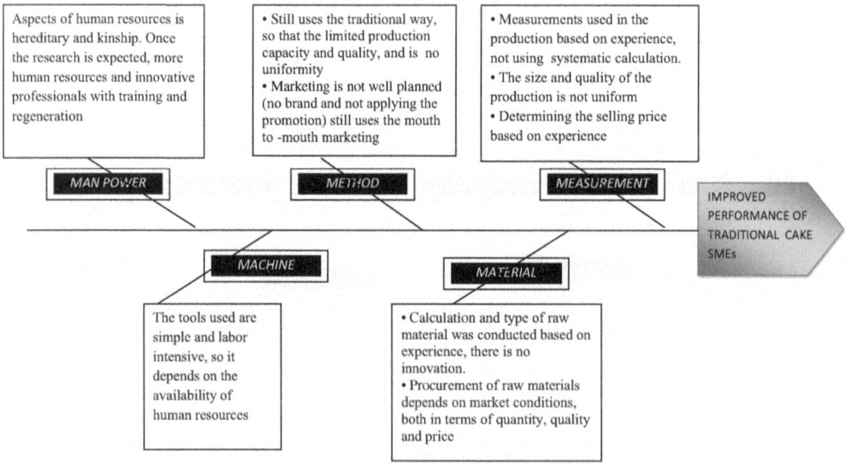

Fig. 3 Ishikawa diagram of SME traditional cake (Dodol, Wajik, Bugis)

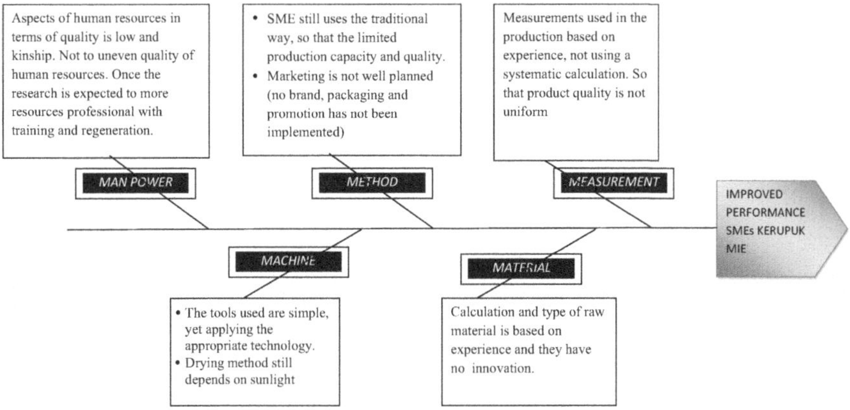

Fig. 4 Ishikawa diagram of SME Kerupuk Mie

6 Analysis of SME Kerupuk Mie

See Fig. 4.

7 Conclusion

Based on the analysis, this research's result shows that marketing is the most difficult problem of them. The solution is building a cluster with Digital Marketing Center (DMC) in the center of cluster (Kecamatan Cibinong).

References

1. Bank Indonesia. 2012. Paper in National Seminar at Soedirman University, Purwokerto, Indonesia.
2. Render, Barry and Jay Heizer. 2001. Principal of Operation Management. Salemba Empat. Jakarta, Indonesia.
3. Tague, Nancy R. Quality Toolbox. 2005. Asq Publisher. USA.
4. Ilyas, Yasnimar, Resista Vikaliana, Asti Andayani. 2013. *Analysis of Performance Improvement Barrier Using Fishbone Diagram (Case Study SME Bamboo Hitam Meubeul).* Proceeding of Call for Paper at Universitas Pendidikan Indonesia. Bandung, Indonesia.

Clustering Residents Based on Perceptions Toward Tourism Growth in Yogyakarta, Indonesia

Harwati

Abstract The development of tourism sector in the region will deliver benefits for local community including the increase sales level, more job openings, the increase in the ability of people to learn foreign languages, encouragement to repair facilities and infrastructure; but on the other hand, it have negative impacts on local residents like congestion, pollution, noise, and rise in criminal cases. Yogyakarta, as a tourism destination, faces the same problem. The tourism growth in that place makes raise the pros and cons among the community. Some people are more pleased with increasing number of tourist arrivals to their area because it gives the direct impact on their economic development, but other people show their objection to its negative impact, especially the traffic. Therefore, a study in Yogyakarta, Indonesia, was conducted to identify and analyze the residents' perception as the variable considered for the local government to arrange the policy toward tourism development. A total of 180 completed questionnaires were obtained from 200 respondents approached. The respondents were drawn from different walks of life by considering various aspects such as age, occupation, distance between home and the nearest tourism object, income, origin. Non-hierarchy clustering method—k-means cluster—is used to analyze and identify the respondents' perceptions of the tourism growth. From the calculation using SPSS 16, it is found that there are three clusters formed based on eight variables: positive residents (47%)—people who are delighted and enjoying the tourism environment; negative residents (21%)—people who do not like the tourism, especially for its impact on traffic; and opportunist residents (32%)—people agree that tourism causes traffic, but they are not care about that because they get benefit impact from tourism activities for their income. In conclusion, from the data clustering, it is informed that the growth of the tourism in Yogyakarta has provided more benefits than costs to the residents.

Keywords Tourism · Resident · k-means cluster · Yogyakarta

Harwati (✉)
Data Mining Laboratory, Faculty of Industrial Engineering,
Universitas Islam Indonesia, Yogyakarta, Indonesia
e-mail: harwati@uii.ac.id

T. Taufik et al. (eds.), *ICoSI 2014*, DOI 10.1007/978-981-287-661-4_16

1 Introduction

Yogyakarta is one of the tourist destinations for domestic and foreign travelers besides Bali, Bandung, Lombok and so on. There are various attractive tourist places in Yogyakarta such as historical tourism temple, sites, beach tourism, travel shopping, nature, hotels. Yogyakarta tourism is potential to grow not only because it is already growing old but now also provide so many new tourism objects. The number of tourist arrivals to Yogyakarta each year continues to increase as shown in the following figure (Fig. 1).

The growth of the tourism sector in a country or region will have a positive impact on its economy, where one indicator is the increase in local government revenue [1]. In line with the impact of tourism industry, the tourism sector in Yogyakarta has become one of the most significant contributors to local government revenue, Sumbarja [2]. Yogyakarta legislators stated that 38% of revenue comes from tourism sectors. Thus, the impact of tourism industry on the socioeconomic development makes this sector the one of Yogyakarta's visions for 2005–2025, namely "Making of Yogyakarta as a leading tourist destination in Southeast Asia"

In addition to providing a positive impact on household incomes and local community, tourist arrivals also provide other positive effects such as more job openings, the increase in the ability of people in mastering of foreign language, encouragement of repair of facilities and infrastructure. But on the other hand, the increase in the number of travelers could give a negative impact such as congestion, pollution, noise, the increase in criminal activities [3].

These trade-off effects certainly shape the perception of its own for local people of Yogyakarta where public perception is very important for the development of the

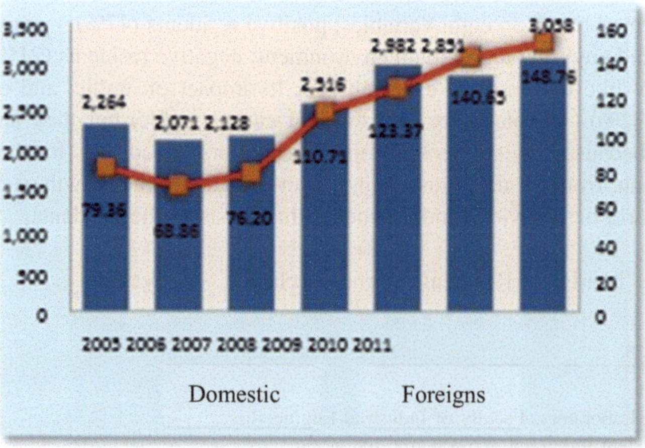

Fig. 1 Graphic representation of tourism arrival in Yogyakarta 2005–2011 (*Source* BPS)

tourism industry [4]. In general, according to Chohen [5] the public perception of travelers formed through the following phases:

- Euphoria. Tourist arrivals were well received, with an abundance of hope. It occurs in the early phases of the development of tourism in a tourist destination.
- Apathy. Communities receive tourists for granted, and the relationship between public and tourists is dominated by commercialization.
- Annoyance. It is already almost reached saturation point, and people are starting get annoyed with the presence of tourists.
- Antagonist. Society has demonstrated his displeasure to the public and sees tourists as a source of the problems.

Related to the tourism industry in Yogyakarta, it is important to assess how exactly the public response to the arrival of tourists is. This information is required as the input for the preparation of the development plan of Yogyakarta tourism industry. It is important to be analyzed how the people of Yogyakarta mapping linier to the phase of public perception. That is the focus of this study so that a planning scheme development of the tourism industry in Yogyakarta will be arranged. Mapping perception in this study will be carried out by any of data mining techniques, namely clustering. Clustering technique is used because it can provide information about groups of objects that have the same profile or certain characteristics. Thus, the resident profile can be obtained based on their perceptions of tourist arrivals in Yogyakarta.

2 Literature Review

Studies on the analysis of local perceptions in tourist arrivals mostly are done in countries that have tourists-attracted objects to be visited. Türker and Öztürk (2013) studied the perceptions of local communities in tourist arrivals in Turkey Mountain National Park Kure (TNGK) [6]. Survey was conducted by the research approaches in three cities (Pinarbasi, Azdavay and Şenpazar) located in the buffer zone TNGK. A two-page questionnaire was conducted, along with structured interviews with 207 residents, local governments, and NGO members between January and March, 2011. Two main conclusions can be reached from the study: first, TNGK population has a neutral or slightly positive perception toward tourism. Second, location affects the community in assessing tourist, with people living in Pinarbasi observe the positive impact of tourism higher than the population Azdavay and Şenpazar. By distributing 299 questionnaires, Marzuki (2012) studied the impact of the main tourist arrivals on the region Pukhet Thailand by using principal component analysis [7]. The results show that there are four major economic impacts: 1. gain, 2. higher cost of living, 3. economic costs, and 4. supporting local economy. Mensah (2012) studied the perceptions of local residents as a result of tourist arrivals in the region Tafi Atome, Ghana. Using the method in the survey of 197 residents aged 18 and over, it is known that tourist arrivals have a positive impact on the local population and the sustainability of endangered animals in the tourist areas. Chan (2011) used cluster analysis to map communities based on their perceptions of the annual activities in Macao [8]. Through the results of 714 questionnaires, approximately the residents of Macao were divided into three groups: the first group has the residents who feel that Macao's annual activities have a positive impact on people's lives in terms of economic and social. The second group believes that in addition, the positive impact of tourist arrivals in Macao also brings negative impacts such as pollution, noise. The last group consists of people who not overly concern about Macao travelers and more focus on his own personal life in Macao than think of a greater impact in Macao.

3 Research Objectives

The purpose of this study is to analyze residents' perception of growth in tourist arrivals in Yogyakarta by using k-means clustering techniques.

4 Method

The first stage of this research is a preliminary study including an analysis of the problem to obtain an overview of the system to be studied. The next step is the distribution of questionnaires. Samples were the people of Yogyakarta randomly starting from communities around the tourism areas to the people who live far from the tourism attractions. A number of correspondents who surveyed approximately 200 people in total spread evenly in every region.

The questionnaire contained questions about what things are perceived by the public as the impact of tourist arrivals in Yogyakarta including positive and negative impacts as well as the expectations of the people associated with the concept of Yogyakarta as an area for tourists. Some variables that need to be known in this survey are divided into two main groups, namely:

1. Profile correspondents include gender, age, occupation, and domicile.
2. Perceptions rating includes a response to the arrival of tourists, gains, losses and so on.

Furthermore, cluster analysis is performed to classify people based on their perceptions of the travelers. This is done to determine the distribution map of the community so that profiling policies related to the tourist industry in Yogyakarta will run maximally. Clustering technique that is used in this research is non-hierarchy cluster (k-means cluster). It is chosen because k-means clustering algorithm is more prominent due to its intelligence to cluster massive data rapidly and efficiently [9].

5 Result

5.1 Respondents' Profile

From 200 distributed questionnaires, only 180 respondents completed the form. This survey consists of 117 males and 67 females. In terms of age, proportion of residents in age between 20 and 30 is 34% and that of residents in age between 30 and 40 is 26%; the next row is people between 15–20 and 40–45 as 15% and the proportion of residents in age over 50 years is 10%. Nearly 67% are the native of Yogyakarta, and the rest 33% are people from other cities who live in Yogyakarta where about 43% from the respondent have been living in Yogyakarta for more than 15 years, 13% in the length of stay between 15 and 20 years, 14% between 10 and 15 years, 12% between 5 and 10 years, and 18% have been living in Yogyakarta for less than 5 years. 51% of them have been living in more than 5 km away from center of the town. In terms of main occupation, the residents are divided into five groups: students 38%, officer 27%, business man 17%, seller 9%, and others 9%. From the survey, more than 67% have annual income less than 2 million IDR. The questionnaire was also asked about the activity of the respondent during holiday; nearly 29% of them said that stay in home is the best activity during the holiday, 21% work as usual, 18% join with the tourist visiting tourism objects, 16% visit their family, 6% visit tourism object outside Yogyakarta, and about 11% do other activities such as shopping, going to the restaurant with family, playing in the city park, and so on.

6 Response to the Tourist

To investigate the general response from the resident, a question was asked. It is about the perception of people to the arrival of the tourists. Nearly 79% of the residents like the growth of the tourism, and only about 21% dislike that (Fig. 2).

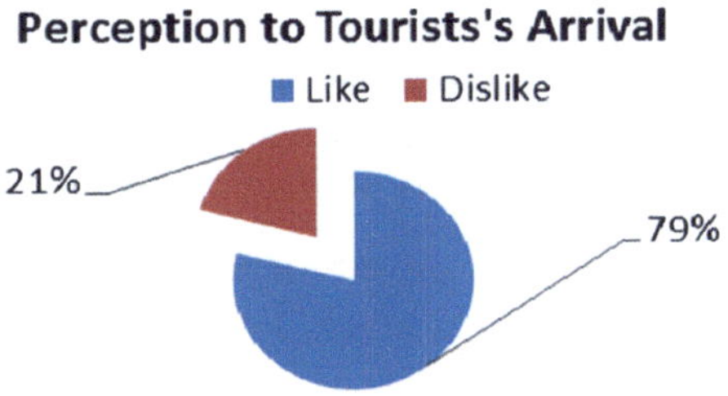

Fig. 2 Graphic representation of residents' perception to tourists' arrival

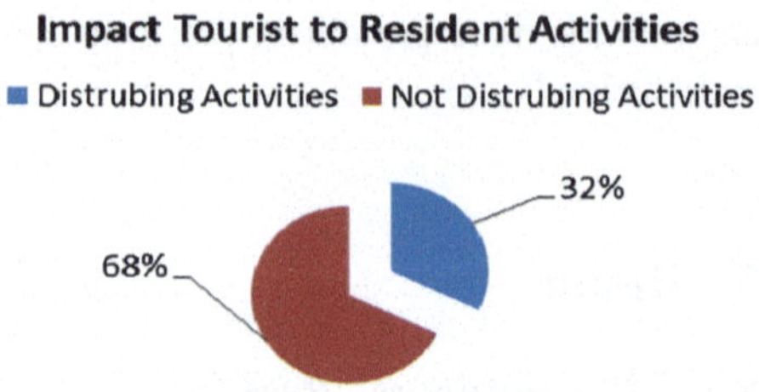

Fig. 3 Graphic of impact of tourists' arrival to residents activities

Specific to the impact of tourists on resident activities, it is only nearly 32% from the respondents who feel that the tourists disturb their daily activities, but most of them (68%) said that they do the activities as well as other days although from the questionnaire its also known that only 23% of the respondents get the direct additional income from tourists' arrival (Fig. 3).

7 Respondents' Cluster

The result from computation process using SPSS 16 is shown in Table 1. It shows that respondents can be grouped in three clusters where profile for each cluster is as follows:

- Cluster 1. Consists of 38 members where the people who belong to this group are the residents who live in Yogyakarta as native and stay near from the center of the city. They have been living for more than 10 years in Yogyakarta where there is no any significant impact of tourism in their activities and their income. Tourism only caused traffic congestion and piles of garbage for Yogyakarta, so that it is better for them to stay at home when the holiday comes. For this cluster, it can be called as **negative cluster** to describe their profile.
- Cluster 2. Consists of 57 members who are the residents who live in Yogyakarta as migrants and stay in place that far from the downtown. They agree that tourism causes traffic, but they are not care about that because they get benefit impact from tourism activities for their income. **Opportunist cluster** can describe their profile.
- Cluster 3. Since it consists of 85 members, this group becomes the biggest cluster of the residents. People who belong to this group are the residents who live in Yogyakarta as native and have been living for more than 15 years. They stay in near the center of the city and accept benefit from the tourism. They not agree that tourism is one of contributors for traffic jam because sometime they enjoy the holiday with the tourist visiting any tourism object. For this cluster, it can be called as **positive cluster** to describe their profile.

These three clusters can be described in a graphic representation as follows (Fig. 4).

Table 1 Result of clustering using k-means algorithm

Initial cluster centers			
	Cluster		
	1	2	3
Zscore(Job)	1.52296	−1.21837	1.52296
Zscore(Native)	−0.69634	1.42809	−0.69634
Zscore(Age)	−0.61979	−1.45234	1.87787
Zscore(Distance_to_center)	1.63666	1.63666	−0.91767
Zscore(Leght_of_stay)	0.94295	−1.59983	0.94295
Zscore(Activities_in_holiday)	1.68447	−0.55320	−1.11262
Zscore(Response_to_tourism)	1.96046	−0.50725	−0.50725
Zscore(Impact_of_tourism_for_income)	−1.80761	0.55014	0.55014
Zscore(Impact_of_tourism_for_activities)	−1.46489	0.67885	0.67885
Zscore(Response_to_traffic)	−0.53304	−0.53304	1.86562

Number of cases in each cluster		
Cluster	1	38.000
	2	57.000
	3	85.000
Valid		180.000
Missing		0.000

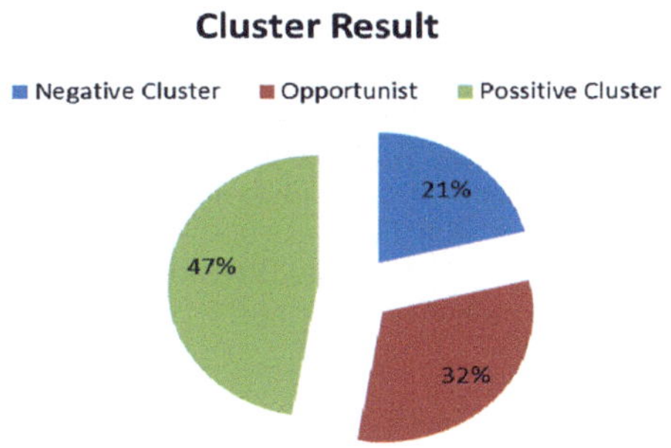

Fig. 4 Clustering result

8 Further Research

This study has identified the resident of Yogyakarta based on their perceptions of tourist arrivals. Most of the natives feel disturbed by the increasing tourist activity in the city of Yogyakarta (21%), others find it is helpful in terms of revenue (47%), and other community groups are not concerned with the tourist activity (32%). In general, the resident of Yogyakarta has a good perception toward tourism because its activities can have a positive impact on the economy of the community. This study has several limitations: the number of respondents who are still few and have not been able to reach all levels of society in Yogyakarta, and grouping variables are limited, relatively short study period, and so on. In future, it has very large opportunities for other similar studies that will be able to know with more details about the public response to the arrival of tourists in Yogyakarta or otherwise for other studies that classify tourists based on their perceptions of resident of Yogyakarta.

References

1. Dinardani Ratrisari, Hadi Wahyono, 2012, Kontribusi Obyek Wisata Di Kota Yogyakarta Terhadap Pendapatan Asli Daerah (PAD), Jurnal Tata Loka; Volume 5; Nomor 1, Universitas Diponegoro, Semarang.
2. Sumbarja Bagus, 2011, Sektor Pariwisata Sumbang 38% PAD Kota Yogyakarta acibanget. blogspot.com, accesed on 7 Juli 2013
3. Sharma Bishnu, Dyer Pam, 2012, A Longitudinal Study of The Residents' Perceptions of Tourism Using Data from The Sunshine Coast Asutralia, PASOS, Vol 10 No 2 p. 37–46, Australia.
4. Cengiz Bulent, 2012, Local Residents' Perceptions of and Attitudes Toward Sustainable Tourism Planning and Management in Amasra (Turkey), Landcape Planning, ISBN 978-953-51-0654-8.In Tech Publisher
5. Chohen E, 1984, The Sociology of Tourism: Approaches, Issues, and Findings, Annual Review of Sociology, Vol. 10: 373–392 (Volume publication date August 1984)
6. Türker Nuray, Öztürk Sevgi, 2013, Perceptions Of Residents Towards The Impacts Of Tourism In The Küre Mountains National Park, Turkey1, International Journal Of Business And Social Science, Vol. 4 No. 2; February 2013, P.45–56
7. Azizan Marzuki, 2012, Lokal residents' perceptions towards economic impacts of tourism, development in Phuket,, Tourism Original Scientific Paper, Vol 60 No 2 p 199–212.
8. Chan S Sandy, 2011, Residents' Perceptions of the Impact of Major Annual Tourism Events in Macao: Cluster Analysis, Journal of Convention & Event Tourism, 12:106–128.
9. Yedla Madhu, Pathakota Rao Srinivasa, Srinivasa T M, 2010, Enhancing K-means Clustering Algorithm with Improved Initial Center, *International Journal of Computer Science and Information Technologies,* Vol. 1 (2), p 121–125

Embedding Accountability Throughout the Innovation Process in the Green Economy: The Need for an Innovative Approach

Andri D. Setiawan, Rajbeer Singh and Henny Romijn

Abstract Innovation in the green economy entails transformative change in society. Vital infrastructure technologies in the fields of energy, water, communication, waste handling, and so on have many interdependencies with other economic sectors. Hence, introducing green innovation in such sectors has complex repercussions and requires many complementary changes to be made elsewhere in the economy as well. Therefore, such innovations come with large uncertainties and risks. They need a long time to be fully developed and assimilated into society, and this is never a smooth journey because many effects were not foreseen at the start. Undesirable social, economic and environmental consequences can easily occur along the way. Therefore, innovation in the green economy introduces the need to foreground issues of accountability more prominently and explicitly throughout the innovation process. This implies that it is necessary to start defining accountability from the beginning of the innovation process—not only toward the end. This can help us to figure out beforehand about what (negative) impacts could possibly emerge and how they can be prevented, in order to lead to satisfactory outcomes for

A.D. Setiawan (✉) · H. Romijn
Eindhoven Centre for Innovation Studies (Ecis), Department of Industrial Engineering and Innovation Sciences, Eindhoven University of Technology, Den Dolech 2, 5612AZ Eindhoven, The Netherlands
e-mail: a.d.setiawan@tue.nl

H. Romijn
e-mail: h.a.romijn@tue.nl

A.D. Setiawan
Systems Engineering Modeling and Simulation (SEMS), Industrial Engineering Department, Universitas Indonesia, L301-L302 Gedung Teknik Industri UI, Depok, Indonesia

R. Singh
Department of Values and Technology, Faculty of Technology, Policy and Management, Delft University of Technology, Jaffalaan 5, 2628BX Delft, The Netherlands
e-mail: rajbasera@gmail.com

T. Taufik et al. (eds.), *ICoSI 2014*, DOI 10.1007/978-981-287-661-4_17

all stakeholders. Therefore, this paper attempts to create an understanding of the importance of embedding accountability throughout innovation process in the green economy by addressing the following questions: How do we define accountability for innovation in the green economy? What kind of approach may help us in embedding accountability throughout the innovation process in the green economy? To answer these questions, the concept of responsible innovation is introduced and the application of the approach is illustrated through a case study. The paper concludes that responsible innovation approach allows the careful scrutinizing of the (emerging) impacts of innovation and provides a clear map of the roles, tasks, and liabilities of innovation actors as the basis for embedding accountability throughout the innovation process.

Keywords Accountability · Green economy · Innovation actors · Responsible innovation · Sustainability

1 Introduction

'Green economy' is evolving as the concept of economy that is geared towards improved human well-being and social equity, while significantly reducing environmental risks and ecological scarcities [1]. It was adopted as the core sustainability concept at the global Rio + 20 United Nations Summit for Sustainable Development in 2012 [2]. Green economy entails economic activities that reduce energy-intensity in production and consumption ('decoupling') and/or improve environmental quality for the society [3]. In this sense, green economy aims to attain economic, social, and environmental sustainability. In the green economy, innovation—in terms of process and product—plays an important role in order to attain such sustainability. One major example is the adoption of renewable energy technologies (RETs), e.g., biomass, hydropower, wind, and solar photovoltaic (PV).

Taking a broader or a systems perspective, the adoption of RETs usually links to other sectors and utilities in which the RETs are used or supported. This attribute makes RETs elements of larger systems [4–6], for example the integration of renewable energy into smart grids and the adoption of solar PV in telecom towers. Such adoption involves various actors or so-called innovation actors—the actors involved in the development, adoption, customization, and assimilation of innovation. Further, the interlinking between different sectors displays systemic and/or sectoral interdependencies, e.g., the telecom sector depends on the energy system and vice versa. However, there is likelihood that such interdependencies also potentially increase the risk for the occurrence of a systems disaster because when one or more elements in the system experience failures, it will most likely affect the operation of the whole system and might create system disruption. And when such disruption cannot be properly resolved, it could lead to a disaster: the collapse of entire systems. Surely, the occurrence of such disaster is highly undesirable since the impact—in society, economy, and environment—will be very severe, and thus, the risk of such major failures becomes a hindrance for the development of the green economy. Further, the occurrence of a disaster often raises an egg-chicken situation; a cyclical situation where what came first between causes and effects is unclear and difficult to recognize [7]. Therefore, what caused the problems will also be difficult to pinpoint. Moreover, since various actors are involved in the adoption of innovation, the egg-chicken situation often gives the appearance of finger-pointing between innovation actors regarding each other's liabilities. Apparently, in such disaster situation it is difficult to locate the accountability of each innovation actor, or in other words, it is difficult to trace who should be responsible for what and to whom. Therefore, it could create uncertainty for the system as the whole, and obstacles for investment in innovation.

The above indicates that innovation in the green economy needs more explicit attention for accountability issues in order to deal with and manage undesirable consequence of innovation (i.e., systems disaster), and therefore, accountability should be present throughout the innovation process. This implies that it is necessary to start defining accountability from the beginning of the innovation process —not toward at the end of the process. This can help us to figure out beforehand about what (negative) impacts could possibly emerge and how they can be prevented—for example, through technological modifications in the innovation itself, or through introducing certain institutional arrangements like use regulation—in order to lead to satisfactory outcomes for all stakeholders. In this sense, if the required accountability is set at the end of the process of innovation, it will be inadequate to deal with the possible negative impacts of innovation. Hereafter, the questions arise: How do we define accountability for innovation in the green economy? What kind of approach may help us in embedding accountability throughout the innovation process in the green economy? By addressing these questions, the paper attempts to create an understanding of the importance of embedding accountability throughout the innovation process in the green economy.

The paper is organized as follows. Section 2 discusses the meaning of accountability and introduces the concept of responsible innovation (RI) [see, e.g., 8–10]. This section will also describe a few examples of the application of such approach found in literature. Next, Sect. 3 will re-discuss a recent case study about the adoption of solar PV in telecom towers in Indonesia by Setiawan and Singh [11]. Thereby, this section illustrates the practice application of RI as the approach that may help us to ensure that accountability of innovation actors could be taken into account and elaborated and embedded throughout the process of innovation. Yet, for this purpose, the paper is not going to delineate all micro-details about the case study, but to take only some examples of stakeholders involved in the adoption of solar PV to illustrate the case. Finally, Sect. 4 concludes and suggests future research for the application of the approach in different areas of innovation in the green economy.

2 Accountability and the Concept of Responsible Innovation

The term accountability is subject to various definitions as it has been widely applied in many disciplines. Accountability is usually associated with the concept of liability, transparency, and some type of authority and commonly implies an assumption of responsibility [12]. Accountability suggests that incentives or disincentives should be paid when one has fulfilled or not fulfilled such liabilities or responsibilities [13]. In the common sense, incentives or disincentives are the sources of motivation for one to be held accountable. Therefore, the likelihood is that people will be motivated to fulfill their tasks and responsibilities when the incentives are comparable (or even bigger) with their tasks and responsibilities. Leat [14] clearly explicated three basic elements as the foundation of accountability. The first is 'accountability to whom?' or in other words 'to whom is accountability owed?' which is related to the persons involved in an issue at different levels—downwards, upwards, and lateral—such as regulator, producer, consumer. The second is 'accountability for what?' which is commonly related to the subject, role, and task of the persons involved in an issue such as accountability for monitoring the process or procedure, program accountability. And the third is 'what type of accountability is required?' for example [cf. 15] accountability for taking into account (responsive accountability), accountability for giving an account (explanatory accountability), and being held to account (accountability with sanctions). These three elements are important to be seriously considered in the innovation process in order to achieve the expected outcomes. However, ensuring such elements to be presented throughout the innovation process becomes challenging. Therefore, in the following paragraph we will further elaborate this issue through the concept of RI.

RI is neither a new nor an old concept. The term itself reflects the relation between innovation and responsibility. The fact that innovations create not only positive impacts but also undesired negative impacts—especially in relation to social and environmental sustainability aspects—is one of the reasons for paying greater attention to responsibility as the theme of innovation practice [10, 11]. After all, innovation for bringing about the green economy must aim to not only achieve sustainability in economic terms but also bring benefits in the societal and ecological domains, thus avoiding trade-offs that often occur between these different sustainability dimensions. Therefore, RI has been evolving as the concept which offers a more explicitly reflexive approach to the development, adoption, and implementation of innovation. Various interpretations of RI can be found in the literature [cf. 8–10, 16–20]. According to von Schomberg [20], RI is a transparent and interactive process involving societal actors and innovators in a mutual and responsive way with a view to the (ethical) acceptability, sustainability, and societal desirability of the innovation process and its products. Stilgoe et al. [10] provided a broader interpretation: '…taking care of the future through collective stewardship of science and innovation in the present.' They also introduced four dimensions of

RI: anticipation, responsiveness, reflexivity, and inclusion. Singh and Kroesen [9] eloquently explained RI as '…being caring or ensuring care for certain values for social, economic, and environmental sustainability by engaging in anticipation, reflexivity, deliberation, responsiveness, and participation for bringing up any change in any idea, product, process, method, way of doing business, technology, etc., in order to bring them into a specific market or use them in a society.'[1]

Further, Singh and Kroesen [9] elaborated that 'being caring' implies being sensitive or considerate, 'ensuring care' espouses responsibility effort to make certain, and 'certain values' pinpoints a list of either universal or culturally specific values. Anticipation expresses the act of looking forward and dealing with risks beforehand and to foresee the plausible impact of innovation. In this regard, the knowledge about foresight and its methodologies [e.g., 21–23] can be used to help in understanding innovation and its impacts. Reflexivity means being reflexive or paying attention to the cause–effect relationships through learning from accumulating experience with the innovation. Reflexivity suggests an iterative process of creating and shaping innovation. Deliberation conveys a rigorous exploration process and a careful consideration of divergent aspects which displays in-depth discussions in order to find agreement about the way forward. Responsiveness refers to being able to readily respond to and cope with certain circumstances caused by diverse requirements, needs, views or perspectives, issues, and values. Participation refers to the need to have broad stakeholder involvement as the condition of RI. Anticipation, reflexivity, deliberation, responsiveness, and participation are considered as the five dimensions of RI [9].

The concept of RI has been applied in a number of studies [cf. 8, 11, 24–26]. Owen and Goldberg [8] used four dimensions: anticipation, responsiveness, reflexivity, and inclusion for analyzing the development of climate geo-engineering technologies project in the UK. Based on the four dimensions, they constructed a set of criteria for assessing the responsibility issues within the project. The criteria are set to pay attention to the following aspects: risks—how they should be identified, managed, and deemed acceptable; regulatory compliance; clear communication of the nature and purpose of the project; applications and impacts of the project/technology; and mechanism to understand publics and stakeholder views. Stahl, Eden [26] developed a framework of responsible research and innovation for studying the ethical issues of emerging information and communication technology (ICT) applications. In that study, they argued that the ethical issues should be addressed by encompassing the following aspects: clear description of subjects of responsibility in ICT (e.g., researchers, professional bodies, funders, policy makers); attribution of objects to subjects (e.g., usability, privacy, sustainability, and transparency); facilitation of discourses on norms, values, and ethical foundations; scheme of sanction and incentive structures; and the reflection on preconditions for

[1]There are also recent approaches that are very closely related to the RI approach, for example the term "responsive research" used by researchers in the UK and "reflexive modernization" used by René Kemp and Jan Peter Voss. (See e.g., https://blogs.nottingham.ac.uk/makingsciencepublic/2014/03/31/responsive-research-which-research-whose-responsibility/.).

successful responsibility relationships. Interestingly, the concept of RI is also adopted in the field of finance. Muniesa and Lenglet [24] introduced 'RI in finance'. In that, they elaborated seven possible perspectives that need to be considered for building responsible financial innovation; function, moral values, internalized values, aggregate consequences, accountability, precaution, and democracy.

3 Embedding Accountability by Engaging the Five Dimensions of RI

The previous section shows that RI has been applied in different disciplines and fields. However, there is still limited number of detailed case studies of where the RI approach has been applied successfully and the applications of RI outside the Western Europe are rarely found. It shows the fact that RI is still a very recently introduced approach—it is not yet a mature research and action field. Therefore, a little more elucidation is still needed to show how the approach can be applied, especially in the green economy context. From this point forward, we will briefly re-discuss a recent case study which we consider can illustrate the application of RI approach in the green economy context; the adoption of solar PV in telecom towers in Indonesia by Setiawan and Singh [11]. Telecom towers constitute the main telecommunication infrastructure in Indonesia. There is a growing concern from the telecom operators in Indonesia to gradually replace diesel generator (DG) by adopting more sustainable and environment-friendly technology such as solar PV for powering the towers (or in a hybrid mode—solar PV combined with DG) [11], especially for the towers which are located in the remote areas with no connection to the main grid. The adoption of solar PV offers several advantages for the operation of telecom towers such as providing supplementary power, less pollution in terms of CO_2 emissions, and lowering the operational cost due to the less usage of diesel. Therefore, such adoption is considered as a sustainable innovation in powering the towers. However, in spite of the advantages of solar PV adoption, the field-related factors such as the local climate condition and improper cleaning regime remain the issues in the post-deployment stage. The temperature and humidity in a tropical country like Indonesia are ideal for rapid accumulation of dust in PV modules. Moreover, this could be worse with the existence of carbon and soil particles, acid components, bird droppings, and other organic matter in PV modules. It is therefore important to periodically clean the PV modules with a good cleaning regime, especially those in the remote areas. But so far the cleaning of PV modules in most telecom towers in the remote areas is only performed once a year, especially in the dry season [11].

So apparently, the field-related factors—local climate condition and the absence of good cleaning regime (e.g., skills of cleaning, the way to transport water to remote areas, technology of cleaning, difficult accessibility of the panel, etc.)—were neither properly handled nor anticipated from the beginning of the adoption process by the actors involved. And as a result, the power production or output of PV modules shows a declining trend. This, however, raises another issue: While the power consumption of the towers remains constant and the power production of PV systems decreases, the towers will have insufficient power for operation and thus can create telecommunication systems disruption. At the end, such condition can lead to the collapse of telecommunication systems, because telecom towers as the main telecommunication infrastructure stop functioning. Further, the collapse of telecommunication systems most likely will give adverse effect to other related systems or sectors that are also dependent on telecommunication systems, such as transport sector, energy sector. Surely, the impact of such disaster will be very severe and therefore such disaster is not expected to occur by telecom operators and the users. Through interviews with some of the actors involved in the adoption of solar PV in telecom towers (i.e., telecom operator, tower company, solar PV manufacturer, solar PV installer), Setiawan and Singh [11] found that if and when such disaster really occurs, no one would have been able to recognize what has caused the problem since they are trapped with the egg-chicken dilemma situation. And not surprisingly, when they were asked about who should be responsible for such disaster and the way to overcome it, most of them were prone to finger-pointing at each other regarding the liabilities in the adoption of solar PV.

Further, in the situation above, it will be difficult to locate the accountability of innovation actors due to lack of clarity about who should be accountable to whom and for what. Such situation also reflects that accountability is not proportionally distributed regarding the tasks that should be performed by each of actors involved. Henceforth, Setiawan and Singh [11] explained that the occurrence of such situation could be prevented beforehand, or at least the risk can be minimized. So in order to ensure that the accountability of innovation actors could be taken into account, elaborated, and embedded throughout the process of solar PV adoption, they proposed the following steps. First, they set a working definition of RI (adopted from [9]) as the frame for analyzing the case; 'RI means ensuring the accountability of innovation actors (the actors involved in the adoption of innovation) through the engagement of anticipation, reflexivity, responsiveness, deliberation, and participation in the adoption of innovation while looking at the impact of innovation on three aspects: environment, social, and economy.' Following this definition, the second step is stakeholder and impact analysis. The actors involved are listed and identified, their interests and motivations are assessed, and their roles and responsibilities are clarified. Further, the possible benefits and impacts of solar PV adoption in terms of environment, social, and economy are also analyzed. Table 1 provides the summary of stakeholder and impact analysis. Next, the result of stakeholder and impact analysis is then further elaborated in the third step with the analysis of the five dimensions of RI; anticipation, reflexivity, responsiveness, deliberation, and participation. The analysis is performed in an iterative way. Table 2 epitomizes this analysis.

Tables 1 and 2 give some inspirations for the provision of a guideline on how to embed the accountability of innovation actors throughout the innovation process in a practical way. Based on Setiawan and Singh [11], at least two possible solutions

Table 1 Stakeholder and impact analysis of solar PV adoption in telecom towers

Actor	Motivation	Task	Impact			Benefit
			Environment	Social	Economy	
Telecom operators	Business profit in telecom sector as the network and service provider	Providing telecom network and services (reliable telecom systems and network infrastructure (i.e., adopting solar PV in telecom tower, providing budget for maintenance cost.))	(+) The adoption of solar PV provides (more) sustainable and environment-friendly power supply for telecom towers, e.g., less CO_2 emission, less noise compare to DG. (−) The failure of the adoption of solar PV may lead to the rejection of using renewable energy technologies, and a setback to the use of unsustainable and non-environment-friendly technology such as DG for powering telecom towers	(+) Sustainable power supply and good tower maintenance lead to reliable telecom systems and infrastructure which keep the customers service level high (reliable use of mobile communication). (−) The absence of sustainable power supply and good tower maintenance will lead to unreliable telecom systems and infrastructure, low customers service level, and thus suffering the users	(+) The adoption of solar PV reduces the cost of energy for powering the towers. This may boost the development of renewable energy sector in general. (−) The failure of the adoption of solar PV creates the risk of telecommunication systems collapse, at the end if and when such disaster occurs, the economic losses will be severe.	Reliable telecom service and infrastructure keep the customers' loyalty, attract new customers, and create profit.
Tower companies	Business profit in telecom sector as the telecom infrastructure provider	Providing tower construction and maintenance services (site maintenance and monitoring systems performance)				Cost reduction in business operation in the long run by reducing the use of diesel generator
Solar PV manufacturers/installers	Business profit from renewable energy development	Providing solar PV modules. (production of reliable solar PV modules (i.e., resistance to climate condition))—PV manufacturer providing solar PV system setup and installation services. (solar PV system operates in the best performance)—PV installer				Increased market for innovation PV technology creates more opportunity for making business profit rising demand for PV as people/society comes to trust in innovation

(continued)

Table 1 (continued)

Actor	Motivation	Task	Impact			Benefit
			Environment	Social	Economy	
Regulator	National economic development through telecommunication sector	Providing rules and regulation for telecom sector (regulation/standard for managing telecom systems (i.e., telecom towers))				High competitive and environmentally friendly telecommunication sector

Source Setiawan and Singh [11]

Table 2 Analysis of five dimensions of RI

Dimension	Form of question	Indicative finding/answer
Anticipation	What (negative) impacts possibly emerge and should be prevented?	Impacts on environment, social, and economy
Reflexivity	What causes the problem and the effects of it?	Causes of the problem: the field-related factors. Effects: declining in power output/production, dysfunction of towers, and risk of telecom system collapse
Responsiveness	What issue needs to be addressed?	Accountability of actors involved in dealing with the field-related factors
Deliberation	What should be done to address the issue? What possible actions can be proposed?	Developing innovative way or (adoption of) technology that considers three aspects: environment, social, and economy
Participation	Who should be involved and contribute to problem solving?	Telecom operators, tower companies, PV manufacturers, solar PV installers, and regulator

Source Setiawan and Singh [11]

could be proposed. The first is a kind of an accountability assessment framework or procedure which can help in assessing and fixing up the responsibility of actors involved in the adoption of solar PV, from the beginning of the adoption until the post-deployment or the monitoring stage. The second alternative is by adopting the self-cleaning PV module technology which can cope with the field-related factors. In this way, the load of periodic cleaning task can be reduced and thus need less maintenance cost. The solutions proposed by Setiawan and Singh [11] reflect that accountability of innovation actors could be embedded throughout the innovation process with 'strategic procedural approach' (accountability assessment framework) and 'technological approach' (self-cleaning PV technology). The first approach seems to be more complicated and less practical, while the second one offers more practical and cost-efficient PV modules maintenance. However, on top of that, both options are aimed to address the power losses problem of solar PV systems due to the local climate condition. Further, both options offer the distribution of responsibility of innovation actors proportionally. This is important in order to deal with the egg-chicken situation. Once the accountability of each of the actors involved can be clearly defined and distributed, it will be less difficult to keep the innovation process running on the right trajectory. Further, it may prevent the occurrence of finger-pointing between actors involved and thus can prevent or minimize the risk of systems disaster.

4 Conclusion

Above we have described that the concept of RI could be used as an approach for embedding the accountability of innovation (actors) throughout the innovation process. The approach works through five dimensions (anticipation, reflexivity, responsiveness, deliberation, and participation) coupled with stakeholder and impact analysis. In the context of green economy, this is particularly useful since the approach allows the careful scrutinizing of the impact of innovation and provides a clear map of the roles, tasks, and liabilities of innovation actors as the basis for embedding accountability throughout the innovation process. Also, the accountability measure can be clarified such as what kind of benefits or detriments could be envisaged if the actors involved fulfill or do not fulfill their tasks and responsibilities. The approach may also serve as a guide to open up a deliberative discussion between innovation actors before deciding or starting the adoption of innovation. Such discussion could be guided by a leading question like how to ensure that the innovation adoption can lead to satisfactory outcomes.

Yet, there is no perfect approach. As for RI approach, one limitation that could be highlighted is that this approach should have been presented not in the middle of the adoption process. It should be in place in the pre-adoption stage or before the innovation project is started, because defining the accountability of innovation actors could be uneasy and problematic after the adoption is started. It is the nature of RI as clearly reflected in the five dimensions, in particular the dimension of anticipation. So in this way, it is apparent that RI approach is currently more suited as a precautionary measure for preventing unsatisfactory outcomes of innovation adoption. And to a large extent, many areas in the green economy need early and strong attention of accountability of innovation actors. This, however, provides opportunity for future research for the application of the approach in different areas of innovation in the green economy such as the development of eco-cities, the integration of renewables to smart grid.

Acknowledgements We would like to thank Indonesia Endowment Fund for Education (LPDP), the Ministry of Finance of Republic of Indonesia for the scholarship support to the first author of this paper.

References

1. UNEP. *What is the "Green Economy"?* 2010; Available from: http://www.unep.org/greeneconomy/AboutGEI/WhatisGEI/tabid/29784/Default.aspx.
2. UNCSD. *The Future We Want. Official Outcome Document of the Rio + 20 United Nations Conference on Sustainable Development, Rio de Janeiro*. 2012; Available from: http://www.uncsd2012.org/index.php?menu=140.
3. Chapple, K., et al., *Innovation in the Green Economy: An Extension of the Regional Innovation System Model?* Economic Development Quarterly, 2011. **25**(1): p. 5–25.
4. Peng, L. and G.-S. Yan, *Clean Energy Grid-Connected Technology Based on Smart Grid.* Energy Procedia, 2011. **12**(0): p. 213–218.

5. Verbong, G.P.J., S. Beemsterboer, and F. Sengers, *Smart grids or smart users? Involving users in developing a low carbon electricity economy.* Energy Policy, 2013. **52**(0): p. 117–125.
6. Phuangpornpitak, N. and S. Tia, *Opportunities and Challenges of Integrating Renewable Energy in Smart Grid System.* Energy Procedia, 2013. **34**(0): p. 282–290.
7. van de Poel, I., et al., *The Problem of Many Hands: Climate Change as an Example.* Science and Engineering Ethics, 2012. **18**(1): p. 49–67.
8. Owen, R. and N. Goldberg, *Responsible Innovation: A Pilot Study with the U.K. Engineering and Physical Sciences Research Council.* Risk Analysis, 2010. **30**(11): p. 1699–1707.
9. Singh, R. and O. Kroesen, *Understanding responsible innovation from developing countries perspectives*, in *The 2nd Conference on Responsible Innovation 2012*. 2012: The Hague, The Netherlands.
10. Stilgoe, J., R. Owen, and P. Macnaghten, *Developing a framework for responsible innovation.* Research Policy, 2013. **42**(9): p. 1568–1580.
11. Setiawan, A.D. and R. Singh, *Responsible Innovation in Practice: The Adoption of Solar PV in Telecom Towers in Indonesia*, in *Responsible Innovation Volume II*, J. Van den Hoven, E. J. Koops, H. A. Romijn, T. E. Swierstra and I. Oosterlaken, Editor. Forthcoming, Springer.
12. Ubois, J., *Responsible Innovation/Sustainable Innovation*, in *Annual Meeting of the American Anthropological Association, November 2010*. 2010.
13. Halle, M., *Accountability in the Green Economy.* Review of Policy Research, 2011. **28**(5): p. 473–477.
14. Leat, D., *Are voluntary organizations accountable?*, in *Voluntary Agencies: Challenges of Organization and Management*, D. Billis, and Harris, M., Editor. 1996: Basingstoke: Macmillan.
15. Taylor, M., *Democracy and Accountability*, in *International Encyclopedia of Housing and Home*, S.J. Smith, Editor. 2012, Elsevier: San Diego. p. 314–318.
16. Hellström, T., *Systemic innovation and risk: technology assessment and the challenge of responsible innovation.* Technology in Society, 2003. **25**(3): p. 369–384.
17. Kemp, R., J. Schot, and R. Hoogma, *Regime shifts to sustainability through processes of niche formation: The approach of strategic niche management.* Technology Analysis and Strategic Management, 1998. **10**(2): p. 175–195.
18. Lee, R.G., *Look at Mother Nature on the Run in the 21st Century: Responsibility, Research and Innovation.* Transnational Environmental Law, 2012. **1**(01): p. 105–117.
19. Sutcliffe, H., *A Report on Responsible Research and Innovation for the European Commission*. 2011: MATTER, London.
20. von Schomberg, R., *Prospects for technology assessment in a framework ofresponsible research and innovation. In: Dusseldorp, M., Beecroft, R. (Eds.), Tech-nikfolgen Abschätzen Lehren: Bildungspotenziale Transdisziplinärer. Vs Verlag,Methoden, Wiesbaden*. 2011.
21. De Moor, K., et al., *Towards Innovation Foresight: Two empirical case studies on future TV experiences for/by users.* Futures, (0).
22. Peter, M.K. and D.G. Jarratt, *The practice of foresight in long-term planning.* Technological Forecasting and Social Change, (0).
23. Sarpong, D., M. Maclean, and C. Davies, *A matter offoresight: How practices enable (or impede) organizational foresightfulness.* European Management Journal, 2013. **31**(6): p. 613–625.
24. Muniesa, F. and M. Lenglet, *Responsible Innovation in Finance: Directions and Implications*, in *Responsible Innovation*. 2013, John Wiley & Sons, Ltd. p. 185–198.
25. Parkhill, K., et al., *Deliberation and Responsible Innovation: A Geoengineering Case Study*, in *Responsible Innovation*. 2013, John Wiley & Sons, Ltd. p. 219–239.
26. Stahl, B.C., G. Eden, and M. Jirotka, *Responsible Research and Innovation in Information and Communication Technology: Identifying and Engaging with the Ethical Implications of ICTs*, in *Responsible Innovation*. 2013, John Wiley & Sons, Ltd. p. 199–218.

Tourist Village for Rural Economic Development

Agus Mansur, Riski Nobriandiro and Yasser Azka Ulil Albab

Abstract The majority of Indonesian people live in rural areas. Most of them are working in the agricultural sector. However, the agricultural sector is not going well so it cannot contribute to the welfare of the society. This factor pushes a massive urbanization, where people in rural areas migrate to the urban areas; they prefer to be a factory worker or a store employee instead of managing the economic potential in their hometown. In fact, the rural areas can be developed for economic growth, such as the development of the tourism sector. Saturation of tourists to the city's attractions gives opportunities to develop a village-based alternative tourism. By using the status of DIY as a world-class tourism city, some districts are trying to attract tourists by developing tourist villages. This study aimed to analyze the factors that influence the development of rural tourism located in Sleman Regency. During the development, the tourist villages are divided into two groups in relation to their growing capability, i.e., middle growing and non-growing groups.

Keywords Tourism · Development · Village

1 Introduction

The Statistics Indonesia (*Badan Pusat Statistik*/BPS) recorded that the percentage of Indonesian people who live in rural areas are bigger than those who live in urban areas, i.e., 50.21% [1]. As generally known, most of the people in rural areas are working as farmers, although the job could not significantly help their economy and

A. Mansur (✉) · R. Nobriandiro · Y.A.U. Albab
Industrial Engineering Department, Universitas Islam Indonesia, Yogyakarta, Indonesia
e-mail: agusmansur.am@gmail.com

R. Nobriandiro
e-mail: riskinobriandiro@gmail.com

Y.A.U. Albab
e-mail: yasserazka@gmail.com

T. Taufik et al. (eds.), *ICoSI 2014*, DOI 10.1007/978-981-287-661-4_18

welfare. It can be known from the BPS's data that the number of poor people in rural areas is six times more than the poor in urban areas [2].

There is an alternative to increase the welfare of those who live in rural areas, and it has already applied in some villages named tourist villages. This is one form of the creative economy that can significantly help the society to earn the economic value of their village by utilizing and accommodating the resources to increase the advantage values that can be exchanged. Special Region of Yogyakarta (DIY) is a well-known tourist destination in Indonesia. Based on the data published by the Statistics Indonesia in 2013, DIY is a province that can significantly raise its amount of foreign visitors [3]. The increasing number of foreign visitors in DIY is shown in the below figure. There are several tourist villages in DIY that have been introduced to public (Table 1; Fig. 1).

Table 1 Tourist villages based on the regions in DIY

Sleman	Bantul	Kulonprogo	Gunungkidul	Yogyakarta
Bokesan	Giriloyo	Kalibiru	Bobung	Dipowinatan
Gabugan	Kasongan	Nglinggo	Nglanggeran	
Grogol	Kembangarum	Glagah		
Nawung	Wukirsari			
Pajangan	Tembi			
Pentingsari	Karangtengah			
Tunggularum	Manding			
Brayut	Candran			
Kinahrejo	Kebonagung			
Tanjung				
Rumah Domes				
Agrowisata Turi				

Sources [4, 5, 6, 7]

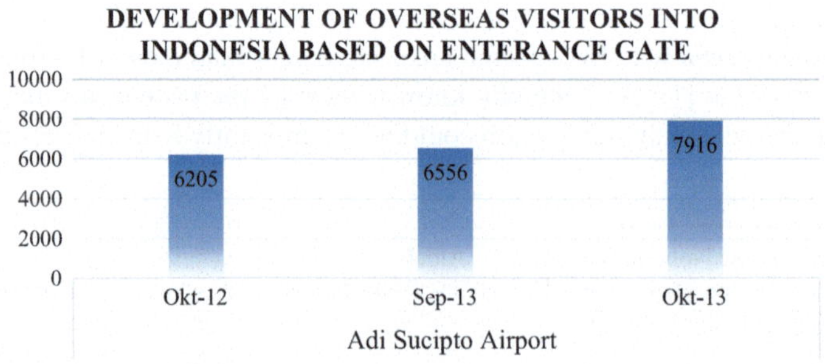

Fig. 1 Development of foreign visitors in Yogyakarta

Although DIY has a lot of tourist villages, there are still many weaknesses and deviations among each other. There are villages that are being highly visited, and there are also villages that are being poorly visited. The ideal condition is that each village has its own uniqueness and competitive values so they can attract tourists to visit their places.

There are three types of tourist villages in DIY, i.e., natural, art and culture, and the adventure attraction based. This research only focuses on art and culture attraction-based tourist village in Sleman Regency, because this type has a large number of potential values and advantages that can be further developed.

2 Literature Review

2.1 *k-Means Clustering*

Cluster analysis is a tool to define the group from many objects based on its characteristics [8]. For that reason, cluster can classify the objects that have similar preference [9]. Distance measurement is an important thing in clustering; it can be done by calculating the difference of its attributes [10–12].

The k-means clustering algorithm [13] is a straightforward and effective algorithm to find clusters in data. The algorithm proceeds as follows:

1. Ask the user how many clusters k the data set should be partitioned into.
2. Randomly assign k records to be the initial cluster center locations.
3. For each record, find the nearest cluster center. Thus, in a sense, each cluster center "owns" a subset of the records, thereby representing a partition of the data set. We therefore have k clusters, C1, C2, ..., Ck.
4. For each of the k clusters, find the cluster centroid and update the location of each cluster center to the new value of the centroid.
5. Repeat steps 3–5 until convergence or termination.

2.2 *Benchmarking*

Benchmarking can be defined as an activity to measure organizational performance against those companies that best-in-class; determine how the best companies are achieving the level of performance; and use that information as the basis for the targets, strategies, and implementation of the organization itself. In another meaning, the benchmarking process can be simplified as an industry's best practice search process that led to superior performance [14]. Among ASEAN, the country that has a lot of benchmarking projects is Malaysia. This is because the implementation of the benchmarking projects requires an understanding of the corporate culture of openness and requires huge costs [15]. There are several steps that should be done to reach the benchmarking goals as follows [16].

1. Planning

Organization should be able to describe the deepest vision of its goal because it has critical success factors among other steps.

2. Search

Organization should be able to find the appropriate partner.

3. Observe

Organization in this process should understand and document the partner's process.

4. Analyze

After all data have been recorded, organization analyzes performance between his own and his partner's that can make a gap between them.

5. Adapt

Organization can arrange such best practice or conduct practice that can be implemented in his own business process.

3 Methods

3.1 Variable Selection

Variable is one important point as a basic analysis to cover the evidence that can increase the performance of tourist village in DIY. The variables used in this research have been discussed with several organizations that concern about tourist development in Sleman. Therefore, the sources of attributes in this research are learnt not only from the expertise in Sleman but also from the various literatures. The variables that will be used in this study are shown in [17–19] Table 2.

3.2 Data Analysis

After collecting the data, analysis is done using clustering method based on the above variables; cluster is used to differentiate the position of every single village. The villages will be classified into three clusters—the villages with highest performance, middle performance, and poor performance. The further analysis will be conducted using the benchmarking approach so we can get the factors that influence the difference among the villages (Table 3).

Table 2 Variables of village tourism

No	Variable	Attributes
1.	Attraction activities	Number of attractions
		Creative supported activities
		Existing tradition
		Routine traditions and show performances
2.	Managerial capacity	Orderliness of meeting administration
		Public transportation management
		Financial management
3.	Community involvement	People's care about tourism activity
		Tourism impact on its environment
4.	Facilities and infrastructure	Lodging availability
		Restroom availability
		Electricity supply
		Clean water supply
		Office and administration room
		Souvenir shop availability
5.	Marketing	Marketing methods
		Involvement of the tourist village itself
		Development of the tourism package
6.	Accessibility	Distance to the city center
		Road condition to the destination
		Comfortability along the road
		Signboards' availability
7.	Visitation	Frequency of visitors
8.	Asset	Assets ownership

4 Result

From the above result, the villages are classified into three clusters—highest performance, middle performance, and poor performance. This classification will help the understanding and analysis of the characteristics of each cluster. The computation is conducted using SPSS with k-means clustering algorithm, and the last result is shown in Table 4.

5 Discussion

5.1 Cluster Profiling

From the cluster data, after analyzing the character of each object at a particular cluster toward the criteria, it can be known that the first cluster consists of objects

Table 3 Result of survey

	Question																							
Villages	1	2	3	4	5	6	7	8	9	10	11	12	13	14	15	16	17	18	19	20	21	22	23	24
Pentingsari	3	3	3	3	3	3	3	3	2	3	3	3	3	2	3	3	3	3	3	2	3	3	3	3
Mlangan	1	2	2	2	2	3	3	3	2	2	2	2	2	2	1	1	2	2	1	2	3	3	1	3
Sukunan	2	3	2	2	3	3	2	3	2	3	2	3	2	2	2	2	2	2	2	2	3	3	2	3
Sendari	2	2	2	2	2	3	2	2	2	3	2	2	2	2	2	2	3	2	2	2	2	3	2	3
Brajan	2	2	2	2	2	3	2	2	1	3	2	2	2	2	2	2	3	2	2	2	3	3	2	3
Gamplong	2	2	2	2	2	3	2	2	1	2	2	2	2	1	2	2	2	3	2	1	3	3	2	3
Sangubanyu	2	2	2	2	2	3	2	2	1	2	2	2	3	2	2	2	3	2	2	1	2	3	2	3
Tanjung	2	2	1	2	2	3	2	3	2	2	3	3	3	2	2	2	2	3	3	1	3	3	3	3
Rumah domes	3	2	3	2	3	3	3	3	2	3	3	3	3	3	2	3	3	3	3	1	3	3	3	3
Sidoakur Godean	3	3	3	3	3	3	3	3	2	3	3	3	3	2	2	3	3	3	3	1	3	3	3	3
Candi Abang	1	1	1	1	1	3	1	1	1	1	1	2	1	2	1	1	1	1	1	1	1	1	1	3
Plempoh	2	2	2	2	3	3	2	3	2	2	2	2	2	2	2	2	2	2	2	2	2	3	3	3
Pajangan	2	2	2	2	3	3	2	2	2	2	2	3	2	2	1	2	2	2	2	2	2	3	2	3
Mlangi	1	1	2	2	2	3	2	2	2	2	2	2	2	2	1	1	2	2	2	1	3	3	2	3
Bokesan	2	2	2	2	3	3	2	3	2	2	3	2	3	2	2	2	2	3	2	2	3	3	2	3
Grogol	3	3	2	2	2	3	3	2	2	2	2	2	3	2	1	2	3	3	3	1	3	3	3	3
Brayut	2	3	2	2	2	3	2	2	2	2	2	2	3	2	1	2	2	2	2	2	2	3	2	3
Pulgading	3	3	2	2	3	3	3	3	2	3	2	2	2	2	2	3	3	3	3	3	3	3	3	3

Table 4 Clustering results

	Cluster 1	Cluster 2	Cluster 3
	Pentingsari	Mlangan	Candi Abang
	Tanjung	Sukunan	
	Rumah domes	Sendari	
	Sidoakur Godean	Brajan	
	Grogol	Gamplong	
	Pulgading	Sangubanyu	
		Plempoh	
		Pajangan	
		Mlangi	
		Bokesan	
		Brayut	
Total	6 villages	11 villages	1 village

that have the highest value for each attribute that has been proposed while the second cluster consists of objects that have attributes character that they tend to be in the middle of the attribute value. The third cluster is indicated to have the poorest grade because of all the attributes of known values; it has the lowest level compared to other objects in the group.

The first cluster consists of tourist villages that have ideal characteristics in which all the tourist villages are expected to have. Unfortunately, this group only consists of 33.33% of the objects observed. Instead, the second cluster, which consists of the villages in midst of other clusters, has the highest percentage, i.e., 61.11%. The third cluster which has the worst value only consists of one village or 5.56%.

5.2 *Developing Tourist Villages*

From the research, it can be concluded that the tourist villages in Sleman which have continuous potential and ideal conditions to attract visitors are only a third from all the objects observed. However, most of the tourist villages are not ready to face the increasing fierce competition. Under the conditions that have been previously described, it is known that the tourist villages are not the first choice for tourist destination.

If the condition is not immediately anticipated, the most feared is that the work diversification in the villages cannot occur and tourist village cannot elevate the economic life of society. In fact, if the tourist villages' management is well organized, it is not impossible that in the future the tourist villages will be the first choice of tourists to visit.

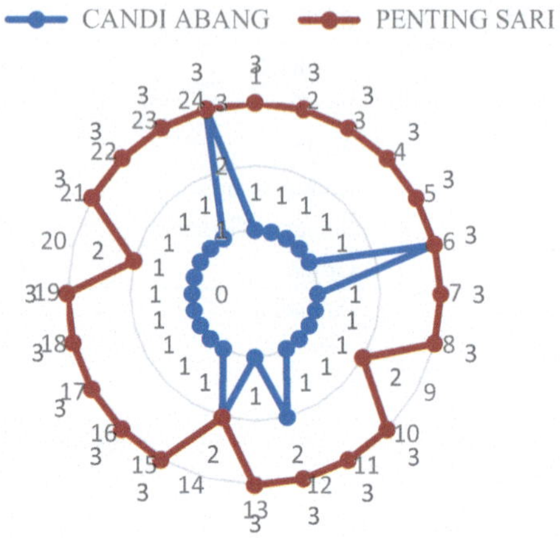

Fig. 2 Comparison between the best and the worst tourist villages

The most interesting issue to discuss is the characteristics of high performance of tourist village cluster and differ it from the poor one, so the management can learn what should be focused to increase the performance of all the tourist villages. One of the tourist villages in the first cluster which has the highest value among the others is Pentingsari. The tourist village which has the worst performance is Candi Abang. The following figure illustrates the relationship that occurs between the two villages (Fig. 2).

From the above figure, benchmarking can be done between both the objects. Although located in a position that does not excel in many aspects, but Candi Abang still has the advantages in the aspect of managerial capacity and ownership of assets. Candi Abang can improve its performance by conducting a proper strategy. In addition, it can be concluded that the characteristics mentioned above (the variables used) are minimum requirements that must be met in order to make a tourist village have a high performance.

Recommendation that can be given to the third cluster (Candi Abang) is adding the facilities and the quality of the attributes proposed in the questionnaire. The future research can be done by observing and analyzing the order of importance of the variables that need to be prioritized.

Acknowledgements This work has been supported by the Industrial Engineering Department, Faculty of Industrial Technology, Universitas Islam Indonesia. This support is gratefully acknowledged.

References

1. Statistics Indonesia. 2010. Data of Population Census.
2. Statistics Indonesia. 2013. Amount and Percentage of Poor Population, Poverty Line, Indexes of Poverty.

3. Statistics Indonesia. 2013. Official Statistics News. No.87/12/Th. XIV, 2 December 2013
4. DIY Tourism Board. 2013. At http://www.visitingjogja.com/?mod=akomodasi& sub = Desa %20Wisata&act = view#tengah and http://www.visitingjogja.com/?mod= akomodasi &sub = Desa%20Wisata&act = view&page = 2#tengah
5. Bantul Tourism Board. 2013. www.disbudpar.bantulkab.go.id
6. Sleman Tourism Board. http://www.tourismsleman.com/dir_detail.php?id=15&PageNo=2 and http://www.tourismsleman.com/dir_detail.php?id=15
7. Jogja Tourism Board. 2013. http://pariwisata.jogjakota.go.id/index/extra.arsip/2
8. Everitt BS. 1994. A hand book of statistical analysis usings-plus 1st ed. London: Chapman&Hall
9. Chinese D,Meneghetti. 2004. A Energy analysis in industrial districts: A statistics-based approach," International Journal of Global Energy Issues 2004; 21 (4): 315 e28
10. Calinski T,Harabasz J. 1974. A dendrite method for cluster analysis. Communications in Statistics 1974;3(1):1e27
11. Jain A K, Dubes R C. 1988. Cluster validity in algorithms for clustering data. Englewood Cliffs, NJ: Prentice Hall; 1988 p. 143e222 [chapter4]
12. Banfield J D, Raftery A E. 1993. Model based Gausian and non-Gausian clustering. Biometrics 1993; 49 (3): 803e21
13. J. MacQueen. 1967. Some methods for classification and analysis of multivariate observations. Proceedings of the 5th Berkeley Symposium on Mathematical Statistics and Probability, Vol.1, pp. 281–297, Berkeley: University of California Press
14. Evans, Lindsay. 2007. An Introduction to Six Sigma and Process Improvement, Jakarta: Salemba Empat
15. Zairi, M. 1996. Effective Benchmarking: Learning from the Best. London: Chapman & Hall
16. Andersen, Bjorn and Pettersen, Per-Gaute. 1996. The Benchmarking Handbook. London: Chapman & Hall
17. Seba, Jaime A. 2012. Tourism and Hospitality: Issue and Developments. USA: Apple Academic Press
18. Uysal, Muzaffer, Perdue, Richard R., and Sirgy, M. Josep. Handbook of Tourism and Quality of Life Research. USA: Springer
19. Echtner, Charlotte M and Ritchie, J.R.Brent. 2003. The Meaning and Measurement of Destination Image. The Journal of Tourism Studies. Vol. 14 No.1 May: 37–49

Innovative Women Entrepreneurs in the Green Economy: Two Case Studies from Mauritius

Satyadev Rosunee and Adeela Peer

Abstract Women make up 51% of the population of Mauritius, but only a very small percentage manages to emerge as entrepreneurs. Contextual inquiry was conducted with two women, both green entrepreneurs, to understand their motivations, business model and the socio-economic impacts of their enterprises. These two women are from completely different social backgrounds, have different levels of formal education, which makes their cases and their business approaches all the more unique and interesting. Both women have lifted other women—the nurturers of future generations—out of deprivation and poverty enabling their families to live with dignity and enhanced financial security. This paper is a narrative of how two women, having espoused sustainable business practices, have successfully grown businesses that are capable of exporting duly certified green products to the most demanding and highly competitive global markets. The concept of sustainability (people, planet, profit) underpinned the key aspects of their business, which proves that sustainability can be a positive driver of innovation and socio-economic change.

Keywords Women entrepreneurs · Role models · Poverty alleviation · Mauritius

1 Introduction

Women are deemed to be transformational leaders while men are generally more effective as transactional leaders [1]. Women are more likely to start businesses with dual goals: social and economic while also ensuring little adverse impact on the environment, making the business more sustainable [2–5]. Female leadership

S. Rosunee (✉) · A. Peer
Department of Applied Sustainability and Enterprise Development,
University of Mauritius, Reduit, Mauritius
e-mail: s.rosunee@uom.ac.mu

A. Peer
e-mail: adeela.peer@gmail.com

T. Taufik et al. (eds.), *ICoSI 2014*, DOI 10.1007/978-981-287-661-4_19

traits that bring societal changes for the better are flexibility, adaptability, collaboration, communication, and empathy [1]. They serve as role models who mentor and empower others. Women entrepreneurs, however, are not usually classified among the top business personalities who inspire people the most [6].

Because women entrepreneurs have a keener eye for the deprivations faced by other women, many across the globe have dedicated themselves to supporting women with the aim to bring a positive change to society. Shiva [7], a world-renowned environmental activist, eco-feminist and academic researcher, has worked assiduously to end the oppression of both women and nature. She believes that the empowerment of women and profound respect for nature is part of the answer to achieving global sustainability [8]. The lack of role models has been reported as a challenge facing women entrepreneurs [9]. The importance of role models should not be underestimated as people learn from those who inspire them and try to imitate them to achieve their own goals [10]. A survey carried out by Bosma et al. [11] revealed that a majority of entrepreneurs would prefer entrepreneurship role models within the same gender, sector and nationality. Personal contact with role models had more impact on future entrepreneurs; during start-up and growth stages of companies [11, 12]. Levie et al. [13] and Hang and van Weezel [14], however, argued that the media played an important part in encouraging an entrepreneurial spirit in the society though it does not directly motivate the majority of future and settled entrepreneurs. Pietilainen [15] observed that it is worth narrating the success stories about female entrepreneurship as a specialty that deserves public attention since there is a general belief that entrepreneurship is a male issue. Hausmann et al. [16], in the Insight Report of the World Economic Forum, noted that in Mauritius, women are 2.5 times more likely to be unemployed than men. Encouraging and supporting women who represent half of the population to become entrepreneurs will increase their contribution to the economy.

The UNDP has emphasized "women empowerment" as one of the main challenges in order to achieve the Millennium Development Goals. Lemmon [17] argued that businesses are important for women because earning an income is earning respect. Pietilainen [15] noted that female entrepreneurship is often projected in the light of gender equality. But overall women participation in the economy remains low [16]. Women are the first educators of future generations and working women contribute significantly to the family budget. Brush [18] found that women entrepreneurs are creating jobs, innovation and contributing to the GNP and are investing back in their community. Matsui [19] reported that educated women contribute to the quality, size, and productivity of the workforce while Kabani [20] pointed out that women are now able to emerge in male predominant sectors.

The motivators for female entrepreneurs are wide ranging: entrepreneurial spirit, ambition and passion, the desire for autonomy, the turn-off of office politics, unhappiness with the gender pay gap, the glass ceiling in corporate life, and the need for a work–life balance and time flexibility [21]. Irrespective of the motivation, women were generally more likely to be running single founder businesses compared to men, indicating a tendency to operate on their own instead of in teams [22]. Women value tremendously the flexibility to take care of their home and family while being their own boss [23], even if many of them are highly educated.

2 Methodology

Two Mauritian women entrepreneurs were chosen as role models for their entrepreneurship initiatives. Contextual inquiry was carried out with the two women during a half-day workshop held at the University of Mauritius in 2013. Online information about their achievements, on-going projects, including videos and articles published in the press were collected. Their profile pages on social networks were searched for information relevant to this article. All information secured online and through social networks was corroborated through phone interviews.

N.B. Throughout the text, both women entrepreneurs are identified as Entrepreneur A and Entrepreneur B, respectively.

3 Results and Discussions

Besides engaging their enterprises on a growth path and successfully exported their products, both entrepreneurs have integrated social and environmental aspects into their businesses. They are constantly working on new products and looking for new markets. In this paper, the success of these two entrepreneurs shall not be analysed from the traditional view point towards successful women entrepreneurs; that is "despite being a women" or compared to the dominant characteristics of male entrepreneurs. Rather, this paper is focussed on the women's own entrepreneurial spirit and their will to pursue a dream as well as bring a positive change in society, for the benefit of future generations and for the common good.

Entrepreneur A launched her venture with a group of women from her immediate, fairly deprived, neighbourhood. The women had lost their jobs when apparel factories closed down massively prior to the end of the multi-fibre arrangement (c.2000). The women's financial situation was precarious and some could not even afford to feed their babies or send their toddlers to pre-primary school. Entrepreneur A's target product was organic herbal teas with medicinal properties, which are known to cure a number of common ailments. The organic herbal teas were prepared from a mix of about 34 medicinal plants that were either grown in-house or purchased from a network of local planters to whom she had earlier supplied plantlets. She initiated a "leaves buy-back' system that provided a regular income to sixteen men and twelve women. After promoting her products in overseas trade fairs, Entrepreneur A started securing export orders to various countries. Although herbal teas may not fall into the category of innovative product, it is the work ethics, mutual trust and collaborative method of value-addition and the ensuing socio-economic benefits it brought to the community that kindled our interest in this small enterprise.

Entrepreneur B is at the helm of a freeport-based horticultural company involved in establishing nurseries, propagation and marketing of quality tropical flora worldwide. Entrepreneur B started her 'mini green revolution' in 1994. Her objectives were: to combat climate change, promote reforestation in African tropical rain forests and to alleviate poverty. She researches endemic plants in various parts of the world, mostly in remote forested areas where the people are generally very poor. She plans expeditions in the forests, identifies the plants, works with the local community there to establish nurseries and cultivate the plants in larger quantities. The plants are then imported into the Mauritius Freeport and re-exported across the globe mainly to first world countries.

4 Family Background and Support

Entrepreneur A and Entrepreneur B were born in Mauritius in 1959 and 1965, respectively. Both women were influenced mostly by their fathers. Since their childhood the foundational platform for their respective enterprises may have been laid in as much as they were both encouraged to be socially and environmentally conscious while striving for the common good. Despite his humble origins, Entrepreneur A's father always helped those in need. The family and close relatives showed affinity for tackling social issues. From her mother, she learned to appreciate the values taught by religions other than her own, which made her tolerant and comfortable in plural societies. Though she had a natural liking for plants, her grandmother transmitted to her the knowledge about medicinal plants. Entrepreneur B's father was a forester (government employee) and he inculcated in her the love for plants and gardening. Her passion for the green world was enhanced when she accompanied her father on trips to forests in Africa. This early experience was an eye opener on the plant biodiversity of those countries but also on the poverty faced by the tribes whose livelihoods depended on the forest. It can be concluded that both women had close relationships with their parents that helped them build character and self-confidence. Entrepreneur A received constant support from her husband and in-laws to set up and implement her projects. Entrepreneur B is a divorcee with a university-going son, that she raised singlehandedly.

5 Self-improvement and Empowerment

Entrepreneur A dropped from school when she was fifteen and was apprenticed to a clothing factory to financially help her family. After a few years, she was promoted to higher posts that encouraged her to pursue higher studies. On leaving the textile and clothing sector she founded a pre-primary school, upgraded herself with child pedagogy courses with the aim of advancing the education of the poor children of her locality. She also built a good rapport with the unemployed women who admitted their children to the school. It was from this association that she launched the herbal tea business in order to provide sustainable livelihoods to these women. To better grow the business, she embarked on a course in the processing of herbal teas in India. Education remained on top of her agenda, not only for the children but for their mothers as well. Twenty mothers who could not find alternative jobs because of their low level of formal education were recycled into learning new agricultural techniques from the Agricultural Research and Extension Unit (AREU), Ministry of Agriculture. They were taught how to process and prepare the herbal teas. They were also trained about hygiene, packaging and marketing techniques that helped them to develop self-confidence. Thus, a novel set-up was created where in the same building children were being educated and formerly unemployed mothers were generating income, thus improving their livelihoods. In addition, flexible working hours (flexi-time) was the norm. For example, if a mum and/or child was sick and had to visit hospital in the morning, she could replace lost working hours either in the afternoon or by instalments over the coming days. This system has up to now given full satisfaction to Entrepreneur A.

As a child, Entrepreneur B was introduced to the world of plants by her father. Her family being financially more comfortable, Entrepreneur B had a normal schooling followed by higher studies in the field of horticulture. Her business idea was to search for rare or little known varieties of endemic plants around the world, nurse and propagate them, with two main objectives: reforestation and improving livelihoods through exports. Her laborious research has indeed contributed to botanical data. With the help of her team and tribal people, Entrepreneur B taught the local communities how to nurse and propagate the endemic plants for export, thus providing alternative means for earning a living. Wherever, she went, she benefitted immensely from the knowledge transmitted by the elders of the tribes about the endemic plants and their virtues.

6 Certification and Greening of Operations

Each country has its quality and safety requirements and norms that exporters must comply with, especially in the agricultural sector. Certification of the herbal teas by the HACCP (Hazard Analysis and Critical Control Point) and Eco-cert has been undertaken. By growing the plants in-house using eco-friendly methods and by supervising the planters supplying leaves, Entrepreneur A is able to ensure consistent quality at all production stages. Entrepreneur A's advice to budding entrepreneurs is not to be discouraged by the investment needed to upgrade in order to meet quality and certification requirements as the return on investment is guaranteed. Each new product is sent to India to be tested and further 'tea-tasting' is carried out for about three more months (in Mauritius) before commercialisation. Tea processing is power intensive. The installation of photovoltaic (PV) solar panels, generating 14360 kWh of electricity per annum, has resulted in a saving of electricity charges of about US$550 per month. The PV project was funded by the SGP-GEF Small Grants Programme of the UNDP [24]. UNDP's belief in the viability of the venture and funding of the renewable energy initiative was a major publicity boost for the venture.

As a privileged witness of the massive deforestation of African rain forests, Entrepreneur B undertook to raise awareness about those endemic plants and encourage reforestation to mitigate climate change. Since the immediate forest community was dependent on the forest, they were part of the deforestation problem. With a lot of persuasion, she managed to convince them to participate in her company's mission. She proceeded by training them to set up a seed bank of the endemic plants for exports, thus reducing their dependence on the forest to earn a living. With the help of her team and tribal members, they could gradually bring a positive change. About 2500 farmers with previously no alternative livelihood options are now benefitting from the project. Sustainably harvested seeds are also used to reforest degraded land, helping to ensure a future income. In order to conquer new markets and offer a wider variety of endemic plants, she constantly has to search for new species. Through her initiatives, more than 300 million plants from the "Seeds of Hope" are now growing in more than 45 countries.

Through their ventures, the women entrepreneurs have demonstrated that the core concepts of sustainability (people, planet and profit) can be a positive driver of innovation and socio-economic change. Although for any business to prosper, profitability is a sine qua non condition, they never diverted from the principle of keeping people at the center of business decision-making, thereby building mutual trust and understanding. They invite others to replicate their methods.

7 Empathy, Mutual Trust and Understanding

Entrepreneur A had worked in the textile sector for many years. She could understand the jobless mothers' difficulties to find employment as they had not developed other skills besides the routine tasks of the factory. Moreover, her contact with the children nurtured sympathy for the mothers whose family lives were being shattered due to financial limitations. Entrepreneur B, as a young lady visiting the forests with her father observed the conditions in which the bushmen were living. Her wish to create awareness about the plants and their commercialisation was closely linked to empowering the tribal people.

Among women entrepreneurs, including the two studied, it may be suggested that they tend to be closer to their employees and to be more understanding and compassionate than men. They think that a woman is more patient and despite difficulties will struggle to continue. Entrepreneur A reported that all the women work wonderfully as a team and there has never been any conflict. During her various expeditions in search of plants across the globe, Entrepreneur B has to set up a team in every country that she visits. The team accompanies her in the forest and they may stay there for a period of 6–11 days. Mutual trust in such conditions is vital.

8 Financing

According to Inverso [25], women-owned business leaders highlighted access to capital among the biggest challenges. Entrepreneur B self-financed all her projects while expanding her business gradually according to her means. Entrepreneur A schooling project run with the help of private donors, fund raising, CSR funds and sponsorships [26] and under this umbrella the herbal tea project was launched. The PV project received financial assistance from the UNDP GEF Small Grants Programme [24]. The venture received other sponsorships as well. Entrepreneur A had to plough back profits generated from the herbal tea business so as to maintain the growth momentum.

9 Marketing

Based on the narrative, it is obvious that initially marketing must have been a key challenge for both entrepreneurs. Indeed, they lacked capital to engage in big scale marketing. However, the deep-rooted social and environmental facets of both enterprises seduced the market. Entrepreneur A first participated in local fairs, then in international B2B fairs with partial financial support from the government. The exposure she obtained in these fairs kick-started her export ventures. Packaging was enhanced in order to make the products more attractive for overseas markets. The herbal teas were placed in local supermarkets as well. As success breeds success, the local and international press have publicised her achievements and she has been approached by several organizations to replicate the project and to present her products at fairs. Entrepreneur B observed that a loud noise is necessary to open deaf ears [27]. She recalls that when she started her business in 1994, online tools were not yet available and communication with overseas clients was expensive. But determined to succeed, she wrote numerous letters and faxed potential clients. Hundreds of clients gradually responded to her call. Over time, as her network enlarged, she benefitted from word of mouth marketing which then increased her client base. Afterwards she developed a full-fledged website for her worldwide clientele. Both entrepreneurs have been using social media for marketing purposes.

10 Conclusion

This work has shown that women entrepreneurs have a more pronounced affinity to empathise with other women and to be more in tune with social and environmental causes. Their entrepreneurial spirit and values make them a good example as role models. Their upbringing was a huge determining factor in their success. Irrespective of their financial situation, the values absorbed during their upbringing, played an essential role in shaping their respective personalities and determination to overcome barriers, to achieve their goals and embarking others in their project. They showed impeccable leadership qualities. Furthermore, the ladies imagination and creativity in perceiving the right business approach and work towards its implementation is highly commendable. Their idea of a successful business would not be whole if people suffered or the planet was being depleted of natural capital. While helping the needy ones or integrating other members in their network, the women entrepreneurs have the innate tendency to develop close relationships with those people either through compassion or enthusiasm to fulfill their mission. They believe to move forward as a big family instead of establishing strict corporate hierarchy.

References

1. Hall (2009). Six Principles of Women's High-Engagement Philanthropy. Tactical Philanthropy. 27 May. Retrieved March 2014 from http://www.tacticalphilanthropy.com/2009/05/six-principles-of-womens-high-engagement-philanthropy/
2. Sankey (2013). Why Social Enterprise Attracts Women. Green Futures. Forum for the Future. Retrieved March 2014 from http://www.forumforthefuture.org/sites/default/files/images/GreenFutures/Futuristas/Futuristas_LOWRES.pdf
3. Meyskens M., Allen I. E. & Brush C. G. (2011). Human Capital and Hybrid Ventures, in G.T. Lumpkin, Jerome A. Katz (ed.) Social and Sustainable Entrepreneurship (Advances in Entrepreneurship, Firm Emergence and Growth, Volume 13), Emerald Group Publishing Limited, pp. 51–72.
4. Dees J. G. (2005). Entrepreneurship in Philanthropy. Center for the Advancement of Social Entrepreneurship. Retrieved February 2014 from http://www.caseatduke.org/articles/0805/corner.htm
5. Spark:Her. Investing In Women Social Entrepreneurs. Foundation for Youth Social Entrepreneurship. Hong Kong. Retrieved March 2014 from http://sparkher.org/about-us/investing-in-women-social-entrepreneurs/
6. Small Business Trends (2011). Women Entrepreneur Role Models: Who Would You Name? Business Insider. 29 March. Retrieved March 2014 from http://www.businessinsider.com/women-entrepreneur-role-models-who-would-you-name-2011-3
7. Shiva V. (2012). Vandana Shiva: Everything I Need to Know I Learned in the Forest. Yes! 05 December. Retrieved February 2014 from http://www.yesmagazine.org/issues/what-would-nature-do/vandana-shiva-everything-i-need-to-know-i-learned-in-the-forest
8. Brinker R. (2009). Conference on Earth Democracy: Women, Justice, and Ecology. 23 October. Oregon State University. Retrieved March 2014 from http://womenjusticeecology.wordpress.com/2009/07/04/dr-vandana-shiva-and-feminist-theory/

9. Sommer C. (2013). Who Are the Most Powerful Women Entrepreneurs? Forbes. 19 September. Retrieved April 2014 from http://www.forbes.com/sites/carisommer/2013/09/19/who-are-the-most-powerful-women-entrepreneurs/
10. Abrams R. (2013). Strategies: Women entrepreneurs need a role model. USA TODAY. 29 November. Retrieved March 2014 from http://www.usatoday.com/story/money/columnist/abrams/2013/09/20/small-business-women-role-models/2840875/
11. Bosma N., Hessels J., Schutjens V., van Praag M. & Verheul I. (2011). Knowing entrepreneur role models personally or through media. Discussion Paper, Tinbergen Institute.
12. Klyver, K., Grant, S., & Hindle, K. (2007). The Influence of Social Network Structure on Entrepreneurial Participation: Gender Differences Studied across 47 Countries. Paper. Third Global Entrepreneurship Monitor Research Conference. 01-03 Oct. Washington, D. C., United States. Retrieved March 2014 from http://new.gemconsortium.org/assets/uploads/1326045332Influence_of_Social_Network_Structure_on_Entrepreneurial_Participation.pdf
13. Levie J., Hart M. & Shamsul Karim M. (2010). Impact of media on Entrepreneurial Intentions and Actions. Global Entrepreneurship Monitor. November. Retrieved April 2014 from https://www.gov.uk/government/uploads/system/uploads/attachment_data/file/32243/11-773-impact-of-media-entrepreneurial-intentions-actions.pdf
14. Hang M., & van Weezel A. (2005). Media and entrepreneurship: A survey of the literature relating both concepts. In 18th Scandinavian Academy of Management Meeting, Aarhus School of Business, Denmark (pp. 18–20). August. Retrieved March 2014 from http://www.personal.psu.edu/amh13/e-ship/hangandweezel.pdf
15. Pietilainen T. (2001). Gender and female entrepreneurship in a pro-entrepreneurship magazine. Working Paper 458, Swedish School of Economics and Business Administration. Retrieved March 2014 from https://helda.helsinki.fi/bitstream/handle/10227/156/458-951-555-691-0.pdf?sequence=2
16. Hausmann R., Tyson L. D., Zahidi S. (2012). The Global Gender Gap Report 2012. Insight Report. World Economic Forum. Retrieved March 2014 from http://www3.weforum.org/docs/WEF_GenderGap_Report_2013.pdf
17. Lemmon G. T. (2012). 'Women entrepreneurs: Example not exception'. Ted. January. Retrieved March 2014 from http://www.ted.com/talks/gayle_tzemach_lemmon_women_entrepreneurs_example_not_exception/transcript
18. Brush C. (2013). How Women Entrepreneurs Are Transforming Economies and Communities. Forbes. 16 February Retrieved March 2014 from http://www.forbes.com/sites/babson/2013/02/16/how-women-entrepreneurs-are-transforming-economies-and-communities/
19. Matsui K. (2013). The Economic Benefits of Educating Women. Bloomberg Businessweek. 07 March. Retrieved March 2014 from http://www.businessweek.com/articles/2013-03-07/the-economic-benefits-of-educating-women
20. Kabani S. (2014). How to get more women in tech and business. Forbes. 10 March. Retrieved April 2014 from http://www.forbes.com/sites/shamakabani/2014/03/10/how-to-get-more-women-in-tech-and-business/
21. McCartney C. & Worman D. (2013). Entrepreneurs: What can we learn from them? Research Report, Part 2 in Inspiring female entrepreneurs. August. Chartered Institute of Personnel and Development. Retrieved March 2014 from http://www.cipd.co.uk/binaries/6280%20Entrepreneurial%20report%202%20WEB.pdf
22. Kelley D. J., Brush C. G. & Greene P. G., Litovsky Y. & Global Entrepreneurship Research Association (2013). Global Entrepreneurship Monitor 2012 Women's Report. Global EntrepreneurshipMonitor. Retrieved March 2014 from http://www.gemconsortium.org/docs/download/2825
23. Caprino K. (2013). Ready to Chuck Your Corporate Job? Solopreneurship Is a Growing Movement for Women. Forbes. 26 February. Retrieved April 2014 from http://www.forbes.com/sites/kathycaprino/2013/02/26/ready-to-chuck-your-corporate-job-solopreneurship-is-a-growing-movement-for-women/

24. The GEF Small Grants Programme (2012). SGP Solar Powered Herbal Tea Production. Retrieved March 2014 from https://sgp.undp.org/index.php?option=com_sgpprojects&view=projectdetail&id=19186&Itemid=205
25. Inverso E. (2014). Why Women Entrepreneurs Are More Optimistic Than Ever. Forbes. 12 March. Retrieved April 2014 from http://www.forbes.com/sites/emilyinverso/2014/03/12/why-women-entrepreneurs-are-more-optimistic-than-ever/
26. CIEL Group (2012). APEDED - Association Pour l'Education des Enfants Défavorisés. ACTogether. Retrieved March 2014 from http://www.actogether.mu/education_details.aspx?id=42
27. News on Sunday (2008). Portrait: Social entrepreneurship award: Our girl among the best. News on Sunday. 1–7 February. Retrieved April 2014 from http://www.tropical-vacances-loisirs.com/modules.php?name=News&file=categories&op=newindex&catid=25

Optimization of Law Enforcement on Reclamation Process in Indonesia

Yordan Gunawan, Andika Putra and Mohammad Hazyar Arumbinang

Abstract Indonesia is blessed with abundant natural and energy resources. The natural and energy resources consist of mine, natural gas, and oil which spread out in all of regions of Indonesia. Nowadays, the development of mining activity is more rapid and useful. Yet, most of the companies possess lack of intention to fulfill the liabilities, in terms of arrangement, maintenance, recovery, and also environmental protection. Many victims got the negative impact because of the activities done by the mining companies which did not conduct the liabilities. One of the problems is the post-mining reclamation process. The paper aims to analyze the effectiveness of the administrative sanctions of post-mining reclamation process and the role of criminal sanctions on it. The method used is a normative legal research with descriptive quality. The legal basis of the paper is the Law No. 4 of 2009 on Mining and Coals, Government Regulation No. 78 of 2010 on Reclamation and Post-Mining, and some particular regulation related to the issues. As a result, this paper finds whether or not the administrative sanctions are effective to solve the environmental damage to post-mining through reclamation. The paper also suggests an idea on the enforcement of criminal sanctions for companies which made the environmental damage and/or did not conduct the reclamation process.

Keywords Mining activities · Environmental damages · Reclamation · Administrative and criminal sanctions

1 Introduction

Indonesia continuously becomes a significant player in the global mining industry with significant levels of productions of coal, copper, gold, tin, and nickel. In particular, Indonesia remains among the world's largest exporters of thermal coal. In global mining companies, Indonesia consistently ranks high in terms of coal and

Y. Gunawan (✉) · A. Putra · M.H. Arumbinang
Faculty of Law, Universitas Muhammadiyah Yogyakarta, Yogyakarta, Indonesia
e-mail: yordangunawan@umy.ac.id

T. Taufik et al. (eds.), *ICoSI 2014*, DOI 10.1007/978-981-287-661-4_20

mineral prospects.[1] However, most of the companies have weak intention to fulfill the liabilities, in terms of arrangement, maintenance, recovery, and also environmental protection. Ignoring such conditions will result in the environmental damages including erosion, formation of sinkholes, loss of biodiversity, and contamination of soil, groundwater, and surface water by chemicals from mining processes. Besides creating environmental damage, the contamination resulting from leakage of chemicals also affects the health of the local population.[2]

Hence, to solve the issue, the Government of Indonesia enacted Law No. 4 of 2009 concerning mineral and coal mining and also promulgated Government Regulation No. 78 of 2010 concerning reclamation and post-mining. But both regulations are not able to force the companies which run the mining activities to fulfill the obligation, which reclaims after the end of mining process. In fact, so many lands used as mining area were not reclaimed by the companies. It means that the companies did not fulfill the obligations. Ignoring the reclamation caused severe environmental damages, which affected the human life and surrounding environmental system. This paper seeks to answer whether or not the administrative sanctions may force the mining companies to conduct the reclamation process and how the role of criminal sanctions is in post-mining reclamation process in Indonesia.

2 Methodology

This type of research is a normative legal research with the Indonesian Law approach through the regulation related to the issue of reclamation. The research uses statute approach, because it elaborates some regulations such as Law on Management and Protection of Environment and other regulation relating to the issue of reclamation which can be applied to the environmental case. The paper also used the following case approach:

1. Legal materials
   The research uses material researches taken from the learning literature consisting of primary legal materials, secondary legal materials, and tertiary legal materials;

   a. Primary legal materials include several regulations as follows:

      (a) Law No. 4 of 2009 on Mining and Coals,
      (b) Law No. 40 of 2007 on Limited Liability Company,
      (c) Law No. 32 of 2009 on Management and Environmental Protection,
      (d) Government Regulation No. 78 of 2010 on Reclamation and Post-Mining, and

[1] Sacha et al. [6, p. 7].

[2] Ref. [19].

(e) Local Government Regulation of Tanah Bumbu Regions No. 12 of 2012 on Reclamation and Post-Mining.

b. Secondary materials consist of several documents as follows:

(a) Books related to the issue of reclamation,
(b) Scientific journals,
(c) Seminar and conference paper related to the issue, and
(d) Other related documents.

c. Tertiary legal materials

(a) Black Laws dictionary,
(b) English dictionary, and
(c) Indonesian dictionary.

2. Method of data collection
The method of data collection in this research is done through library research. The data are collected from reading, analyzing, and making conclusion of all the related documents, namely law books, legal journals, and other related data.

3. Data analysis
The data are analyzed systematically through juridical thinking. The term "systematically" means that the research data are analyzed based on Indonesian Law, especially relating to the issue of reclamation.[3]

3 Discussion

3.1 The Effectiveness of Administrative Sanctions

Reclamation is reclaiming the land, that is, restoring land that was once used for mining to make a park, nature sanctuary, or some other beneficially use; in other words, reclamation is the process of returning land to its originals or in better condition after the mining is completed.[4] Basically, the Indonesian government has prevented the companies which run the mining activities but did not conduct the reclamation process. According to article 50 (2) of Government Regulation on Reclamation and Post-Mining, there is an administrative sanctions that shall be given to the company which has done the mining activity but does not conduct reclamation process.[5] Article 50 (2) stated the administrative sanctions referred to

[3]Ibrahim [3, pp. 301–302].
[4]Ref. [15, in Article 1, Paragraph (1)].
[5]Ref. [14].

paragraph (1) may include the following: a written warning, temporary cessation of activities, and/or revocation of IUP, IUPK, or IPR.[6]

With the Government Regulation on Reclamation and Post-Mining that has been enacted by the government, hopefully, the mining companies are encouraged to conduct the reclamation process. However, article 50 (3) Government Regulation No. 78 of 2010 stated "holders of *IUP, IUPK, or IPR imposed to administrative sanction in the form of revocation of IUP, IUPK, IPR as the meant in paragraph letter c, are not eliminated as of their obligation to perform reclamation and post-mining activities.*"[7] This regulation confirms that the reclamation is obligation for the companies, and there are no reasons to neglect it. On the other hand, paragraph 4 in article 50 stated "the administrative sanction refers to paragraph (1) will be imposed by the minister, governor, or region/mayor in accordance with their authority."[8]

It means that there is still a bargaining position that will happen between the companies, which are the main actors of mining activities, and government as policy makers, to impose administrative sanctions for companies that did not conduct the reclamation process as stated in the Government Regulation.

Article 51 stated "(1) further provision on the procedure for the imposing of administrative sanctions to holders of IUP and IUPK as meant in article 50 shall be regulated by Minister Regulation. (2) Further provisions on procedure for the imposing of administrative sanctions to IPR holders as meant article 50 shall be regulated with regency/city regional regulation."[9]

Ali Masykur Musa stated that the sanctions, especially administrative sanctions on the Government Regulation on Reclamation Process, are very weak.[10] On the other hand, Minister Regulation on Post-Mining and Reclamation in article 69 stated "administrative sanctions in form of a written warning referred to article 68 (1), letter a given at most three (3) times with the 30 (thirty) days period and in article 70 (1) in the case of holders of IUP Exploration, IUP Production Operation, IUPK Exploration, IUPK Production Operations, after the imposition of the sanction refer to this article, has not been carrying out their liability, imposed by temporary cessation of a part or all mining activities referred to article 68 (1) letter b. And in paragraph (2), the administrative sanctions in the form of temporary cessation of a part or all mining activities referred to paragraph (1), not more than 90 (ninety) days".[11]

[6]Mining Permit, referred to Mining Business License (IUP) and Special Mining Business License (IUPK), People Mining License (IPR) holders of Production Operation Mining Business Licenses (Production Operation IUP) and Production Operation Special Mining Business Licenses (Production Operation IUPK).

[7]Ibid.

[8]Ibid.

[9]Ibid.

[10]Musa [5].

[11]Minister Regulation of Energy and Mineral Resources 2014, Op. Cit.

Then in chapter 71 administrative sanctions in form of revocation of IUP, IUPK as referred to article 68 (1) letter c imposed to the holders of IUP, IUPK which did not conduct reclamation process until the end of the imposement of temporary cessation of a part or all mining activities referred to article 70.[12] The regulation also has no power to encourage the companies to conduct the reclamation process as the liability or the responsibility to protect the environment.

Because when the companies did not conduct the reclamation process, it results in the environmental damages including erosion, formation of sinkholes, loss of biodiversity, and contamination of soil, groundwater and surface water by chemicals from mining processes, and also, health and well-being of the people living in and around the mining complexes get affected due to the pollutants in the air, water, noise, and vibrations. In fact, the society in the mining complexes has to bear the various costs of abating the affects of environmental pollutions in various ways. The people working in the mines and associated facilities also get affected by the work place environment, which can cause various problems, e.g., skin problems, lung diseases, and deafening.[13]

The issue needs the role of government and the law to enforce criminal sanctions. Because it is associated with a company that runs their business or activities on the natural resources, it means that the activities are parts of environmental law issues. Therefore, the company is one of the subjects of criminal sanctions which refer to the Law No. 32 of 2009 on the Protection and Management of the Environment.

In this case, the reclamation and post-mining is a liability that has to be fulfilled by the mining companies at the end of the activities, because reclamation is the process of returning land to its originals or in better condition after mining is completed. This effort tries to prevent the environmental damage and as a part of sustainable development.

3.2 The Criminal Enforcement on Post-mining Reclamation Process

In facts, there are many mining companies that did not conduct post-mining and reclamation activities correctly and precisely or even had not done it at all. It creates pollution and environmental damage on a massive scale in that area.[14] It is because of the weakness of Government Regulation No. 78 of Reclamation and Post-Mining and Law No. 4 of 2009 on Mineral and Coal that did not state criminal provisions on companies which did not conduct the reclamation process as their liability.

---

[12]Ibid.

[13]Saxena [10, p. 4].

[14]Mahadli [4].

In terms of environmental protection, criminal law has always had a place in regime of all jurisdictions in Indonesia, as well as civil law and international law.[15] Therefore, when the administrative sanctions are not effective to force the mining companies to conduct the reclamation as their liability, the state needs the existence of legal framework to force the companies and give the deterrent effect for companies that did not conduct reclamation. Criminal sanctions are the best solution, because the purposes of criminal law are to prevent wrongdoings and punish wrongdoers.[16]

Sanction, according to Van Wijk, is a tool of public power (*publiekrechtelijke machts-middelen*) which is used by the government in response to the violation of administrative law.[17] Correspondingly, P de Han describes a similar point of view that the administrative law is often interpreted as the application of administrative sanctions. The characteristic of administrative sanctions is coercion.[18]

Criminal sanctions in protection of environmental issues are not new idea. It is also regulated on the Law No. 32 of 2009 on the Protection and Management of the Environment. The company can be subjected to criminal sanctions which refer to the law on the Protection and Management of the Environment. Paragraphs in article 54 state:

1. Everybody polluting and/or damaging the environment shall be obliged to restore the environmental function.
2. The environmental function as referred to paragraph (1) shall be restored by phases as follows:
   (a) Discontinuation of source of pollution and cleaning of pollutant;
   (b) Remedy;
   (c) Rehabilitation;
   (d) Restoration; and/or
   (e) Other methods in accordance with the developments of science and technology.

In this case, reclamation is the rehabilitation of land after the coal mining operations have stopped, and reclamation is one of the liabilities of the companies. When the companies did not conduct the reclamation process and cause the environmental damage and pollution, then criminal sanctions can be imposed based on Law No. 32 of 2009 in Chapter "A Cluster Model for Increasing Performance of Small and Medium-Scale Enterprises (A Case Study in Bogor, Indonesia)" of the penal provisions of article 98:[19]

---

[15]Silalahi [9, p. 3].

[16]Samaha [8, p. 3].

[17]Hadjon [1, paper was presented at the Faculty of Law Universitas Islam Indonesia, Yogyakarta, p. 1].

[18]Saleng [7, p. 175].

[19]Ref. [11].

1. Anybody intentionally committing action causing standard quality of ambient air, water, and sea water or standard criteria for environmental damage to be surpassed shall be subjected to the imprisonment for 3 (three) years at the minimum and 10 (ten) years at the maximum, and a fine amounting to IDR 3,000,000,000 (three billion rupiahs) at the minimum and IDR 10,000,000,000 (ten billion rupiahs) at the maximum;
2. In the case of the action as referred to in paragraph (1) wounding people and/or endangering human health, the said person shall be subjected to the imprisonment for 4 (four) years at the minimum and 12 (twelve) years at the maximum, and a fine amounting to IDR 4,000,000,000 (four billion rupiahs) at the minimum and IDR 12,000,000,000 (twelve billion rupiahs) at the maximum; and
3. In the case of the action as referred to in paragraph (1) causing serious injury or death to people, the said person shall be subject to imprisonment for 5 (five) years at the minimum and 15 (fifteen) years at the maximum, and a fine amounting to IDR 5,000,000,000 (five billion rupiahs) at the minimum and IDR 15,000,000,000 (fifteen billion rupiahs) at the maximum.

Furthermore, Law No. 40 of 2007 on Limited Liability Company at the article 74 paragraph (1) stated "a corporation operating business activity in the field and/or related with natural resources shall be obliged to implement social and environmental responsibilities" and at the paragraph (3) stated "a corporation that does not implement its obligation as referred to in paragraph (1) shall be imposed with sanctions in accordance with prevailing laws and regulation."[20]

Mining is an important industry for the economic development, but operation of the mines may harm surrounding environment as well as the population. Therefore, those laws aim at balancing between the mining activities and the sustainable development in terms of protecting the environment. Law No. 40 of 2007 has given the provisions of article 74 that the company has social and environmental liability. When the companies did not conduct their liability, it can be punished with the criminal sanctions referring to Law No. 32 of 2009 on Protection and Management of the Environment.

It is in accordance with the theory proposed by Hadjon that "sanctions is repressive actions to force the obedience."[21] Although the criminal provisions of environmental laws can be used in the reclamation process, in fact, the criminal provisions were never used. Therefore, it results in the environmental damage because the companies did not fulfill their liability. In this issue, not only the government's role needs to enforce the law, but also the role of the legal enforcement and society needs to supervise and guarantee the law enforcement toward the environmental protection, because supervision also supports the law enforcement (*handhaving*) toward all the regulations.[22]

---

[20]Ref. [13].

[21]Hadjon [1, Op. Cit, p. 174].

[22]Hamzah [2, p. 61].

As we know, not only law legal basis, but also local government regulation reduces environmental damage caused by the mining activities. i.e., Local Government Regulation (*Perda*) Tanah Bumbu Region No. 12 of 2012 on Reclamation and Post-Mining, as stipulated in the provisions in article 52 (1) stated "anybody or agency which is not implement the provisions in this regulation shall be punished with maximum 6 (six) months imprisonment or a fine of IDR 50,000,000 (fifty million rupiahs)."[23]

From these regulations we may conclude that the regulation enacted because no national law that may used as an umbrella act to punish the companies that did not conduct post-mining reclamation process. While few regions have promulgated their own regulations relating to reclamation and criminal provisions, they have a similar problem as like as Tanah Bumbu Region. Therefore, for uniformity law, new legal provisions are needed which are able to harmonize the criminal provisions and the company that did not conduct the reclamation process.

It is clear that Law No. 4 of 2009 on Mineral and Coal Mining and Government Regulation No. 78 of 2010 on Reclamation and Post-Mining did not mention the provision of criminal law when the mining companies did not conduct their liability and created environmental damage. Yet referring to the Law No. 40 of 2007 on the Limited Liability Company and Law No. 32 of 2009 on the Management and Environmental Protection, the criminal sanctions may be imposed to the companies itself.

4 Conclusion and Recommendation

4.1 Conclusion

Based on the discussion above, it may be concluded as follows:

(a) Administrative sanctions are not effective to force the mining companies to conduct their liability in terms of reclamation process.
(b) Government Regulation No. 78 of 2010 on Reclamation and Post-Mining as well as Law No. 4 of 2009 on Mineral and Coals do not formulate the penal provisions against companies that ignored the reclamation; moreover, the enforcement of the criminal law can be still applied by referring to the Law on Limited Liability Company and Law on the Management and Environmental Protection, which mention penal provisions to the companies that did not conduct the reclamation process.

[23]Ref. [16].

4.2 *Recommendation*

With regard to the issue of criminal punishment, the author may suggest two options as follows:

(a) Imposing the effectiveness of criminal law enforcement against companies that did not conduct the reclamation process, by using the provision referring to the Law No. 40 of 2007 on Limited Liability Company and Law No. 32 of 2009 on the Management and Environmental Protection, or
(b) Amending or creating a new law which provides criminal sanctions to the actor or companies who do not conduct the reclamation process and make the environmental damage in the end of mining activities.

References

Books and Journals

1. Hadjon, Philiphus M, 1995, *Penegakan Hukum Administrasi dalam Pengelolaan Lingkungan*, Yogyakarta, Makalah FH-UII.
2. Hamzah, Andi, 1995, *Penegakan Hukum Lingkungan*, Jakarta, Arikha Media Cipta.
3. Ibrahim, Johnni, 2006, *Teori dan Metode Penelitian Hukum Normatif*, Malang, Bayu Media.
4. Mahadli, Zulkifli, 2013, *Simposium Nasional Pertambangan*, Jakarta, Asosiasi Pemerintah Kabupaten Indonesia (APKASI).
5. Musa, Ali Masykur, 2013, *Keynote Speech at International Seminar and Workshop on Wetlands Environmental Management*, Banjarmasin, Universitas Lambung Mangkurat.
6. Sacha, Winzanried *et al.*, 2012, *Mining in Indonesia Investment and Taxation Guide*, Indonesia, PwC (Price Water House Coopers) Mining Publications.
7. Saleng, Abrar, 2004, *Hukum Pertambangan*, Makassar, UII Press.
8. Samaha, Joel, 1999, *Criminal Law* 6th Edition, United State, Wadsworth Publishing Company.
9. Silalahi, M. Daud, 1996, *Penegakan Hukum Lingkungan di Indonesia Melalui Peningkatan Kesadaran Hukum*, Bandung, Journal of Padjajaran University, Nomor 3 Vol. 14.
10. Saxena, N C, *Centre of Mining Environment*, Dhanbad (Jharkhand), Indian School of Mines.

Legislations

11. Law No. 32 of 2009 on Management and Protection of Environment.
12. Law No. 4 of 2009 on Mining and Coals.
13. Law No. 40 of 2007 on Limited Liabilty Companies.
14. Government Regulation No. 78 of 2010 on Reclamation and Post-Mining.
15. Minister Regulation of Energy and Mineral Resources No. 7 of 2014 on the Implementation of Post-Mining Reclamation and Operations Mineral and Coal Mining.
16. Local Government Regulation of Tanah Bumbu Regions No. 12 of 2012 on Reclamation and Post-Mining

Internet

17. http://www.tempo.co/read/news/2013/09/20/206515340/BPKi-196-Ribu-Hektare-Bekas-Tambang-Menganga, last viewed on April 11th, 2014, at 5.13 pm.
18. http://www.majalahtambang.com/detail_berita.php?category=18&newsnr=8326, last viewed on April 10th 2014, at 10.12 pm.
19. http://ngm.nationalgeographic.com/2009/01/gold/larmer-text/6, last viewed on April 9th 2014, at 9.22 pm.

Planning Livable Community with Social Systems Approaches: Medan, Indonesia

Dwira Nirfalini Aulia

Abstract Livable Community is an ideal condition desired by all people. The perfect condition has variety specification based on the need of the community. Social system approaches the resident used as the approach of planning strategy of a livable community. The social system approaches are community profile, social fragmentation, and social dynamics. The aim of the research is to find the planning concept of the livable community in Indonesia with the approach of the local community system. The components of planning livable community are land use, social and cultural opportunity, economic resources, access and mobility, safety and aesthetic, environmental consideration and cumulative effects. The research is done by survey method in three social housing developed by Perumnas in the city of Medan such as Perumnas Helvetia, Perumnas Martubung and Perumnas Simalingkar. Community residents of Perumnas (public housing in Indonesia) represent the character of the local community in Indonesia. Perumnas is also the form of the urban fabric which built to meet housing needs in Indonesia.

Keywords Planning · Livable community · Social system approaches

1 Introduction

Livable community is an ideal condition desired by all people to the neighborhood in which they live. The perfect condition has various specifications based on the need of community. Planning the good-quality neighborhood when the occupant feels comfort now and the future is the main purpose of the architect. Based on [1], designing livable community required the following: (1) Some criteria are based on residential satisfaction; (2) the scope of design area is neighborhood; and (3) basic criteria of design can be measured by the indicator. Social characteristic of the resident is used as the approach of planning strategy of a livable community.

D.N. Aulia (✉)
Department of Architecture, University of Sumatera Utara, Medan, Indonesia
e-mail: dwira.nirfalini@usu.ac.id; dwira_aulia@yahoo.com

T. Taufik et al. (eds.), *ICoSI 2014*, DOI 10.1007/978-981-287-661-4_21

Research finding will be useful for planning livable community in Indonesia's formal housing for the future. Several previous researches connect the social aspect and the quality of build environment design. Dempsey [1] said that the social interaction, social network, sense of community, sense of place attachment act as a social binder in an urban neighborhood.

The study is conducted with the prediction that the livable community will be formed when effectively accommodated the needs of residents and they involve in the management of the settlement. Residential satisfaction researches have done with the aim to determine the occupant's perception about their neighborhood. Residential satisfaction criteria determine the components of livable community planning. These elements are land use, social and cultural opportunity, economic resources, access and mobility, safety and aesthetic, environmental consideration, and cumulative effects as the criteria for planning livable community. The study was conducted in three social housing in Medan, namely Perumnas Helvetia, Perumnas Simalingkar, and Perumnas Martubung. The structure of the paper consists of background research, literature review, research methodology, data analysis, discussion, and conclusion.

2 Livable Community

The component of a livable community in this research is classified by Deverman [2] into seven components as follows: (1) land use, where the community has diverse land use to meet their needs; (2) social and cultural opportunities; (3) economic resources, there are dynamic of economic resources can be a catalyst for raising the level of cultural, recreational, and artistic opportunities; (4) access and mobility means efficient land use and transportation linkages that give residents access and mobility; (5) safety and aesthetics means safe community, safe neighborhoods, and safe streets; (6) environmental considerations such as good water quality, air quality, and noise; and (7) cumulative effects can affect a variety of human and ecological resources. Livability in transportation refers to leveraging the quality, location, and transportation facilities and services that are available to help achieve broader community goals [3]. The effort to create a livable community has been done all over the world. Stevens [4] compare the effort of planning livable community in USA and UK. In USA, livability embraces a wide range of long-term efforts by increasing environmental sustainability, by increasing densities and land use mix, and by improving transport, whereas in UK, target livability focuses on the management and redevelopment of spaces in existing neighborhood rather than the planning in the new one.

Some studies show an interest in 'place shaping' as a way to create a livable community. Allen and Crookes [5] said there are two aspects to consider when exploring the 'place shaping.' The features are as follows: *epistemic* that means what kind of understanding do planners and resident have of the place in particular and *aesthetic* which means how do planners and resident relate to transforming the look of places. To achieve the goal, livable community should pay greater attention to the social aspects of housing occupants.

Livable community planning criteria translated to several indicators. Indicators of changes in the population of residents, number of residents who had a home first, and change in housing supply policies are used to measure livable community [6]. Livable community indicators were also used by Christian [7] in designing housing projects. These indicators refer to livable neighborhood guidelines interpreted from New Urbanism movement. The indicators are as follows: (1) a community design, (2) a movement network, (3) public parklands, and (4) lot layouts. Residential satisfaction criteria as finding by Aulia et al. [8] are physical features (location, housing facility, and housing typology) and nonphysical aspects (social interaction, security, housing estate management, and legal status) into consideration in determining the criteria of a livable community.

3 Components of Livable Community

3.1 Land Use

Planning a livable community is about the planning and development control. In other words, drawing up land use plans and policing development [9]. Land use planning can produce or inhibit the opportunity, for the formation of a livable community depends on the circumstances. Land use planning requires linkages with the aspirations of the community regarding their needs in the use of urban space.

3.2 Social and Cultural Opportunities

Communities can be classified into different types [10]. There are four classifications of community types, namely first is the community of elected official whose position of power makes they are important and unique, second is the community of engaged publics who are made by the principle of participatory, third is communities of cultural differences, and fourth is professional communities. Each community possessed different criteria. Understanding the community needs as a way to planning a livable community.

3.3 Economic Resources

In preparing the scenario of comprehensive urban planning, officials and residents have the objective economic vitality as one of their top goals for achieving high quality of urban life. The dynamic of economic resources always becomes a catalyst to upgrade the level of cultural, recreational, and artistic opportunities of the community offers [2]. There is a direct relationship between successful economic conditions and the perceived level of livability and quality of life.

3.4 Access and Mobility

Access and mobility in the neighborhood connects with the transportation system, pattern of road users, and residential movement pattern. Access can be classified as access with street structure (street types such as the speed limit and design layout) and social structure (such as age, gender, nationality, social status, and households) [11]. These components are affecting the access and mobility in the neighborhood. One method used by Bell et al. [12] is organizing the crossroad of neighborhood's main corridor to planning livable community. Walkability of the occupants of the neighborhood can be increased by measuring walking behavior of the occupant firstly [13].

3.5 Safety and Aesthetics

Globalization and urbanization have changed the social structures of the city. The changes parallel to the rise of crime rate. The role of public participation should be strengthened to achieve safe and livable cities. Like the traditional way to create attachment and sense of belonging, public participation is one way to create safe and secure neighborhood [14].

3.6 Environmental Consideration

Environmental damage is getting worst if did not doing the effort to improve environmental condition. Air pollution and greenhouse gas emissions are causing environmental damage. Lam and Mullen [15] doing the strategy to improve environmental condition by (1) encouraging environmentally sustainable medium to high-density residential and mixed-use development and (2) facilitating government regulation that support increased environmentally sustainable development.

3.7 Cumulative Effects

Cumulative effects are the impacts upon the environment from a proposed action [2]. Cumulative effects can affect variety of human and ecological resources.

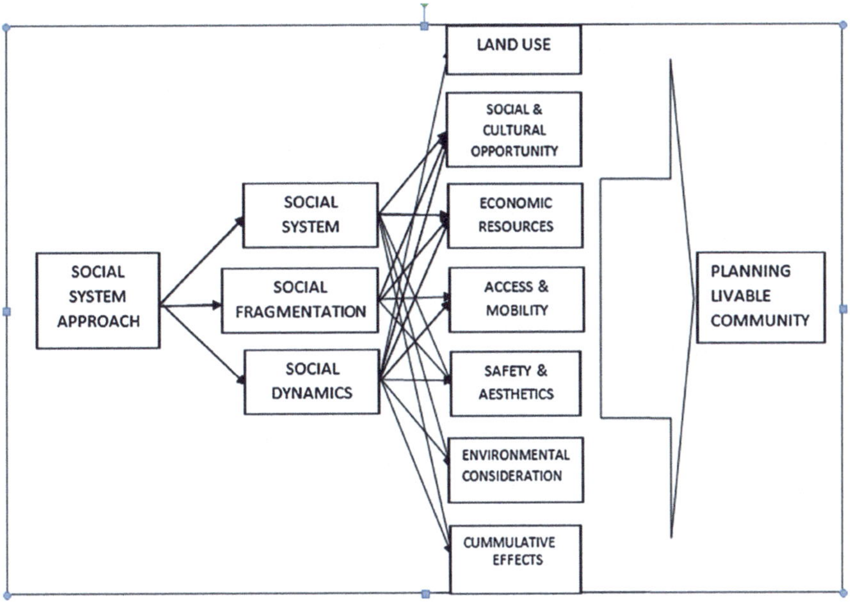

Fig. 1 Theoretical framework

According to Dizdaroglu et al. [16], there are six main targets to make urban ecosystem more sustainable which are as follows: (1) establishing a hydrological conservation, (2) providing ecological conservation, (3) improving environmental quality, (4) creating sustainable mobility, (5) sustainable design of urban environment, and (6) using a renewable resources. Theoretical framework is shown in Fig. 1.

4 Research Methodology

Collecting data method is done by questionnaire distribution, field observation, and photograph documentation. Research conducted is a descriptive study. The research is done by survey method in three social housing developed by Perumnas in the city of Medan such as Perumnas Helvetia, Perumnas Martubung, and Perumnas Simalingkar. Residential community who live in Perumnas housing represent the image of the low-income community. The spread of the questionnaire carried out at random sample to get a picture of the characteristics of Perumnas community occupants. The number of respondents from the three community occupants is 388 respondents.

To get an idea of the community characteristics of the Perumnas occupants, the tabulations related questionnaire about the background of the occupants. Socioeconomic condition of residents can be grouped into two portions: (1) profile of residents described by age, sex, marital status, and number of home contents [17]; and (2) the condition of social fragmentation is described by income, occupation, and education the head of household [18]. Following [19], the characteristics of the occupants will affect the social dynamics that will occur in the neighborhood. There are three aspects that affect the social dynamics as follows: (1) homeownership status, (2) social identities in residential areas, and (3) balance between incoming residents moved into the area and residents who came out of the residential area.

5 Data Analysis

Social system of residential community is a complex system of interdependence between the parts, components, and processes surrounding the rules of relationship between the residential community. Residential community is a set of people who lived in a residential location with a variety of status and role in an organization of society. By understanding the social system within a residential community, the role and involvement of each member of the community in achieving a livable community can be planned. This approach is used in planning livable community.

Community profile describes that the community with young families and middle-age group is in productive age. This community is still possible to evolve by added family members and the improvement of economic conditions. In the human life cycle stages, this community still allows more and the improved quality of life. This condition causes that the community is dynamic and changing rapidly [20]. This condition describes the characteristic of modern society community. This community requires the following: the ability and the right of people to develop choices, recognition as individuals than as members of society, and social structure is open and voluntary.

Social fragmentation in these communities illustrates the low- to moderate-income groups with different composition. This community represents a heterogeneous group of people from the type of work, education, and income. This condition indicates that the community social capital can be used optimally. Community social capital is [21] the potential of groups and patterns of relationships among the individuals within a group and between groups with the chamber's attention on social networks, norms, values, and beliefs within the group. High social capital will tend to be more efficiently and effectively carried out various policies for the welfare and advanced the group. A group of people who have high social capital will open up the possibility of completing the complexity of the problem more easily.

Social dynamics in these communities illustrate that the mobility rate is high enough. It can be seen that the majority of residents had inhabited the area for 1–5 years although the age of housing is already more than 10 years. Although the mobility of residents is quite high, the majority of residents are homeowners. This condition describes that the transfer of homeownership is common. This condition describes a dynamic community, and the mobility of residents who moved in and out of this community is quite high. Perumnas as a public housing provided by government for low- to moderate-income communities often become a temporary home in accordance with the family's ability at a certain life cycle stages. This is also confirmed by Winstanley et al. [22] which conducted an evaluation of public housing in Venezuela. It was found that the respondents choose public housing because the price is cheaper than the other and because of the concept design 'Core House' allows the development of the house when necessary according to the stage of the life cycle. Result survey is shown in Table 1.

6 Result and Discussion

Discussion of the result is conducted by connecting the social aspect with the components of livable community. The analysis is shown in Table 2.

Social system in young families and middle-age group describes the dynamic social structure. In this stage of human life cycle, the community is still possible to grow and to increase the quality of life. The high mobility of residents will be created the basis of social unity based on the togetherness in the neighborhood. This is will facilitate community empowerment in planning livable community. The arrangement of status and the role of the community member in the social system of Perumnas's resident must be clear so that management system of the livable community running well. This community needs the ability and the right of people to develop options, acknowledgment as individual than as a member of society, and an open social structure. The dynamic of potential economic development is more possible because the group of productive age will produce the economic productive. This community requires livable street planning at the component of access and mobility. According to Park [23], livable streets are the streets for everyone that has been planned, designed and operated which enable safe access for all users. Safety and aesthetic components are rather difficult to implement because of the high mobility of occupants. Creating strong attachment and sense of belonging can be done when the community has an obvious social system. The community has neighborhood leader that respected and honored by the residents. Likewise the components of environmental consideration and cumulative aspects, the obvious structure of the social system will form the community that consistently maintain the environmental quality. All the residents better have the same perception about the importance of maintaining the environment quality. It will require continuous socialization about livable community.

If social fragmentation of the community is middle and low income, livelihoods are civil government employee and private employee. The component of social and cultural opportunity and economic resources has potential to develop in planning livable community because of the social interaction more closely. Sense of brotherhood in improving the socioeconomic conditions makes the interaction more closely and, similarly, the possibility of cooperation in opening more business

Table 1 Result

Household age	Helvetia	Martubung	Simalingkar	Total	Percentation
Profile					
21–30 years	7	7	7	21	5.41
31–40 years	56	48	35	139	35.82
41–50 years	50	40	30	120	30.93
51–60 years	31	28	21	80	20.62
61–70 years	11	10	7	28	7.22
Gender					
Male	132	101	60	293	75.52
Female	23	32	40	95	24.48
Household size					
2 peoples	0	0	4	4	1.03
3 peoples	19	9	13	41	10.57
4 peoples	41	45	30	116	29.90
5 peoples	52	54	26	132	34.02
6 peoples	39	24	22	85	21.91
>6 peoples	4	1	5	10	2.58
Marital status					
Not married	7	7	4	18	4.64
Married	148	126	92	366	94.33
Widower	0	0	1	1	0.26
Widow	0	0	3	3	0.77
Social fragmentation					
Occupation					
Civil government employee	28	24	17	69	17.78
Private employee	53	43	41	137	35.31
Businessman	68	51	31	150	38.66
Retired	5	11	8	24	6.19
Unemployment	1	4	3	8	2.06
Education					
Primary school	0	0	0	0	0.00
High school	72	9	4	85	21.91
Bachelor	46	91	10	147	37.89
Graduate	28	33	66	127	32.73
Postgraduate	9	0	18	27	6.96
Income					
<Rp. 500.000	0	0	0	0	0.00
Rp. 500.000 –Rp. 750.000	0	0	0	0	0.00
Rp. 751.000–Rp. 1.000.000	37	0	1	38	9.79
Rp. 1.001.000–Rp. 2.000.000	53	29	15	97	25.00
Rp. 2.001.000–Rp 2.500.000	43	104	65	212	54.64

(continued)

Table 1 (continued)

Household age	Helvetia	Martubung	Simalingkar	Total	Percentation
>Rp. 2.500.000	0	0	17	17	4.38
Social dynamic					
Ownership					
Owner	122	97	74	293	75.52
Tenant	33	36	24	93	23.97
Official	0	0	1	1	0.26
Other	0	0	0	0	0.00
Length of stay					
<1 year	10	12	11	33	8.51
1–5 years	87	76	58	221	56.96
5–10 years	45	28	20	93	23.97
>10 years	13	17	11	41	10.57

Table 2 Discussion of analysis data

Components of livable community	Social system	Social fragmentation	Social dynamic
Land use	–	–	Social potential
Social and cultural opportunities	Social potential	Social potential	Social potential
Economic resources	Social potential	Social potential	Social potential
Access and mobility	Livable street	Good public transportation	–
Safety and aesthetics	–	Social potential	–
Environmental consideration	Social potential	–	Socialization of livable community
Cumulative effects	Social potential	–	Socialization of livable community

opportunity. However, it may also occur the possibility of competition in the effort of economic improvement. Planning livable community will open the opportunity of the development of some micro-businesses. The business opportunity will be functioned as the urban catalyst in developing the neighborhood. The component of access and mobility requires attention to adequate public transportation planning. It should be maintained that they do not switch utilizing private vehicle in mobility everyday because, based on the income level, using public transport will help the household allocate the expenditure for transportation cost. Planning the component of safety and aesthetic should be conducted by '*gotong royong*' like the implementation of '*siskamling*.'

Social dynamic is quiet high in the research area and caused that the changing of land use also is dynamic. The planning has to anticipate the dynamic of land use change. Development control should be done in a short term because of the changing. Majority Perumnas's resident is the community of engaged publics and communities of cultural. The cluster of ethnic and religious is observable in neighborhood. The condition will affect the planning of component social and cultural opportunity. The diversity of ethnic and cultural has to become a social potential in planning livable community. The component of economic resources will grow because majority Perumnas's resident is the owner. Sense of belonging in the neighborhood is quite high because the ownership status is obvious. The enhancement of economic condition will improve the sense of belonging that will stabilize livable community. The component of access and mobility is not affected the social dynamic. The determination of Perumnas location has considered accessibility factor. Social dynamic changes happened based on the improvement of living standard the occupant. Based on the study of Dahl and Sorenson [24] that the economic factor is more affecting the mobility when it is compared with social factor in neighborhood attachment, living in Perumnas still be a springboard before they can afford to buy a better quality of neighborhood. Planning the component of safety and aesthetic will be difficult because of the high of the social dynamic. The changing of occupant frequently will affect neighborhood attachment. The neighborhood leader has to anticipate the changing of occupant by enforcing required to report for the new resident. The environmental consideration and cumulative affects' components would be difficult to implement because of the high of the social dynamic. The strategy of continuous socialization should be implemented to maintain the environment sustainability.

7 Conclusion

Assessment of the community and identification of social potential is indispensable for planning livable community. Social system, social fragmentation, and social dynamic Perumnas's resident are used to approach to planning livable community. If the planning of the livable community is successfully applied in the neighborhood, then the same method can be used in another neighborhood. So it can be achieved the planning of sustainable housing. With these limitations and conditions, commiting to create sustainable housing will be a good step ahead, at the very least, to achieve environment sustainability for the next generation.

References

1. Dempsey, Nicola (2009) Are good-quality environments socially cohesive? Measuring quality and cohesion in urban neighborhoods, *TPR*, 80 (3), pp 315–345.

2. Deverman, Ron (2007) Perspective on solid ground: Preserving the quality of space, *Environmental Practice,* 9(1), 1–4.
3. Kenneth, Patty (2011) An introduction to the Livability in Transportation Guidebook, *Institute of Transportation Engineers, ITE Journal* 81,4, pp 59–61.
4. Stevens, Quentin (2009), "Broken public spaces in theory and in practice, *TPR* 80 (4–5), pp. 371–391.
5. Allen, Chris and Crookes, Lee (2009) Fables of the Reconstruction. A phenomenology of 'place shaping' in the North of England. *TPR* 80 (4–5), pp 455–480.
6. Wong, Cecilia and Watkins, Craig (2009), Conceptualize spatial planning outcomes. Towards an integrative measurement framework. *TPR* 80 (4–5), pp. 481–509.
7. Christian, Hayley; Knulman, Matthew; Bull, Fiona; Timperio, Anna; et all (2013) A New Urban Planning Code's Impact on walking; The Residential Environment Project, *American Journal of Public Health,* pp 1–11.
8. Aulia, Dwira Nirfalini and Ismail, Abdul Majid (2013) Residential Satisfaction of Middle Income Population: Medan City, *Procedia Social and Behavioral Sciences* 105, pp. 674–683.
9. Gallent, Nick and Wong, Cecilia (2009) Introduction Place shaping, spatial planning and livability, *TPR* 80 (4–5) pp. 253–258.
10. Holden, Meg (2009) Community interest and indicator system success, *Social Indicator Research,* 92, pp. 429–448.
11. Sauter, Daniel and Huettenmoser, Marco (2008) Livable streets and social inclusion, *Urban Design International,* 13 pp 67–79.
12. Bell, Stanford Garson; Johns; Anthony David (2006) A Strategic approach to developing Livable and sustainable Arterial Corridors in Auckland City, *Institute of Transportation Engineers Journal* 76.7 pp 30–37.
13. Ewing, Reid; Handy, Susan (2009) Measuring the Unmeasurable: Urban Design Qualities related to Walkability, *Journal of Urban Design* 14(1) pp. 65–84.
14. Dola, Kamariah and Mohd Noor, Khairul Baharein (2012) Managing Towards Sustainable City: Public Participation for safety and security, *British Journal of Arts and Social Science,* 4 (1) pp. 116–122.
15. Lam, Alven H and Mullen, Brianne M (2012) Comparative analysis of Best Practices of Sustainable Communities: Adelaide, Australia Case Study, *Cityscape: A Journal of Policy Department and Research,* 14(3) pp. 235–242.
16. Dizdaroglu, Didem; Yigitcanlar, Tan; Dawes, Les (2012) A micro level indexing model for assessing urban ecosystem sustainability, *Smart and sustainable Built Environment* 1 (3) pp 291–315.
17. Zhang, Xia, (2008) Social Diversity in the local Residential Environment, A comparative socio economic profile of Dalhousie community.
18. Van Kempen, R., Schutjens, A.J.M., Van Weesep, J (2000), Housing and Social fragmentation in the Netherlands. *Housing Studies,* 15(4), 505–531.
19. Morrison, Nicola, Neighbourhoods and Social Cohesion: Experiences from Europe, *International Planning Studies,* Vol. 8, No. 2, 2003: pp. 115–138.
20. Kartasasmita, Ginanjar, *Karakteristik dan struktur masyarakat Indonesia Modern,* disampaikan pada ujisahih Penyusunan Konsep GBHN 1998, Jogyakarta, 29 Juni 1997.
21. Witrianto, (2007) *Modal sosial dan Pembangunan Manusia Melayu: Kasus Indonesia dan Malaysia*, Universitas Andalas, Padang.
22. Winstanley, Ann; Thorns, David C & Perkins, Harvey C, Moving House (2002) Creating Home: Exploring Residential Mobility, *Housing Studies,* 17(6) pp. 813–832.
23. Park, Michael K (2013) Livable streets: Lee's Summit (Part 1 of 2), *Institute of Transportation Engineers,* 63, 11 pp 35– 40.
24. Dahl, Mikhael S and Sorenson, Olav (2010) The Social Attachment to Place, *Social Forces,* 89 (2) pp. 633–658.

Poverty, Its Measures and Determinants: Evidence Across Provinces in Indonesia

Masyhudi Muqorobin, Venia Prissi Ramadhani and Agus Tri Basuki

Abstract This paper attempts to examine the measurement concepts and problems of poverty and provides empirical evidence, by analyzing the influence of population, gross domestic regional product, share of agriculture, share of industry, and mean years of schooling to poverty in the case of 33 provinces of Indonesia period 2006–2012 by using random effect model. The empirical evidence suggests that population is positively and significantly influence number of poverty in Indonesia, while gross domestic regional product, share of industry, and mean years of schooling are negatively and significantly influence the poverty in Indonesia. In addition, this study also finds that share of agriculture is having no significant influence to poverty reduction in Indonesia. This finding is acceptable as agricultural productivity in Indonesia remains low so that it is not capable enough to be an adequate source of income for the society.

Keywords Poverty · Gross domestic regional product · Share of agriculture · Share of industry and mean years of schooling

1 Introduction

Poverty is among the debate-provoking issues in economics. Some economists argue that poverty is no longer a serious problem [1], though, some others such as George and Howards [2] suggest, it deserves the highest priority. As a matter of fact, poverty (and the poor as well) has been defined and redefined over the years, including its measurement concepts [3] The paper attempts to examine the measurement concepts of poverty and provides an empirical evidences in Indonesia.

It is thought that measurement of poverty becomes an endless issue demanding for further investigation. Nevertheless, a study on measurement is meaningless without discussing definitions of the subject in question, in order to get clear-cut

M. Muqorobin (✉) · V.P. Ramadhani · A.T. Basuki
Universitas Muhammadiyah Yogyakarta, Yogyakarta, Indonesia
e-mail: masmubin@yahoo.com

T. Taufik et al. (eds.), *ICoSI 2014*, DOI 10.1007/978-981-287-661-4_22

meanings (not necessarily clear-cut definitions) of such a subject, because of their close relationship. Consequently, before examining the concepts of measurement, definitions and meanings of poverty from both perspectives deserve an adequate delineation.

Poverty is usually refers to a condition where people live below certain income threshold. Hence, many economists considered economic growth as one way to eliminate poverty. However, there is an interesting fact as reported by the World Bank on World Development Report 2003, mentioning countries with quite impressive economic growth and even take place consistently in the past two decades. However, the economic growths of those countries are not necessarily reducing poverty.

In Indonesia, economy continues to grow steadily despite the global economic downturn in recent years. Steady economic growth considered has led to a gradual reduction in overall poverty in Indonesia. A study by World Growth (2004) showed that in case of Indonesia, agricultural GDP growth is beneficial in reducing poverty, particularly in rural areas. In rural areas, agriculture contributed to almost half of rural household income (wages and farming income). In addition, the next study by World Growth (2009) noted that expansion of industry in Indonesia has been a significant source of poverty alleviation through farm cultivation and downstream processing.

However, an Indonesia overview by World Bank indicates that out of a population of 234 million, more than 32 million Indonesians currently live below the poverty line and approximately half of all households remain clustered around the national poverty line set at 200,262 rupiahs per month ($22). Further, International Fund for Agricultural Development (IFAD) indicates that 16.6% of rural people are poor compared with 9.9% of urban populations.

Actually, national wealth of a country is considered highly correlated with the education of its population in which human capital is one of the most influential factors in economic growth. Hence, countries with a highly educated work force considered able to achieve higher economic growth rates and at the same time wealthy countries have the financial resources to invest more in education. According to data by BPS-Statistics Indonesia, the number of labor force in Indonesia that formally nine years educated is only reached about fifty percent.

2 Definition and Meaning

Definition of poverty should include the answers to three questions of how a person's requirements should be satisfied and of the quantity as well as the quality of such requirements [2]. In terms of these three-dimensional requirements, poverty may be defined to have four varying depths of want: *starvation, subsistence, social coping,* and *social participation.*

Starvation is "the characteristic of some people not *having* enough food to eat," [4] or not having enough money or enough resources to obtain the amount of food required to avoid undernourishment [2]. This definition may fall under Sen's [4: 11–14] category of *biological approach*. This term is usually used to represent the very low level of societies in terms of economic conditions, such as in the earlier stage of industrialization of the advanced industrial economies, and currently in the third world economies, under George and Howards's [2] definition. However, Sen's biological approach also includes other types of necessities such as clothing, housing, developed from a famous study of poverty in York by Rowntree[1] and thus may also cover *subsistence* definition of George and Howards [2]. It is clear, because Sen [4] perceived that starvation is not merely dependent on food supply but also on distribution. He on the other hand maintained that subsistence–level of poverty definition provides the irreducible core of *absolute* deprivation in the poverty concept.

"Absolute" definitions of poverty, thus, refer to the *items* included in the list of the necessities. In the absence of these items, people may be defined as living below the subsistence level. Hence, Sarlo [5] used the term "absolute approach" interchangeably with "necessities approach," while "relative approach" is used to have equal meaning with "social amenities approach," because relative poverty conveys the impression of lack of both physical and "social" needs. The latter is considered more subjective.

Nearly all definitions state that poverty is a matter deprivation. However, since this term is ambiguous, it needs further elaboration. The concept of *relative* (as opposed and thus alternative to *absolute*) *deprivation*, where people possess less of some desired attribute (i.e., income, favorable employment conditions, or power) than what others do is among the answers. Defining poverty in terms of deprivation should restrict to *economic deprivation* as suggested by the Panel on Poverty and Family Assistance (PPFA) to the US Government [3]. A narrowed definition, they argued, is required partly to arrive at a specific concept or measurement.

On the other hand, the word "relative" also connotes the meaning that different group of people may possess (or perhaps, need) different attribute. This may lead to confusion about the difference of poverty from the issue of "inequality," of which an approach to poverty concept has also been developed to accommodate broader social concerns. One may disagree with either conceptually separate or equivalent treatment of the two, still their presence is interrelated.[2] George and Howards [2] advocated that broader considerations of poverty definition have accommodated social aspects of mankind above the subsistence level of poverty, whether based on

[1]This study is worth mentioning in these two books discussed above by George and Howards [2] and Sen [4].

[2]Compared to Sen [4: 14–15 and 22–23].

working-class living standard as defined in terms of *social coping* or that of *social participation* based on living standard of the whole society. These two definitions are in dispute, however. There are some other definitions proposed by economists, including Hagenaars [6]. Nonetheless, limitation of the paper does not warrant their inclusion.

3 Measurement Problems

3.1 Concepts and Objectives

Nearly all scientific activities invariably necessitate measurement, without which any progress in the modern sciences would not be significantly achieved. Measurement means comparison of a particular object and assignment of numerical values on it according to a certain method or procedure. Kemeny suggested that a procedure should provide stages commencing from classification, arrangement to measurement scale in numbers, and each stage may have some detailed procedures.

Poverty measurement assigns numerical values to the poverty phenomenon using a certain procedure so as to allow for possible aggregation or, otherwise, disaggregation, of such a phenomenon. Sen [4] admitted this meaning of measurement by stating "that the measurement of poverty can be split into two distinct operations, viz. the *identification* of the poor, and the *aggregation* of their poverty characteristics into an overall measure."

3.2 Basic Criteria for Poverty Measures

It is worthy of note discussing in a brief criteria for a poverty measure the Panel on Poverty and Family Assistance of the USA deliberates three recommended criteria: *public acceptability; statistical defensibility;* and *operational feasibility*. Public acceptability talks about capability of a measure to be accepted by people in the society. This criterion conditions public access through government policies, intellectual discourses, or dissemination with the help of mass media. However, for acceptability it also necessitates that such a measure should be reasonably valid and understandable as well.

As a device in the decision making and intellectual or academic purposes, the measure must also be statistically defensible to get its calculations accepted by both analysts and statisticians. It must be logically consistent and welcomes the reasonable comparative analyses across time, places, types of families, and across population groups as well. To be operationally feasible, the measure must provide possibility of data availability and their collections.

The focus of the discussion of poverty measurement shall be on selected conventional approaches to measurement. It seems that these approaches may be formulated into two basic categorizations in terms of the following:

(a) The extent to which the amount and quality of needs should be satisfied in order to set up the *poverty line*; and
(b) The bases on which the needs should be measured.

3.3 Specifying the Needs: Clothes, Shelter, and Others

Apart from foods, there are some kinds of goods appealing for inclusion into minimum needs. These include minimum requirements of clothes, housing, schooling, and health. Fulfillment of such necessities is obligatory for human life for survival, or otherwise, their shortage leads to human deprivation.

3.4 Consumption or Income: Direct Versus Indirect

Among the questions addressed in poverty measurement is the choice between consumption and income as a yardstick. Does poverty indicate a lack of consumption or of income? Of the advantages of using consumption, which is also known as *direct method*, is that poverty measurement needs not any assumption of consumption, so that one can directly examine the level of the needs below which poverty happens. "It does not involve the use of any income notion," including that of poverty line income [4]. The difficulty, as Sarlo [5] admitted, arises when some households consume non-necessities while having lack of basic necessities.

On the other hand, *indirect method* employs income as a proxy of welfare indicator. An income cutoff, being the poverty line, would judge whether an individual is considered as a poor or not depending upon his income, below the line or not. There are some difficulties of this method, two of which are: (a) income and living standard among the population groups may be quite different; and (b) household and other non-income-earning activities may not be covered in this method. The major advantage of this indirect method is, as Sarlo [5] pointed out, its large coverage of information about income and income distribution.

3.5 Distribution and Inequality Issue

It is argued, according to relative approach that economic growth, by itself, is unable to eradicate poverty, if there is no change in income distribution. Consequently, inequality among population groups in a society remains unchanged. In contrast to the absolute approach, given the distribution of income, alleviation of poverty can take place in a society where there is a significant economic growth, though inequality remains the same, since the minimum basic needs do not considerably change. As many people's income increases exceeding the poverty line, these people are no longer poor, so that the poverty accordingly reduces. On the other hand, according to the relative approach, they are still poor as compared to the riches whose income may double over time; they are relatively less well off than the rest of the society.

Meanwhile, it is also argued, distribution or redistribution, by itself, does not guarantee the reduction of (especially relative) poverty. In a society where distribution sufficiently takes place, inequality may decrease, given the stagnant economy, enabling all the members of the society to approach the poverty line from below and above, respectively, for the poor and the rich. Finally, measurement of poverty comes up with a number of determinants of the poverty that require more elaboration. Empirical evidence of the variables influencing the poverty is given in the following sections.

4 Determinants of Poverty in Indonesia

In Indonesia, poverty line can be defined as the expenditure value in rupiah (Indonesian currency) to purchase the minimum requirement for foods and non-food per capita. The minimum standard for food is based on the recommendation of the National Workshop on Food and Nutrition in 1978, which is stated that a person have to consume 2100 kilo calories per day in order to stay healthy. Similar to food standard, a number of non-food items were selected based on value judgment. The poverty line based is then defined by simply summing up these minimum expenditure values both in urban areas and in rural areas.

According to data from Badan Pusat Statistik, Indonesia's poverty rate is at 11.37% in March 2013, or equivalent to 28.07 million Indonesians. This number shows a steady downward trend and it is assumed that the number will continue in a slower pace in the future as the inflationary pressure that faced by Indonesia nowadays as the increment of oil fuel prices.

Table 1 Variables and data sources

1.	Poverty	Statistical Yearbook of Indonesia, BPS, various years
2.	Population	
3.	Gross domestic regional product	Gross Regional Domestic Product of Provinces in Indonesia by Industrial Origin, BPS, various years
4.	Share of agriculture	
5.	Share of industry	
6.	Mean years of schooling	Trends of Selected *Socio-Economic Indicators* of Indonesia, BPS, various years

5 Data and Methods

The data used for this study consist of data on poverty, population, gross domestic regional product, share of agriculture share of industry and educational attainment of labor force of 33 provinces in Indonesia periods 2006–2012. All of the data are extracted from BPS as shown in Table 1.

This study also obtains relevant information from many books, journals, newspaper, and others used as reference. Analysis of these balanced panel data which used in this study is done by using program EViews 6, in which the dependent variable (poverty) assumed influenced by four independent variables (population, gross domestic regional product, share of agriculture, share of industry, and mean years of schooling). The models that will be used are as follows:

$$\begin{aligned}\mathrm{POV} = {} & \beta_0 + \beta_1 \mathrm{X1}_{it} - \beta_2 \mathrm{X2}_{it} - \beta_3 \mathrm{X3}_{it} - \beta_4 \mathrm{X4}_{it} - \beta_5 \mathrm{X5}_{it} \\ & + \gamma_1 D_{1i} + \gamma_2 D_{2i} + \gamma_3 D_{3i} + \gamma_4 D_{4i} + \gamma_5 D_{5i} + \gamma_6 D_{6i} + \gamma_7 D_{7i} + e_{it}\end{aligned}$$

where

POV	Number of poor people (thousand people),
X1	Number of population (thousand people),
X2	Gross domestic regional product (billion Rps),
X3	Share of agriculture (percent),
X4	Share of industry (percent),
X5	Mean years of schooling,
D	Dummy variables,
i	Intercept, and
t	Time.

6 Results and Discussion

Based on the Hausman test, probability of cross-sectional random effect is 0.3259, which means that cross-sectional effects are not significantly correlated with at least one repressor in the model and thus the random effect model is not problematic. Hence, period random effect model is better than the period fixed effect, in which the panel data regression result is as follows:

$$\begin{aligned} POV = {} & 964.024 + 0.17\ X1_{it} - 0.0014\ X2_{it} - 0.74\ X3_{it} - 5.62\ X4_{it} - 111.39\ X5_{it} \\ & + \gamma_1 D_{1i} + \gamma_2 D_{2i} + \gamma_3 D_{3i} + \gamma_4 D_{4i} + \gamma_5 D_{5i} + \gamma_6 D_{6i} + \gamma_7 D_{7i} + e_{it} \end{aligned}$$

6.1 The Strong Influence of Human Capital

Human capital can be defined as to include the size of population and their education and skill so as to develop the capacity of the society in advancing technology and thus technological progress of the country. Though an increase in population size may contribute to increasing poverty, surely, the quality of the population which is defined as human capital consequently can substantially reduce the poverty and conversely increase economic growth. The evidence of this study suggests positive and significant effect of population to poverty in Indonesia. It is in line with the hypothesis, in which population growth assumed has positive effect to poverty in Indonesia. Just like the theory, Malthus [7] states that population growth exerted on the availability of food so that poverty increases. History also already shows that the downward pace of Indonesian population followed by the downward pace of Indonesian poverty headcount ratio. The coefficient of population is 0.17, means that an increment or decrement of population in the amount of 10 thousand people, the number of poor people will increase or decrease in the amount of 1.7 thousand people by assuming the other variables are constant.

On the other hand, mean years of schooling representing improvement in human capital through education has negative and significant effect on poverty in Indonesia. It is in line with the hypothesis, in which mean years of schooling assumed have negative effect to poverty in Indonesia. It supports the argument of Hanushek et al. [8] who states that there is a stronger impact of education quality and of years of schooling in open economies. History also shows that the upward trend of mean years of schooling in Indonesia is followed by the decrement of poverty in Indonesia.

The coefficient of mean years of schooling is −111.39, means that an increment or decrement of mean years of schooling in the amount of a year, the number of poor people will decrease or increase in the amount of 111.39 thousand people by assuming the other variables are constant.

6.2 The Influence of Income on Poverty in Indonesia

Provincial income in Indonesia represented by GDRP for each province has negative and significant effect to poverty. It is in line with the hypothesis, in which GDRP assumed has negative effect to poverty in Indonesia. Just like the argument by Dollar and Kray [9], policies to raise average incomes are likely to be central to successful poverty reduction strategies. Then, by looking at the trend of poverty headcount ratio and GDP growth, the economic growth seems having negative relationship with poverty headcount ratio in the case of Indonesia.

The coefficient of GDRP is −0.0014, means that an increment or decrement of GDRP in the amount of 10,000 billion Rupiahs, the number of poor people will decrease or increase in the amount of 14 thousand people by assuming the other variables are constant.

6.3 The Sectoral Shares of Agriculture and Industry

Share of agriculture has negative but insignificant effect to poverty in Indonesia. It is in line with the hypothesis, in which share of agriculture assumed has negative effect to poverty in Indonesia. Just like the finding of World Growth (2004) which is showed that in case of Indonesia, agricultural GDP growth is beneficial in reducing poverty, particularly in rural areas. History also shows that negative relationship almost exists in case of agricultural sector. The coefficient of share of agriculture is −0.74, means that an increment or decrement of share in agriculture in the amount of 10 percent, the number of poor people will decrease or increase in amount 7.4 thousand people by assuming the other variables are constant.

Nonetheless, the finding of this study shows that the share of agriculture has no significant effect in reducing poverty in Jawa Tengah. This result is possible as agricultural productivity in Indonesia remains low so that it is not capable enough to be an adequate source of income for the farmers, particularly in area where the agriculture is more primitive. In Indonesia, as elsewhere in the developing world, most people who depend on agriculture for their incomes are poor and; otherwise, most of those who are poor depend on agriculture for a living.

Meanwhile, the share of industry has negative and significant effect to poverty in Indonesia. It is in line with the hypothesis, in which share of industry assumed has negative effect to poverty in Indonesia. Just like the finding of World Growth (2009) which is noted that expansion of industry in Indonesia has been a significant source of poverty alleviation. History also shows that negative relationship almost exists in case of industry sector.

The coefficient of share industry is −5.62, means that an increment or decrement of share of industry in the amount of 10%, the number of poor people will decrease or increase in the amount of 56.2 thousand people by assuming the other variables are constant.

7 Concluding Remarks

Based on the results of regression estimates, population is positively and significantly influenced number of poverty in Indonesia, while gross domestic regional product, share of industry and mean years of schooling are negatively and significantly influenced number of poverty in Indonesia. In addition, this study also finds that share of agriculture is having no significant influence to poverty reduction in Indonesia.

The advice can be given as follows: (1) As there is a significant positive relation between number of population and number of poor people, hence, it is necessary to control the number of population. Government intervention is needed in order control the population. Besides, it is rather to increase the productivity of population so that the increment of population may create more wealth. (2) As there is a significant negative relation between gross domestic regional product and number of poor people, hence, having more income is then important. Therefore, government intervention in creating high-quality human resources is needed in order to stimulate the economy. (3) As there is an insignificant negative relation between share of agriculture and number of poor people, hence, it is necessary to increase the agricultural productivity in Indonesia so that it could be capable enough to be an adequate source of income for the farmers. If agriculture could be capable enough to be an adequate source of income, share of agriculture possibly works in eliminating poverty. (4) As there is a significant negative relation between the share of industry and number of poor people, hence, it is better to increase the share of industry. Therefore, government intervention in adding industrial productivity is needed in order to stimulate the industrial sector. (5) As there is a significant negative relation between mean years of schooling and number of poor people, hence, it is better to increase the mean years of schooling. Therefore, government intervention in increasing the mean years of schooling is needed.

References

1. Schiller, Bradley R., 1984, *The Economics of Poverty and Discrimination,* Prentice-Hall Inc., Englewood Cliffs, N.J.
2. George, Vic and Irving Howards, 1991, *Poverty Amidst Affluence,* Edward Elgar, Hants, England.
3. Citro, Constance F. and Robert T. Michael (editors), 1995, *Measuring Poverty: A New Approach,* National Research Council: National Academy Press, Washington.
4. Sen, Amartya Kumar, 1982, *Poverty and Famines: An Essay on Entitlement and Deprivation,* Clarindon Press, Oxford.
5. Sarlo, Christopher A., 1996, *Poverty in Canada,* 2nd ed., the Fraser Institute, Vancouver, B.C.
6. Hagenaars, Aldi J.M., 1986, *The Perception of Poverty,* North-Holland-Elsevier Science Publishers B.V., Amsterdam.
7. Malthus TR. *An Essay on the Principle of Population,* New York: W. W. Norton, 1976, p. 132.

8. Hanushek, et al. (2006). "The Effects of Education Quality on Income Growth and Mortality Decline". Working Paper 12652. National Bureau of Economic Research.
9. Dollar, David., and Aart Kraay. (2002). "Growth is Good for the Poor". *Journal of Economic Growth*. Development Research Group, The World Bank.

Quality Improvement of Product–Service Package of "Mangunan" Agrotourism Using Quality Function Deployment Approach

Nafis Khuriyati, Wahyu Supartono and Rahmat Hidayat

Abstract Mangunan is one of growing agrotourism in the district of Bantul Yogyakarta. Basically, Mangunan had many criteria to support the implementation of product–service package sustainability agrotourism that included product, service, and others. Product–service package is a correlation between the products and services offered simultaneously. Mangunan has a variety of recreational and educative facilities and is supported by a natural landscape as a special attraction. Although it has a wide range of potential, Mangunan has not been able to attract tourists to visit. The study aimed to formulate technical requirements for improving quality of "Mangunan" agrotourism using Quality Function Deployment approach.

Quality function deployment (QFD) provides a comprehensive, systematic approach to ensure whether new products meet or exceed customer expectations. QFD can be a powerful tool because it can reduce time to market, improve quality, and increase customer satisfaction. The techniques that were employed in the study included consumer surveys to collect the consumer preference in product, price, place, promotion, people, process, physical evidence (7P) of Mangunan Agrotourism and convert consumer and market information into detailed technical requirements.

Based on the 7P Tourism Marketing, dimensions that must be corrected immediately were as follows: Product, which included services and product sales of fruit and a comfort place to enjoy the view. Place dimension such as ease to reach the locations. Promotion dimension included the granting bonuses for certain package orders, the appropriate price with the service received for price dimension. People dimension such as attractive and good appearance employee. Process dimension was recognizing the loyal customers, and the physical evidence such as transportation facilities in the garden without having to walk away. Technical requirements for Mangunan quality improvement included education and training for employees and establishing Standard Operating Procedure (SOP) of services.

N. Khuriyati (✉) · W. Supartono · R. Hidayat
Department of Agroindustrial Technology, Faculty of Agricultural Technology, Gadjah Mada University, Yogyakarta, Indonesia
e-mail: nafiskhuriya@yahoo.com

T. Taufik et al. (eds.), *ICoSI 2014*, DOI 10.1007/978-981-287-661-4_23

Keywords Mangunan · Agrotourism · Product–service package · Quality function deployment

1 Introduction

The high population growth rate that exceeds the capacity of the earth's productivity and globalization makes the lives of the world more complex and unmanageable. Sustainable development is needed in order to meet the needs of the present without ignoring the needs of future generations. Referring to the concept of sustainability, agrotourism has the potential to be developed into sustainable tourism. Agrotourism is the notion of travel activities with no damaged or polluted nature with a view to admire and enjoy the beauty of nature, wild animals, and plants in their natural environment as well as educational facilities [1].

The Mangunan fruit orchard is one agrotourism located in Dlingo, Bantul, Yogyakarta, which is 35 km from the city of Yogyakarta. The government built Mangunan in 2003 with an area of 23.5 ha at an altitude of 150–200 m above sea level. The vision of the Orchard Mangunan is the leader in the field of Agro tour, to become an international orchard and establish employment opportunities for the people of Indonesia. Its mission is to distinguish with other businesses to develop agrotourism by involving people around the tourist sites, providing entertainment agro creative and education, providing educational services, training to communities in need, and creating employment opportunities for the surrounding community. Mangunan agrotourism is a hilly region with a typical mountain landscape that has cool air. Tourists can enjoy the view from Mangunan, i.e., Parangtritis beach, Bantul City, as well as the beauty of the very steep Oyo River. Mangunan cultivates a variety of tropical fruits such as mango, durian, rambutan, guava, orange, sapodilla, 'duku,' and mangosteen. There are also scarce fruits like matoa, longan, carambola, 'cempedak,' and various organic vegetables. Tourism potential of Mangunan is packaged in a product–service package but it has not been able to attract tourists. Data from Department of Culture and Tourism showed that in 2009, 1,338,112 tourists visited Bantul with the percentage nearly 67% dominated coastal attractions, 18% cave attraction, and the rest is occupied by tourist attractions such as the tourism village, tourism history, and so forth.

The number of travel alternatives requires Mangunan to pay attention to the feedback of the customer as a basis to improve its quality. One way to do it is to generate products–services package to meet consumer need using quality function deployment (QFD) approach. Quality function deployment is a structured approach to defining customer needs or requirements and translating them into specific plans to produce products to meet those needs [2]. Some of the benefits of adopting QFD have been documented as reduced time to market, reduction in design changes, decreased design and manufacturing costs, improved quality, and increased customer satisfaction [3]. The QFD has been used to improve the quality of some products [4, 5] and services [6, 7]. This study aims to identify the needs of consumers towards agrotourism, which includes 7Ps marketing and formulate the technical requirements to meet the needs of consumers taking into account the competition and the ability Mangunan.

2 Methods

This study proposed integration of the 7Ps marketing and QFD approaches for quality improvement of Mangunan Agrotourism. The 7Ps is a marketing model that adds to the aforementioned 4Ps, including product, price, place, promotion, people, process, and physical evidence. It was used as product of Mangunan Agrotourism as a product–service package, not merely a physical good. The customer needs were expressed in 7 Ps. This understanding of the customer needs was then summarized in a product planning matrix or "House of Quality: HOQ" due to its house-like shape (Fig. 1). These matrices were used to translate higher level "what's" or needs into lower level "how's"—technical requirements to satisfy these needs.

2.1 Customer Needs

Interview (59 panelists) was carried out to explore numerous consumer issues associated with the agrotourism and especially Mangunan agrotourism. This information was then used to obtain detailed consumer needs for Mangunan agrotourism. The needs were further grouped into 7Ps marketing.

Fig. 1 The house of quality

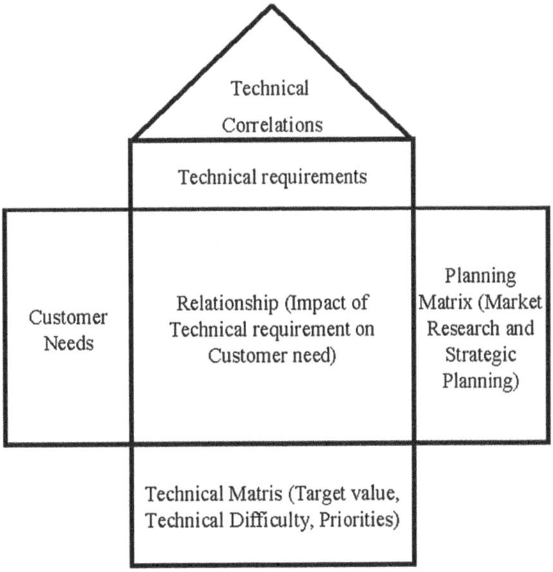

2.2 *Planning Matrix*

The planning matrix is the tool that helps to prioritize customer needs. The data placed in the planning matrix allows to make strategic decision about the product or service.

2.2.1 Importance to Customer

The importance to customer show how importance each need is to the customer. The consumer survey obtained the views of 100 respondents who were randomly recruited at Mangunan agrotourism areas. They were asked to complete the questionnaire of the importance of consumer need. A 5-point Likert scale, anchored with the terms: not very important (1), not important (2), neutral (3), important (4), and very important (5), was used to indicate the consumer's opinion.

2.2.2 Customer Satisfaction and Competitive Satisfaction Performance

The customer satisfaction performance is the customer perception of how well the current product service package is meeting the customer needs. As for benchmarking the competition satisfaction performance, a competitive satisfaction performance was also measured. The method for estimating this value was by asking the customer, via survey, how well he or she feels the Mangunan agrotourism's product–service package has met each customer need compared to that of the Kusuma agrotourism. Kusuma agrotourism is located in Batu-East Java and were formed in 1991. It has 17 hectares wide and is located 1000 masl. Kusuma agrotourism is one of the pioneer Agro Tourisms in Indonesia which offer orchard picking apples, oranges, guava, dragon fruit, strawberries, and pesticide-free hydroponic vegetable. Tourists accompanied by tour guides who explain about the cultivation of plants and related matters about the fruits. In addition to picking tour, Kusuma Agro also offers outbound, the apple house and strawberry house where they present menus using apples or strawberries. Twelve respondents who had visited Mangunan and Kusuma recruited through purposive sampling method. The satisfaction level was expressed as a grade or a performance level. Grades were given on a five-point scale: very poorly (1), poorly (2), neutral (3), well (4), and very well (5).

2.2.3 Goal and Improvement Ratio

By comparing importance to customer, customer satisfaction performance, and competitive satisfaction performance, Mangunan managers in the Department of Agriculture and Forestry of Bantul district decided what level of customer performance they want to aim for in meeting each customer need (the goal). The performance goals were expressed in the same numerical scale as performance level. The goal, combined with current rating–customer satisfaction, was used to set improvement ratio. The improvement ratio is one of the most multipliers of importance to customer, and thus, setting the goal is a critical strategic step in QFD.

2.2.4 Sales Point

The sales point contains information characterizing the ability to sell the product or service, based on how well each customer need is met. In this study, sales point was determined by Mangunan manager. The values assigned for sales points were 1 (no sales point), 1.2 (Medium sales point), and 1.5 (strong sales point).

2.2.5 Raw Weight

The raw weight contains the overall importance to the manager of each customer need, based on its importance to the customer, the improvement ratio, and the sales point value. The value of the raw weight for each customer need was

$$Raw\,weight = (importance\,to\,customer) \cdot (Improvement\,ratio) \cdot (Sales\,point).$$

2.3 Technical Requirements and the Relationship/Linkage to the Customer Need

The technical requirement represents a translation from the customer's language (voice of the customer) into the organization's technical language (voice of the developer). The strength of the linkage between one technical requirement and one customer need was classified into four levels: not linked (0), possibly linked (1), moderately linked (3), and strongly linked (9).

2.4 Technical Matrix

2.4.1 The Level of Difficulty Fulfilling the Technical Requirements

Basically, the technical requirements often cannot be met due to the limited ability of firms to comply. For it performed an analysis of the company's degree of difficulty in meeting the technical needs expressed by the value from 1 (low difficulty) to 5 (high difficulty).

2.4.2 Target Values for Technical Requirements

Targeted technical requirements were the parameters of the technical requirements needed to meet the customer needs. This determination was based on field observations, literature review, and discussions with managers and employees' Mangunan involved in the operation.

2.4.3 Technical Requirement Scale (Priorities)

Once all the linkages of technical requirement were determined, some simple arithmetic provides one of the key results of QFD: the relative contributions of the technical requirements to overall customer satisfaction. These represent the priorities of technical requirements.

2.5 *Technical Correlation*

Technical correlation represents interrelationships and interdependencies between technical requirements. The potential positive and negative interactions between technical requirement were determined using symbols for strong or medium, positive, or negative relationships: ● (strong positive impact), ○ (moderate positive impact), blank (no impact), □ (moderate negative impact), and ■ (strong negative impact). Too many positive interactions suggest potential redundancy in technical requirements. The negative interactions focused on considering technology to overcome these potential trade-offs or considering the trade-offs in establishing target values.

3 Results and Discussion

Quality function deployment is the comprehensive quality system aimed specifically at satisfying the customer. It concentrates on maximizing customer satisfaction (positive quality). The 39 items of customer needs grouped into 7 Ps marketing were listed as shown in the Customer Needs column (Fig. 2). A questionnaire based on 39 items was designed and used to identify the customer expectations and needs through personal interviews. A summary of the personal information of all respondents: as many as 52% of respondents were male and 48% were women, between the ages of 16–25 years. Respondents as much as 78% were students, 18% were employees, and 4% others. They come from Bantul (57%), Yogyakarta (32%), and outside Yogyakarta (11%). The importance of each customer need is shown in the Importance column of Fig. 2. The highest and lowest importance value is the conformity price with the service received (3.58) and the openness company to the

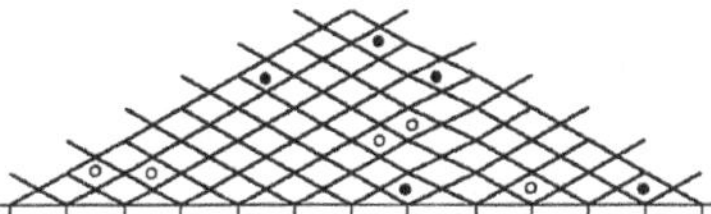

		Customer Needs \ Technical Requirements	In coop. with Transportation Dept	Recruitment of employees	Development of parking facilities	Recruitment tour guide	The addition of a gazebo	Regular hygiene checks	Addition of Customer Services	Promotions, discounts, souvenirs	Provision of vehicles in the garden	Construction of the waiting area	The use of uniforms for employees	Training for employees & SOP	Importance	Mangunan Performance	Kusuma Agro Performance	Goal	Improvement Ratio	Sales Points	Raw weight	Normalized Raw weight
Product	1	Fruit sales		9											3,20	2,83	4,33	4,00	1,41	1,2	5,43	2,1
	2	The comfort place					9	3							3,20	2,92	4,17	4,00	1,37	1,5	6,58	2,5
	3	Cleanliness of tourist spots					3	9							3,48	2,50	3,67	4,00	1,60	1,5	8,35	3,2
	4	Seeds sales		9											3,32	2,75	3,92	4,00	1,45	1,2	5,79	2,2
	5	Greenhouse arrangement		3											3,39	1,67	4,50	4,00	2,40	1,2	9,74	3,8
	6	Plantation arrangement		3											3,45	1,92	4,50	4,00	2,08	1,2	8,63	3,3
Place	7	Location is easy to reach	3												3,47	2,67	3,83	4,00	1,50	1,2	6,24	2,4
	8	Public transport access to the site	9												3,20	2,33	3,25	4,00	1,72	1,5	8,24	3,2
	9	Access private vehicles (signposts)	9												3,32	3,42	3,67	4,00	1,17	1	3,88	1,5
Promotion	10	Giving bonuses								9					3,30	2,58	3,58	4,00	1,55	1,5	7,67	3,0
	11	Discounts for special order								9					3,35	2,58	3,83	4,00	1,55	1,5	7,79	3,0
	12	Giving a souvenir as a token of gratitude								9					3,10	2,58	3,50	4,00	1,55	1,5	7,21	2,8
	13	Providing correct and accurate information												1	3,30	2,58	3,58	4,00	1,55	1	5,12	2,0
Price	14	Price product - service package								1					3,55	2,50	3,75	4,00	1,60	1,2	6,82	2,6
	15	Conformity price with the services received		3	3	3	3	3	3					3	3,58	2,58	3,33	4,00	1,55	1,2	6,66	2,6
	16	Variation in a single package product package offered								1					3,56	2,58	3,83	4,00	1,55	1,5	7,81	3,0
People	17	Tour guide				9								3	3,25	2,17	4,25	4,00	1,84	1,5	8,99	3,5
	18	Guiding skills in interacting with visitors				3								9	3,09	2,33	4,42	4,00	1,72	1,5	7,96	3,1
	19	Communicative guides				3								9	3,21	2,25	4,50	4,00	1,78	1,5	8,56	3,3
	20	Hospitality empoyees												9	3,57	2,50	3,92	4,00	1,60	1,2	6,85	2,6
	21	Attractive employee											3		3,21	2,42	3,92	4,00	1,65	1,2	6,37	2,5
	22	The employees are always ready to help												3	3,29	2,67	3,92	4,00	1,50	1	4,93	1,9
	23	Knowledge and operational expertise of employees												9	3,14	3,00	4,00	4,00	1,33	1	4,19	1,6
	24	Understanding customers' expectation												3	3,10	2,75	3,67	4,00	1,45	1,2	5,41	2,1
Process	25	Customer service							9					3	3,33	2,58	3,67	4,00	1,55	1,2	6,20	2,4
	26	Company's response to complaints of customer							9					3	3,39	2,50	3,67	4,00	1,60	1	5,42	2,1
	27	Secure parking area			9										3,52	2,75	3,92	4,00	1,45	1,2	6,14	2,4
	28	Convenience parking			9										3,19	2,75	3,92	4,00	1,45	1,2	5,57	2,2
	29	Ease of access to purchase tickets tour							9						3,37	2,75	3,67	4,00	1,45	1,5	7,35	2,8
	30	Alertness of responding to the needs of consumers												3	3,14	2,58	3,67	4,00	1,55	1	4,87	1,9
	31	The speed and accuracy of handlingcomplaints							3						3,21	2,92	3,75	4,00	1,37	1,5	6,60	2,5
	32	Recognizing the regular customers												1	3,07	3,08	4,00	4,00	1,30	1,2	4,78	1,8
	33	The company open with customers												1	2,59	2,83	3,83	4,00	1,41	1,2	5,07	2,0
Physical evidence	34	Large parking area			9										3,50	2,50	4,00	4,00	1,60	1,2	6,72	2,6
	35	Transport facilities around the garden									9				3,42	2,08	3,67	4,00	1,92	1,5	9,87	3,8
	36	Reservations and information by telephone							9						3,29	2,42	3,83	4,00	1,65	1,2	6,53	2,5
	37	Comfortable waiting area										9			3,17	2,75	3,83	4,00	1,45	1,2	5,53	2,1
	38	Complete bussiness facilities											9		3,17	2,42	4,00	4,00	1,65	1	5,24	2,0
	39	The distinctive and attractive bussiness facilities											9		3,16	2,42	4,00	4,00	1,65	1,5	7,83	3,0
			Consumers easily towards the location	Ease in purchasing fruit and vegetables	Safely leave the vehicle	Visitors are not confused	Consumers can relax enjoy the view	Consumers feel comfortable	Quick respond to complaints	Increased consumer loyalty	Not tired around the orchard	Consumers are comfortable at rest	Employees are easily recognized	Skilled employees								
			3	2	4	2	2	5	4	3	2	2	3	3								
			42	68	72	58	40	44	104	84	34	19	53	151								
			5,4	8,8	9,2	7,5	5,2	5,7	13,4	10,9	4,4	2,5	6,8	19,4								

Fig. 2 The house of quality for Mangunan agrotourism

customer (2.99), respectively. The importance of each 7Ps can be seen in Fig. 3. It is known that the price has the highest level of importance to customer. This suggests that based on tourist perception the price is the most important component in the implementation of an agrotourism business. The six other of 7Ps have the same relative importance.

How well is Mangunan Agrotourism doing in meeting the needs today? The answer of this question is expressed in Mangunan Performance column (Fig. 2). Fulfillment of almost all consumer needs by Mangunan (Mangunan Performance) assessed by the respondents under a score of 3, in fact even under a score of 2 for the greenhouse arrangement and plantation arrangement, while Kusuma Agro as a benchmark rated about a score of 4. From Fig. 4 it can be seen that Mangunan has

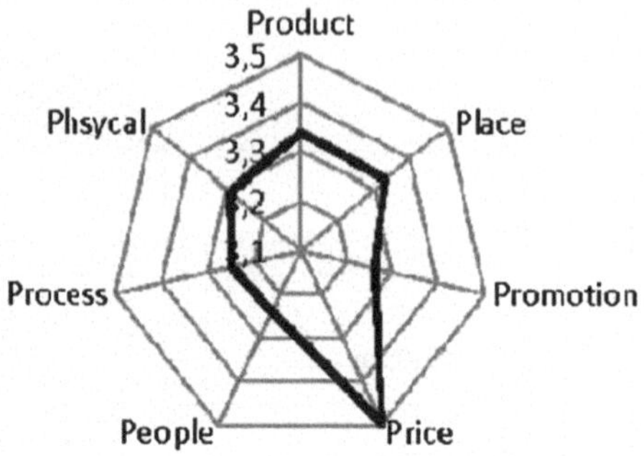

Fig. 3 Importance of customer needs based on 7Ps marketing

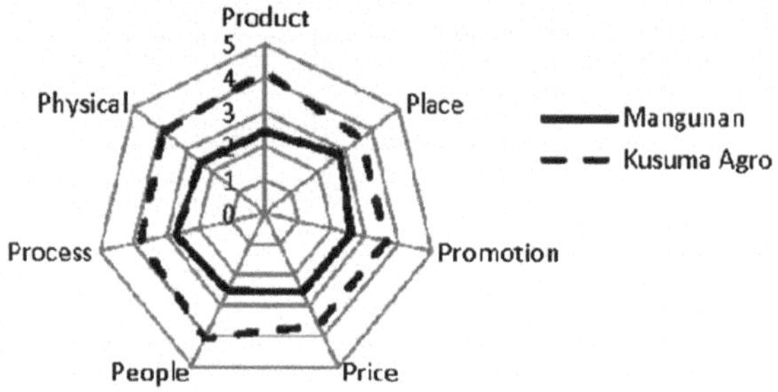

Fig. 4 Customer satisfaction of Mangunan and Kusuma agro based on 7Ps marketing

disadvantages compared to Kusuma Agro, for example, in the green house and plantation arrangements. There are some greenhouses in Kusuma Agro which are grouped based on the type of plant and the technology used while in Mangunan, there is only one greenhouse for many types of plants. Moreover, Kusuma Agro provides fruit picking tour that is not offered by Mangunan. However, there is a Unique Selling Point owned by Mangunan, i.e., beautiful life view that attracts tourists to visit Mangunan. Based on the importance to the customer, the improvement ratio, and the sales point value, the raw weight for each customer need was calculated and then normalized as shown in the rightmost column in Fig. 2.

Quality function deployment focusing on delivering value by seeking out the needs, translating these into actionable services, and communicating this throughout the organization, further QFD allows the customer to prioritize their requirement, tell how the Mangunan are doing compared to the (Kusuma Agro), and then direct Mangunan to optimize those aspects of organization service that will bring the greatest competitive in agrotourism. To meet the 39 consumer needs a number of technical requirements formulated as shown in Fig. 2. In the House of Quality for Mangunan also demonstrates the strength of the linkage between one technical requirement and one customer need. For example, the fruit sales of customer need had a strong linkage (9) to the recruitment of employees because Mangunan does not always sell the fruits due to the limited number of employees.

By considering the components that have been arranged in a matrix of HOQ, the priority order of the technical requirements to improve the quality of the product–service package was as follows: training for employees and developing standard operating procedure (SOP) for service, adding customer services, making promotions, giving discounts, providing souvenirs, developing parking facilities, recruiting more employees, recruiting tour guides, providing uniforms for employees, cooperating with Transportation Department, performing regular hygiene checks, building more gazebos, providing vehicles in the garden, and constructing the waiting area.

4 Conclusion

Quality function deployment (QFD) is the dominating approach which assures that customer requirements actually do guide the quality improvement and offer an approach that supports structuring and the representation of information on customer requirements, and especially how to link customer requirements to design technical requirements. The HOQ analysis has demonstrated that the price has the highest level of importance to customer. However, after considering the Mangunan performance compared to competitors, as well as future goals and sales point, all items of 7Ps marketing had almost similar importance value. The technical requirements prioritized to meet the needs of consumers among others were training for employees and developing standard operating procedure (SOP) for service, adding customer services, making promotions, giving discounts, providing souvenirs, developing parking facilities, and recruiting employees.

References

1. Anonymous. (2007). *Pengembangan Agrowisata Indonesia*. Agricultural Depertment of Indonesian Republic.
2. Bossert, J. L. (1991). *Quality Function Deployment: A practitioner's approach*. New York: ASQC Quality Press.
3. Rudolph, M. J. (1995). *The Food Product Development Process*. British Food Journal, 97(3), 3–11.
4. Fertiasari, R., Khuriyati, N., Ushada, M. (2011). *Product Development of Aloe vera Chinensis Gel Coated with Chocolate as West Borneo's Local Potential*. Proceeding Seminar on Agricultural Reform towards Food Sovereignty. Surabaya Indonesia. October 20, 2011.
5. Prihasti, G., Khuriyati, N., Falah, A.F. (2011). *Quality improvement of Nata de Banana-Skin Through Application of Technical Requirements in QFD*. Proceeding Seminar of Indonesian Society of Agroindustrial Technology II. Yogyakarta. Nopember 23–24, 2011.
6. Mazur, G.H. (1999). *QFD for Service Industries-From Voice of Customer to Task Deployment*. The Fifth Symposium of Quality Function Deployment. Michigan USA. June, 1999.
7. Nurlela, E., Khuriyati, N., Wahid, I. (2011). *Application of Quality Function Deployment Method for Improvement of Service Systems in Fish Market (Case Study in Depok Beach, Bantul, Yogyakarta)*. Proceeding Seminar on Agricultural Reform towards Food Sovereignty. Surabaya Indonesia. October 20, 2011.

Spatial Use and Satisfactory Level of the High-Density Housing Residents

Sri Astuti Indriyati

Abstract This research presents evidence on how the space used by the residents and their satisfactory level towards space of living. The study proceeds by looking at what the occupants' perceptions towards their space. Various variables are used to foresee the possible factors that affect their satisfactions. Variables used for assessment are some of possible dimensions occupants' feel of necessary for the space. Physical, psychological, and perceptual performance of space is assessed in this study as indicated by Murtha's Criteria on Dimensions of User Benefits, such as behaviour facilitation (functional conformance and spatial conformance), physiological maintenance (climate conformance and hazard regulation, perceptual maintenance (sensory initiation and sensory conformance), and social facilitation (social initiation and social isolation) (Fisher in Environmental psychology, 2nd edn. Holt, Rinehart and Winston, New York, 1984, [1]). The primary data, collected from respondents of particular type of high-density housing units in Jakarta, were analysed and interpreted by qualitative–quantitative methods using the Statistical Package for Social Science (SPSS), and results were further presented in descriptive form. This study concluded that the perceptions and satisfactorily level of residents towards their space are variously found. From the previous research, the implication for planning has mentioned that the minimum standard of 36 m^2 is advised for the high-density housing (Indriyati in Low-cost housing in Jakarta: toward a more humane architecture. Penerbit Pustaka Sutra, Bandung, 2009, [2]), and it now further investigates whether this minimum standard has also met residents' satisfaction for the extended dimensions of user benefits. The findings will provide architects the potential criteria of space that need to be highly concerned. They should develop design ideas based on a more profound knowledge of potential users' physical and perceptual needs of expected design for the high-density housing. This research is also resulting in an advice that users are always to involve in the design process.

Keywords Perception · Satisfaction · Housing

S.A. Indriyati (✉)
Faculty of Engineering, University of Persada Y.A.I, Jakarta, Indonesia
e-mail: indriyati@yai.ac.id; sriastutiindriyati@yahoo.com

T. Taufik et al. (eds.), *ICoSI 2014*, DOI 10.1007/978-981-287-661-4_24

1 Overview

The research was inspired by Bell et al. [3, p. 44] and Boutourline [4] in Holahan [5, p. 3] who defined that the dominant situation in modern life is that individuals live in a setting which was not built for them and reality shows that, for many centuries, human beings have modified their environments by creating a human-built world to augment their natural settings. This study seeks to oversee the use of space for the activities, perceptions that ends up with their satisfactory level of the occupants living in high-density housing in Jakarta, particularly those living in 36 m^2 unit type. This unit type focused for 36 m^2 unit type in relation to the implication for planning in the previous research said that the minimum standard of 36 m^2 is advised for the high-density housing [2]. For that, the study is continued to investigate whether this minimum standard of 36 m^2 unit type has also met residents' satisfaction?

In regard to building design, it is useful for understanding the meshing between the human physical characteristics of architectural form as defined by Holahan [5, p. 3]; he said that "The buildings containing both people and objects need to be more humanized and residential buildings are one example to be considered in this light". The study that modern settings are not built for people is borne out by the fact that the current design processes still neglect some of the important needs of users, in particular designing for a typical high-density housing in which users were not involved for discussion of their needs with designers [6, p. 33]. When the failure of architecture occured for building designs around the world, one real issue is urban housing which now becomes a global problem and remains the worst in undeveloped countries, such as Jakarta, capital city of Indonesia [7, pp. 31–33]. The local government of Jakarta responded to this problem through the development of high-density residential in the inner city of Jakarta. High transport costs expensed by disadvantaged people, compelling them to live closer to their places of work which are located in the inner city of Jakarta, although for that, the people remains in slums or in very dense areas with very limited or no basic facilities and other basic infrastructures [8, p. 2].

While the future housing design is required, the issue of existing housing and surroundings in Jakarta was also addressed. The facts indicate that more than 50% of existing housing and surroundings were classified as substandard, with those areas located in the urban centre requiring high investment for facilities and infrastructure to bring them up to a minimum standard [7, p. 11]. The study has indicated that the basic psychological and physical needs of housing and daily living have not yet been reached, such as the poor provision of floor space and the minimum standards for adequate housing needs [9, p. 299]. Previous research on high-density housing in Jakarta has found that there were unidentified matters towards socio-psychological effects on occupant [10]. What this also demonstrates is how the occupants of high-density housing may perceive and rating their satisfactory level towards the environments created and available for them. An important part of the quality of how individuals live is dependent upon their levels of satisfaction with the world and conditions they experience [11, p. 22]. As perception is directly related to satisfaction, this study also looks at an occupants' sense of satisfaction with their home environment, and to what extent their degree of satisfaction was affected by their perception of the spatial and socio-psychological conditions of their housing complexes, housing unit, and rooms.

This paper comprises a response to this question by concentrating on an analysis of the users' perceptions towards the standards of physiological and social well-being, and the degree the occupants' perceptions towards the architectural spaces considered in the design process. The research results indicate that the failure of architecture was occured since the design interpretation was taken inappropriately of which the occupants' needs to what they expects for being a users were ignored. What arises from this research was variety: a poor perception rating and poor sense of satisfaction rating of their (occupants) use of various rooms in the 36 m^2 unit types considered in someways. The continuing of 36 m^2 unit types should require more considerations of points of design development concepts aiming for the better quality of occupants living there.

2 Research Objectives and Hypothesis

The main hypothesis of this research was that the occupants of existing high-density housing in Jakarta have negative perceptions of the rooms and leading to the dissatisfaction of occupants and the advent of various coping[1] strategies.

The objective of this research is therefore focused on investigating the occupants' perceptions of the physical environment and their satisfactory level living within it. As an outcome of the various issues identified, examined, and assessed in respect of the case studies, strategies for the generation of possible alternative/modified design ideas will be conceptually proposed for future high-density housing provision in Jakarta.

3 Theories

Perceptions are created when humans and environments (physical objects) contact each other through the medium of the senses (the sensation process). The perception process starts once the object has stimulated the five human senses, which is then followed by the process of cognition. Once the cognitive process is completed, it then returns to the human senses, producing a so-called response or perception

[1]Coping with the environments refers to the space adjustment occupants make to fit the occupants' behaviour and the occupants do adaptation of their behaviour to fit the existing environments [3, p. 224].

[12, p. 111]. Human perception towards the environment includes cognitive, affective, interpretive, and evaluative components, all of which operate at the same time [11, p. 21]. The physical environment of a residential development should be able to provide opportunities for humans to meet their perceived needs and desires. So defined, the housing environment embraces both social and physical elements [13].

How perception towards the physical environment of residential settings indicates so far? In most situations, housing has not been well designed according to the physical and social criteria. Study taken place in low-cost housing located in Indonesia [14, pp. 26–28] indicated that 60% of occupants living in these two locations were either disappointed or very disappointed with their housing conditions. Various aspects of dissatisfaction towards their living space were identified from the occupants' feelings on physical and psychological condition. Steidl [15] found in his study regarding housing units that the size and floor plans of rooms were often mentioned as sources of difficulty or ease in performing household tasks. Not having enough room to work, having too many rooms to clean, and being too close to noisy areas of the house were often mentioned as features difficult to adapt to, and thus as sources of dissatisfaction as clarified in the book written by Bell et al. [3, p. 231].

Jakarta still has issues related to high-density housing complexes and its physical, social, and psychological conditions, which are pointed out by Komarudin [9, p. 175] in his book, such as the minimum floor space available for most rooms within units such as kitchen, bathroom, and toilet, and types of room available. Problems were also found for psychological and technical disturbance. The occupants' perception further affects the psychological feelings, namely lack of available space that creates poor air circulation system in the room, and it feels stuffy; it is difficult to arrange space and problems related to building conditions, and maintenance was also mentioned.

Other than perceptions of rooms, the occupants also have perceptions towards physical environment of a residential development. Their perceptions are supposed to be able to provide opportunities for humans to meet their needs and desires. In most situations, housing has not been well designed as one was found that the occupants of low-cost housing in Jakarta tend to cope with their residential setting [9, pp. 167, 175]. The dwelling layout design, including required facilities within a residential complex; building design, including problems varying resulting from the building's appearance; and building orientation and building utilities are often inadequately considered. On top of that, the unit size, layout of rooms, and types of rooms within a unit in public housing are subject to more demanding constraints, as a result of the lack of building maintenance. Power [16] has concluded that design problems were classed as problems stemming directly from the shape of the buildings, the nature of the construction, and the layout of the estates (a survey taken from a twenty European countries). The clumsy structure of the estates, blocks, and the environment; unused open spaces and communal areas; and the lack of security were common problems found in most of estates. The size, number, and layout of rooms in public housing are subject to more demanding constraints than for detached housing. With increased density, site configuration and planning controls (boundary setbacks, height restrictions, open-space requirements, and road and parking standards) are more restricting, and the attachment of dwellings limits the access to light and ventilation. The facts indicate that lower socio-economic status communities in general have limited access to quality housing and live in neighbourhoods that do not facilitate outdoor activities or provide sufficient access to healthy food. Effective residential site planning requires placement of buildings on the site in such a way that each dwelling will be provided with usable outdoor space for daily family needs and that both interior and exterior living space will have satisfactory daylight, sunlight, air circulation, quiet, and safety. The neighbourhood is a physical concept: it is the area within which residents may all share the common services, social activities, and facilities required in the vicinity of the dwellings [17].

Dilapidated housing conditions will no doubt impact on the health of inhabitants, and the research indicates that the burden of illness is greater among minorities and low-income communities [18, p. 3]. With inequalities in the construction and maintenance of low-income housing, these communities may experience greater rates of respiratory disease, developmental disorders, obesity, chronic illness, and mental illness [19, p. 1446] and even create a stress factor that also affects the human immune system [20, p. 7]. Moreover, from a large study undertaken to evaluate housing and health in seven European cities [21, p. 5], it was found that housing conditions affect the residents' mental health. Even more than health, dilapidated housing conditions can also be associated with higher rates of crime, making neighbourhoods less safe for walking, and in some cases resulting in greater social isolation [22, pp. 343–367]. Davidson [23] claimed that factors other than design are also important in the occurrence of crime and vandalism. Wider social factors such as class, social mix, and social environment also seem to be influential.

The social environment is another dimension of housing estates. The housing environment embraces both social and physical elements. It should be able to provide opportunities for humans to achieve their needs and desires, including the physical quality of the dwelling unit itself, which directly affects the level of resident satisfaction with occupancy of that dwelling unit. Important points which lack full consideration in residential design include the quality of physical safety, health, living in high-rise building, noise, privacy, and vandalism. However, availability and proximity to local social services and facilities are no longer an issue for most housing complexes, which seem to meet the needs of occupants, especially those in housing complexes which are located in the city centre [13]. The good quality of neighbourhood should be well planned. These include physical environment conditions; locational characteristics; local services and facilities; and sociocultural environment [24]. Other that those, there are four aspects of neighbourhood quality, such as physical safety; health risks (concerned with the atmospheric, water, or soil pollution); noise (a number of possible sources can seriously diminish quality of life); and personal security [25]. Careful design and layout can reduce the opportunities for crime and can therefore encourage crime- and vandalism-free environments [26].

The psychological effect of a street layout, street lighting, accessibility, fire-fighting and fire rescue purposes, fire equipments, and access for garbage collection provides a variety of outlooks [27]. Privacy is considered very important as argued by Franklin [28, p. 23], including this visual and aural privacy between housing units. Space required for a wide variety of activities includes tenant meetings, preschool programs, teenage activities, and senior citizens' activities. Space that can be used for social gatherings is important [29, p. 1]. Neighbourhood community facilities need to be provided, such as educational—elementary, kindergarten, nursery school, and adult education; outdoor recreation—playgrounds and park; indoor culture—church, library, assembly, and recreation space; neighbourhood shopping—food and drug stores, and miscellaneous services; and health services—access to medical and dental services, and participation in official and voluntary health services [17].

In the case of high-density housing estates in Indonesia, the residents' perceptions towards the social environment directly impact on the physical environment, social, and psychological problems. A high-rented housing price has not been considered and unaffordable. Minimal places for communal and community facilities are limited provided in the complex. Noise from stairs and corridor is overly disturbing. Socio-psychological disturbances experienced by the occupants, such as not feeling free for their own activities, not free to have guests visiting them at home, discomfort, difficulty with concentration and creativity, noise, unable to work or study at home, limited spoken ways, and neighbourly relationships, often being disturbed. Physiological and technical disturbance includes difficulty in moving freely and difficulty in fulfilling their own interests. In addition, buildings and environment are not well-maintained: any damage that occurs in buildings and their facilities are not usually well responded to, in terms of repairs [9, pp. 175–177].

Environment plays a role for individuals. This role is explained in the following ways: (1) the environment may change individual behaviour, so that an individual has to act against the environment in order to protect himself/herself from behaviour changes; (2) the environment is changed and whose positive alteration would stimulate people to react and participate within it; (3) the environment is an object with which individuals must cope. All points of highlighted issues described above are then used for the research variables on the questionnaires finding out the perceptions of residents towards their space/room and housing complex. A need to cope with the physical environment is initiated upon the perception of the environment by an individual, in interrelation to physical objects. If this perception cannot be accepted by the human, it will then be followed by stress in an effort to cope with the environment [3, p. 89]. Successful coping strategies involve adaptation and adjustment. Adaptation is the way in which we change our behaviour to fit environmental demands, while adjustment is the way in which we change the environment to fit our behaviour [3, p. 224].

4 Research Design and Survey Methodology

A Post Occupancy Evaluation (POE) is employed in this study. It represents the most appropriate method for this study. POEs have primarily been used for the study of environments and users who are constrained or regulated, and often in the public sectors such as hospitals, schools, the elderly, dormitories, and public housing [30, p. 14]. In-depth field work was carried out at two different housing complexes located nearby each other in the central of Jakarta, all of which were built by the National Urban Development Corporation of Indonesia *(Perum Perumnas)*. Semi-structured interviews were conducted using questionnaires. Data obtained from the interviews were in quantitative form. Part of the field observation concerned the taking of photographs to illustrate the existing housing units and complexes. The quantitative data were analysed and interpreted using the Statistical Package for Social Science (SPSS). Parallel mixed analysis was also employed. One of the scenarios for combining these qualitative and quantitative methods was to use qualitative methods to explain quantitative results [31, p. 135].

4.1 Selecting the Case Study Area

The case study area taken was Tanah Abang Housing complexes (TAB) which were located next to one another. It has only one unit type of 36 m^2 with two different designs, referred to as 36-A and 36-B. Both 36-A and 36-B unit types have an open-plan room for living space and dining, two bedrooms and a combined kitchen, and bathroom space and balcony. A balcony is provided in all type 36-A units, but not in 36-B units. The layout of both unit types is detailed below (Figs. 1 and 2).

4.2 Sample Employed and Sample Size

One of the most crucial issues to consider when designing a survey is the sample size for the study. Type of simple random sampling was employed. Proportionate random sampling is used to the extent that it is more representative than simple random sampling. A typical sample size is 10% of a total population [32, pp. 101–102]. The proportionate random sample determined for the research represented 10% of the total units available. Given that the total number of units within the specified housing complexes was 960 units; therefore, the samples taken are a total of 96 units and consist of 54 units from the TAB-A complex and 46 units from TAB-B complex.

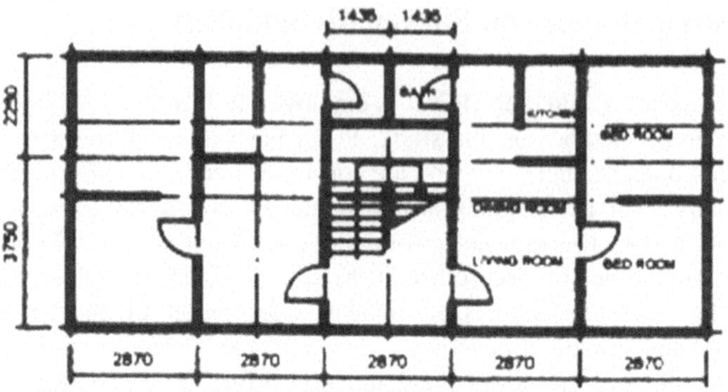

Fig. 1 Floor plan of 36 m² units block A *Tanah Abang* HC Block A (TAB-A)

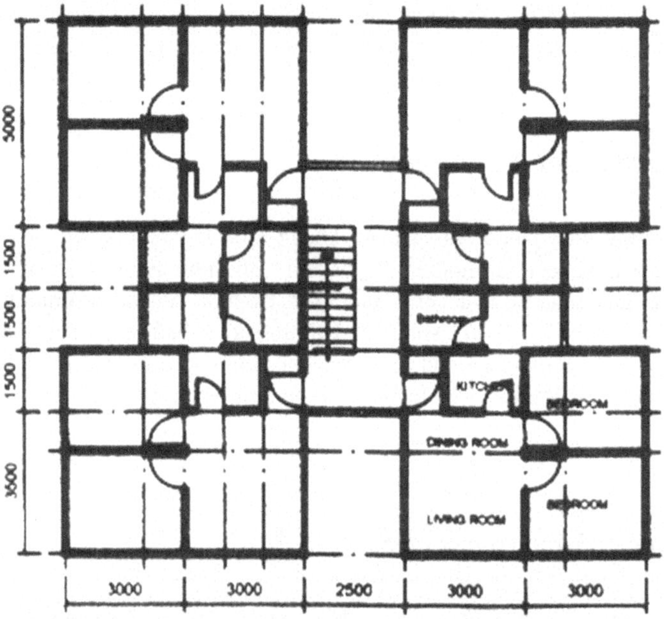

Fig. 2 Floor plan of 36 m² units block B *Tanah Abang* HC Block B (TAB-B)

4.3 *Method of Data Collection and Analysis*

The data collection was taken by conducting the semi-structured interviews using face-to-face questionnaires. Face-to-face interviews were chosen because they can be used with people who might not otherwise be able to provide information, such as those who are illiterate, blind, bedridden, or very old [33, p. 230]. The interviews covered the pre-coded or fixed-alternative questions employed for asking socio-demographic, general information, and perceptions. Those questions were gained from interviewing an adult member of each household. Open-ended questions were used for constructing a diary of occupants' coping strategies. For a total family structure of less than 4 (four) family members per household, the respondents fill in diary on behalf of all family members. For a total family structure of more than 4 (four) family members per household, the respondent to fill in diary on behalf of 4 (four) members with various ages/occupations.

4.4 Analysis for the Socio-economic and Demography Data

While open-ended questions were analysed statistically using SPPS software, the responses were first analysed before transforming the qualitative data to numerical form (quantifying the qualitative data). The qualitative and quantitative data were then analysed using frequency tables. Cross-tabulation techniques were also used to run some statistical analyses for some related variables. Quantitative analysis of qualitative data was undertaken for the type of socio-economic, demography data, perception value, and coping strategies.

4.5 Analysis for the Occupant's Perception

The closed-ended responses were analysed to calculate the descriptive statistics for central tendencies, such as averages or means. For this study, a 5-point semantic scale was employed for the types of perceptions data outlined below. The means of 3 above resulted from the analysis were considered positive (positive perceptions of occupants).

5 Analysis of Findings

Within the TAB complex, a two-storey multi-function-building provides public, social, and commercial facilities for the residential community. These housing complexes—TAB-A and TAB-B—are also characterized by extensions built on the ground floors. Compared to the residents upstairs, the occupants of ground-floor units have the advantage of adjacent land next to their units for expansion of dwelling space. The extensions range from permanent to semi-permanent and include rooms for private use (terraces, balconies, and gardens with fences) or for their own commercial business operations (see Figs. 3 and 4). The nature of this situation offers clear economic advantages to those on the ground floor over those living in upper-floor units.

Fig. 3 Extensions to ground-floor units (TAB-A)

Fig. 4 Extensions to ground-floor units (TAB-B)

The households living in the 36 m^2 units of TAB-A and TAB-B complexes were found to differ. The socio-demographic survey and analysis showed that there was on average 4–5 residents among all 36 m^2 units: TAB-A has 3.76 pp/hhs, and TAB-B has 4.24 pp/hhs. These result in space density of TAB-A of 9.57 m^2/pp and TAB-B of 8.49 m^2/pp.

The proportion of male and female occupants living in the TAB-A and TAB-B housing complexes is about equal, with an average age between 28 and 34 years.

The predominant ages of residents in this unit type across all locations were similar, within the range of 25–55.9 years (productive ages). Residents were of Moslem, Christian (including Catholic), Buddhist, Hindu, and Kong Hu Chu faiths. Of these, there were more than 73% of Moslem residents in two locations. The majority of residents in the all locations were originally from Jakarta. The majority of occupants were found to have achieved a senior high school level of education. The residents in TAB-A and TAB-B possessed the average education level (diplomas and university degrees). The typical[2] occupations of residents were defined as private employee, entrepreneur, housewife, and student. Of these, private employee and student were the most common. The average family income of residents was above the provincial minimum wage per month in 2013, which was about Rp. 2.400.000—(DKI Jakarta 2013). Households who had an average monthly income above that figure are found 50.0% for TAB-A and 52.4% for TAB-B).

6 Occupants' Perceptions to the Physical and Social Environments of the Residential Settings

On the question of perceptions, these were found to concur on certain aspects and diverge in others. The living room, bedrooms, kitchen, bathroom, and balcony were typical for the design of 36 m^2 units in TAB-A and TAB-B complexes. The findings reveal that occupants had a positive response to the living rooms in many respects (see Table 1). The shape of the rooms, natural lighting, glare from windows, air freshness and circulation (room's stuffiness and air circulation), and humidity were all positively perceived by residents. However, the availability of floor space and extension of rooms was perceived negatively by all inhabitants. Residents of TAB-A and TAB-B perceived the amount of living space provided as adequate. Temperature control in the living room was also thought too hot by those in TAB-A.

The overall satisfaction of the living room space was affected by some negative relations as shown in the table above. The quantitative analysis found a satisfaction value of 6.9% for TAB-A and 20.9% of TAB-B. The variable of floor space availability was statistically significant at a 5 per cent level of significance as concluded from the *t*-test affecting the occupant's perception in TAB-B.

The perception of TAB-A and TAB-B occupants towards the dining room zone was found to be similar. The occupants of TAB-A and TAB-B recorded positive responses towards the dining room in relation to its flexibility of layout, the shape for usage, natural lighting, glare coming from windows, freshness or stuffiness of the air, air circulation, and humidity level. However, negative perceptions towards the minimum number of rooms, the availability of floor space, and restriction to extended use of the space were also recorded (see Table 1). The satisfaction of

[2]Typical means type of resident's occupation found in all four complexes.

Table 1 Occupants' perceptions of living room and dining room within unit

No.	Criteria	Room type—housing complex			
	How the occupant feels about the spatial, visual, heating, and air conditions of the living room or dining room	Living room		Dining room	
		TAB-A	TAB-B	TAB-A	TAB-B
1	The number of rooms	3.11	3.00	**2.98**	**2.90**
2	Availability of floor space	**2.67**	**2.62**	**2.54**	**2.55**
3	Extension of the room	**1.30**	**1.43**	**1.30**	**1.45**
4	Flexibility of the room for its functions	3.26	**2.95**	3.24	2.98
5	The shape of room for usage	3.67	3.74	3.39	3.60
6	The natural lighting of the room	4.04	3.88	3.78	3.81
7	The glare from windows	3.69	4.00	3.87	3.95
8	Heating conditions of the room	**2.63**	3.00	**2.70**	3.00
9	Freshness/stuffiness of the room air	3.74	3.71	3.69	3.71
10	The room's air circulation system	3.96	3.90	3.85	3.88
11	The room's level of humidity	3.83	3.83	3.85	3.81
	Satisfaction with the living room/dining room	3.37	3.45	3.24	3.48

Bold fonts Negative perceptions (less than 3 of the scale of 5)
For TAB-A and TAB-B units Living room/dining room zones all within one open-plan unit area

occupants towards their dining rooms was found to be affected by certain negative aspects. The quantitative analysis found that satisfaction values of 6.7% of TAB-A and 23.1% of TAB-B occupants were negatively impacted on various reasons. The variable of number of rooms was found statistically significant at a 5 per cent level of significance as concluded from the *t*-test affecting the occupant's perception living in TAB-B for their dining room.

On the question of bedroom space (see Table 2), in general, bedroom one, shape of room, natural lighting, glare, air circulation system, humidity conditions, and visual and conversational privacy were all positively perceived by residents. Although they were generally satisfied with the bedroom, negative perceptions touched on several criteria. Three of the 16 criteria were typically viewed negatively by occupants: availability of floor space, extension of the room, and perceiving the room as an uncomfortable place to be alone.

Residents of both locations portrayed negative perceptions towards the heating of the bedroom and its flexibility for other functions. It should be noted that the unsatisfactory nature of heating conditions correlated with whether or not the rooms were air-conditioned. The figures indicate that between 33.3 and 48.3% of units in the locations had air-conditioned rooms. The satisfaction value of 36 m^2 unit occupants towards bedroom one was affected by some negative aspects. The quantitative analysis found that the satisfaction values of 26.2% of TAB-A and 21.9% of TAB-B occupants towards bedroom one are also explained by negative responses. The variable of availability of floor space was found statistically significant at a 5% level of significance as concluded from the *t*-test affecting the occupant's perception living in TAB-A for their bedroom 1.

Table 2 Occupants' perceptions of bedroom(s) within unit

No.	Criteria	Room type—housing complex			
	How the occupant feels about the spatial, visual, heating, and air conditions of the bedroom 1 (one)/bedroom 2 (two)	Bedroom 1		Bedroom 2	
		TAB-A	TAB-B	TAB-A	TAB-B
1	The number of rooms	**2.98**	3.24	**2.94**	3.02
2	Availability of floor space	**2.59**	**2.67**	**2.63**	**2.67**
3	Extension of the room	**1.28**	**1.67**	**1.19**	**1.33**
4	Flexibility of the room for its functions	**2.80**	**2.81**	2.96	**2.71**
5	The shape of room for usage	3.59	3.88	3.54	3.74
6	The natural lighting of the room	3.56	3.69	3.54	4.02
7	The glare from windows	3.94	4.07	4.02	3.88
8	Heating conditions of the room	**2.50**	**2.88**	**2.46**	**2.83**
9	Freshness/stuffiness of the room air	3.39	3.60	3.24	3.69
10	The room's air circulation system	3.67	3.55	3.67	3.95
11	The room's level of humidity	3.78	3.74	3.65	3.83
12	Visual privacy of room	3.39	3.52	3.37	3.40
13	Conversational privacy of room	3.50	3.71	3.30	3.50
14	Having a comfortable place to be alone	**1.06**	**1.00**	**10.6**	**1.00**
15	Having place to store possessions	3.09	3.55	3.09	3.52
16	Having a lock for the room	4.31	4.12	4.02	3.67
	Satisfaction with the bedroom 1 or 2	3.24	3.45	3.13	3.40

Bold fonts Negative perceptions (less than 3 of the scale of 5)

For bedroom two (as shown in Table 2), the shape of the room, natural lighting, glare, room's stuffiness, air circulation system, humidity, and visual and conversational privacy were all typically found to be positively received by occupants. Although they were generally satisfied with bedroom two, certain typical negative perceptions were found in relation to several criteria. Three of 16 criteria were typically valued negatively by occupants, such as the availability of floor space, the possibility of extending of the room, and whether they found it a comfortable place to be alone. The satisfaction value of 36 m^2 unit occupants towards their second bedroom was affected by some negative aspects. The analysis found that satisfaction values of 18.4% of TAB-A and 30.5% of TAB-B occupants towards their bedroom two are explained by negative responses. The variables of availability of floor space and extension of the room were found statistically significant at a 5 per cent level of significance as concluded from the *t*-test affecting the occupant's perception living in TAB-B for their bedroom 2.

The bathroom generally seen in positive terms from many responses to the 6 criteria (see Table 3). However, in TAB-A and TAB-B, there were also negative responses towards the availability of the floor space. It was also ascertained that satisfaction value of 36 m^2 unit occupants towards their bathroom was affected by some negative aspects perceived by the occupants. The quantitative analysis found

Table 3 Occupants' perceptions of bathroom/kitchen/balcony within unit

No.	Criteria	Room type—housing complex					
	How the occupant feels about the spatial, visual, heating and air conditions of the bathroom, kitchen and balcony?	Bathroom		Kitchen		Balcony	
		TAB-A	TAB-B	TAB-A	TAB-B	TAB-A	TAB-B
1	The number of rooms	3.30	3.57				
2	Availability of floor space	**2.69**	**2.86**	**2.04**	**2.24**	**2.20**	**2.26**
3	Extension of the room			**1.63**	**1.40**		
4	Flexibility of the room for its functions			**2.44**	**2.62**	**2.20**	**2.67**
5	The shape of room for usage	3.69	3.79	3.33	3.62	3.17	3.62
6	The natural lighting of the room			3.72	4.12		
7	The glare from windows			4.02	4.07		
8	Heating conditions of the room	3.37	3.40	3.28	3.43		
9	Freshness/stuffiness of the room air			3.61	3.90		
10	The room's air circulation system	3.46	3.79	3.78	3.95		
11	The room's level of humidity	3.35	3.64	3.78	3.83		
	Satisfaction with the Bathroom/Kitchen/Balcony	3.19	3.40	**2.91**	3.50	**2.69**	3.48

Bold fonts Negative perceptions (less than 3 of the scale of 5)

that the satisfaction values of 11.1% of TAB-A and 21.9% of TAB-B occupants towards their bathroom are explained by negative responses. The variable of floor space availability was found statistically significant at a 5% level of significance as concluded from the *t*-test affecting the occupant's perception living in the locations —TAB-A and TAB-B.

The perception of occupants towards their kitchen (as seen in Table 3) was found to be typically negative in several respects. The lack of floor space, possibilities of extending, and flexibility of the room were general to all residents. In this sense, the kitchen space received negative responses. Next to the above, residents were generally positive in terms of the kitchens shape, natural lighting, glare, air freshness/stuffiness and air circulation systems, and humidity levels. Satisfaction of the 36 m^2 unit occupants with their kitchen was affected by some negative responses. The satisfaction values of 25.5% for TAB-A and 20.6% for TAB-B occupants towards their kitchen are explained by identified negative responses (perceptions). The variable of extension of the room was found statistically significant at a 5% level of significance as concluded from the *t*-test affecting the occupant's perception living in TAB-A towards their kitchen.

The balcony space was disliked in relation to lack of floor area and flexibility of space to accommodate its intended functions (see Table 3). The satisfaction value of 36 m^2 unit occupants towards their balcony was found affected by certain negative aspects. In the quantitative analysis, the satisfaction values of 41.8% of

TAB-A and 11.8% of TAB-B occupants towards their balcony are explained by negative responses. The variables of availability of the floor space and the flexibility of the room for its functions were found statistically significant at a 5% level of significance affecting the TAB-A occupants, based on the *t*-test.

Issues concerning residents' perceptions towards their building were also considered. Their level of satisfaction with the units and building complex as a whole in TAB-A and TAB-B housing complexes was positive. However, some negative responses were recorded, in terms of insufficient floor space, drying space for clothes, and availability of lift. Corridor design was negatively perceived by TAB-B occupants. TAB-A and TAB-B were mostly negative about the exterior building design. Fire protection systems and means of escape were seen as inadequate in TAB-A and TAB-B housing complexes.

The social environment including neighbourhood quality was also measured. The 36 m^2 unit occupants were typically found to be satisfied with the social environment of their housing complexes. However, the physical environment in terms of lack of available drying space and landscaping/streetscaping conditions was seen in negative terms. The provision of parking areas and stairs or lift services for the high-density population in case of fire was also deemed inadequate by TAB-A and TAB-B occupants. Cleanliness and building maintenance (improvement) were also portrayed negatively by TAB-A residents. TAB-A and TAB-B residents saw the special areas for teenage activity negatively, whereas only TAB-B residents felt special spaces for senior citizen activities was lacking.

7 Coping Strategies Towards the Living Environment

This paper has a little touch on the coping strategies. It shows how the occupants would use their spaces in the units and outside the units to cope with the living environment. In the case of these case studies, where the negative perceptions have been found clearly in many ways, the coping behaviours are presumed definitely occured in these complexes. This condition is to refer since as one of main theories of coping behaviour has defined that when people are dissatisfied with the way they use the physical environment, they start to cope in order to meet their needs. Where perceptions of 36 m^2 unit occupants were found to be negative, their coping behaviour was studied. Two types of coping behaviour were found in this unit type: adjustment towards the space (space adjustment) and adaptation to the space. It is the degree to which behaviour and environment or space are interrelated and impact on each other that is investigated here. The results show a significant number of households coped with the environment both through physical and functional adjustments to the space provided and through adaptation to the space.

The study found that both physical and functional adjustments were undertaken. There were 13 types of space adjustment made by TAB-A occupants, 10 related to TAB-B residents (see Table 4). The findings show various types of space adjustment for all households.

Table 4 Coping behaviour/strategies: space adjustment

No.	Types of space adjustment made by households	TAB-A	TAB-B
Functional space adjustments			
1	Bedroom used for having meals on the floor	3.7	–
2	Bedroom used for praying purpose with praying carpet provided	64.8	69.0
3	Living room used for praying purpose with praying carpet provided	16.7	4.8
4	Living room used for preparing food to sell using sofa, chair and table	3.7	–
5	Living room used for sleeping purpose with bed provided	1.9	–
6	Living room used for sleeping purpose with mat (on the floors)	20.4	11.9
7	Living room used for sleeping purpose on the floor (without mat)	1.9	2.4
8	Living room used for ironing with ironing table provided		2.4
9	Living room used for ironing using sofa/table set	5.6	4.8
10	Chairs and small table provided in the balcony	1.9	–
11	Balcony used for cooking purpose with stove and cooking tools	7.4	–
12	Balcony used for washing with washing machine provided	–	2.4
13	Corridor used for children's space to play	3.7	2.4
14	Corridor used for drying space with wire/string provided		
15	Corridor used for eating, chatting and relaxing purpose with chair/sofa/small table provided	7.4	4.8
16	Corridor used for gardening activities with flowerpots provided	1.9	2.4
Physical space adjustments			
17	New space built on the spare ground-floor area	1.9	–
18	New space created in the ceiling area for sleeping	–	2.4
19	New space created in the ceiling area for study	–	2.4

For TAB-A and TAB-B units Living room/dining room zones all within one open-plan unit area

Typical[3] functional space adjustments towards bedrooms and corridors made by the occupants of 36 m^2 units in TAB-A and TAB-B are shown in Table 4. Functional space adjustments of bedrooms were made for the purposes of praying with a proper mat provided and for having meals. Functional space adjustments to the living and dining room areas were made by occupants of TAB-A and TAB-B. The living room was functionally changed by doing ironing using sofas and tables or an ironing table. The living room was also used for sleeping with a mat or bed provided. Balconies were functionally changed for washing and cooking activities.

The nature of the physical space adjustments is recorded in Table 5. The main type of permanent physical space adjustment was performed by fitting partitions; this was only found in TAB-B (9.5% of the units visited). The use of high furniture (more than 1.80 m high) was found used in most units in TAB-A and TAB-B.

[3]Typical means type of form of functional space adjustment found in all four complexes.

Table 5 Physical space adjustment: the use of partitions in the unit

No.	Types of partitions	TAB-A	TAB-B
1	Full-partition (plywood)	–	9.5
2	High furniture used to slit the rooms (>1.80 m)	7.4	11.9
3	No partition used in the unit	68.5	42.9
4	No response	24.1	35.7
	Total	100.0	100.0

These types of physical space adjustment suggest occupants, particularly of TAB-B divided the room to create a sense of more rooms and private areas.

Another made of physical space adjustment was found where occupants created new permanent spaces for sleeping. TAB-A occupants only undertook this on ground-floor units. It was on the top floor that TAB-B residents extended into the ceiling area for extra room. Although there were only a few households in TAB-A (1.9%) who constructed extra space on the spare ground-floor area, and 2.4% of the households in TAB-B which created new space in the ceiling area, these findings show that occupants use coping behaviour through physical space adjustment to meet their behavioural needs.

Other than space adjustments (physical and functional), residents were found to cope with their environment through the way they adapted to the space. Rather than making physical alterations to the available space, they also attempted to change their behaviour to fit their allocated space. Adaptation of behaviour by occupants was found to have occurred in TAB-A and TAB-B with some seemed as improving lives and others detrimental. The study found that five types of typical[4] behavioural change were perceived as improving the quality of life of occupants, while another four types seen as necessary in order to live in the complex were considered as impacting poorly on their quality of life. Of these considered to improve the quality of life, these included becoming more sociable and increased play time for children due to the provision of public facilities in the complex; being able to commute to work later because of nearer workplaces; increased availability of ways to earn a living within the immediate neighbourhood of the complex; and cooking more comfortably within a smoke-free kitchen. On the other hand, there was a range of behavioural alterations that were seen to lessen resident's quality of life. These included poor socialization in relation to inferior neighbourhood conditions; choice of fewer social activities in neighbourhood area; having to get up earlier because work and school locations being further from home; and the use of stairs being

[4]Typical means type of occupant's adaptation behaviour identified in all four complexes.

tiring and reducing going out. The findings show how space has impacted on behaviour in all locations. In terms of adaptation behaviour, 19 types were found in occupants of TAB-A with 1.9–25.9%, and TAB-B with 2.4–19.0%.

In conclusion, the behaviour and activities of residents within 36 m^2 units have impacted significantly on the way they use their space: 51.7–76.3% of households have changed the space in permanent or improvised ways. The space provided has also been shown to impact significantly on the occupants' behaviours: 19.0–51.7% of occupants in the households living in the 36 m^2 unit type have been affected as described.

8 Conclusion: Implications of the Design for High-Density Housing in Jakarta

This study has reported the physical and socio-psychological aspects of living condition of the 36 m^2 unit of high-density housing in Jakarta. It is mentioned that these residential buildings have not been planned and designed in ways that provide for basic human needs. This paper comprises a response to this question by concentrating on an analysis of their perceived standards of physiological and social well-being. In answer to these issues, the outcome of this research indicates that the behavioural relations of occupants have not yet been adequately considered in the design.

As perception is directly related to satisfaction, this study also looks at an occupants' sense of satisfaction with their home environment and to what extent their degree of satisfaction was affected by their perception of the spatial and socio-psychological conditions of their housing unit and rooms. What came up from this was a poor satisfaction rating based on cramped very dense living space, too few rooms for the number of occupants and functions. Other findings further reinforce the contention that space significantly impacts upon an occupants' sense of their use of various rooms in this 36 m^2 unit types considered.

This research has resulted that occupants of 36 m^2 unit type are unsatisfied towards their rooms within units for the kitchen and balcony. The variables below were found statistically significant at a 5 per cent level of significance as concluded fro the *t*-test affecting the occupant's perception living in TAB-A and TAB-B, as follows:

- Availability of the floor space for the living room, bedrooms, bathroom, and balcony;
- number of rooms for the dining room;
- extension of the room for the kitchen; and
- flexibility of the room to its function for balcony.

Successful coping strategies involve both adaptation and adjustment. This research confirms that occupants cope in a variety of ways with their physical setting, to the extent that both functional and physical space adjustments were undertaken by the occupants, as well as adaptations to their behaviour. The findings thus clearly demonstrate that space and behaviour are interrelated with high-density housing and that occupants have to cope with their physical setting in various ways.

Implications of design for high-density housing in Jakarta are developed/pointed out as follows:

1. Architects should be encouraged to develop new design ideas, by taking the Post Occupancy Evaluation on the Occupants' Perceptions. This is needed particularly for users who are not involved in the process of design, as is the case for most high-density housing complex.
2. It is recommended to provide the 36 m^2 unit type with modifications design in order to response what most of occupants feel lack on. Specific types of room provision, improvement, and modifications are required to be made to accommodate the functioning space with its users' needs (for the fulfilment of the occupants' physical, socio-psychological expectations).
   - The provision of one multi-function room within the unit or an enlarged living room is required to accommodate various activities in this space for working, physical exercise/gymnastics, sleeping, having meals, reading newspapers or magazines, reading books, playing with friends, relaxing by sitting and drinking, and chatting with neighbours.
   - An improved floor space standard for the kitchen is necessary to allow the option of cooking food to sell to take place.
   - A sink should be fitted in the kitchen for occupants to wash the dishes.
   - A laundry area should be located adjacent to the bathroom for occupants to wash the clothes (instead of doing so in the bathroom) and to do the ironing.
   - Space should be provided within the units or in the communal space outside them for drying clothes; it can be done by reviewing a floor space standard for the service balcony.
   - A small private room or corner of a room (screened off) within the unit is required to allow Moslem occupants to conduct their frequent private activity of praying five times daily.
3. In terms of development of the new high-density housing design ideas, it is crucial to address users' needs for 36 m^2 unit and its building quality, such as the needs for drying space, lift provision, and improved design for the exterior and interior buildings. Including this, architect should develop new design ideas that address the users' needs for better neighbourhood quality, concerning the availability of drying yards for occupants to dry their clothes in the sun.
4. The design alternative (e.g. space layout with a non-open plan) is needed by occupants. It provides rooms with a variety of private and separate functions in order to eliminate the need for physical space adjustments (such as putting partitions up in the unit).

References

1. Fisher, JD 1984, *Environmental Psychology,* 2nd edn, Holt, Rinehart and Winston, New York.
2. Indriyati, SA 2009, *Low-Cost Housing in Jakarta: Toward a More Humane Architecture,* Penerbit Pustaka Sutra, Bandung.
3. Bell, PA, Fisher, JD & Loomis Ross, J 1978, *Environmental Psychology*, W.B. Saunders Company, Philadelphia.
4. Boutourline, S 1970, 'The Human Dimension in Architecture. In C.J. Holahan', in *Environment and Behaviour: A Dynamic Perspective*, Plenum Press, p. 3–4, New York.
5. Holahan 1978, *Environment and Behaviour: A Dynamic Perspective*, Plenum Press, New York.
6. Heimsath, C 1977, *Behavioural Architecture: Toward An Accountable Design Process*, MCGraw-Hill Book Company, New York.
7. Provincial Government of DKI Jakarta 1991, *Jakarta Master Plan for 2005*, Provincial Government of DKI Jakarta, Jakarta.
8. Ministry of Settlements and Regional Infrastructure 2004, *Source of Funds and Assets for the National Program in 2004*, Ministry of Settlements and Regional Infrastructure. from http://www.indonesia-house.org/Humanitarian/Hum1103/110503Govt_to-build_200_.
9. Komarudin 1997, *Menelusuri Pembangunan Perumahan dan Permukiman (Development of Housing and Settlement: An Investigation)*, P.T. Rakasindo, Yayasan Realestat Indonesia, Jakarta.
10. Subroto, E, P 1984, 'Aspek Sosial-Psikologis pada Pemukiman Masyarakat Berpenghasilan Rendah', in *Sejumlah Masalah Permukiman Kota. In Eko Budihardjo (ed)*, Alumni, Bandung, p. 58.
11. Ittelson 1978, 'Environmental Perception. In J.D. Fisher, P.A. Bell & A. Baum', in *Environmental Psychology*, 2nd edn, Holt, Rinehart and Winston, New York, p. 22.
12. Widayatun, TR 1999, *Ilmu Prilaku,* 1 edn, CV. Sagung Seto, Jakarta.
13. Grigsby & Rosenburg, L 1975, *Urban Planning Policy*, APS Publications, Inc., New York.
14. Buchori, M, & Supradono, G. 1978, 'Evaluasi Penghuni Perumnas Semarang terhadap Tempat Kediaman Mereka.' in *Lokakarya permukiman bagi Penduduk Berpenghasilan Rendah*, Jakarta, pp. 26–8.
15. Steidl, RE 1972, 'Difficulty Factors in Homemaking Tasks: Implications for Environmental Design', *Human Factors,* vol. 14, pp. 471–82.
16. Power, A 1997, *Estates on the Edge: The Social Consequences of Mass Housing in Northern Europe*, Macmillan Press Ltd., London.
17. American Public Health Association 1960, *Standards for Healthful Housing: Planning the Neighborhood*, Public Administration Service, Chicago.
18. Bashir, SA 2002, 'Home is where the harm is: Inadequate Housing as a Public Health Crisis', *American Journal of Public Health,* vol. 92, pp. 733–38.
19. Shobha, S, Liam, R, O'Fallon & Allen, D 2003, 'Creating Healthy Communities, Healthy Homes, Healthy People: Initiating a research agenda on the Built-environment and Public Health.' *American Journal of Public Health,* vol. 93, no. 9, p. 1446.
20. Rauh, VA, Chew, GL & Garfinkel, RS 2002, 'Deteriorated Housing Contributes to High Cockroach Allergen Levels in Innercity Households.' *Environmental Health Perspect.* vol. 110(suppl), pp. 323–27.
21. Brody, DS, Hahn, SR & Spitzer, RL 1998, 'Identifying Patients with Depression in the Primary Care Setting: a More Efficient Method.' *Arch Intern Med,* vol. 58, no. 22, pp. 2469–75.
22. Kuo, F & Sullivan, WC 2001, 'Environment and Crime in the Inner City: Does vegetation reduce crime?' *Environment and Behaviour,* vol. 33, pp. 343–67.
23. Davidson, R. N, 1981, *Crime and Environment*, Croom Helm, London.

24. Huttman, E & Vliet, WV 1988, *Handbook of Housing and the Built Environment in the United States*, Greenwood Press, New York.
25. King, A 1994, *Towards Indicators of Housing Stress*, Canbera.
26. Poyner, B & Webb, B 1991, *Crime Free Housing*, Butterworth-Heinemann Ltd., Oxford.
27. Judd, B 1993, 'Designed for Urban Living: Recent Medium Density Group Housing in Australia', in *The Royal Australian Institute of Architects (RAIA)*, Canbera.
28. Franklin, EM 1980, *Use of Housing Space by Families with preschool Children: Their Needs as Future Design Criteria*, University of Minnesota.
29. Abrams, ED & Blackman, EB 1973, *Managing Low and Moderate Income Housing*, Praeger Publishers, New York.
30. Rabinowitz, HZ 1989, 'The Use of Boundaries of Post Occupancy Evaluation: An Overview. In Preiser, W., F.E.' in *Building Evaluation*, Plenum Press, New York, p. p. 14.
31. Mason, J 1996, *Qualitative Researching,* 1st edn, Sage Publications LTD., London.
32. Nasution, S, Prof., DR. 2002, *Metode Research*, P.T. Bumi Aksara, Jakarta.
33. Bernard, H, Russell 2000, *Social Research Methods: Qualititative and Quantitative Approaches*, Sage Publications, Inc., California.

The Development of DIY's Southern Coastal Communities: Issues and Challenges

Imamudin Yuliadi

Abstract The reality shows that Indonesia is one of the countries that have the longest beach in the world. This fact gives implication that the economic potential of people life at the beach is important factor to economic growth indoors welfare for improving economic development and equity both national and regional scope. Research methodology that is applied for the investigative descriptive obtained the fact about the economic problems of people at the beach especially beach economy potential for improving people's economic welfare. Analytical method at this research is location quotient (LQ), shift-share, and Klassen typology. The output of this research is to come up with a planning system and model of integrated investment promotion which can result in more equitable economic development at the DIY's coastal areas.

Keywords People empowerment model · Integrated investment · Coastal area

1 Introduction

Development in coastal areas requires different treatments in comparison with other regions considering the peculiarities of coastal areas including the aspects of geography, geomorphology, anthropology, economics, and social. Changes in natural conditions which are relatively more extreme than the land area encourage people to do economic activity with a greater degree of risk with the possibility of a loss which is also greater. It needs a synergistic cooperation between local communities and government agencies to resolve issues related to the economic development of the people living in coastal areas. Southern coastal region of the Province of Special District of Yogyakarta (DIY) is an area that holds great nature potential to sustain local development and improvement of food security for the local community and other areas. This issue is becoming increasingly prominent

I. Yuliadi (✉)
Faculty of Economics, University of Muhammadiyah Yogyakarta, Bantul, Indonesia
e-mail: imamudin2006@yahoo.co.id

T. Taufik et al. (eds.), *ICoSI 2014*, DOI 10.1007/978-981-287-661-4_25

since government plans to build an international airport in the area of Congot, district of Kulon Progo, and the province of DIY. It is expected that the realization of the development plan not only impacts on promoting economic development perceived by middle and upper class society with large capital but can also be felt by the lower layers more evenly. Integrated economic development and strategies of coastal area of DIY are a form of alternative models of economic development to answer the challenge in carrying out economic development and the development of the region/area in an integrated and comprehensive approach through the development of potential of south coastal areas of Yogyakarta in an integrated manner. The number of fishermen in DIY is not relatively too much but it gives a pretty good contribution in increasing the level of welfare of fishermen and also in meeting the needs of marine catches for DIY and surrounding communities. The fishermen in DIY live in areas directly faced to the Indian Ocean coastal region which are in the district of Bantul, the district of Kulon Progo, and the district of Gunungkidul. Most of the activities of people living in coastal areas in addition to fish they also conduct fish catches processing and other kind activity such as processing of agricultural land.

2 Research Method

To maintain the credibility and validity of the results of this study, the data is not only taken directly from the object through survey research in the field but it is also taken from a credible reference both of the books (references) published in international and national publications with ISBN. The data is taken from credible sources such as data from related agencies such as Bappeda, SKPD, and related agencies, which are the Department of Industry and Trade (Diperindag), Department of Agriculture (Dipertan), and also from the IFS, BPS, BI, and Bappenas. Technically the research site is in DIY's southern coastal locations, but procedurally the data and the information about the study site can be accessed through an online medium that can be easily downloaded.

The method of analysis in this research is done using several methods of analysis, which are analysis of location quotient (LQ), shift-share analysis, and analysis of society's economic empowerment. Shift-share analysis also compares the difference in the rate of growth of the various sectors (industry) in areas (regency) with regional/national areas. Shift-share method is relatively more specific than the LQ method because it can explain the cause of the change of several variables. Shift-share analysis can also determine the projection of economic sectors in the coming period. Shift-share analysis aims to determine the contribution of each subsector to the regional gross domestic product (GDP). Shift-share analysis can be used to describe the trend of aggregate statistics. Shift-share analysis clarifies the GDP change at any time in the region, as compared with the three categories. The components in the shift-share include specific sector GDP (i) in the regional level, the GDP growth rate in national level (rn), the rate of GDP growth of certain sectors (i) in national level (rin), and the rate of GDP growth in certain sectors (i) in the level of the region. Analysis of Society's Economic Empowerment is an approach used in the analysis of economic empowerment by assessing the economic state of participatory rural appraisal or PRA which is a set of techniques and tools that encourage coastal communities to participate in enhancing and analyzing their knowledge about life and their own condition so that they can make plans and actions. Another approach to the analysis of economic empowerment in participatory society is participatory learning methods or PALM that emphasize on participatory learning processes in rural communities in facing economic problems and challenges. Identification of problems and the formulation of strategies in problem solving are based on the public participation process. Location quotient (LQ) is an analytical tool to determine whether there is an area of specialization for the specific sector (industry).

$$\mathrm{LQ} = (E_{ij}/E_j)/(E_{in}/E_n)$$

in which

E_{ij} Job opportunities in I sector in j region
E_j Job opportunities in j region
E_{in} Job opportunities in I sector in n country
E_n Job opportunities in n country.

The interpretation of the results of the calculation with the formula is as follows:

(a) If the value of LQ > 1, then there is specialization j region in I sector (The level of specialization areas > national specialization level)
(b) If the value of LQ = 1, then there is specialization j region in I sector (The level of specialization areas = national specialization level)
(c) If the value of LQ < 1, then there is specialization j region in I sector (The level of specialization areas < national specialization level)

LQ analysis showed that the all cities/regencies both located in key region or not have greater LQ than one in some subfield of business. It means that all regencies/cities have featured subsector and the establishment of key region based on those requirements for superior sector can be said right.

In Klassen typological analysis, area can be divided into four categories: fast-growing developed areas, fast-growing areas, potential areas, and relatively underdeveloped area. In Klassen typological analysis, area can be divided into four categories, which are as follows:

(a) Fast-Growing Developed Areas
It is an area that has characteristics of the regional GDP contributions of regency/city/the average of regional GDP contributions of province $\geq$ 1 and the average growth of regional GDP (regency/city)/the average of regional GDP $\geq$ 1.

(b) Faster Growing Areas
It is an area that has characteristics of the regional GDP contributions of regency/city/the average of regional GDP contributions of province $\leq$ 1 and the average growth of regional GDP (regency/city)/the average of regional GDP $\geq$ 1.
(c) Potential Areas
It is an area that has characteristics of the regional GDP contributions of regency/city/the average of regional GDP contributions of Province $\geq$ 1 and the average growth of regional GDP (regency/city)/the average of regional GDP $\leq$ 1.
(d) Relatively Underdeveloped Areas
It is an area that has characteristics of the regional GDP contributions of regency/city/the average of regional GDP contributions of province $\leq$ 1 and the average growth of regional GDP (regency/city)/the average of regional GDP $\leq$ 1.

3 Results and Discussion

The structure of the economic and social structure of each district in the province differs from each other based on the Klassen typology. Based on the rate of growth of regional GDP and the regency's regional GDP contribution, the formation of provincial regional GDP can be grouped into several categories. The regency's/-city's economic development towards special region of Yogyakarta year 2007–2011 can be seen in Table 1.

From the analysis above it is shown that the area which has been developed and is growing fast is the city of Yogyakarta, and this finding is not surprising because the city of Yogyakarta is the economic driver of DIY, while the Bantul district is in the category of fast-growing area, which means that it has considerable development potential to become a more developed area. Sleman district is in the category of potential areas that require huge investment and development strategies to encourage its economic growth, while Gunungkidul and Kulon Progo districts are in the category of underdeveloped areas in the context of DIY's economic development considering their relatively low economy potency and minimum centers of economic growth.

Table 1 The development of regional economic DIY by Klassen typology

Criteria	Area
Fast-growing developed areas	City of Yogyakarta
Faster growing areas	Regency of Bantul
Potential areas	Regency of Sleman
Relatively underdeveloped areas	Regency of Gunungkidul Regency of Kulonprogo

Sources DIY in figures 2007–2011 (processed)

Location quotient (LQ) is an analytical tool to determine whether specialization exist in an area or not in a specific sector (industry).

$$\mathrm{LQ} = (E_{ij}/E_j)/(E_{in}/E_n)$$

in which

E_{ij} Job opportunities in I sector in j region
E_j Job opportunities in j region
E_{in} Job opportunities in I sector in n country
E_n Job opportunities in n country.

LQ analysis showed that the all cities/regencies both located in key regions or not have greater LQ than one in some subfield of business. It means that all regencies/cities have featured subsector and the establishment of key region based on those requirements for superior sector can be said right (Table 2).

From the data above it can be concluded that each regency/city has a particular leading subsector in the sector which has a value of LQ > 1. Regency of Sleman has a leading sector in the sectors of processing industry, building, trade, restaurants and hotels, finance, renting and business services, and other services. It can be seen from the analysis that the value of LQ > 1, while other sectors such as agriculture, mining and quarrying, and transport and communication have value < 1 so that they are not dominant sectors. Economic aspect to note from the results of the LQ analysis is on the agricultural sector, because until nowadays the agricultural sector includes sectors that became the pride of the community such as *salak pondoh* (*Pondoh* zalacca) and agrotourism. This phenomenon becomes the indication that a significant land use occurred to non-agricultural use because of promising economic necessity, for example, for housing (real estate), shop, and business

Table 2 The value of location quotient (LQ) DIY's economy

Economic sector	The average of location quotient (LQ)				
	Yogyakarta	Bantul	Kulonprogo	Sleman	Gunungkidul
Agriculture	0.01852	1.33	1.52	0.9025	2.18
Mining and quarrying	0.00714	1.36	1.30	0.7095	2.57
Processing industry	0.78396	1.21	1.11	1.0969	0.81
Electricity, gas, and clean water	1.04143	0.74	12.54	0.7571	0.46
Buildings	0.82524	1.23	12.54	1.1960	0.88
Trading, restaurant, and hotel	1.19414	0.94	0.82	1.0736	0.70
Transportation and communication	1.92354	0.69	0.98	0.5754	0.67
Financing, renting, and business service	1.50558	0.67	0.68	1.1161	0.51
Services	1.22672	0.79	1.04	1.0321	0.79

Sources DIY in figures 2007–2011

infrastructure development, whereas the opposite condition occurred between the cities of Yogyakarta and regency of Bantul, regency of Gunungkidul, each of which has superiority in certain sectors. City of Yogyakarta and Regency of Sleman also have superiorities in the secondary and tertiary sectors of the economy, while Regency of Gunungkidul and Bantul have superiorities in the primary sector.

Shift-share analysis aims to determine the contribution of each subsector towards regional gross domestic product (GDP). Shift-share analysis can be used to describe the trend of aggregate statistically. Shift-share analysis clarifies the GDP change at any time in the region compared to three categories. The components that form the shift-share are (i) regional GDP in the specific region, the growth rate of national GDP (rn), the growth rate of GDP in certain sectors (i) at the national level (rin), and the growth rate of regional GDP in certain sectors (i) in the level of the region (rij). The components of change nationally present components of national distribution for the calculation where the regional growth rate has changed appropriately followed changes in the national level for all subsectors in the national level for all subsectors in the assessment period. If the regional growth is in contrast to the national level (in the form of a positive or negative shift in Regional GDP), the total shift consists of a structural shift and also the shifts of the proportional division. The impact of change in regional GDP calculated at the regional level changes according to the type of GDP in certain subsectors (including the fast or slow than the national growth rate). The shifts consist of differences in the shift and knowledge about the impact in the regional impacts where the calculation of the regional GDP changes along with location factors in each region. The following are the results of the calculation of shift-share analysis (Table 3).

From the result analysis table above, seen in 2011, there was a difference between the characteristics of the economic development of the regency/city in DIY, meaning that each region has their own economic superiority and uniqueness. Besides, it also shows the development among sectoral shift phenomenon which can be seen from the significant rate growth. It can also show a tendency of character changes between economic sectors.

In the city of Yogyakarta, it is seen that the competitive superiority decline in the financial sector, renting and business services as well as other services (due to the European and American crisis). From the shift-share analysis in 2011 as a whole and by looking at the shift-share analysis, regional GDP has improved compared to 2010. Overall, all sectors' competitive superiorities are increased compared to Regional GDP in 2010 within the city of Yogyakarta. The competitive superiorities sectors that are increased are trade, restaurants and hotels, transport and communication, and services. These sectors are relatively stable in influencing the competitive superiorities of the city of Yogyakarta. Changes in shift-share of the 2009–2011 experience a fluctuation in total and in general experienced a decline in performance in 2011 compared to 2009. The analysis of shift-share can also be a material of evaluation of the development planning process in the city of Yogyakarta, which experience changes in the characteristics of the primary economic sectors to the service sector.

Table 3 The value of DIY's economy shift-share

Economic sector	The value of shift-share year 2011				
	Yogyakarta	Bantul	Kulonprogo	Sleman	Gunungkidul
Agriculture	416	21.809	11.768	23.788	30.212
Mining and quarrying	15	2.023	80	1.815	3.376
Processing industry	30.93	35.620	13.833	50.657	20.032
Electricity, gas, and clean water	3.23	1.746	55	2.872	910
Buildings	24.19	26.864	5.553	42.825	16.536
Trading, restaurant, and hotel	101.81	59.768	23.467	109.087	36.903
Transportation and communication	174.22	44.868	27.188	58.941	35.598
Financing, renting, and business service	−295.34	−98.176	−41.329	−249.477	−61.959
Services	70.26	33.203	19.825	67.965	27.569

Sources DIY in 2007–2011 figures

Meanwhile, for other regions such as Regency of Bantul, it shows a different tendency. From the shift-share analysis conducted in 2011, thoroughly by looking at the shift-share analysis on PDRB, the characteristics have improved compared to the ones in 2010. However, thoroughly, the decreasing of competitive excellence compared the fact in 2010 specifically on PDRB in Regency of Bantul. The analysis shows that the biggest influence of province's growth (Nij) in 2011 is in agricultural sector, trading sector, restaurant, and hotel. The restaurant's and hotel's growths are 46,195 and 42,157. This result is in line with Klassen typology and the economic structure analysis in Bantul. Objectively, Regency of Bantul is a fertile and potential area to develop the agricultural and farming sectors.

The analysis of empowerment on the fishing communities in southern coastal of Yogyakarta shows the improvement tendency as the number of the catch and also the quality of the fishery products are increased and improved. This condition happens as the active involvement of the fishermen to enhance their skill and establish a good cooperation network including their efforts to increase the economic value of the fish catches. The modernization of equipment, such as boats and fishing nets, and also the improvement of management and catch processing techniques are the efforts done by the fishermen to improve the marketability of the product. In addition, the efforts are their real action to increase their incomes and welfare. The improvement of fish catches and fishermen' income is considered essential as it also increases collaboration between groups of fishermen through organizations. The aim of the organizations is to develop the economic value of the fish catches and the prosperity of the fishermen's family. Besides forming the group of fishermen, there is also a joint venture (KUB) which processes the fish and also mediates the fish traders. The member of this organization is the fish merchants themselves. The fish production produced by Pandan *TPI* (an auction center) Mino

Table 4 The fish catches' product of fishermen in fish auction center TPI Pandan Mino, Bantul

Types of fish	Numbers (kg/year)
Bawal	13.692
Layut	9.108
Kakap	1.368
Tengiri	88
Hiu	1.156
Pari	3.200
Keting	8.284
Gatho	208
Tongkol	80.000
Ekor Kuning	728
Teri	6.024
Campur	7.944
Surung	156
Samanganti	3.3572

Source Profile of fish auction (TPI) DIY, 2013

reaches 375 kg/day or 11,200 kg/month. Here is the catch data of TPI Pandan Mino (Table 4).

The table shows that the commodity consists of various kinds of fish which have high economic value. Those products are processed into fried fish, sweet-sour fish, and other products. However, sometimes the fresh fish products are imported from other areas outside Yogyakarta. It can be concluded that the potency of fishery sector's development in Yogyakarta is high including the needs of the fresh fish to support fish processing industry or to fulfill the increasing of fresh fish needs. In addition, the fish sellers also have a quite significant contribution to increase the economic value of the fish catches by directly processing the raw fish at their stalls. It is also seen that the development of fishery sector also supports the development of marine and beach tourism as the existence of the auction center. This phenomenon attracts the foreign and domestic tourists to visit the beach enjoying the scenery and also the culinary. They often bring the fish home or enjoy them there. The famous marine and tourism resort that they visit often are Baron, Depok, Congot, Glagah, Krakal, and Parangtritis. The development of beach tourism in Yogyakarta supports the fishery sector development and also increases the income and prosperity of fishermen.

4 Conclusion

From the research, it can be concluded that

1. The development of fishery sector in Yogyakarta requires comprehensive and integrated plan and development from all of the stakeholders including the government, fishermen, society, and also business;
2. The development of fishery sector has to be focused on the increasing of fishermen's family prosperity. It can be done by increasing the catching value;
3. The increasing of caught-fish's value is done by increasing the quality and capacity of catching equipments or by providing modern ship and net equipment. These efforts will possibly make the catching activity reach more distant sea;
4. The improvement of income and prosperity of the fishermen is also can be done by improving the economic value of the caught fish especially in storing and processing fish so that the product will be able to fulfill the market's demands;
5. The integrated development of the fisheries' sector by developing the tourism sector and agricultural coast land gives more economic opportunities for fishermen;
6. It is necessary to arrange a roadmap of fishery sector development in the medium and long-term period toward the establishment of new international airport in Kulonprogo. It is expected that the efforts can have the bigger roles in fishery and economic sector in Yogyakarta.

The Effect of Leadership Style, Work Motivation, Organizational Culture and Job Discipline on Employees Performance

Retno Ulfayatun Hidayah and Ietje Nazaruddin

Abstract The purpose of this study is to analyze the influence of leadership styles, motivation, and organizational culture and work discipline on employee performance. This study is conducted to see whether there are differences in the performance of employees in the conventional and Islamic banking. The benefits obtained in this study are to add to knowledge about organization policy related to employee performance. Questionnaires were used to obtain related data and analysed by using SPSS. The results obtained in this study are the variables of leadership style, organizational culture and work motivation, and discipline of work have a significant effect on employee performance, motivation of work gives the most impact among other standardized coefficient variable that is equal to $\beta = 0.392$ in islamic banking and $\beta = 0.395$ in conventional banking and no difference performance of the conventional and islamic banking employees.

Keywords Leadership style · Motivation · Organizational culture · Work discipline · Employee performance

1 Introduction

Practices that occur over time in the organization employees might feel bored on the job, or there are other factors that make the employee's performance declined. This is a problem for the organization because it can degrade the performance of the organization. Leader in organization is one of the important factors that affect how the employee's performance will have impact on organization performance. To have the employees to achieve the organization's goals must first understand the motivation of employees working at the organization. Motivation will determine how the behavior of employees to work productively for the organization. Culture influences the way employees act in an enterprise culture is related to the way one

R.U. Hidayah · I. Nazaruddin (✉)
Department of Accounting, Universitas Muhammadiyah Yogyakarta, Yogyakarta, Indonesia
e-mail: ietje_effendi@yahoo.com.sg

T. Taufik et al. (eds.), *ICoSI 2014*, DOI 10.1007/978-981-287-661-4_26

considers the work, in collaboration with colleagues and looking at the organization's goals. Conscious employees comply with laws and regulations applicable in an organization are a work discipline that will benefit the organization.

Formulation of the problem in this study is whether the leadership style, motivation, organizational culture, and work discipline effect on employee performance and whether there are no differences in performance between conventional banks and Islamic banks.

The purpose in doing this study was to determine the effect of variable styles of leadership, motivation, organizational culture, and work discipline on the performance of employees at conventional banking and Islamic banking.

2 Literature Review

2.1 Employee Performance

Performance describe how far the implementation of a program to establish a corporate policy goals, the purpose of the vision and mission of the enterprise, or in general an enterprise performance achieved in a given time period based on the strategic plan of the organization. According Syanputri [9] the performance is the output of a process. If the output is derived and employee or as a result of the case was named employee performance According Sentono (1999) in Chairunnisah [2] is the result of the performance that can be achieved by a person or group within an organization in accordance with the authority and responsibility of each individual in order to attempt to achieve the purpose of the enterprise concerned is legally not break the law and in accordance with moral or ethics.

2.2 Styles of Leadership

According Sutrisno [8] leadership is an activity of a process to mobilize others to lead, guide, influence others, to do something in order to achieve the desired results. The style of leadership is the process of influencing or moving employees to want to do what is desired and expected leadership. According Sutrisno [8] the leader of the enterprise has an important role both internally and externally synchronize enterprise where both are intended to improve the organization's ability to achieve its objectives. According to Ralph and Ronald in Purwoko et al. [5] leadership style is divided into three kinds, namely:

(a) Autocratic leadership style
    Autocratic leadership style is the style of leadership in which the leader much influence or determine the behavior of their subordinates.

(b) Democratic Leadership Style
Is a leadership style that a lot of emphasis on the participation of members. This leadership style is not using its authority to make the final decision and to provide specific guidance to employees, but to find a variety of opinions and ideas from employees about decisions to be taken.
(c) Leadership Style Laissez Faire (free)
Is a leadership style that emphasizes more group decisions. Leaders will submit decisions on the wishes of the group and is responsible for the execution of such work on the employee.

2.3 Work Motivation Theory

People need motivation to achieve his desire. Motivation questioned how morale boosts. According Gitosudarmo (2001) in Sutrisno [8] motivation is a factor that drives a person to do a particular activity, motivation is often interpreted as a factor driving behavior. When someone wants something, he will be tempted to do something to get what they want. According to Banni et al. [1] motivation is the key benefit of as an effort to improve the quality and increase the quality and productivity at an individual. For enterprise motivation will increase profits from increased productivity of employees. Beneficial for employees to enhance skills and enthusiasm to work harder later on employee and revenue performance will be better in their work. And to society will get satisfaction from what is offered and produced an enterprise. The purpose of this motivation is to improve employee morale and job satisfaction, increase employee productivity, improve the stability of the organization's employees, improving employee discipline, streamline procurement employees to create an atmosphere and a good working relationship between employees increase loyalty, creativity, and employee participation to improve the welfare employees reinforce a sense of responsibility to the task, improve the efficiency of the use of tools and raw materials

2.4 Organizational Culture

According Wibowo (2011) in Tintami et al. [10] organizational culture is organization policies that download philosophy confidence, norms, and values with a characteristic core of how to do something in the enterprise. Confidence, norms, and values to be holding all the human resources (employees) of the organization in implementing the performance. According Nurjanah [4] organizational culture is the values, symbols are understood and complied with, which belongs to an organization so that members of the organization feel a family and create a situation that is different from other organizations. Organizations that have a culture will have something unique or a distinctive feature that will provide traction for individuals to join. Culture itself is a strong beneficial to direct the behavior of individuals or employees. With the advent of culture will help employees to do a better job. Luthans (2006) in Nurjanah [4] stated organizational culture has a number of important characteristics. Among them:

(a) Rules of conduct are observed when members of the organization which interact with one another, they use the language, terminology, and common rituals associated with respect and how to behave.
(b) Norms are standards of behavior, including guidelines on how much the job done.
(c) Where the dominant enterprise support and hope participants shared core values, such is the high product quality, a little absent and high efficiency.
(d) The philosophy there is an organization that builds trust policies on how employees or customers are treated.
(e) There are strict guidelines related to enterprise performance. Newcomers should learn the techniques and procedures that have to be accepted as a member of a growing group.
(f) Organizational climate is the overall feeling conveyed employee with the physical nature of the rules, the way employees interact and how members relate with clients and individuals from outside.

2.5 *Job Discipline*

Regularity is one of the key features of the organization and discipline of work is one method to preserve the orderliness. Its main purpose to increase efficiency by preventing the waste of time and energy and, in addition discipline will prevent damage or loss of property when caused by caution, amusement or theft. Discipline try to resolve errors and negligence caused because less attention, the inability of employees or employee tardiness. According Sutrisno [8] the discipline of employees is an appropriate behavior with rules, existing work or discipline is the attitude, behavior, and act in accordance with good industrial rules written or unwritten. Disciplines seek to protect good behavior by getting the desired response. According Moekijat [3] discipline comes from the latin word meaning "diciplina" training or education modesty and spirituality and behavioral development. This emphasis on employees to develop proper attitudes toward work and the way in making supervisory role in relation to the discipline. According Singodimedjo (2002) in Sutrisno [8] there are several factors that affect employee discipline, namely:

(a) The size of the compensation
(b) Presence or absence of exemplary leadership within the organization
(c) Whether or not there are definite rules to hold on
(d) Leadership courage in taking action

(e) Presence or absence of supervisory leadership
(f) Presence or absence of attention to employees
(g) Created habits that support the establishment of discipline

Based on the above theoretical formulation of the problem and then derived the following research hypothesis:

Nurjanah [4] conducted a study with the results of the leadership style has a positive effect on employee performance. A leader must apply leadership style to manage subordinates in the work so that the leadership style of a leader will affect the organization's success in achieving its objectives through employee performance.

H_1: leadership style has a positive effect on employee performance

Banni et al. [1] conducted a study with the results of work motivation influence performance. Motivation is one of the factors that a person works better. Employees who have high motivation to work will have a peak performance. Motivation is necessary in order to encourage employees to continue working with better.

H_2: work motivation has a positive effect on employee performance

Prasetyono and Nurul Kompyurini [6] doing research with the results of the organizational culture affects employee performance. Priyahayanto also said employees are constantly surrounded by the culture of the organization which forms the backdrop of this culture relating to the environment, working conditions, attitudes, behaviors, and perceptions of employees in the organization where they work.

H_3: organizational culture has a positive effect on employee performance

Reza [7] conducted the study with the results of the work discipline effect on employee performance. Reza also says discipline is the loyalty and devotion of a person or group of people for rules both written and unwritten, as reflected in the form of behavior and actions.

H_4: work discipline positive effect on employee performance

2.6 Research Method

The population in this study is employees of conventional and islamic banking in Yogyakarta the sample are 60 employees with 30 employees on conventional banking and 30 employees on islamic banking. Data used in this study is primary data using questionnaires given to employees as research respondents. Test equipment used is the classical assumptions, the analysis used in this study is chow test analysis with the performance of the employee as the dependent variable, and leadership style, work motivation, organizational culture, and work discipline as the independent variable. Test the validity of using the corrected item and uses the Cronbach Alpha to reliability test.

2.7 Result and Discussion

Classical Test Multicollinearity Assumptions in this study resulted in VIF values below 10 and the tolerance value above 1 thus concluded multicollinearity does not happen. Normality test in this study resulted in the P-plot graphs were spread and follow the direction of the diagonal line so that it can be said not happen auto-correlation. Autocorrelation test in this study at conventional banking DW value of 1.89 > 1.5 and Islamic banking of 2.16 > 1.5 so it does not happen autocorrelation inferred.

By using SPSS 13.0 can be seen the results of multiple regression analysis as follows:

$$\text{Regression result conventional and Islamic banking}$$
$$Y = 0.319X_1 + 0.347X_2 + 0.236X_3 + 0.216X_4$$

$$\text{Regression result conventional banking}$$
$$Y = 0.271X_1 + 0.395X_2 + 0.237X_3 + 0.241X_4$$

$$\text{Regression results Islamic banking}$$
$$Y = 0.231X_1 + 0.392X_2 + 0.287X_3 + 0.243X_4$$

From the regression equation above can be explained that the variables of leadership style, motivation, organizational culture and work discipline has a coefficient that is positive towards the performance of employees either at conventional banking and Islamic banking. Based on the results of the statistical test using the F test F conventional banking known count of 37.57 with a 0.000 sig, and islamic banking known F calculated at 28.98 with a 0.000 significance values below 0.05.

Based on the results of the statistical calculations it can be seen the value of adjusted R^2 conventional banking of 0.835 this shows the percentage contribution of the effect of independent variables on the dependent variable was 83.5% and the adjusted R^2 islamic banking of 0.794 so that donations influence of independent variables on the dependent variable was 79.4%. The results of hypothesis testing in conventional banking leadership style variable indicates t value of standardized coefficients $\beta = 0.271$, $p < 0.05$ so that H_1 is accepted motivation to work with of $\beta = 0.395$, $p < 0.01$ thus H_2 is accepted. Organizational culture was $\beta = 0.237$, $p < 0.05$ successfully receives H_3 and variable labor discipline $\beta = 0.241, p < 0.05$ received the H_4. While the leadership style Islamic banking $\beta = 0.231$, $p < 0.041$ work motivation $\beta = 0.392$, $p < 0.01$, organizational culture $\beta = 0.287$, $p < 0.01$ and work discipline $\beta = 0.243$, $p < 0.05$, thus explained that the $H_1 - H_4$ is accepted.

3 Conclusion

Based on the validity and reliability of the data indicates that the data used to conduct the study is valid and reliable. Based on the classic assumption test variables can be inferred in this study can be used for regression models. Based on the results of the F test is known that variable leadership style, motivation, organizational culture, and discipline work simultaneously on variables affect the performance of employees. Based on the results of the regression coefficient of determination unknown R^2 conventional banking for the remaining 83.5, 16.5% influenced by other variables not included in the research model. While at Islamic banking by 79.4% and the remaining 20.6% influenced other variables. Based on the t test result that H_1, H_2, H_3, and H_4 is accepted. Work Motivation gives the most impact among other variable that is equal to 39.75% in conventional banking and 43.26% in islamic banking there are difference performance between islamic banking and conventional banking.

Suggestion

1. Those companies to further improve their performance in work motivation so that employees will achieve better performance again.
2. For further research in order to increase the sample and extend the research by looking at the performance of private sector employees and public employees. R^2 test results show there are still other variables that can affect the performance of the employee.

References

1. Banni, Korompot, dan Robiansyah,(2012), *Pengaruh Disiplin dan Motivasiter hadap Kinerja Pegawai PT PLN (Persero) Wilayah Kalimantan Timu rdan Area Samarinda,* journal: Mulawarman University
2. Chairunnisah Siti,(2012), *Pengaruh Gaya Kepemimpinan, Komunikasi Internal dan Motivasi Kerja Terhadap Kinerja Pegawai (studi kasus di kantor kecamatan Grogol Petamburan Jakarta Barat),* Journal: Gunadarma University
3. Moekijat.(1999). *Manajemen sumber daya manusia (manajemen kepegawaian).* Bandung: mandar maju
4. Nurjanah,(2008), *Pengaruh Gaya Kepemimpinan dan Budaya Organisasi terhadap Komitmen Organisasi dalam Meningkatkan Kinerja Karyawan (Studi pada Biro LingkupDepartemen Pertanian),* Thesis: Diponegoro University
5. Purwoko, Hidayati and Robiansyah, (2012), *Pengaruh Gaya Kepemimpinan dan Disiplin Kerja Terhadap Kinerja Karyawan pada PT. Tri Mandiri Selaras Samarinda,* Journal: Mulawarman University
6. Prasetyono and Nurul Kompyurini, (2008), *Analisis Kinerja Rumah Sakit Daerah Berdasarkan Budaya Organisasi, Komitmen Organisasi, dan Akuntabilitas Publik (Survei Pada Rumah Sakit Daerah di Jawa Timur),* Simposium Nasional Akuntansi 11, IAI, Pontianak.

7. Reza Regina Aditya,(2010), *Pengaruh Gaya Kepemimpinan, Motivasi Kerja dan Disiplin Kerja terhadap Kinerja Karyawan PT Sinar Sentosa Perkasa banjar Negara),* Essay: Diponegoro University
8. Sutrisno Edy.(2009). *Manajemen sumberdaya manusia.* Jakarta: prenanda media group
9 Syanputri Firstri, (2009). *Analisis Pengaruh Kepuasan Kerja Terhadap Kinerja Tenaga Kependidikan (Studi Kasus: Sembilan Fakultas Institut Pertanian Bogor, Darmaga Bogor),* Thesis: Fakultas Ekonomi dan Manajemen, Institut Pertanian Bogor, Bandung.
10. Tintami, Pradhanawati, and Susanto,(2012), *Pengaruh Budaya Organisasi dan gaya kepemimpinan transformasional terhadap kinerja Karyawan melalui Disiplin Kerja pada Karyawan Harian SKT Megawon II PT. Djarum Kudus,* Journal: Diponegoro University

The Study of Snack Purchasing Behavior and Foreign Tourists' Eating Habits at Taiwanese Night Markets

Shu-Hsien Liao and Pei-Yuan Hsiao

Abstract Taiwan is well known for its reputation cuisine, and annually thousands of tourists come to Taiwan to taste authentic traditional cuisine, in addition to provide the residents' life needs, taking into account the meaning of leisure culture and tourism. Taiwan has many night markets in different cities and counties, and more than several hundreds of thousands of street vendors make a living from their stalls, and many of them, including the night markets, are in fact recently established. The night market become pooled these delicacies field. "Night market" has become a nighttime leisure activity in Taiwan. Taiwan's night market continued to flourish, more diversified, combined with leisure, entertainment, and cuisine, becoming the first choice for domestic and foreign tourists and attractions. It provides consumers the night leisure and recreation place, and demographics they cover and consumers who are not limited to regional groups, including the more exotic visitors from Taiwan to the world famous gourmet snacks; attracting international tourists, Taiwan Tourism Bureau has been the promotion of policy, according to Tourism Bureau statistics. In 2014 the number of people reached 9.91 million in Taiwan. The research focuses on night market industry in detail. Total frequency for going to night market per month and different annual incomes, ages, occupations, types of consumption and past experiences of consumption, and finally deliver a conclusion on the frequency and content of consumers' behavior in each respective target group. Taiwan's population of overseas tourists are divided into two types, "Europe, USA and Canada" and "Japan and South Korea," and must have shopped in Taiwan night market of expatriates to conduct analysis, analytical method using factor analysis, and correlation analysis. The purpose of the consumption is to focus the main factors in Taiwan night market.

S.-H. Liao (✉) · P.-Y. Hsiao
Department of Management Sciences, Tamkang University, Tamsui, Taiwan, ROC
e-mail: michael@mail.tku.edu.tw

P.-Y. Hsiao
e-mail: tenoraa@gmail.com

T. Taufik et al. (eds.), *ICoSI 2014*, DOI 10.1007/978-981-287-661-4_27

Keywords Purchasing behavior · Eating habits · Night market · Data mining

1 Introduction

In the early market fair of Taiwan, Taiwanese snacks on behalf of Taiwan local food and geographical and cultural characteristics have driven one of the main elements of Taiwan's tourism industry, to create Taiwan's unique charm, reviving the economy and business opportunities. The night market provides the consumer unique life experience at night. It combines with local culture, shopping, dining, entertainment, and street performers. It is the very important place for the consumers who spent night time for the activities. It also becomes the overseas tourist spot to tour the place. According to Hsieh and Chang [7], the values of night markets are highly related with the local people, demonstrating the reality of lifestyle and local culture. About 2015 Taiwan Tourism Bureau for inbound visitors spending and travel trends survey trend analysis reports that, "Night Market" is ranked first foreign tourists to Taiwan for sightseeing attractions. There is a gourmet paradise known as Taiwan, especially the most famous pearl milk tea, as well as a special challenge to foreigners and pig's blood cake flavored stinky tofu in Taiwan, no matter how discerning whether foreign tourists, will be able to find co-appetite in Taiwan snack. According to the finding, the night market in Taiwan foreigner's imagery is equipped with authentic cuisine specialties, high visibility, and good public facilities [9]. Night market in Taiwan has a unique culture, bright lights, a variety of food, inexpensive goods, high visibility, and excellent public facilities [4, 5, 9].

The aim of this study is to investigate whether the tourist from difference group of countries would affect their purchasing behavior to enjoy night markets in Taiwan. This study also will make a comparison between night markets in Taiwan such as Taiwan's Shilin, Raohe street, Shida, Huaxi street, Gongguan, Fengjia, and Liuhe Night Market and then compares their differences. Survey data from 316 tourists indicate that Taiwan night market stands for the characteristic culture of Taiwan with high reputation.

The research questions are as follows:

1. Are there any difference between the night market back ground and culture characteristic?
2. What are the motivations of the consumers purchasing behavior in spending time in Taiwan night markets?
3. What are the consumers experience and intention of spending time in Taiwan night market?
4. What are the eating habits of the tourist in Taiwan night markets?

2 Literature Review

A night market is defined as a trading place during evening and includes small business for a variety of cheaper products, cooked food, and game playing. Night market was defined by Huang et al. [8] as a trading place during the evening, where small businesses offered a variety of cheap products and cooked food. People like to enjoy "freedom," "liberation," or temporarily alienated from society, so they come to the night market and produce a mass of sense of fusion. Therefore, customers do not care in the crowded, sweltering night market, the common dining tables with strangers, and many vendors have adopted a low table, low chair, so friends and family can work together to the night market, symbolized in the low table, low chair under the intimate dining atmosphere. In this way, they have become a crucial index of both local culture and tourism.

2.1 Purchasing Behavior

Consumer purchasing behavior is that people to meet the needs and desires to find, select, purchase, use, evaluation and disposal of products, processes and services activities involved, including consumers subjective and objective psychological activities both materials activities [10]. Consumer purchasing behavior is complex, resulting in buying behavior which is influenced by internal factors and external factors. There is a widespread recognition that consumer behavior is the success key factors of contemporary marketing [6]. Consumer behavior has been legitimized in marketing for which it provides the conceptual framework and strategic thinking for carrying out successful segmentation of markets [11]. For excellent sales staff, to master the customer buying decision process and understand the impact of the customer to make a very important factor in purchasing decisions.

Belch and Belch [1] explains that consumers undertake both internal and external searches. The possibility of alternative options is meaning whether there is another better or cheaper product available. Many internal factors influence consumer purchasing behaviors, and there are individual factor and psychological factor consumers. The purchaser's age, gender, income, and education level factors have a great impact on the purchasing behavior of consumers.

2.2 Eating Habits

Foreign tourists consider Taiwan to attract them the most important reason that is "snacks," such as Xiaolongbao, steam-fried baozi, Tempura, Wheel pie, Grilled Corn, Stinky tofu, vermicelli with chitterlings, Oyster omelet, Pigs blood cake, Braised food, Chicken feet, Coffin, Salty rice pudding, Ducks blood cake, Rice tube pudding, Taiwanese Meatballs, Snake soup, Green bean paste cake, Fried white radish patty, Sausage, Sesame oil chicken, and Pork chops in Chinese medicine soup. Eating has become to attract foreign tourists to Taiwan's biggest incentive. It is not only an art but also a manifestation of culture to promote to Taiwanese people. Snack allows foreign tourists a unique experience in the tourism process [3].Therefore, foreign tourists will also be able to come to Taiwan by "eating" journey deeper understanding of Taiwan.

Dietary culture in different ethnic groups in the world is an important part of the overall cultural dimension. In Taiwan, the hybridization of Taiwan and China ways fused to the point that the eating habits and cuisines transplanted by Chinese immigrants were assimilated into Taiwan urban culture along with the creation of a hybridized Taiwan identity and behavior patterns.

Taiwanese snacks were from the folk, with a deep folk character. Taiwanese snacks refer to Chinese provinces hometown tastes, as well as parts of Taiwan (Taiwan, Penghu, Kinmen, and Matsu including) the local snacks. Folk snacks are divided into eight major cuisines above Anhui, Cantonese, Fujian, Hunan, Jiangsu, Shandong, Szechuan, and Zhejiang cuisines. Due to the geographical character of Taiwan, Cuisine distinctive mix will be made in these eight Taiwanese snacks.

3 Methodology

The research areas covered Shihlin, Raohe Street, Shida, Huaxi Street, Gongguan, Fengjia, and Liuhe Night Markets in Taiwan. In an attempt to explore the Foreign Tourists' Eating Habits, related consumer purchasing behavior, and the perceived existing problems, a survey method was employed. The survey was conducted daily from December 1, 2013 to January 1, 2014. The survey time was from 5:00 pm to 3:00 am, the peak business hours for the night markets. A total of 400 questionnaires, 316 valid questionnaires were found; the effective recovery rate was 79.00%. The questionnaire was composed of four parts: (1) the foreign tourists regarding purchasing behavior in tourist night markets; (2) the tourists' preferred eating habits when they shop in tourist night markets; (3) the perceived existing problems of tourist night markets; and (4) background information about respondents.

4 Results Analysis

The socio-demographic characteristics of the sampled foreign tourists came to Taiwan were profiled and presented in Table 1. Of the 400 returned questionnaires from seven tourist night markets, 316 were valid and usable. The profiles of the foreign tourists are as follows: Descriptive analysis of the sample showed that the mean age of the respondents was about 18–25 years. The ratio of female and male respondents was nearly 1–1. Their occupations are predominately students (48.1%), followed by employee (21.5%), Homemaker (8.9%), and others.

This study implements ODBC bridge, into SPSS modeler data, to establish the analysis process, which are classified using cluster analysis to identify different cluster groups. SPSS modeler provides a different classification of clustering in the modeling node, then the data analysis process and the main set of nodes are linked together to complete the analysis of the data stream processing, as shown in Fig. 1.

Clustering analysis data characteristics according to the characteristics of the study proposed, 316 consumers via the three clusters, so called for the cluster-1 (208 records) named for the cursory tourists. Cluster-2 (39 records) called the in-depth tourists, the cluster-3 documents (69 records) called for the gap year tourists, such as shown in Table 2.

Table 1 Socio-demographic profile of Taiwanese senior travelers

Characteristics	Descriptions	Statistics	Percentage
Age	Below 17	14	(4.4)
	18–25	153	(48.4)
	26–35	69	(21.8)
	36–45	30	(9.5)
	46–55	34	(10.8)
	Above 56	16	(5.1)
Gender	Male	152	(48.1)
	Female	164	(51.9)
Area	Asia	136	(43)
	Europe	37	(11.7)
	America	89	(28.2)
	Canada	41	(13)
	Oceania	13	(4.1)
Occupation	Employee	68	(21.5)
	Homemaker	28	(8.9)
	Retiree	6	(1.9)
	Volunteer work	19	(6.0)
	part time jobber	17	(5.4)
	Student	152	(48.1)
	Freelancer	26	(8.2)
Spending in night market	Under NT$100	16	(5.1)
	NT$101 to NT$300	143	(45.3)
	NT$301 to NT$500	101	(32.0)
	More NT$501	56	(17.7)

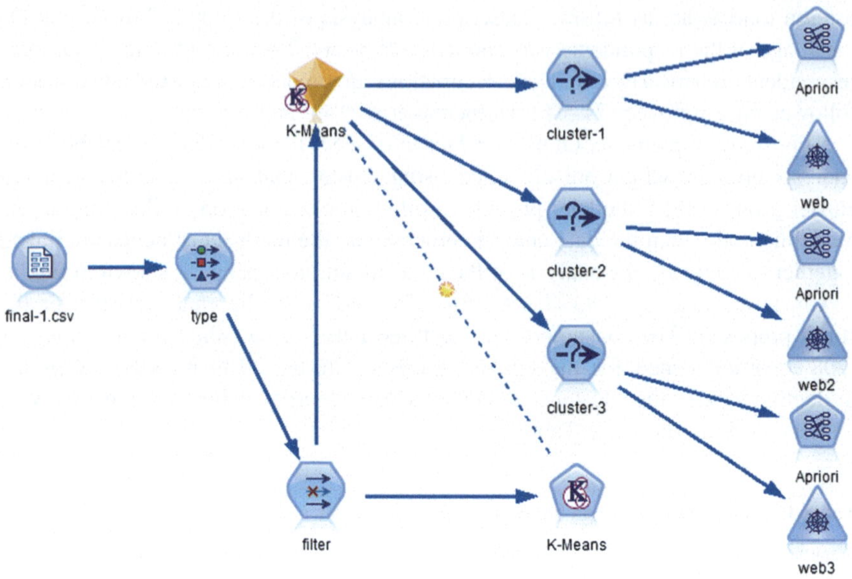

Fig. 1 Analysis process model

Table 2 K-means analysis

ID	Cluster-1	Cluster-2	Cluster-3
Name	Cursory tourists	In-depth tourists	Gap year tourists
Sample (%)	208 (65.8%)	39 (12.3%)	69 (21.8%)
Sex	Male (55.3%)	Male (53.8%)	Female (76.8%)
Age	18–25 (53.4%)	26–35 (53.4%)	18–25 (44.9%)
Area	Asia (56.7%)	Asia (35.9%)	America (44.9%)
Spending in night market	NT$301 to NT$500 (39.9%)	More NT$501 (47.8%)	Under NT$100 (56.4%)
Shopping	(39.4%)	(36.2%)	(43.6%)

For example, the cluster-1 cursory tourists, Nationality mostly in Asia, such as Korea, Japan, the amount of consumption between NT $101 and $300 which is very small, mostly buy snacks or buy one or two pieces of clothing, jewelry, night markets are looking for something to eat or to buy used (such as clothing accessories), all week to 2–3 times! It seems that foreigners feel very interesting night market; these ethnic eateries are all interested to mostly go with friends; mostly 19:00–21:00 to the night market, there may be a dinner, to accept snacks as ranked in Fig. 2 (Table 3).

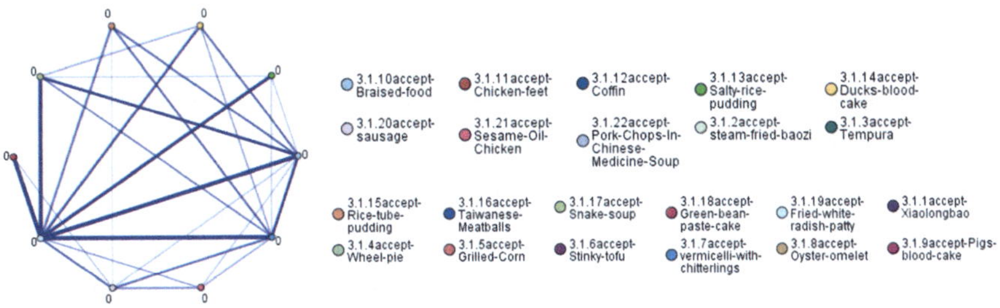

Fig. 2 Cluster–cursory tourists web model

Table 3 Accept snakes' type

Snakes name	Accept member	Ranking
Xiao Long Bao	83	1
Steam-fried baozi	60	2
Tempura	54	3
Stinky tofu	54	3
Oyster omelet	47	4
Grilled corn	44	5
Chicken feet	44	5
Wheel pie	41	6
Taiwanese meatballs	40	7
Coffin	33	8
Sausage	30	9
Green bean paste cake	29	10

5 Conclusion and Suggestion

According to survey results, the foreigner's view on Taiwan's night market is too dirty and crowded, I hope you can plan a good night market route, do not eat or drink with clothes and jewelry classes put together. In terms of food, mostly foreigners can accept tofu, pig blood cake unacceptable, and snake soup Taiwanese folk snacks. The next night market should be more intensive development of local specialties, and if you want to promote Taiwan's night market culture to the international stage, the Tourism Bureau can organize similar "International Competition snacks"-related activities "Taiwan snacks Expo", etc., to improve the visibility of Taiwanese snacks.

You can even collaborate with television, producing a series of Taiwanese snacks tourism programs, or at home on their own DIY show Taiwanese snacks, let foreigners know more about our snack culture. It may also be in cooperation with the logistics industry, with a home delivery service for Taiwanese snacks spread to every corner of the world; more can enhance the development of fresh frozen technology, and after going through a long journey, consumers can get the freshest Taiwanese snacks.

The study found that foreigners in the planning and management of the store partition blocks quite ideas; foreigners understand that the "Night Market" is Taiwan's culture, lively, bustling night in the night market is also a major feature, but foreigners cannot understand the greasy oil. Why food and colorful costumes or shoes placed in the same block; selling the same street, many respondents believe that foreigners snacks, clothing, or entertainment segment should be done separately, looking directly to reduce the time. So in addition to the future government it should strengthen the night market health treatment, and it should also take every night market re-planning division and the operating line.

References

1. Belch, G.E., and Belch, M.A. (2007), Advertising and Promotion: An Integrated Marketing Communications Perspective, Seventh Edition. New York: McGrawHill/Irwin.
2. Blackwell, R.D., Miniard, P.W. and Engel, J.F. 2001, Consumer Behavior, 9th Edition, South-Western Thomson Learning, Ohio, USA.
3. Dittermer, D. (2001). Dimensions of the hospitality industry: An introduction (3rd ed.). New York: Wiley.
4. Bloom, D. (2002). Dan Bloom Taiwan's night markets. Taipei: Tongyo Cultural Affairs Inc.
5. Huang, C. H. (2012). Service quality of night markets in Taiwan. International Journal of Marketing Studies, 4(3), 36–44.
6. Hawkins, Del I., Best, R., & Coney, K., 2003, Consumer Behaviour: Building Marketing Strategy, Boston, Massachusetts: Irwin McGraw-Hill, 775.
7. Hsieh, A. T. and Chang, J. (2006) Shopping and Tourist Night Markets in Taiwan. Tourism Management, 27(1), 138–145.
8. Huang, S. O., Liou, Y. H., & Tzeng, G. H. (2009). Development strategies for improving the service of tourist night markets through hybrid MCDM technique.
9. Lee, S. H., Chang, S. C., Hou, J. S., & Lin, C. H. (2008). Night market experience and image of temporary residents and foreign Visitors. International Journal of Culture Tourism and Hospitality Research, 2(3), 217–233.
10. Philip Kotler, Marketing Management, Interbooks publications, 2000.
11. Schiffman, L. G. & Kanuk, L. L., 2000, Comportamento do Consumidor, 6ª Edição.Rio de Janeiro: Editora LTC.
12. Wang, C. (1999). The study on the planning of Taipei Tourist Night Markets. Taipei: Taipei City Government.
13. Zhang, QL. Taipei night market literature symposium report. Taipei City Literature, vol.148, pp. 1–90, 2004.

The Sustainability of Bayt Al-Maal Wa Al-Tamwil's Contribution to Yogyakarta Citizen

Masyhudi Muqorrobin, Yuli Utami and Abdiel Fadhil Ridho

Abstract This paper aimed at identifying the influence of selected BMT's products (Pr), empathy of Islamic microfinance institution to the member (G), and profit-sharing ratio determination (Ps) on their member's welfare (Y). This welfare is the increment of a combination of member's income (y), consumption (C), and investment (I) after being BMT's partner. By using Slovin formula and multiple linear regression method, this study evaluates 95 BMTs members who are in their second (or above) financing periods. The finding indicates that the improvement in their welfare has a positive relationship with the products offered, the empathy of BMTs to their members, and profit-sharing ratio they agreed. In the long run, this will promote the sustainability of contribution of BMTs for small entrepreneurs. Therefore, paying consistent attention to the members and being committed on the track as BMTs, with social services as their platform of services, will contribute to solving the problems of micro- and small enterprises as the grassroot economy level who are unable to secure access to Islamic banks.

Keywords *Bayt al-Maal wa al-Tamwiil* · BMT · Empathy · Profit sharing · Yogyakarta · Slovin formula · Multiple linear regression

M. Muqorrobin (✉) · Y. Utami · A.F. Ridho
International Program for Islamic Economic and Finance, Economics Department, Universitas Muhammadiyah Yogyakarta, Bantul, Indonesia
e-mail: masmubin@yahoo.com

Y. Utami
e-mail: yuliutami@umy.ac.id

A.F. Ridho
e-mail: abdiel_banjary@plasa.com

T. Taufik et al. (eds.), *ICoSI 2014*, DOI 10.1007/978-981-287-661-4_28

1 Introduction

1.1 Research Background

Post-economic 1977 crisis in Indonesia has proved sustainability of the majority Islamic financial institutions. However, micro- and small enterprises (MSEs) as a grassroot-level economy are almost unable to get a light procedure to connect to the Islamic banks. Based on these facts, Bait al-maal wal-Tamweels or BMTs have started to mushroom and provided financial assistance for those who are considered standing in this level and given a green light to them to improve their capacity such as incomes, consumptions, and investments, after approval of their membership in BMTs. However, whether these improvements were experienced by majority of BMT's member or not needs further in-depth analysis. Since BMT today tends acting likely an Islamic bank, careful supervision will takeover by the Financial Service Authority (OJK) for the coming year 2015; therefore, the existence of BMTs is under challenge and therefore needs to be evaluated immediately. This study is focused on the response of the citizen of Yogyakarta, where micro-business (98.79%) and small business (1.11%) are considered the most developed companies/businesses, whereas the number of medium and large enterprises covers only 53.9 thousand or less than one percent of all companies/businesses.

The manufacturing sector during 2012 contributed 13.35% of the value added to the economy of Yogyakarta Special Territory Province. The structure of industrial enterprises in Yogyakarta based on the results of the 2006 Economic Census was dominated by micro-scale industry (90.62%) and small industries (8.49%). Meanwhile, the business population of medium and large scale only reached 0.89%.[1] This fact gives rise to mushrooming BMTs to strengthen the unbankable grassroots and to fight against moneylenders that torment grassroots with high rates of interest or usury. This study investigates how strong the contribution of BMTs especially in Yogyakarta city as measured by the increasing welfare of the MSEs under BMTs supervision. Therefore, the study comes up with a research question: "Do BMTs contribute to the welfare creation of Yogyakarta citizen?"

1.2 Purposes and Location

This research will be held in five areas of Kota Yogyakarta, aiming at achieving the following objectives:

[1]BPS Catalog: 1101002.34 [28, p. 56].

(a) To identify the influence of a number of variables such as BMTs' products, empathy, and profit-sharing ratio determination toward MSEs' funding
(b) To analyze whether the improvement of income, consumption, and investment of the members are influenced by the contribution of BMT.

1.3 Target and Contribution

It is expected that this study contributes to the following:

(a) Policy recommendations to support the existence and growth of BMTs, especially to enhance their role to promote MSEs in *Kota* Yogyakarta. The result will also be used as a report and presented to the Mayor of *Kota* Yogyakarta for endeavoring of the development of MSEs which could be next strides to empower them as well as to put it as a target of local strategic policy and a proposal for their next program purposes.
(b) Evidence to support in convincing the MSEs that Islamic financial institutions (BMTs) are able to act as intermediator and facilitator to assist their businesses in compliance with the Shariah principles.
(c) Methodology in analyzing the welfare which is represented by the improvement in income, consumption, and investment of *Kota* Yogyakarta's citizen. Therefore, the study is also prepared for an eligible international or local journal submission.

2 Literature Review

Satisfaction level and welfare are the two definitions which are related to each other. Satisfaction level refers to individual or party's condition, whereas welfare refers to community or society conditions where they are satisfied with their life due to fulfillment of their needs. Welfare is an aggregate condition from individual satisfaction. According to Social Welfare Enactment of the Republic of Indonesia No. 11 of 2009, the social welfare is the fulfillment of the condition of the material, spiritual, and social life of citizens in order to be viable and able to develop themselves, so that it can carry out its social function. Public welfare of middle and lower class can be represented from their living standard, such as alleviation of poverty, acquisition of higher education levels, and the level of productivity of the community.[2] Indonesian Central Bureau of Statistics or BPS (2000) explains that the level of household wealth of an area can be measured by several indicators such as the level of family income, the composition of household expenditures by

[2]Todaro [1, p. 46].

comparing expenditures on food and non-food, the level of education, the level of health, and housing conditions and facilities owned by the household.

Therefore, a person can be said to have been prosperous if he gained increased revenue, since it will influence to an increase in consumption or an increase in investment. The variables of income, consumption, and investment are interrelated to each other. In micro-scale, income is the sum of investment and consumption. If income decreases given the consumption constant, the investment or saving will decrease. If both the income and consumption decrease, investment or saving will decrease proportionately, vice versa.[3]

Income is a monetary payment received in exchange for providing goods or services, or from other sources such as rent or profit from investments. The basic economic model of circular flow diagram shows the flows of production, income, goods, services, and expenditure between firms and households. In macroeconomic perspective, households' consumption decisions affect the behavior of the economy as a whole, both in the long run and in the short run. Short-run analysis argues that consumption decision is crucial due to its role in determining aggregate demand, whereas the long-run analysis believes the importance of the consumption decision for its impact on the economic growth. Consumption is two-third of GDP, so fluctuations in consumption are a key element of ups and down or booms and recessions of an economy.[4]

In 1936, Keynes introduced a theory of economic fluctuations, where consumption plays a key role in macroeconomic analysis. Instead of relying on statistical analysis, Keynes makes conjectures where consumption is related to behavior of individuals in the economy, creating a tendency of all individuals in consumption in what he terms as marginal propensity to consume (MPC) which lies between zero and one. Ratio of consumption to income, called the average propensity to consume (APC), falls as income rises. Income is the primary determinant of consumption. In the analysis of Keynes, interest rate does not have any important role. According to the theory, households with higher income who consume more have the higher marginal propensity to consume, which confirms that MPC is greater than zero. On the other hand, households with higher income but save more will confirm that the marginal propensity to consume decreases. In addition, the higher-income households who save a large fraction of their income confirm that the average propensity to consume falls as income rises.[5]

During the years when income was unusually low, such as in the depths of the Great Depression, both consumption and saving were significantly low, and the marginal propensity to consume remains to place between zero and one. During those years of low income, the economy was facing what Keynes defines as the second conjecture, being the ratio of consumption to income, which was high. Finally, because the correlation between income and consumption was so strong,

[3]Mankiw [2].

[4]Mankiw, N.Gregory. *Op. Cit.* Page 456.

[5]Ibid Page 459.

thus, no other variable appears to be important to influence consumption. This is called the Keynes's third conjecture that income is the main variable by which people determine to consume or not. Investment plays a key role not only for the long-run growth but also the short-run business cycle since investment is the most volatile component of GDP. When expenditure on goods and services falls, especially during a recession, most of the falls are associated with a decrease in investment. The 1992 US recession is also a good example of this relationship, when the US real GDP fell to $105 billion from its peak in the third quarter of 1981 to its trough in the fourth quarter of 1982. Investment spending over the same period fell to $152 billion, accounting for more than the entire falls in spending.[6]

Driven by a sense of deep concern to many poor people who are caught by moneylenders, Indonesian government supports the existence of the *Bayt al-Maal Wa al-Tamwils* (BMTs) as micro-financial institutions which enhance the economic empowerment programs, especially among micro-, small, medium, and cooperative enterprises through the Islamic system. The term *Baytal-Maal* means house of finance and *Bayt al-Tamwiil* means financing house.[7] These terms already existed since the era of the Prophet *Rasulullah* then established as a formal economic institution suggested by jurist named Walid bin Hisham.[8] Since this period and during the triumph of Islam (especially under the Dynasties of Abbasid and of Umayyad), *Bayt al-Maal* has become an institution that had an important role for the country life. At that time, the *Bayt al-Maal* had been handling various matters such as *zakat, ghanimah, infaq, shadaqah,* or building public facilities such as roads, bridges, hiring soldiers and officials countries, and other public interests. So, the *Bayt al-Maal* since the earlier Islam, particularly starting from the period of the Caliph Umar ibn al-Khattab until that of the triumph of Islam had acted as what we have today as the Ministry of Finance, Directorate General of Taxation, Ministry of Social Affairs, Ministry of Public Works, or combination of those ministries for a small country.

Nevertheless, the development of civilization has transformed the meaning and roles of *Bayt al-Maal* in Indonesia to a narrower scope, namely as financial institution or services to a small community. BMTs play a strategic role to reduce poverty in response to the fact that the banking institutions have been unable to touch the grassroots. The strategic role of BMTs in reducing poverty can be seen by its combined economic activities, which are composed of social (*Bayt al-Maal*) and business activities (*Bayt al-Tamwil*). Socioeconomic activities are carried out such as *zakat, sadaqah,* and *infaq,* whereas business activities are taken in the form of financing in productive economic activities.[9] According to the Ministerial Decree of

[6]Ibid Page 487.

[7]Andriani.*Baitul Maal Wa tamwiil (konsep dan mekanisme di Indonesia).* (Empirisma. Volume 14 No. 02 July 2005) page 249.

[8]Ibid Page 250.

[9]Dra. Siti Zubaidah, MM, AK. 2010. *Pengembangan Model Sistem Pengendalian Aliansi "Alkamil" di Malang.* (Malang: Universitas Muhammadiyah Malang). Page 2.

the of Cooperatives and SMEs Number 91, 2004 about Activities Guidelines for Shari'ah Financial Services Cooperatives (*Koperasi Jasa Keuangan Syariah*) Chap. 1 Article 1, "BMTs as *Shari'ah* microfinance institutions and incorporated Islamic financial cooperatives, their implementation should follow the rules and guidance of *Shari'ah* financial services, namely business activities engaged in financing, investment, and savings based on profit and loss sharing."[10]

Bayt al-Tamwil has 3 principles in its practices: First, loss and profit-sharing system using two types of contracts, i.e., *Mudharabah* (trustee partnership) and *Musyarakah* (joint venture) contract. The former involves a combination of entrepreneurship and capital while the latter involves a partnership in both entrepreneurship and capital. These equity-based products are unique to Islamic banking and in some sense account for its superiority over conventional banking on the grounds of ethics and efficiency. Therefore, these two financing products are the niche products for Islamic financial institutions because of its departure from conventional one.[11]; Second, sale and purchase system using two general selling prices, i.e., exchange goods for money (*Murabahah, Bay' Bithaman 'Ajil, Istisna',* and *Salam*) and exchange services for money (*Ijarah, Wakalah, Kafalah*); Third, non-profit contract (*Qard Hasan).*This research will focus on the niche products in BMT that are only used in Islamic financial institutions.

As also conventional financial institutions that act as fund provider and borrowers, likewise, Islamic financial institutions have double roles, being financier (capital provider) and *mudharib* (entrepreneur). In *Mudharabah* case, profit is shared according to a pre-agreed ratio. If *mudharib* experiences losses, the losses are entirely absorbed by the capital provider (Islamic Financial institutions). *Mudharabah* may be of two types—restricted or unrestricted. In a restricted *mudharaba* (*mudharaba al-muqayyada*), the bank or the financier may specify a particular business in which investments may be undertaken. *Mudarabah* may also be an unrestricted one (*mudharabah al-mutlaqa*); in this case, the *mudharib* may invest the capital provided in any business he deems fit. The *robbulmal* or financiers own the business's net worth, while the *mudharib* or the managers control the business's assets. Moral hazard problem may arise when the *mudharib* tends to understate profit in financial reporting. Hence, to determine the proper utilization of the *mudarabah* fund by the *mudharib,* a detailed (and may also costly) monitoring system is an important factor to consider.[12] A joint venture *musharakah* shares the profit between the financier and the customer in the pre-agreed ratio. Losses are shared strictly in proportion to their respective capital contributions.[13]

---

[10]Hendro wibowo, "*BMT sebagai Corporate Social Entrepreneurship*". http://forumriset.iaei-pusat.org/downlot.php?file=BMT%20sebagai%20Corporate%20Social%20Entrepreneurship%20(revisi).pdf.

[11]Obaidullah, Mohammed. 2005. *Islamic Financial Services.* Jeddah, Saudi Arabia: Islamic Economic Research Center, King Abdul Aziz University Page 57.

[12]Rosly, Saiful Azhar. Op. Cit. page 193.

[13]Obaidullah, Mohammed. Op.Cit.Page 59.

3 Research Methodology and Findings Discussion

3.1 Research Subject

This study is focused on micro- and small enterprises of BMT members in Jogja or Jogjakarta City, with sample estimation using Slovin formula as follows:

$$n = \frac{N}{1 + Ne^2}$$

where

n: sample size

N: population size

e: margin error that are tolerated based on statistic (1, 5, 10%)

Based on data in http://umkm.jogjakota.go.id/, the number of UMKM (medium-, small- and micro-enterprises) in Yogyakarta city covers about 2082 MSMEs. Following Slovin formula, the percentages of error tolerated are 1, 5, and 10%,[14] the study considers 10% of error that can be tolerated.

Sampling: Due to limitations, only BMTs under PUSKOPSYAH (Center for Sharia Cooperatives) in Yogyakarta city are considered in this study, with 95 respondents randomly taken from 5 BMTs Jogjakarta city, such as BMT Beringharjo (Central Jogjakarta), BMT Bina Ihsanul Fikri (West Jogjakarta), BMT Bina Sparta Mandiri (North Jogjakarta), BMT Al-Ikhlas (South Jogjakarta), and BMT BAROKAH Padi Melati (East Jogjakarta).

Variables and Data Collection Method: The study considers BMT members' welfare as dependent variable (Y), explained by independent variables of the value of Islamic financial institution product offered toward their member (X_1), their empathy toward members (X_2), and profit-sharing ratio in their schemes of financing (X_3). Primary data are collected using simple random sampling method, by distributing questionnaire for several BMT members.

Validity Test: Validity test is used to measure the validity of the data. A questionnaire is considered as valid if the questions in the questionnaire can show the result measured in it.[15] Validity test can be done by using SPSS program. A trial of a data can be said as valid if r is greater than r-table by using the formula given below:

$$\Gamma xy = \frac{\mathrm{N}\sum xy \pm (\sum x)(\sum y)}{\sqrt{\left\{\mathrm{N}\sum x^2 - (\sum x)^2\right\}\left\{\mathrm{N}\sum y^2 - (\sum y)^2\right\}}}$$

[14]Danang Sunyoto. 2013. *Metode dan Instrumen Penelitian (Untuk Ekonomi dan Bisnis).* Yogyakarta: CAPS (Center for Academic Publishing Service). Page 16.

[15]Azwar, Saifuddin.2013. *Reliabilitas dan Validitas Edisi 4.* Yogyakarta: Pustaka Pelajar. Page 40.

where

Γxy = correlation coefficient between variable x and variable y
$\sum x$ = sum of x item
$\sum y$ = sum of y item
N = population size
$\sum xy$ = sum of combination of x and y
$\sum x^2$ = sum of x square
$\sum y^2$ = sum of y square

From 64.2% female and 35.8% male respondents, the validity findings this test found are summarized in Table 1.

Table 1 Validity findings

Validity test	Findings
Paired before and after Financing	Average income of respondents before financing is 1.74 and after financing is 2.05. However, by using significant level of 5%, the correlation between two variables is 0.141 with its significance of 0.17. The result shows that there is no correlation between two income averages before and after financing. This is because the revenue of traders in Yogyakarta is more seasonal. At peak season, they can get higher revenue than usual. On the other hand, at low season their revenue is stable. T-test value gives -6.584 with significant value of 0.000, which means less than 0.05. Therefore, in conclusion, their income average before and after financing is different which indicates that financing from financial institutions especially BMTs influence their incomes. Based on macroeconomic theory, the increase in the factors of production such as capital and labor can increase the output of production and consequently increases the income of the traders
Selected financing type	*Musharakah* is likely to be higher demanding contract compared to *mudharabah* by having 87.4% of total profit-sharing schemes, especially for working capital. Surprisingly, *mudharabah,* which is expected to be the majority for business contract for respondents whose jobs are trader, covers only 12.6% of total contracts. It may be associated with a higher market risk and costly contract of *mudharabah* as compared to that of *musharakah.* Though it is a niche product of Islamic financial institutions and uses friendly ratio portion or *nisbah,* yet, BMTs prefer to recommend *musharakah* contract to their members to cover profit and loss sharing together
Level of financing	43.2% of the respondents took financing from the range of Rp 6 millions to Rp 10 millions, 38.9% took the financing on the range of Rp 1 million to Rp 5 millions, and the rest (17.9%) took the financing around the range of Rp 10 millions to Rp 25 millions

Mankiew [2], p. 46

Reliability test: It is a tool to measure the questionnaires being indicators of variables to construct the researches. A questionnaire is considered as reliable if the answers are consistent or stable over time.[16] Cronbach alpha's formula is employed and found that all the variables are reliable: BMT member's welfare 0.836; product value 0.709; BMT's empathy 0.753; profit sharing 0.651.

$$\propto = \frac{k - \bar{r}}{1(k-1)\bar{r}}$$

where: $\bar{r}$ = correlation mean among items

k = number of item

Descriptive Variables Analysis: Each respondent was given 20 questions, and each question was measured using a Likert scale. Questionnaires were given to respondents in the form of questions asking respondents to give their opinions or experiences after using the products and services offered by the BMTs. The full sequence is worth five for strongly agree response, score four for agree, score three for neutral, score two for does not agree, and score one for strongly disagree. The dependent variable or variable Y consists of three questions, while the independent variables X1, X2, and X3 consist of four questions. The respondents' **welfare** increase indicator can be seen from the increase in their **investments, incomes,** and **consumptions** which are summarized from 26 respondents on average (27.37%) who really agree and strongly agree. The total increase in **investment** based on the positive change in **capital formation** (18.26%), **saving** (9.21%), and **household furniture** (6.16) which significantly got the highest response. The positive change in **average income** after financing is summarized based on the change in **revenue** (6.63%**), product sold** (4.74%), and **product ordered** (2.53%) those have highest response as well. The positive change of respondents' **consumption** that concise from the **basic needs** (11.11%), **business cost** (4.26%), and **operational cost** (1.58%). **Product** offered, **empathy** given by BMT, determination of **profit-sharing ratio**, all together are answered that the majority agree and strongly agree to give positive advantage to their business.

4 Data Analysis Method

1. **Multiple Linear Regression**: The study uses the following regression equation:

$$Y = a + bX_1 + cX_2 + dX_3 + e$$

[16] Ibid. Page 28.

where Y = BMT members' welfare
X_1 = value of BMTs' product offered to their members
X_2 = empathy of BMTs toward their members
X_3 = profit-sharing ratio in BMTs
e = error term

2. **Determination coefficient (R^2)**: This R^2 determination is used for measuring the proportion of dependent variables which is explained by independent variables.[17]

R^2 value is placed as $0 \leq R^2 \leq 1$. The closer to 1 indicates the better the model obtained.

$$R^2 = 1 - \frac{SSerr}{SStot}$$

3. **Statistic test**: This statistic test is used for measuring goodness of fit which consists of:

   (a) **T-test** which is done for finding out the influences of independent variables toward dependent variable individually. The used hypotheses are: *Ho:* $\beta 1 = \beta 2 = \beta 3 = 0$ (it means that there is no relationship between the value of BMTs' product, empathy of BMTs toward their members, and profit-sharing ratio in their BMTs with the welfare of BMT members). *Ha:* $\beta 1 \neq \beta 2 \neq \beta 3\ 0$ (it means that there is relationship between the value of BMTs' product, empathy of BMTs toward its member, and profit-sharing ratio in their BMT with welfare of BMT members)
   (b) **F-test** which is done for finding out the influences of independent variables toward dependent variable jointly. The conclusion of F-test can be found by comparing between F-test statistic with F-table at certain level and certain degree. This test is done using the below formula:

$$F = \frac{R^2/k}{(1 - R^2) - (n - k - 1)}$$

where R^2 is the determination coefficient;
k the number of parameters; and
n the sample size.

[17]Qudratullah, Mohammad Farhan.2013. *Analisis Regresi Terapan: Teori, Contoh Kasus, dan Aplikasi dengan SPSS*". Yogyakarta: CV. Andi Offset. Page 71.

4. **Classical assumption test**

   (a) **Multicollinearity Test**
   Multicollinearity test is made available to check the correlation among independent variables. Correlation indicates the presence of multi-collinearity problems. Therefore, a good regression model will have no a correlation problem among independent variables.[18]

   (b) **Heteroskedasticity test**
   Heteroskedasticity test is done to examine the presence of inequalities of regression model residual variance from one observation to another (Table 2).[19]

5 Conclusion

Based on field study of BMTs in Jogjakarta City, it is suggested that BMTs have positive impact on the increasing of Yogyakarta's citizen welfares. The increasing indicators can be seen from the increasing of their investments, consumption, and incomes. Most of respondents stated that BMTs are helpful in supporting their business growth. From product value aspect, BMTs have simple procedure for financing product offered toward their members. Hence, BMT members can increase their business growth because of BMTs ability to help them in fulfilling working capital needed by micro-enterprises. From empathy aspect, BMTs have no burden for installments payment or some penalties for those who are unable to pay the installment in time like conventional financial institutions. In addition, proactive systems done by BMTs can help simple traders to pay installments as well as to save their funds. From profit-sharing aspect, BMTs apply the fairness in deciding profit-sharing ratio without burdening their members by conducting discussion and agreement before signing the contract.

Given contribution of BMTs to increase social welfare today and in the future, it is expected that the government can intensify its support toward Islamic microfinance institutions, especially BMTs rather than pouring economic incentive to the people without strong basis and rationale intense. Government can channel a proportion of public budget through BMTs, and then BMTs can manage this fund for social interests. This study also expects to be a reminiscence to the society especially grassroots people who should understand that Islamic financial institutions, such as BMTs, emerged in order to assist them and release them from the bonds of money lenders.

[18]Ibid. Page 179.

[19]Ibid. Page 192.

Table 2 Discussion of findings

<table>
<tr><th>Classical assumption test</th><th>Findings</th></tr>
<tr><td>Multicollinearity test</td><td>The result from the multicollinearity test for the variable of product offered, empathy of BMT's staff, and profit-sharing determination generally have VIF value less than 10 and tolerance value of greater than 0.1. Therefore, the multicollinearity problem does not occur in this model</td></tr>
<tr><td>Heteoskedasticity test</td><td>The heteroskedasticity problems in the independent variables do not occur as well</td></tr>
<tr><td>Multiple linear regression:
$Y = 2.188 + 1.111 X_1 + 0.211 X_2 + 0.207 X_3 + e$</td><td>
<table>
<tr><th>Variable</th><th>Coefficient</th><th>T Statistic</th><th>Sig</th></tr>
<tr><td>C</td><td>2.188</td><td>1.344</td><td>0.182</td></tr>
<tr><td>X_1</td><td>1.111</td><td>2.843</td><td>0.006*</td></tr>
<tr><td>X_2</td><td>0.211</td><td>3.689</td><td>0.000*</td></tr>
<tr><td>X_3</td><td>0.207</td><td>2.846</td><td>0.005*</td></tr>
</table>
*significance level 1%</td></tr>
<tr><td>T-Test</td><td>• t count > t Table (2.843 > 2.63094), reject H0. This means that there is a positive relationship between BMTs' products offered to their members with BMT members' welfare
• t count > t Table (3.689 > 2.63094), reject H0: means that there is a relationship between empathy of BMT toward its member with BMT members' welfare
• t count > t Table (2.846 < 2.63094), reject H0: means that there is a positive relationship between profit sharing in their BMT with BMT members' welfare</td></tr>
<tr><td>F-Test
F-stat value 17.054
df1 = k − 1, so 4 − 1 = 3; df2 = n − k, so 95 − 1 = 91
Sig.level at 1%, = 4</td><td>H0: There is no relationship between understanding of BMT members toward the product offered, empathy of BMT toward its member, and profit-sharing ratio in their BMT with MSEs welfare of BMT members jointly. Since F count is greater that F-Table (17.054 greater than 4.00), H0 is rejected. It means that there is a relationship between understanding of BMT members toward the products offered, empathy of BMT toward its members, and profit-sharing ratio in their BMT with BMT members' welfare jointly</td></tr>
<tr><td>R^2 Test = 0.360 or (36%)</td><td>The percentage contribution of the effect of independent variables (product, empathy, and profit sharing) on the dependent variable (BMT members' welfare) is equal to 36%; or variations of the independent variables used in the model (product, empathy, and profit sharing) were able to explain 36% of variation in the dependent variable (BMT members' welfare). Meanwhile, the remaining 64% are influenced or explained by other variables that are not included in this research model</td></tr>
</table>

References

1. Michael P. Todaro. 2003. *Economic Development 8th Edition.* London: Pearson Education Limited.
2. N. Gregory Mankiw.2007. *Macroeconomics 6th Edition.* New York: Worth Publisher.
3. M. Obaidullah, Hajah Salma Haji Abdul Latiff.2008. *Islamic Finance For Micro And Medium Enterprises.* Islamic Research & Training Institute Islamic Development Bank, Centre for Islamic Banking, Finance and Management Universiti Brunei Darussalam
4. Rosly, Saiful Azhar. *Critical Issues on Islamic Banking and Financial Markets.* Kuala Lumpur-Malaysia: Dinamas Publishing.
5. A. Parasuraman, V.A. Zeithaml and L Berry. 1988, *SERVQUAL: a multiple-item scale for measuring member perceptions of service quality.* Journal of Retailing, Vol. 64, Spring.
6. Mohammed Obaidillah, 2008. *Introdution to Islamic Microfinance.* India: IBF Net: the Islamic Business And Finance Network
7. Saiful Azhar Rosly. *Critical Issues on Islamic Banking and Financial Markets.* Kuala Lumpur-Malaysia: Dinamas Publishing.
8. Saifuddin Azwar. 2013. *Reliabilitas dan Validitas Edisi 4.* Yogyakarta: Pustaka Pelajar.
9. Danang Sunyoto. 2013. *Metode dan Instrumen Penelitian (Untuk Ekonomi dan Bisnis).* Yogyakarta: CAPS (Center for Academic Publidhing Service)
10. Sadono Sukirno. 2008. *Makroekonomi Teori Pengantar Edisi Ketiga.* Jakarta: PT. Raja Grafindo Persada.
11. Sadono Sukirno. 2006. *Mikro Ekonomi Teori Pengantar.* Jakarta: PT Raja Grafindo Persada.
12. Muhammad Ridwan. 2004. *Manajemen Baitul Maal wa Tamwil.* Yogyakarta: UII Press.
13. Djazuli, 2002.*Lembaga-lembaga Perekonomian Umat*, Jakarta: PT. Raja Grafindo Persada.
14. Pusat Pengkajian dan Pengembangan Ekonomi Islam (P3EI) Universitas Islam Indonesia Yogyakarta. 2008. *Ekonomi Islam.* Jakarta: PT. Raja Grafindo Persada
15. Iskandar Putong, S.E., MMSI. 2003. *Pengantar Ekonomi Mikro & Makro Edisi 2.* Bogor: Ghalia Indonesia
16. Irwin Ananta. 2012. *Tinjauan Kritis Praktek Mudharabah Pada Perbankan Syariah.* Paper on national seminar, BSI Univercity. Bandung
17. Neni Nurul Haida. *Konsep-Konsep Pelayanan Prima.* Pelayananprimaneni.blogspot.com accessed on 23th of January 2014.
18. Mohammad Farhan Qudratullah. 2009. *Analisis Regresi Terapan: Teori, Contoh Kasus dan Aplikasi dengan SPSS.* Yogyakarta: CV. Andi Offset
19. Dra. Siti Zubaidah, MM, AK. 2010. *Pengembangan Model Sistem Pengendalian Aliansi "Alkamil" di Malang.* Malang: Universitas Muhammadiyah Malang.
20. Lepi T Tarmidi. *Krisis Moneter Indonesia. Sebab, Dampak, Peran IMF, dan Saran.* Revision and updating from grand teacher inaugural speech at Economic Faculty Indonesia University under title "*krisismoneter tahun 1997/1998 dan peran IMF*"
21. Andriani.2005. *Baitul Maal Wa tamwiil (konsep dan mekanisme di Indonesia).* Empirisma.
22. M. Agus Khoirul Wafa and Nasrodin. 2008. *"Bait Al-Maal Wa At-Tamwiil"*, Yogyakarta: Universitas Islam Indonesia
23. Ridha Nugraha. *Manajemen Pembiayaan: Panduan Untuk Koperasi Syariah.* SDM Kementrian Koperasi
24. Huda Nurul & Muhammad Heykal. 2010. *Lembaga keuangan islam: tinjauan teoritis dan praktis.* Jakarta: kencana
25. Dahlan Siamat. 1995. *Manajemen lembaga keuangan.* Jakarta: intermedia
26. Ahmad Sumiyanto, SE,M.SI. 2008. *BMT Menuju Koperasi Modern: Panduan Untuk Pemilik, Pengelola, dan Pemerhati Baitul Maal wat Tamwiil dalam Format Koperasi.* Yogyakarta: ISES Publishing.
27. *Sejarah Bank Indonesia: Perbankan Periode 1997–1999.* Unit Khusus Museum Bank Indonesia: Sejarah Bank Indonesia

28. BPS Catalog: 1101002.34, "*Statistik Daerah IstimewaYogyakarta 2003*". *Badan Pusat Statistik* special region of Yogyakarta Province.
29. Annual Meeting Member of BMT BAROKAH Padi Melati at 2012

Traffic Jam in Tuanku Tambusai Road, Pekanbaru City

Lucky Prawira, Abdul Kudus Zaini and Puji Astuti

Abstract Traffic jam in urban road in big cities has been the main issue in developing countries such as Indonesia. It is so easy to own motor vehicles today. This is one of the causing factors of dense traffic. In Riau Province, particularly Pekanbaru City, the number of motor vehicle ownership is growing sharply. This growth is in line with the economic growth and people's mobility. The characteristic of the usage of Tuanku Tambusai Road Part is mixed-use area. The objective of the research is to (1) identify the causing factors of the traffic jam incidence in Tuanku Tambusai Road and (2) to explore the traffic attraction and traffic generation. The method used in this research is a deductive approach based on a general theory which has to be proved with fact finding. In addition to the deductive approach, a quantitative method is used to identify the performance of Tuanku Tambusai Street, its traffic attraction, and traffic generation. The result shows that the causing factors of the traffic jam in Tuanku Tambusai Street among others are (1) the road capacity is about 1978.40 passenger car unit per hour, and (2) the traffic flow volume of Tuanku Tambusai Street from SKA Shopping Center for on the most dense day, Thursday, is 4340.2 passenger car unit, whereas that of from the flyover on the most dense day, Thursday, is only 815.4 passenger car unit. This indicates that the traffic flow volume from SKA Shopping Center (east) has been over the capacity of the road. On the other hand, the traffic flow volume from the flyover (west) is still under the capacity of the road. The saturation level of the road, from west to east, is categorized unstable with the point of 2.19 passenger car unit. Meanwhile, from east to west, it is categorized stable with the point of 0.41 passenger car unit, categorized into an F-level service. The traffic attraction and generation of the road are that correlation value is 0.991. Theoretically, a strong

L. Prawira (✉) · A.K. Zaini · P. Astuti
Departement Urban and Regional Planning, Riau Islamic University, Pekan Baru, Indonesia
e-mail: lucky_prawira@yahoo.com

P. Astuti
e-mail: pajiastutiafrinal@yahoo.co.id

T. Taufik et al. (eds.), *ICoSI 2014*, DOI 10.1007/978-981-287-661-4_29

correlation will occur if the value is close to 1. Based on summary model, the r-square value is 0.975; ANOVA value is 39.930 with a significance level of 0.049; and coefficient value is 304.940. As a conclusion, according to the regression formula, the number of mobility of the traffic attraction and generation is influenced by the increasing use of land.

Keywords Road performance · Traffic attraction · Traffic generation · Traffic jam · Land use

1 Introduction

Traffic jam in urban road has been the main issues in developing countries such as Indonesia. Generally, there are three factors that cause congestion: increase number of vehicle ownership; limited resources for highway construction; and not optimal operation of the existing public transport facilities [1]. It is very easy to own motor vehicles today. This is one of the causing factors of dense traffic. This leads to the uncontrolled growth of the number of vehicles. Congestion problems and poor public transport services trigger an increase in the use of private vehicles in Indonesia.

The number of motor vehicle ownership in Pekanbaru City is growing sharply, the growth in line with the economic growth and people's mobility. Based on the data held by the section numbering Book Vehicle Ownership (BVO), in Regional Directorate of Traffic Police in Riau Province, in October 2011, as many as 9432 units of two wheelers are registered to get numbering BVO, while for four-wheeled vehicles, based on the same data, as many as 2641 units registered. In September 2011, a number of two-wheeled vehicles are listed in the reg numbering much more. During September 2011, there were 9.809 motorcycles in Pekanbaru released to the market, while the number of car type vehicle in November was not as much in October. Only 2091 recorded cars were released to the market. The highest number of vehicles was released to the market occurred in August, a total of 10,956 units of motorcycles and 2941 units of the car (Tribun Pekanbaru, Tuesday, November 22, [2].

Although the distribution of motor vehicle purchases for the entire county, but most of the residents purchased Pekanbaru. The impact of street in Pekanbaru could be compact in certain hours, such as hours go to work or school, breaks, and hours after work or school. This situation has led to reduced levels of user comfort vehicle, timeliness, in addition to air pollution caused by exhaust fumes and combustion vehicles. And on the other side, public transportation cannot accommodate and serve as these routes are shaping up a lot, so people tend to use private transport in order to move faster trip without the need to depend on the public transport system which is still irregular.

The usage characteristic of Tuanku Tambusai Street is the mixed-use area which consists of the region's economy, education, office, residential, worship, and health.

Characteristic of land is the movement of traffic puller so that the tendency of traffic jams often occurs with day-time span and more frequently. This is compounded by the behavior of road users of disorderly segment.

The problems that occur in Tuanku Tambusai Street consist of side barriers form on-street parking, activities of street vendors who sell on the sidewalk, the presence of the crowd at any given time, narrowing the width of the road, current number of vehicles coming and going from the direction of the road environment, and no passage of legislation banning large vehicles the road crosses Tuanku Tambusai in the morning until the afternoon.

There have been many steps taken overcoming the problem of congestion on the Tuanku Tambusai Street such as the construction of the bridge crossing, stop the manufacture, parking prohibition in place, and not the other, but still not able to overcome the congestion problem. Even the facilities are not optimal and switching functions, as the stop and the bridge crossing are now often found flattened, to the resting place of the homeless and the insane.

2 Research Methods

This study uses a deductive research approach with a quantitative analysis method. Data sources use primary data and secondary data. Data analysis method is used in the form performance roads analysis, and attraction and generation analysis of traffic movements. Analysis includes road performance the road capacity, traffic volume, degree of saturation, and level of service. Benchmark analysis of the road capacity is Indonesian Highway Capacity Manual (IHCM [3]). Analysis attraction and generation of traffic movements performed to estimate traffic movements as a result of land use. Analysis uses methods of correlation and simple linear regression analysis.

3 Discussion

3.1 The Characteristics and Traffic Movement Patterns

Tuanku Tambusai Street in Pekanbaru City included in classification of urban roads and function in the form of secondary arterial roads. The length of roads (in research) is meticulous 3.54 miles with varying width 6–12 feet. The road consists of two lanes and a divided two-way movement and bounded by the road median, with different widths on each side. This is one of the main roads in Pekanbaru which has a spiral pattern or veins. Based on survey, results are obtained along the road corridor Tuanku Tambusai; there are 24 points of intersection between the Tuanku Tambusai Street and the neighborhood streets. The number of intersection resulted in frequent delays due to vehicles turning on to or off of the road environment, resulting in delays and reduced vehicle speed of the vehicle.

3.2 *The Characteristics of Land Use*

The observation develops land along Tuanku Tambusai Street among others: trading and services activities, office activities, education activities, health activities, and settlement activities. Tuanku Tambusai Street activities dominated the center of trade and services, but this is not supported with adequate parking facilities. The centers of activities along the street most of the land do not provide parking for visitors. This causes a lot of vehicles parked on the street, which disrupt the flow of traffic and resulted in high rates of side friction, thereby reducing the capacity of the road.

3.3 *The Factors Causing Traffic Jam in Tuanku Tambusai Street*

3.3.1 Analysis of Road Capacity

The data is calculated and analyzed with reference to the Indonesian Highway Capacity Manual (IHCM) [3], and road capacity is determined by the following factors:

1. A basic capacity (Co) and capacity of urban roads in Tuanku Tambusai road part consist of two lanes and divided into two directions, so it has a value of 2900 passenger car units per hour.
2. Width adjustment factor of road traffic (FCw); Tuanku Tambusai Road has a width varying, ranging from 6 to 12 m. To shorten the time of the study, the width of the path chosen is 6 m with values FCw of 0.87.
3. Adjustment factor due to the direction of traffic separation study site is worth 0.97, due to the varied way traffic separation.
4. Adjustment factor aside barriers and road shoulder width (FCsf) determined based on the type of road, beside the class barriers, the effective width of the shoulders. For a two-lane two-way street and the effective width of the shoulder is of one (1) meter. Side friction conditions such as the location of the commercial area of research have high activity side of the road. Indonesian Highway Capacity Manual (IHCM) [3], adjustment factor aside barriers and road shoulder width valuable research sites of 0.86.
5. Capacity adjustment factor due to city size is determined by the population of the city. The population of the city of Pekanbaru is close to 1 million by Indonesian Highway Capacity Manual (IHCM) 1997; value for the population of the city size is $0.5 - 1.0$ million worth of 0.94.

To obtain the value of capacity Tuanku Tambusai road, the value adjustment factor of urban road capacity by IHCM 1997 in mathematics is as follows:

$$\mathbf{C = Co \times FCw \times FCsp \times FCsf \times FCcs}$$

$$C = 2900 \times 0.87 \times 0.97 \times 0.86 \times 0.94 = \mathbf{1978.40}\ \text{pcu/h}.$$

3.3.2 Volume of Traffic Flow Analysis

Collecting data traffic volume or the number of vehicles passing on a line observation is done by record of all vehicles passing through a transverse tread of the line, at the observation post during the period of observation aided by the use of counter. Calculations carried out at three times during one hour, held at 06.30 am–07.30 WIB; lunch held at 12.30–13.30 WIB; and evening held at 17.30–18.30 WIB. The position of the observation point is at two points. The observation of traffic volume is shown in Tables 1 and 2.

The number of each volume of the above vehicles is converted into passenger car units (pcu), which are grouped by types motorcycle (MC = 0.2), light vehicles (LV = 1), and heavy vehicles (HV = 1.3) according to the standard Indonesian Highway Capacity Manual (IHCM) 1997. The results of the volume of traffic flow in Tuanku Tambusai road Passenger Car Unit (SMP) are done using the following formula (Tables 3 and 4):

PCU (Q) = Total vehicle per classification x Equivalent vehicles by type.

Passenger car unit for the volume of traffic flow Tuanku Tambusai Road of direction SKA Mall intersection is greater than the capacity of the existing road, while the passenger car unit for volume flow of traffic from the intersection toward Fly Over is smaller than the capacity of the road. Road capacity is about 1978.40 passenger car unit per hour. Here the difference is in traffic volume roads in the study site in units of passenger cars with a capacity of road (Table 5).

Volume difference of traffic flow in units of passenger cars due to the western part of Tuanku Tambusai road part and a suburb east form the city center. The movement of the vehicle from the suburbs to the city center passes Tuanku Tambusai road part l savings due to the large ore route, distance and travel time to the city center, while the movement of the vehicle from the opposite direction is smaller, which is because the route, distance and time varied as desired road users.

3.3.3 Analysis of Degree of Saturation

Analysis of the degree of saturation (DS) was performed using the data traffic flow (Q) in pcu/h and then divided by the road capacity (C) of 1978.40 pcu/h, so that the degree of saturation results of the two-way flow of vehicles enters the following (Table 6).

3.3.4 Level of Service (LOS) Analysis

Level of service based on Indonesian Highway Capacity Manual (IHCM) in 1997 is characterized by the degree of saturation. From the calculation of the degree of saturation (DS) obtained the highest score in the study site was 2.19 pcu/h (Table 7).

Table 1 Traffic volume roads intersection Tuanku Tambusai of direction SKA Mall

No	Type of vehicle	Number of vehicles (units)												
		Vehicle flow path towards Tuanku Tambusai of horse monument/SKA Mall												
		Monday				Thursday				Sunday				Total
		Morning	Afternoon	Eve ning	Total	Morning	Afternoon	Eve ning	Total	Morning	Afternoon	Eve ning	Total	
1	Motorcycle	1183	1160	1086	3429	1164	1176	1130	3470	991	1018	923	2932	9831
2	Private car	1017	1076	1089	3182	1001	1099	1119	3219	809	973	701	2483	8884
3	Car public transport													
	Oplet	79	64	66	209	65	52	74	191	50	64	58	172	572
	Taxi	30	25	19	74	28	23	20	71	17	17	13	47	192
	Transmetro	1	1	1	3	1	1	1	3	1	1	1	3	9
4	Freight cars													
	Colt diesel	18	19	16	53	19	17	8	44	18	11	11	40	137
	Pick up	24	23	25	72	26	24	19	69	20	18	11	49	190
	Truck	9	10	7	26	12	9	6	27	6	3	4	13	66
Total		2361	2378	2309	7048	7048	2401	2377	7094	1912	2105	1722	5739	19,881

Source Survey of traffic volume, 2013 (Monday, June 24, 2013; Thursday, June 27, 2013, and Sunday, June 30, 2013). Observation point located at postal horse monument

Table 2 Traffic volume roads Tuanku Tambusai of direction junction flyover

No	Type of vehicle	Number of vehicles (units)												
		Vehicle flow path towards Tuanku Tambusai of Jendral Sudirman street												
		Monday				Thursday				Sunday				Total
		Morning	Afternoon	Eve ning	Total	Morning	Afternoon	Evening	Total	Morning	Afternoon	Eve ning	Total	
1	Motorcycle	307	310	323	940	321	332	397	1050	217	276	265	758	2748
2	Private car	132	137	125	394	152	143	128	423	71	83	79	233	1050
3	Car public transport													
	Oplet	28	29	18	75	26	30	24	80	19	14	15	48	203
	Taxi	14	11	7	32	17	9	7	33	3	9	4	16	81
	Transmetro	1	1	1	3	1	1	1	3	0	2	1	3	9
4	Freight cars													
	Colt diesel	10	7	3	20	9	5	2	16	2	1	1	4	40
	Pick up	11	13	5	29	14	10	9	33	3	8	2	13	75
	Truck	7	6	6	19	5	3	1	9	2	2	1	5	33
Total		510	514	488	1512	545	533	569	1647	317	395	368	1080	4239

Source Survey of traffic volume, 2013 (Monday, June 24, 2013; Thursday, June 27, 2013, and Sunday, June 30, 2013). Observation points located at the front of the bridge crossing Vanhollano Bakery Tuanku Tambusai Road

Table 3 Volume of traffic flow Tuanku Tambusai road of direction SKA mall

According to vehicle type classification	Total of vehicles					
	Monday	Value of PCU	Thursday	Value of PCU	Sunday	Value of PCU
Motorcycle	3429	685.8	3470	694	2932	586.4
Light Vehicles	3537	3537	3550	3550	2751	2751
Heavy Vehicles	82	106.6	74	96.2	56	72.8
Total	7048	4329	7094	4340.2	5739	3410.2

Source Result of analysis, 2013

Table 4 Volume of traffic flow Tuanku Tambusai road of direction junction fly over

According to vehicle type classification	Total of vehicles					
	Monday	Value of PCU	Thursday	Value of PCU	Sunday	Value of PCU
Motorcycle	940	188	1050	210	758	151.6
Light vehicles	530	530	569	569	310	310
Heavy vehicles	42	54.6	28	36.4	12	15.6
Total	1512	772.6	1647	815.4	1080	4772

Source Results of analysis, 2013

Table 5 Comparison of total volume of traffic flow two-way (Q)

No.	Flow in vehicle	Day	Value of Q	Difference Q and C	Category
1	Direction of the flow of vehicles from Ska Mall (west–east)	Monday	4329	2351	Exceed capacity
		Thursday	4340.2	2361.8	Exceed capacity
		Sunday	3410.2	1431.8	Exceed capacity
2	Direction of the flow of vehicles from Simpang Fly Over (east–west)	Monday	772.6	−1206	Lack of capacity
		Thursday	815.4	−1163	Lack of capacity
		Sunday	477.2	−1501.2	Lack of capacity

Source Results of analysis, 2013

Table 6 Degree of saturation (DS) Tuanku Tambusai road

No.	Flow in vehicle	Day	Value of Q	Degree of saturation (DS)
1	Direction of the flow of vehicles from Ska Mall (west–east)	Monday	4329	2.18
		Thursday	4340.2	2.19
		Sunday	3410.2	1.72
2	Direction of the flow of vehicles from Simpang fly over (east–west)	Monday	772.6	0.39
		Thursday	815.4	0.41
		Sunday	477.2	0.24

Source Results of analysis, 2013

Table 7 Level of service (LOS) category

No.	Flow in vehicle	Density levels	Value degree of saturation (DS)	Category level of service (LOS)
1	Direction of the flow of vehicles from Ska Mall (west–east)	Solid	2.19	F
		Low	1.72	F
2	Direction of the flow of vehicles from Simpang Fly Over (east–west)	Solid	0.41	C
		Low	0.24	B

Source Results of analysis, 2013

LOS B < LOS C < LOS F, traffic flow conditions ranging unstable, decreased relatively fast operating speed, due to obstacles that arise during the time and freedom of movement are relatively small. This condition can increase to LOS F, when the traffic volume keeps getting bigger, and almost close to or equal to the capacity of the road, when it becomes low speeds, changing, stopped at any time. From the data analysis of the degree of saturation obtained LOS F value of 2.19 over the edge of the peak capacity of roads by a margin of 1.19 where the road conditions are no longer able to accommodate the traffic flow.

3.3.5 The Traffic Attraction and Traffic Generation

The traffic attraction and traffic generation Tuanku Tambusai road can be seen from the movement of vehicles entering the study site. This is in line with the assumption that the output from the calculation of the traffic attraction and traffic generation such as the number of vehicles, transportation of people or goods per unit time, such as vehicles per hour [4]. The traffic attraction and traffic generation depend on two aspects of land use; types of land use and the amount of activity and the intensity of land use. Types of different land uses (residential, educational and commercial) have characteristics different traffic generations.

The pattern of land use in the study site is a mixed area, land use not only for the region but also trade and settlement services, education, health, social facilities, and public facilities. This study examines four types of land use, namely settlements, trade and service area, region, and area health education. Each type of land use will result in towing and different generations.

Land use in the study site is dominated by trade and services (45%), residential areas (40%), the area of education (10%), and health areas (5%). To predict and foresee the possibility of the traffic attraction and traffic generation influenced by land use activities in the location of research, it analyzed the effect of land use on pull and trip generation to obtain a model that is the basis for predicting and forecasting the possibility of resurrection and pull movement on the future, in analyzing the effect of land use on the movement used Linear Regression analysis. The variables used are divided into two (2) which is the dependent variable (the number of which occurs each land use) and the independent variables (broad land use), as shown in the following table (Table 8).

Table 8 Movement of vehicles per each land use Tuanku Tambusai road (timeline one day, date November 14, 2013)

No	Land use type	The number of vehicle movements (pcu/h)		Percentage (%)	Peak hour	Influence area of land use (Ha)
1	Trade and services	2125		45	12.30–13.30	19.85
	Generation		1016			
	Attraction		1109			
2	Settlement	2096		40	06.30–07.30	19.15
	Generation		1138			
	Attraction		958			
3	Education	504		10	06.30–07.30	0.40
	Generation		19			
	Attraction		485			
4	Health	168		5	12.30–13.30	0.25
	Generation		23			
	Attraction		145			
Total		4893		100		39.65

Source Results of analysis, 2013

Table 9 Correlation results

		Total movement (pcu/h)	Area of land use (Ha)
Person correlation	Total movement (pcu/h)	1.000	0.991774
	Area of land use (Ha)	0.991774	1.000
Sig. (1-tailed)	Total movement (pcu/h)	.	0.023
	Area of land use (Ha)	0.023	

Source Results of analysis, 2013

Table 10 Anova test applications using Microsoft Excel 2007

	df	SS	MS	F	Significance F
Regression	1	2,069,682,832	2,069,682,832	39,930	0049
Residual	1	51,831,834	51,831,834		
Total	2	2,121,514,667			

Source Results of analysis, 2013

Table 11 Coefficients

Model		Coefficients	Standard error	t Stat	P value
1	(Constant)	304.940	163.809	1861	0313
	area of land use	93.594	14.811	6319	0049

Source Results of analysis, 2013

Area of land (ha) per each land use in the study site with a width of 200 m and a length of 3.54 km is obtained using the Google Earth application with tools area. Results of linear regression analysis using Microsoft Excel 2007 are as follows (Table 9).

The relationship between the movements of the vast number of variables land use is very strong which is indicated by a correlation coefficient of 0.991. Theoretically, the correlation would be stronger if the value of the correlation coefficient is close to 1 (one) (Table 10).

ANOVA or F test value obtained F count was 39.930 with a significance level of 0.049. Due to this, the probability (0.049) is smaller than 0.05, and the linear regression model can be used to predict the movement of a variable or variables can be considered where broad land use influences the amount of movement generation and attraction of the location study (Table 11).

From the table of coefficients obtained by the regression equation as predictors of seizure occurrence and movement of the pull of the future due to the influence of land use, the regression formula is as follows:

$$\mathbf{Y = 304.940 + 93.594\, X_1}$$

Note:

Y = Total Movement Journey

X_1 = Area of Land Use (Ha)

Constants as large as 304.940 state that if land use has not increased or fixed (mathematically $X_1 = 0$), the amount of movement is of 304 passenger cars units (pcu). From this equation it can be concluded that the higher the extensive use of land means, the higher the amount of movement of attraction and generation of trips produced.

4 Conclusion

Based on the research it can be concluded that

1. Identifyng the causing factors of traffic jam in Pekanbaru City is as follows:
a. Capacity of roads: Tuanku Tambusai road has a capacity of 1978.40 road passenger car units per hour (pcu/h);
b. The research result indicated that the traffic flow volume from SKA Mall (east) has been over than the capacity of the road. On the other hand, the traffic flow volume from the flyover (west) is still under the capacity of the road.
c. The saturation level of the road: from west to east is categorized unstable with the highest point of 2.19 passenger car unit. And from east to west is categorized stable with the highest point of 0.41 passenger car unit,
d. The service level on the basis of Indonesian Road Capacity Manual (MKJI), 1997, is marked with the saturation level. From the calculation, the highest saturation level in Tuanku Tambusai Road is 2.19 passenger car unit per hour. According to the service level based on MKJI, the saturation level ≥ 1 can be categorized into an F-level service. This category has been over the peak limit of the road capacity with a difference of 1.19 passenger car unit. Finally, this road condition cannot hold the traffic flow. As a result, a heavy delay occurred, which is called traffic jam.
2. The traffic attraction and generation of the road are as follows:
a. The correlation between the amount of mobility and the land use area is very strong. It is indicated by a correlation coefficient of 0.991. Theoretically, a strong correlation will occur if the value is close to 1.
b. Summary model: R-square value is 0.975. Means of approximately 97.5% of the movement can be explained by the extensive use of land, while the remaining $(100 - 97.5 = 2.5\%)$ is explained by other factors (side friction).
c. ANOVA value: According to the ANOVA test, the calculated F value is 39.930 with a significance level of 0.049. Because the probability (0.049) is lower than 0.05, the linear regression model can be applied to predict the variable of mobility number. Or it can be said that the variable of land use area has an influence on the amount of traffic attraction and generation in the area.
d. Coefficient value: According to the regression formula, the number of mobility of the traffic attraction and generation is influenced by the increasing use of land. The higher the land use, the higher the number of the traffic attraction and generation mobility will be.

References

1. Wahyuni, R. (2008). Pengaruh Parkir Pada Badan Jalan Terhadap Kinerja Ruas Jalan. Tugas Akhir. Medan : Jurusan Teknik Sipil Universitas Sumatera Utara.
2. Koran Tribun Pekanbaru. (2011). Melonjaknya Pertumbuhan Kendaraan Kota Pekanbaru. Pekanbaru : Selasa, 22 November.
3. Direktorat Jenderal Bina Marga. (1997). Manual Kapasitas Jalan Indonesia (MKJI). Jakarta.
4. Tamin, O.Z. (2000). Perencanaan dan Permodelan Transportasi. Bandung : Penerbit ITB.

Work in Progress—Using Transnational Education Program to Inculcate Social Innovation and Sustainability Mindset

Noel Kristian

Abstract With the rapid progress of globalization, there is a need to innovate the way educational institutions in developing students to be world ready to face current sustainability issues. Learning Express (LeX) program is Singapore Polytechnic (SP) answer to tackle this issue. The objective of this program is to give opportunities for students from SP and partner institutions in using their technical skills and knowledge to conceive and co-create with rural communities on social projects in sustainable ways. In this paper, the detailed structure and pedagogy of LeX is explained using the implementation trip at Yogyakarta, Indonesia in March 2013 as an example. It was found that a number of key success factors like local educational institution partners, buddy system, and methodology used are important for successful implementation of this program.

Keywords Education · Social innovation · Service learning · Design thinking

1 Introduction

Engineering education has traditionally focusing on the technical advancement [1]. However, with the rapid progress of globalization, there is a need to instill the mindset of social awareness and a bias toward sustainability [2]. In Singapore Polytechnic (SP), initiatives to infuse this mindset have been put recently and one of the examples is the institutional Social Innovation Project (SIP) module. This module is aimed to instill social awareness mindset to help community in need. Although SIP module has been implemented successfully for the past 2 years in SP, it is believed that this module could be implemented more effectively overseas for the low-income community referred to bottom of the pyramid [3]. With this idea in mind, Learning Express (LeX) program was conceived and implemented.

N. Kristian (✉)
School of Chemical and Life Sciences Singapore Polytechnic,
500 Dover Road, 139651 Singapore, Singapore
e-mail: noel_kristian@sp.edu.sg

T. Taufik et al. (eds.), *ICoSI 2014*, DOI 10.1007/978-981-287-661-4_30

LeX is a transnational program that aims to give opportunities for students from SP and partner institutions in using their technical skills and knowledge to conceive and co-create with rural communities on social projects in sustainable ways. This is different from the typical community service where participants are mainly expected to contribute their labor on doing pre-determined work with less sense of ownership throughout the duration of the program. While in social innovation project, participants are required to develop deep understanding of the community and unearth their unmet needs and develop and implement ideas to meet those needs.

In this paper, the detailed implementation of LeX program at Yogyakarta, Indonesia in March 2013 is described. The framework used and some key success factors are described. Finally the findings on the students experience and learning in terms of process used, the insight gained and solutions are explained.

2 Structure and Pedagogy of the Program

LeX program consists of two parts which are LeX 1 and LeX 2 that take place in two semesters consecutively. In LeX 1, students undergo a 2-week expedition in selected villages of host countries where the participants immerse themselves into the lives of the community to understand their lifestyle, culture, and issues they face. Each SP student is paired with another student from a partnering institute and they are divided into three groups that work in three different villages. The main objective of LeX 1 is to allow students to experience the living conditions in the village and build empathy and collaboration with the people that they will work with and finally ideate possible solutions. In other words, the activities are mainly in conceiving the needs of the society and designing concepts to satisfy those needs.

In LeX 2, students build working prototypes based on the concepts that they have conceived in LeX 1 and at the end of the semester they implement and operate their prototype in the same village. Besides that students also transfer their know-how in terms of building, operating and maintaining the prototype so that the villagers could sustainably use it and expand the use of the prototypes to other villages. In this paper, the author focuses on the pedagogy and implementation of LeX 1, while the pedagogy and implementation of LeX 2 will be reported in separate publication.

To conceive the needs of the community and design appropriate solutions, Design Thinking (DT) framework is adopted whereby four main stages are applied throughout LeX 1 namely sense and sensibility, empathy, ideation, and prototyping [4].

Although represented in Fig. 1 in a linear fashion, the DT process is best thought of as a system of three overlapping spaces of inspiration, ideation, and implementation, rather than a sequence of orderly steps [5]. In this program, students learn by doing starting from defining the issues. They start in the field, where they develop empathy for people they serving for, uncovering real needs of the

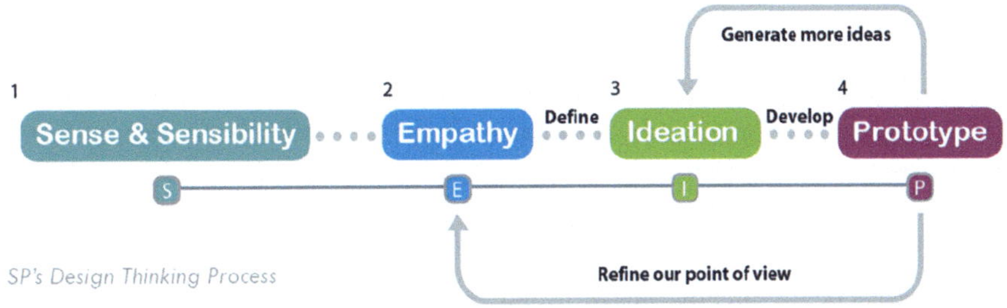

Fig. 1 SP design thinking framework

Table 1 Implementation of DT in LeX 1

Time	Program
Day 1	Arrival in the host country
Day 2–Day 3	**Sense and sensibility** by learning the local language and culture
Day 4–Day 7	**Empathy** by staying in the village and conducting observation and interview to develop deep understanding about the issues faced by the community
Day 8–Day 9	**Ideation** to generate plausible solutions to meet the needs of the community
Day 10–Day 11	**Prototyping** with simple and quick prototypes to gather feedback from the community
Day 12–Day 13	Fine tune prototypes and deliver final presentation for the stakeholders of the projects
Day 14	Departure from the host country

community. They then iterate to develop possible solutions and create rough prototypes to take back out into the field and test with real people [6]. The detail implementation of DT into this program is described in Table 1.

Using design thinking framework, students develop ideas and co-create together with the local community that consists of the villagers and local university students. The main requirement of the final ideas that developed into various solutions is the use of sustainable local resources and technology. The key to this pedagogy approach is to maintain the sustainability of the projects that students and local community conceive because local community can only continue the development of these projects onsite using available resources [7].

3 Project Scopes

One of the key aspects in this program is to identify sustainable projects. In this program, faculty of different expertises such as engineering, business, food science, social science, and industrial design agreed upon projects that have engineering/product development opportunity to increase the livelihood of the community without neglecting their social needs. Before deciding on the projects, a reconnaissance trip is conducted together with the local partners. Around five possible projects are provided by the partners in the host country and throughout the reconnaissance trip SP staff and local partners scope the projects and select on three different projects that are located on three different villages. The criteria used to select the projects in this program are as follows:

1. The projects are directly related to the community.
2. The projects have potential technological solutions that involve social needs.
3. The projects are supported by local education institution partners.
4. The communities are keen to collaborate with SP and local partners.
5. The communities are able to provide authentic homestay experience for international students.

For example, for the LeX 1 program at Yogyakarta Indonesia in March 2013, three of the projects are as follows:

1. To explore effective and environmentally friendly ways of making cassava crackers in cottage industries.
2. To extract and develop low-cost syrup derived from Sorghum in cottage industries.
3. To tap on green identity of eco-village to develop products through up-cycling waste and developing a business model for products.

4 Implementation of the Program in Yogyakarta, Indonesia

In March 2013, a group consisting of 24 SP students, 2 Duy Tan (Vietnam) university students, and 4 Kanazawa Technical College (Japan) students underwent 2-week LeX 1 trip in Yogyakarta, Indonesia. The students consist of students from engineering, business, digital media, chemical, and life sciences school. This trip is the result of the collaborations among SP, University of Muhammadiyah Yogyakarta (UMY) and University of Pembangunan Nasional (UPN) "Veteran" Yogyakarta. The students were grouped into three teams that worked on three different projects and each team was accompanied by one faculty from SP and one from UMY or UPN. As part of the curriculum, credit units were given in-lieu of a compulsory course named Social Innovation Project (SIP) for SP students and Kuliah Kerja Nyata (KKN) for UMY and UPN students, while for Japanese and Vietnamese students, this program is part of their out-of-classroom enrichment program.

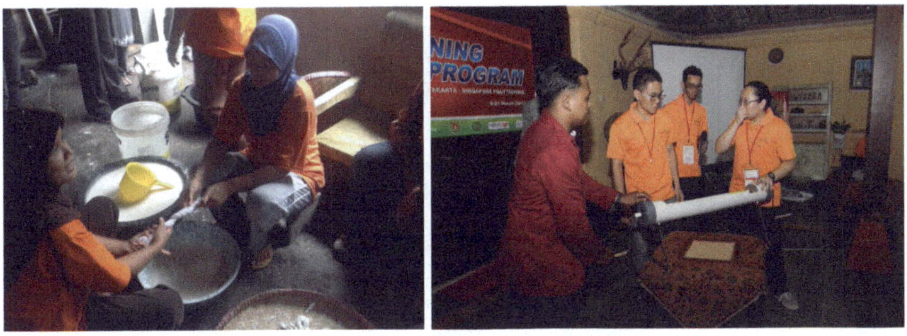

Fig. 2 Women in Polengan village squeezing cassava juice (*left*). An efficient machine to squeeze cassava juice (*right*)

During the expedition itself, each SP student was paired with another student from UMY or UPN. This buddy system has two objectives. First, it is meant for local students to teach the local language and culture to the international students. Second, the international students teach the Design Thinking framework to local students that would be used throughout the 2-week program.

In one of the projects, students were faced with the laborious and tedious process of converting cassava roots into crackers in Polengan village. This cracker business is run by the housewives in the village and is meant as a side income for them. After experiencing the whole process, students identify the most unpleasant process for the workers who were the squeezing of the cassava to obtain the unfermented cassava juice as shown in Fig. 2. In normal operation, two people are needed to do this and they spend up to 5 h daily in doing this. While the process itself is tedious and consume lots of energy, the workers still enjoy doing this due to the presence of the other co-worker doing the same thing. With these insights, students created a manually operated squeezer for 2 people to help the workers reduce the strength needed in this process [8].

In another project, a team worked with Banjarum community to improve syrup production from sorghum stalks. After extracting the juice from the stalks, traditional stove made of stone is used to boil the juice to make syrup. The students immersed themselves into the process and they experienced difficulty in breathing and eye irritation because of the smokes generated by burning the biomass fuel as shown in Fig. 3. On top of that they also encounter difficulties in replenishing the fuel due to high air temperature nearby the stove. With these insights, students proposed a solution to enhance the existing stove by optimizing the design in terms of air intake, circulation and heat transfer and ensure the wood as the fuel can be replenished easily and safely [8].

Fig. 3 Boiling sorghum juice using traditional stove (*left*). A stove designed to produced less smoke (*right*)

5 Students Feedback on the Learning Experience

At the end of the program, a feedback session was held with the students. Overall, the students felt that the trip had made them appreciate the communities and their environment as a whole. Some of the interesting comments written by students in their reflection are listed below:

- *"I learnt not to stereotype and be judgemental to the people just because they live in a village."*
- *"I have learnt that people in the village are caring and willingly let me enter their homes."*
- *"Initially I was not really getting the idea why we needed to use Design Thinking, but when we apply this method in real life situation just like in the village, it really did make sense and helped a lot in getting the almost perfect solution."*
- *"I realised even though the villagers do not have luxuries like air conditioner, they are not bit considered disadvantage, once we could realise that all those luxuries are just superficial objects, they actually are living what could be an even more fruitful lives in comparison to many city dwellers."*
- *"I have learnt how we can impact the lives of the villagers using our engineering skills and knowledge."*
- *"The fact that I am able to serve the community with a group of committed people has made me realize that I enjoy doing this. Also I have learnt that I actually enjoy village life more than city life."*
- *"It was truly an unforgettable experience. Everyone just worked as one despite our differences in nationalities and cultures. We worked hard, and played hard. We felt so proud of ourselves when we got the job done! The Indonesian buddies were extremely friendly and offered help in every way they could. This journey has definitely changed my perception of life greatly."*

Students also felt that their experience make them more socially aware on what is happening outside Singapore and they particularly mentioned how they appreciate learning to work with people with different cultures even though they have different ways in doing things [8].

6 Reflections on Success Factors

Finally, some of the initial critical success factors and pedagogy of the program such as local partner, buddy system, and the used of DT are identified as important for successful implementation of this program. A close working collaboration with local education institutions such as UMY and UPN is really needed for the successful implementation of this program because education institutions have similar mission which is to serve the society and educate people. With this same mission, the main objective of the program which is to give opportunities for students in using their technical skills and knowledge to conceive and co-create with rural communities on social projects in sustainable ways could be achieved. Buddy system whereby local university students who are proficient in English are paired with SP students is essential for the program because the communication and cultural barriers can be reduced. Finally, the use of DT is as essential as the other factors because DT allows students in finding the important issues faced by the people and brainstorm possible solutions without neglecting the physical and emotional being of the community.

7 Conclusions

Learning Express is a transnational program that provides students with opportunity to experience real project for real people, learn new skills, make meaningful new friendships, and rediscover their social awareness as world citizen. This program aims to promote social innovation and sustainability mindset by bringing students from various disciplines and various countries collaborating on projects in developing world. This exciting program provides the opportunities for students to make sense of what they have acquired in the classroom through putting/linking the academic knowledge into practical execution via various methodologies. The various stages of this discovery/application process are conducted in a real-world situation and are properly structured. As a consequence of this immediacy of experience, this learning activity is more likely to be personally meaningful to students and to generate emotional consequences, to challenge values as well as ideas, and hence to support social, emotional, and cognitive learning and development of the students.

Acknowledgements The author is grateful to University of Muhammadiyah Yogyakarta (UMY) and University of Pembangunan Nasional (UPN) "Veteran" Yogyakarta for the successful preparation and implementation of the LeX 1 at Yogyakarta Indonesia.

Remarks All the figures in this paper are publically available in database.

References

1. Crawley, E.et. al. 2007.*Rethinking Engineering Education: The CDIO approach*, Springer.
2. Wade, R. 2008.*Education for sustainability: Challenges and opportunities*, Policy & Practice: A Development Education Review, Vol. 6, Spring, pp. 30–48
3. Amadei, B. 2004. *Engineering for the developing world*, The Bridge, Vol. 34, No.2.
4. KF, Soh. 2011. *An Observational Study of Infusing Design Thinking into the CDIO framework*. Proceedings of the 7th CDIO conference.
5. Brown, T. 2009. *Change by Design*, Harper Business.
6. Kelley, T. 2013. *Creative Confidence*. Crown Business.
7. Brown, T. and Wyatt, J. 2010. *Design Thinking for Social Innovation*, Development outreach, World Bank Institute.
8. Kristian, N. 2014. *Social innovation in South East Asia*. The chemical engineer magazine, issue 874, April 2014, pp 26–27.

Part III
Technology

2-D and 3-D Subsurface Liquefaction Potential Profiling Using Tomography Surface Waves Method

Sri Atmaja P. Rosyidi

Abstract A 6.3 Mw earthquake struck Yogyakarta region in 2006 causing many geotechnical damages, e.g., ground cracks, surface displacement, landslides and local liquefactions and soil billings occurred in some regions. From field observations, it was shown that most minor to major structural damages in buildings and bridges were identified near to liquefaction locations. Consequently, a site investigation and advanced analysis for providing the subsurface liquefaction potential information plays important rule in infrastructure design related to structural damage analysis. The aim of this paper is to use of the tomography surface waves method in order to investigate the liquefaction potential in 2-D and 3-D subsurface profile. These seismic surveys were conducted on deep loose sand and sandy soil deposit located inside Universitas Muhammadiyah Yogyakarta (UMY)'s campus, Indonesia which consists of. The liquefaction potential profile was analyzed and generated from combination of the shear wave velocity and soil properties information. Two earthquake scenarios, 6.3 and 8 Mw, were used to simulate the sensitivity of 2-D profile for identifying the liquefaction potential in each soil layer. The results show that the liquefaction potential widely occurs in observed sandy soil and sand deposit layer for stronger earthquake. Finally, the tomography surface waves method is becoming an effective tool for observing liquefaction potential profile in the site investigation, particularly for the purpose of geohazards analysis in the concept of sustainable environmental development.

Keywords Surface waves method · Liquefaction potential · Earthquake · Tomography

S.A.P. Rosyidi (✉)
Department of Civil Engineering, Universitas Muhammadiyah Yogyakarta,
Yogyakarta, Indonesia
e-mail: atmaja_sri@umy.ac.id

T. Taufik et al. (eds.), *ICoSI 2014*, DOI 10.1007/978-981-287-661-4_31

1 Introduction

Nowadays, a comprehensively system of soil dynamic analysis is becoming significant in order to provide suitable information for earthquake hazards. This analysis system can be used to evaluate the strong-ground motion parameters, microzonation map, liquefaction potential and ground settlements at observed site influenced by earthquakes. Moreover, the results from comprehensive soil dynamic analysis are employed in earthquake resistant design, seismic safety assessment and estimation of possible economical losses as well. Therefore, comprehensible surveys and site investigations relating to subsurface soil dynamic properties, i.e., shear wave velocity and soil modulus of sites play important rule in earthquake hazards analysis. Many studies conducted by Imai and Tonouchi [1] and Craig [2] which observed compressional (P)- and S-wave velocities in an embankment, and also in alluvial, diluvial, and tertiary layers, shows that shear (S)-wave velocity is one of the key parameters in seismic hazards analysis and S-wave velocities have a strong correlation to an index value of formation hardness used in soil mechanics and foundation engineering.

The aim of this study is to describe the use of surface wave technique for generating shear wave velocities profile and based on S-wave velocity, 2-D and 3-D mesh profile of soil liquefaction potential mapping can be developed. A case study was conducted at several observed locations in Universitas Muhammadiyah Yogyakarta (UMY) campus, Indonesia, where several buildings in study area suffered minor to moderate damage level caused by liquefaction. A 2-D shear wave velocity profile was produced based on the multichannel of surface wave survey. Thus, a liquefaction potential map was calculated by using cyclic stress ratio (CSR) and cyclic resistance ratio (CRR) analysis based on the shear wave velocity profile.

2 Study Area

2.1 Geological of Yogyakarta

Yogyakarta region and the vicinity are located as part of the Mount Merapi area and Yogya-Bantul plain that extends toward the south coast [3]. The geologic formation in this region is dominated by recent deposits of Merapi Volcano namely young volcanic deposits of Merapi Volcano and sedimentary and inter-bedded volcanic rock deposit namely Sentolo Formation. The young volcanic deposit can be divided into two formations. The lower part of the young volcanic deposits is Sleman formation and consists of sand and gravel inter-bedded by andesite boulders. Whereas, the upper part of the volcanic deposits, i.e., Yogyakarta formation, consists of sand, gravel, silt and clay inter-bedded. The thickness of these formations is identified from several bore log data obtained at different location. The

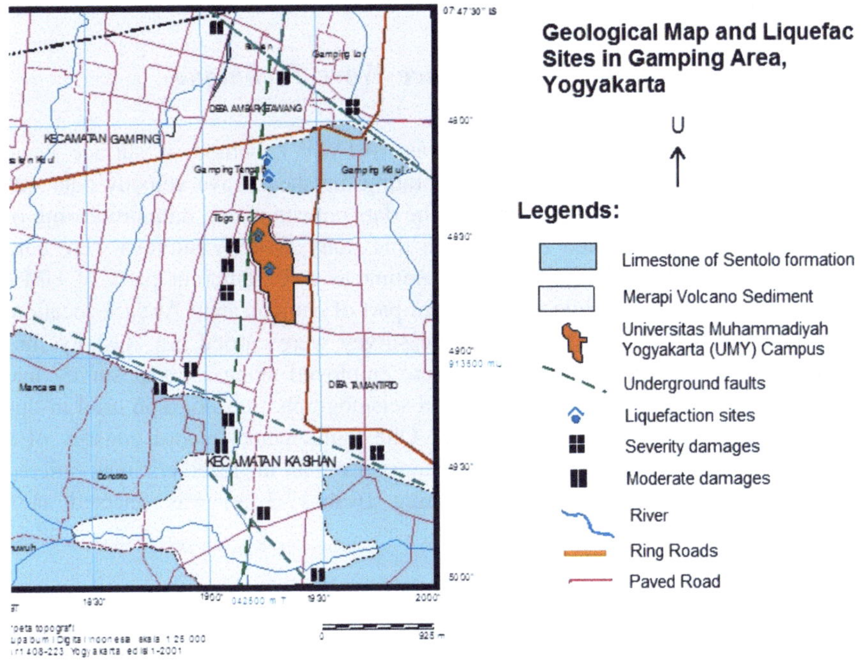

Fig. 1 Geology structure and fault lines locations close to study area Rosyidi et al. [4]

Sentolo Formation is found at depth of approximately 58 m. This formation consists of limestone and marly sandstone. The stratigraphy of Yogyakarta region can be seen in Fig. 1.

2.2 *Location of Study*

The location of study was chosen within the Universitas Muhammadiyah Yogyakarta (UMY) campus. Rosyidi et al. [4] conducted visual observations on structural and geotechnical damages in the UMY campus after the Yogyakarta earthquake, 27 May 2006. Based on the geological condition, i.e., rock formation (Fig. 1) and compilation of visual observation of the structural and geotechnical damages, a geological map and estimated underground faults for location of the study at UMY campus and surrounding area were reported by Rosyidi et al. [5]. From Rosyidi et al. [4], some predicted underground cracks and faults close to the damaged area were drawn with the direction of north to south (NS) located in the western part of the UMY campus. Consequently, a soil investigation by drilling test was carried out in this area by LKPT UMY [6], and underground faults analysis of selected site was conducted by Rosyidi et al. [5].

3 Research Method

3.1 Field Measurement of Surface Wave Technique

The multichannel analysis of surface wave (MASW) method as surface wave technique developed by Park [7] for generating the shear wave velocity data was used in this study. All the procedures in the data collection and data processing of MASW method are adopted from [8]. In this study, MASW surveys were conducted at three different locations representing as three different parts of UMY campus, i.e., northern, middle and southern part of campus area. At each location, three different measurement points of the survey were carried out. Twenty four receivers of 4.5 Hz vertical geophone were employed along a linear survey line with receivers connected to a multichannel seismograph. Seismograph used in this measurement was OYO Mac Seis 24ch. One multichannel record consists of a multiple number of time series or seismic traces from all the receivers in an ordered manner. A heavy sledgehammer weighing 16 lb (7.3 kg) was selected as a mechanical source.

3.2 Data Processing of MASW

Data processing consists of three steps, i.e., (1) preliminary detection of surface waves, (2) constructing the dispersion image panel and extracting the signal dispersion curve, and (3) back-calculating shear wave velocity (Vs) variation with depth. The preliminary detection of surface waves examines recorded seismic waves in the most probable range of frequencies and phase velocities. Consequently, construction of the image panel is accomplished through a 2-D (time and space) wavefield transformation method.

The experimental dispersion curve is then associated with surface wave energy in the frequency–wavenumber (f–k) domain from a 2-D Fourier transform. This curve shows the relationship between phase velocity and frequency for seismic waves including fundamental and higher modes propagated horizontally and directly from the impact point to the receiver line Park and Miller [9]. Consequently, the extracted fundamental-mode Rayleigh waves from an experimental dispersion curve is used as a reference to produce the 1-D Vs with depth in the inversion process. In the inversion process, a profile of a homogeneous layer extending to infinity in the horizontal direction is assumed. Based on the initial profile, a theoretical dispersion curve is then calculated using Knopoff method [10]. The theoretical dispersion curve is ultimately matched to the experimental dispersion curve of the lowest RMS error based on an optimization technique.

3.3 2-D Shear Wave Velocity (V_s) Mapping

A 2-D Vs map is constructed from the acquisition of multiple records of 1-D Vs profile with a fixed source–receiver configuration and a fixed increment of the configuration. A small increment is required when horizontal variation plot is expected. In most soil-site applications, total receiver spread length is set up in the range 10–30 m that will give an optimal fixed increment in the range of 5–15 m. A 2-D Vs map is then obtained from multiple 1-D Vs profiles with a particular fixed increment of the configuration by using tomography interpolation. In this study, all the seismic records are processed using SurfSeis version 2.01 that was developed by Kansas Geology Survey (KGS), USA.

3.4 2-D and 3-D Liquefaction Potential Analysis Based on V_S

Evaluation of liquefaction resistance involves two important calculation, i.e., cyclic stress ratio (CSR) and cyclic resistance ratio (CRR). A simplified CSR calculation was proposed by Seed and Idriss [11] based on the peak ground acceleration. CRR value can be calculated from the shear wave velocity using the method developed by Andrus and Stokoe [12]. Gridding algorithms, such as kriging, minimum curvature, etc., may be used to generate a 2-D liquefaction potential of a vertical section. With density information, a stiffness section can be generated simultaneously. 2-D data processing techniques, such as regression analysis, could be easily applied to a vertical section of liquefaction potential to enhance local anomalies of liquefaction. The calculation procedure is shown in Fig. 2. The summarized equations are described as follows.

1. *Cyclic Stress Ratio*

$$CSR = \frac{\tau_{av}}{\sigma'_v} = 0.65\left(\frac{a_{\max}}{g}\right)\left(\frac{\sigma'_v}{\sigma_v}\right) r_d \tag{1}$$

where τ_{av} is the average cyclic shear stress, $a_{\max}$ is peak ground acceleration, σ_v and σ'_v are the total and effective vertical overburden stresses, respectively, g (9.81 m/s^2) is the gravity acceleration and r_d is a stress-reduction factor.

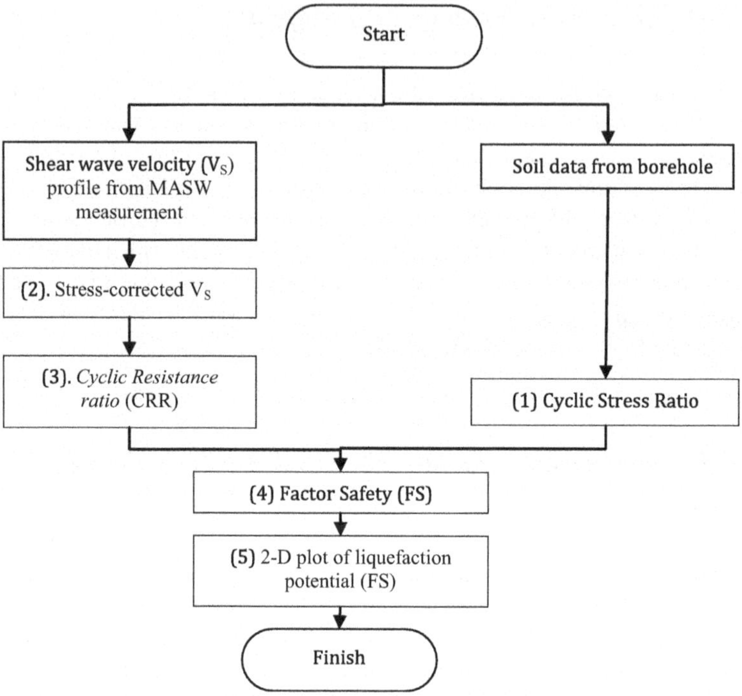

Fig. 2 Flowchart of liquefaction potential analysis

2. *Stress-Corrected Shear Wave Velocity*

$$V_{S_k} = Vs\left(\frac{P_a}{\sigma'_v}\right)^{0.25} \tag{2}$$

where Vs_k is corrected shear wave velocity, and P_a is referenced stress (100 kPa).

3. *Cyclic Resistance Ratio, CRR*

$$CRR = \left\{a\left(\frac{V_S}{100}\right)^2 + b\left(\frac{1}{V^*_{S_k} - V_{S_k}} - \frac{1}{V^*_{S_k}}\right)\right\}MSF, \tag{3}$$

where V^*_{Sk} is upper limit of corrected shear wave velocity for liquefaction events, *a*, *b* are curve fitting parameters and $MSF = \left(\frac{M_w}{7.5}\right)^n$ is the magnitude scaling factor, with $n = -2.56$ and M_w = earthquake magnitude.

4. *Factor of safety, FS*

$$FS = \frac{CRR}{CSR}. \quad (4)$$

4 Results and Discussion

4.1 Shear Wave Velocity Profile

For shear wave velocity analysis, series of seismic data for nine measurement locations of MASW survey were transformed from time to frequency domain. By f–k analysis, the phase velocities of surface waves for each frequency and their amplitudes were calculated and then were plotted in the experimental dispersion curve. In Fig. 3 an example of dispersion curve results from MASW surveys at observed location is shown. From these curves, the seismic wave events, i.e., fundamental mode, higher modes and interferenced body wave are clearly visualized. However, construction of shear wave velocity profile from these dispersion curves only considers the fundamental mode of surface wave.

1-D shear wave velocities (Vs) profile is then obtained by the inversion process. Therefore, the inversion of the dispersion of subsequent movement of the source to

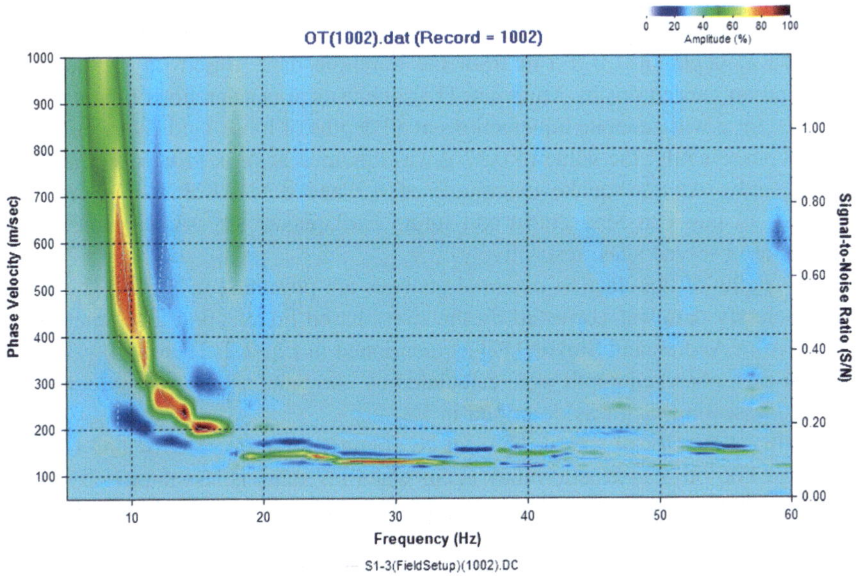

Fig. 3 Example of dispersion curve generated for the southern part of UMY campus

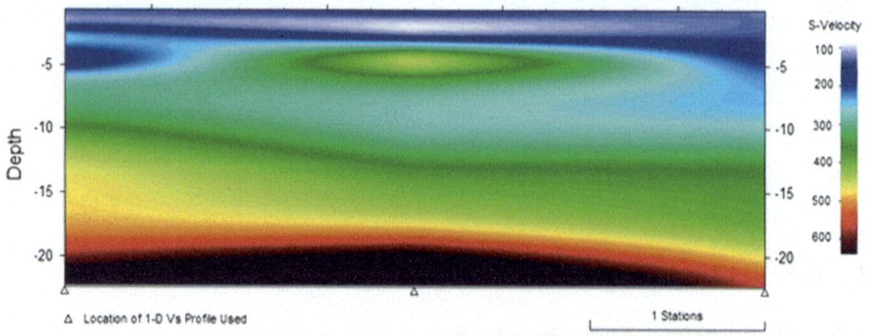

Fig. 4 A plot of 2-D shear wave velocity profile for southern part of UMY campus

geophone array then produces the final 2-D shear wave velocity profile of Fig. 4. The 2-D Vs profile was calculated by tomography correlation on minimum three different V_S profiles with the fixed increment of 5 m.

4.2 Liquefaction Potential Analysis Based on Shear Wave Velocity Values

Based on the shear wave velocity data, evaluation of liquefaction potential was carried out using the procedure mentioned in Eq. 1–4. Soil data of sites, i.e., density, particle distribution, water table and fines content, were collected from laboratory tests on the borehole samples. In addition, the ground water level was observed at depth of 1.0–1.5 m from the borehole data. Based on the previous investigation carried out by Muntohar [13], the critical accelerations ranging from 0.34 to 0.69 g will generate liquefactions at all depths of loose sand layers in hazard areas. Consequently, the value of 0.47 g was obtained as peak ground acceleration in this study. An earthquake magnitude of 6.3 and 8 M_W was also selected as scenarios of past (26 May 2006) and future earthquakes for assessing the liquefactions at observed sites, respectively.

The results of liquefaction potential analysis are presented in Fig. 5. The shear wave velocity data was corrected by the stress-based liquefaction assessment procedure from Andrus and Stokoe [12] as mentioned in Eq. 2.2. Therefore, values of CRR at each observed depth were calculated by corrected shear wave velocity and magnitude scaling factor as described in Eq. 2.3.

Finally, the factor of safety (FS) can be obtained by ratio of CRR and CSR as shown in Fig. 5d. Liquefaction may occur when FS value is less than 1. By using 2-D shear wave velocity data, the 2-D and 3-D mesh plot of FS can be generated as presented in Figs. 6, 7 and 8, respectively. This figure can provide understandable information of critical parts in 2-D and 3-D subsurface profile that may have liquefaction potential during 6.3 and 8 Mw earthquakes at observed locations.

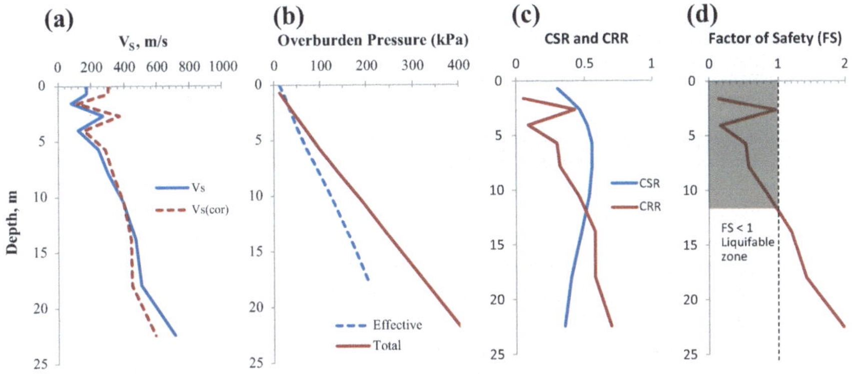

Fig. 5 Liquefaction potential analysis based on the shear wave velocity profile for first scenario of 7.5 M_W Yogyakarta's earthquake at southern part of UMY campus: **a** overburden stress-corrected shear wave velocity **b** total and effective overburden stresses **c** cyclic stress ratio (CSR) and cycle resistance ratio (CRR) and **d** factor of safety for predicted liquefaction

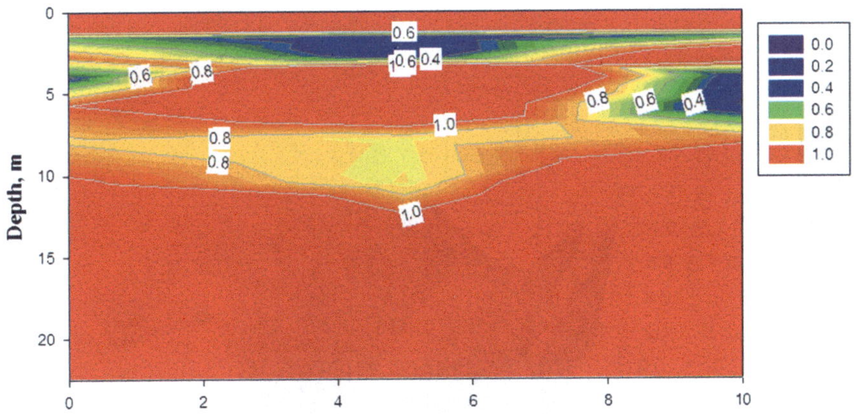

Fig. 6 A plot of 2-D liquefaction potential for first scenario of 6.3 Mw Yogyakarta's earthquake at southern part of UMY campus

A 2-D display of liquefaction potential can easily and quickly be done by contouring the FS data. It provides an efficient way to map FS liquefaction in a vertical section. The 2-D display reduces non-uniqueness of liquefaction potential and the potential for misinterpretation based on 1-D liquefaction potential versus depth plot. 2-D and 3-D data processing techniques, such as regression analysis, could be easily applied to a vertical section of FS to enhance local anomalies of subsurface profile.

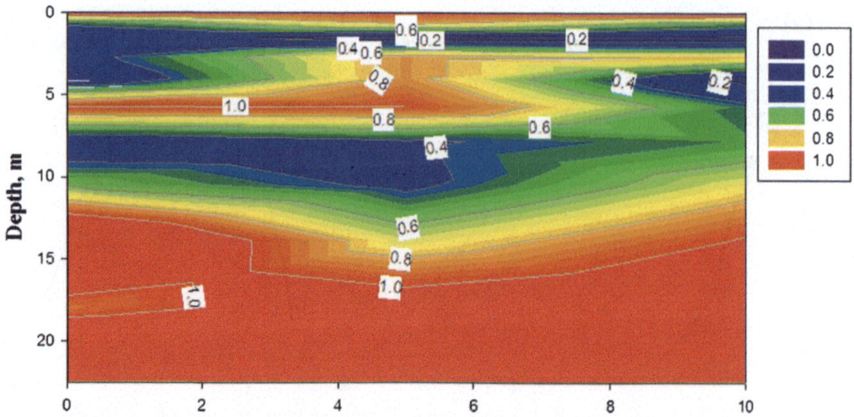

Fig. 7 A plot of 2-D liquefaction potential for first scenario of 8 M_W Yogyakarta's earthquake at southern part of UMY campus

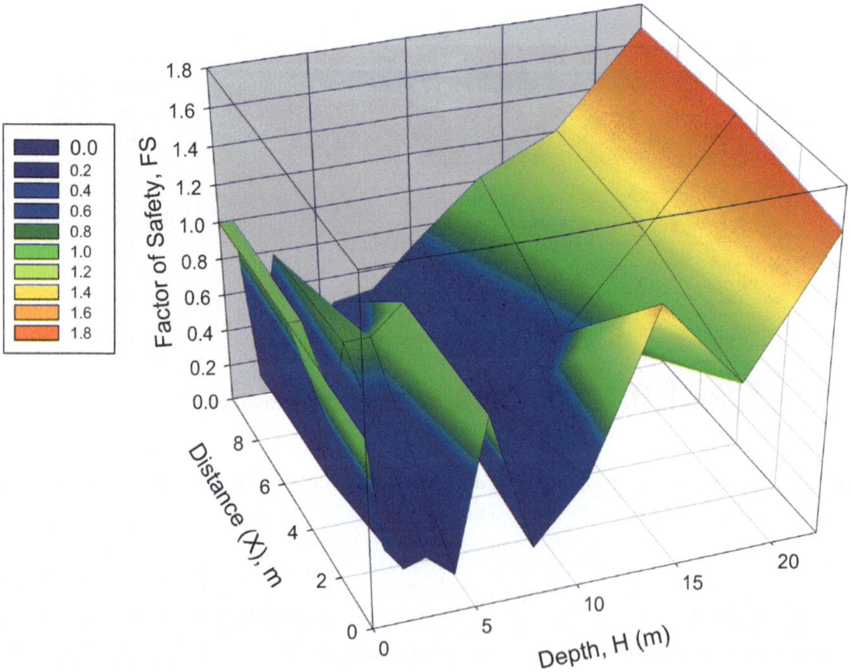

Fig. 8 A plot of 3-D liquefaction potential for first scenario of 8 M_W Yogyakarta's earthquake at southern part of UMY campus

5 Conclusions

The compilation and analysis between seismic (MASW) method and geotechnical site investigation successfully generates the 2-D shear wave velocity of study area. 1-D and 2-D shear wave velocities profiles can be efficiently estimated by MASW method. Based on the 2-D shear wave velocity profile, lateral variation in dynamic properties of soil layer can be clearly obtained. The use of MASW method as an alternative non-destructive method is found to be one of the foremost cost-effective options to estimate in situ shear wave velocities. By using the shear wave velocity, physical soil characteristics and seismicity data, the liquefaction potential of sand layers deposit at observed sites can be analyzed using procedure described in [12]. 2-D and 3-D liquefaction maps are also generated in order to provide understandable information of subsurface liquefiable sections during the earthquake. This analysis is useful and effective in presenting the main data for seismic hazards analysis for the sustainable development.

Acknowledgements This study is supported by Hibah Bersaing 2009–2010 from the Ministry of National Education, Indonesia. Authors would also like to appreciation to Ms. Susy K. Ariestyanti, Ms. Anita Widianti and Mr. Surya Budi Lesmana for their assistance in the surface wave data processing and field measurement.

References

1. Imai, T., and Tonouchi, K. (1982). Correlation of N-value with S-wave velocity: Proc. of 2nd Europ. Symposium on Penetration Testing, 67–72.
2. Craig, R. F. (1992). Soil mechanics (fifth edition): Chapman & Hall, New York.
3. Rahardjo, W., Sukandarrumidi and Rosidi, H.M.D. (1995). Geological Map of the Yogyakarta Sheet, Jawa, 2nd edition, Geological Research & Development Centre, Bandung.
4. Rosyidi, S.A., Taha, M.R., Lesmana, S.B., Wintolo, J. and Adi, A.D. (2008a). Some lessons from Yogyakarta earthquake of May 27, 2006. *Sixth International Conference on Case Histories in Geotechnical Engineering and Symposium in Honor of Professor James K. Mitchell at Arlington, Virginia*:1–8.
5. Rosyidi, S.A., Lesmana, S.B., Wintolo, J., Chik, Z. and Taha, M.R. (2008b). Geo-resistivity surveys for faults identification in geotechnical damages area from Yogyakarta earthquake of May 27, 2006. *Fourteenth World Conference on Earthquake Engineering, Beijing, China*:1–8.
6. LKPT UMY (2006). Driling and Georesistivity Measurement in Universitas Muhammadiyah Yogyakarta Campus. Final Report, Lembaga Pelayanan dan Konsultasi Teknik, Universitas Muhammadiyah Yogyakarta.
7. Park, C.B. (1995). Characterization of geotechnical sites by multichannel analysis of surface waves. *95th Annual Meeting, Korean Ground Society*:15–21.
8. Penumadu, D. and C.B. Park, (2005). Multichannel analysis of surface wave (MASW) method for geotechnical site characterization. *Geo-Frontiers Conference, Austin*:1–8.
9. Park, C.B. and Miller, R.D. (2005). Seismic Characterization of Wind Turbine Sites Near Lawton, Oklahoma, by the MASW Method. Open File Report, Kansas Geological Survey, University of Kansas.
10. Schwab, F.A. and Knopoff, L. (1972). Fast Surface Wave and Free Mode Computations, in B. A. Bolt Ed., Methods in Computational Physics, Academic Press, 87–180.

11. Seed, H.B. and Idriss, I.M. (1971). Simplified procedure for evaluating soil liquefaction potential. *Journal of Soil Mechanics and Foundation Engineering* **97:9**, 1249–1273.
12. Andrus, R.D. and Stokoe, II K.H. (1999). A liquefaction evaluation procedure based on shear wave velocity. *Joint meeting, 31st Technical Memorandum of PWRI 3653.* **Wind and Seismic Effects**:469-474.
13. Muntohar, A.S. (2009). Evaluation of peak ground acceleration using CPT data for liquefaction potential. *Fourth Annual International Workshop & Expo on Sumatra Tsunami Disaster & Recovery 2009*:1–5.

An Analysis of Biogas Productivity with Fixed Dome Type for Supporting Household-Base Energy in Nongkojajar, East Java-Indonesia

Aminatus Sa'diyah, Katrin Klingenberg and Ridho Hantoro

Abstract Indonesia is a developing country endowed by various potential renewable energy resources. Among them is biogas which basically can be produced from the decomposition of biomass by the biochemical processes. Java is the most densely populated island in Indonesia and consumes almost 67% of the national energy supply. This study was carried out under supervised government and founded by IDBP (Indonesia Domestic Biogas Program) for developing sustainable rural renewable energy supply. Many in-ground type of fixed dome biogas reactors have been established in the dairy production field owned by the Cooperative (KPSP) Setia Kawan Nongkojajar, Pasuruan East Java. An individual household of cooperative member was provided by a reactor of 8 m^3 of subtract. Every reactor was assembled by financial sharing mechanism between a farmer and a financial body. The main objective of the study is to determine the effectiveness of the bioreactor in biogas production for knowing the consumption and electrification in rural area community. A batched feeding mechanism was applied using a feedstock composed proportionally from cow dung and water. The reactor was optimized in order to produce sufficient biogas production in terms of its quality [methane purity] and quantity [daily volume] to meet the energy requirement in the surrounding rural area. The result shows that the optimum productivity of biogas reached if the value of TS 16.9 kg/m^3, HRT in 15 day, bioreactor temperature 23 °C, volume of biogas 3.6 m^3, 67.6% CH_4 and 29.4% CO_2 with 0.8 of bioreactor effectiveness.

A. Sa'diyah (✉) · R. Hantoro
Engineering Physics Department, Sepuluh Nopember Institute of Technology, Surabaya, Indonesia
e-mail: aminatus12@mhs.ep.its.ac.id

R. Hantoro
e-mail: hantoro@ep.its.ac.id

K. Klingenberg
Sustainability Economics and Management, Carl von Ossietzky Universitaat Oldenburg, Oldenburg, Germany
e-mail: katrinklingenberg86@hotmail.com

T. Taufik et al. (eds.), *ICoSI 2014*, DOI 10.1007/978-981-287-661-4_32

Keywords Biogas productivity · Monod-type kinetic · Bioreactor effectiveness · Economic impact

Abbreviation and Acronyms

BIRU	*BiogasRumah*
BSP	Biogas Support Program
GC	Gas Chromatography
HRT	Hydraulic Retention Time
IDBP	Indonesia Domestic Biogas Programme
KPSP	*Koperasi Peternakan Sapi Perah*
LPG	Liquid Petroleum Gas
SNV	*Stichting Nederlandse Vrijwilligers*

Units

Vs	Specific yield (volumetric capacity of methane gas production, m^3/day/m^3 reactor)
Bo	The highest methane production in capacity, in m^3 methane produced/kg volatile solid (VS) added
So	Volatile solid (VS) concentration on the input material, [kg/m^3]
HRT	Hydraulic retention time, day
Eff (t)	Effectiveness, []
$V_{reactor}$	Reactor volume [m^3]
P_{CH_4}	Methane content [%]
P_{max}	Maximum methane content [%]

1 Introduction

Indonesia develops and disseminates biogas technology and it is managed by Ministry of Energy and Mineral Resources which founded IDBP in collaboration with SNV and Hivos [1]. SNV is a technology development organization owned by Dutch government engaged in renewable energy such as biogas in developing countries, especially in Asia–Africa [2], whereas social organization Hivos for special support in financial aspect to renewable technology development process [3]. In addition to work with the two organizations, IDBP also formed a program initialed BIRU in charge to assemble the reactor with SNV in the construction process as well as the dissemination of biogas into the community [4]. The reactor establishment is funded by stakeholder with a subsidy from Embassy of the Netherlands Government in Indonesia [5].

This study is expected to obtain new information on the increased productivity of biogas, quantitative and qualitative, as well as new solutions on the modeling of fixed dome type bioreactor (biogas reactor) without compromising the environmental and economic impacts on society [2, 6–8]. It is done in the work area of KPSP Setia Kawan Nongkojajar East Java hereinafter referred to as Cooperative.

2 Materials and Methods

2.1 *Raw Materials of Biogas*

Biogas is derived from the anaerobic fermentation of the substrate in the form of cattle waste by microbes in the bioreactor chamber as digester [7]. The process of methane formed by anaerobic bacteria and carried by a biochemical reaction; hydrolysis, acidogenesis, acetogenesis, and methanogenesis [9]. The dominant component of biogas is methane (CH_4) with compounds of CO_2, H_2S, N and other gas in less than 0.1% [4].

2.2 *Setting—Biogas Reactor Model*

Biogas reactor that is assembled in Nongkojajar-East Java was fixed dome type. This is developed by SNV and preinstalled in Nepal country through BSP Program since 1994 [10].

As shown in Fig. 1, the inlet consists of two pipelines, each of which stem from the mixer to the input of animal waste and latrine (toilet) for input from human waste. The body of reactor or commonly called the digester tank is tubular cylinder with a diameter of 4–6 m and 3–4 varies in depth depending on the size of the biogas reactor installation. Gas holder is a gas storage chamber for microbial decomposition of methanogens and the site of decomposition which consists of a mixture of liquid and solid substrates. Exodus (outlet) consists of a manhole and sludge output as lines. Biogas output paths are at the top of the dome and a main gas pipe valve, which serves to channel the biogas produced by the digester towards household appliances [11]. The formula that describes the performance of biogas reactor is shown in Eq. 1.

$$\mathrm{Vs} = \frac{\mathrm{Bo} \times \mathrm{So}}{\mathrm{HRT}} \cdot \left[1 - \frac{K}{(\mathrm{HRT} \times \mu\mathrm{m} - 1 + K)}\right]. \tag{1}$$

In Eq. 1 is given Monod-type kinetic which is generally used in many grounds of bioenergy research to calculate and simulate a formation process of biogas in

Fig. 1 IDBP biogas reactor (bioreactor) design

reactor (digester tank) [12]. Then, the effectiveness of reactor was determined to get a value of reactor performance and a ratio between output and input of materials or substrate [8].

$$\mathrm{Eff}(t) = \frac{1}{V_{\text{reactor}}} \int_{V_{\text{reactor}}} \frac{P_{CH_4}(t)}{P_{\max}(t)} \mathrm{d}x. \tag{2}$$

There are conditions that have to be fulfilled by stakeholder (under guidance of technical consultant from SNV/BIRU) before assembling the biogas reactor [4]. It is important for preventing a malcondition of reactor which may affect in biogas productivity, or technical humanly error in setting initial condition of materials. The substances that must be considered for an initial condition apply as the preliminary to the stakeholder, it is shown in Table 1.

Table 1 The size of biogas reactor

Digester capacity (m^3)	Gas production/day (m^3)	Cow's manure needed (kg)	Water needed (l)	The number of cows
4	0.8–1.6	20–40	20–40	3–4
6	1.6–2.4	40–60	40–60	5–6
8	2.4–3.2	60–80	60–80	7–8
10	3.2–4.2	80–100	80–100	9–10
12	4.2–4.8	100–200	100–200	11–12

3 Methods

3.1 Simulation of Model

iDynoMiCS is a special program for presenting the simulation of the microorganisms growth based on a mathematical model such as Monod-type kinetics, chemostat, and the growth response factor [13]. The result of this program is in 2D or 3D graph and contains a set of circles with different colors that indicates the species of microbes in the reaction. Each circle represents a microbial colony with a ratio of $1:10^5$ species. Every 1 species has a mass of 1 femtogram (fg) equivalent to 10–15 g. The data collected from Cooperative (KPSP) and BIRU archives.

3.2 Economic and Ecological Aspects

The information about biogas technology and its use is obtained for an interview with stakeholders and the technicians. Interview emphasized on how the utility of household-scale biogas, and of course how biogas, can support their life in the economical aspect. As well as the utilization of biogas residue is used as fertilizer for crops [6]. The nutrient content contained in the sludge adequate to fertilize the soil and avoid the agent pollute of water quality so that it remains clean. Microeconomic regulation on household biogas users are fully supported by Cooperative, which operates by applying the principle of the welfare of members. Furthermore, the various environmental impact assessments will be discussed in detail in Chapter "Development of Breakfast Meals from Local Taro Using Extrusion for Food Security".

4 Result and Discussion

4.1 Effectiveness of Bioreactor

The first step is to determine the number of microbes and microbial growth rate, and perform using Monod-type kinetic model using iDynoMiCS v 1.2. The resulting images are shown in Fig. 2.

Figure 2 shows the results of microbial populations. The blue-circular units are active microbial colonies were afford to degrade the substrate and convert it into methane. Each of units represents the amount (colonies) in increments of 1×10^5 microbial species. There is a significant difference between iterations for day 0, 15, and 21. This is the number of active microbes that affect the productivity of biogas in the reactor. The more microbial population produces more of gases. The reactor produces biogas at day 15–21 (called HRT, Hydraulic Retention Time) though and

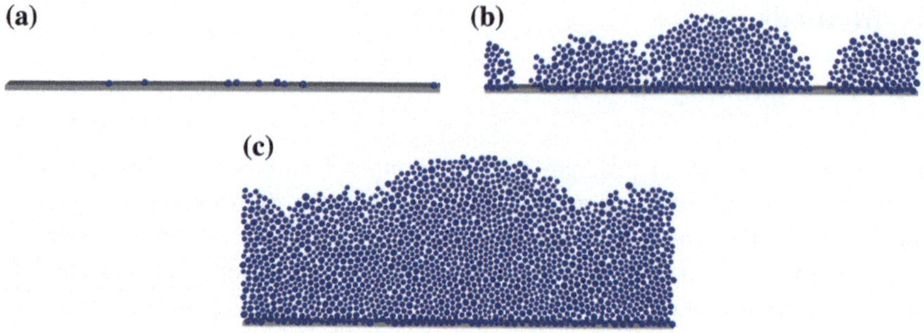

Fig. 2 **a** The populations of active microbes in day 0, **b** The population of active microbes in day 15, **c** The population of active microbes in day 21

it is still in the little amount [about 1676.8×10^5–2546×10^{15} microbial species] with biogas apparently 1–3.5 m^3 in volume/day.

The biogas effectivity calculations (Eq. 2) were performed using MATLAB modeling especially Monod-type kinetics. The result is shown in Fig. 3.

Figure 3 shows that, during 60 days, the effectiveness of reactor was decreased. This is influenced by the levels (input) and the amount of methane produced (output) by the bioreactor. The highest effectiveness produced by bioreactor on day 5 with the value 0.8 [calculated after HRT; 15 days]. On this day, the samples of methane content of the gas tested using GC worth 67.6% CH_4, 29.4% CO_2. Methane content continues to decline until reaches 3.34% at day 60, with 22.87% CO_2 and other gas substances almost 70%. The presence of inhibition factors such as reduction in the total absence of solids and ash in the bottom of reactor is what leads to reduced productivity and effectiveness in the bioreactor.

Fig. 3 Biogas reactor effectivity

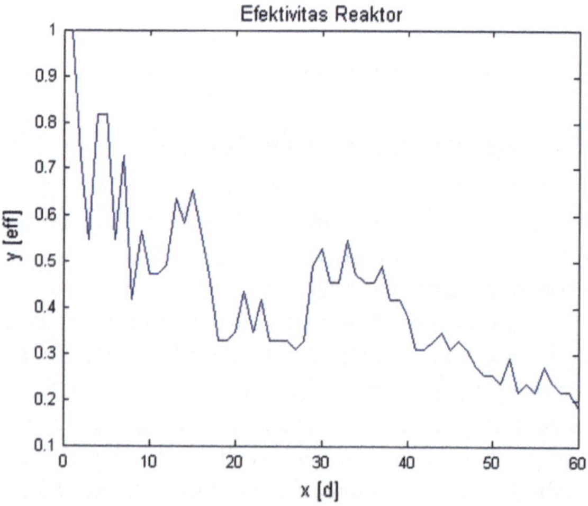

Fig. 4 Biogas lamp in Nongkojajar, Indonesia [6]

The highest volume of biogas is about 3.6 m^3/day which is enough to meet an energy consumption in household-scale. Most of biogas is used to conserve LPG for cooking (Fig. 4a) and as a fuel gas lamp (Fig. 4b). On average, the households can use the biogas 7 h in a day for cooking, and 7 h of a lamp gas with power equivalent to 24 W.

4.2 Economic–Ecological Impact

The microeconomic analysis is done after conducting interviews with users and collecting the written records of the Cooperative archives. The result is known that the use of biogas households is adequate to meet the energy requirement, especially for cooking and electricity. People no longer use LPG and firewood after using biogas in their house. Meanwhile, it can reduce air pollution caused by burning firewood that is harmful to health.

Throughout 2009–2013, the amount of the biogas reactor in Nongkojajar is more than 1000 units with 75% of them is use 8 m^3 in size. The payment of biogas reactor construction is charged to stakeholders with half subvention from the Netherlands Embassy in Indonesia through IDBP that works together with Cooperative to facilitate its realization to the community. The amount of payment shows in Table 2.

The total cost of construction offered by BIRU ranged from 4.7 to 7.4 million Rupiah depending on the size of the biogas reactor. Subsidies are provided by Hivos of 2 million Rupiah for each biogas reactor size. Hence, the amount to be paid by the user community is the difference between the costs of the subsidy. However, the Cooperative has a rule to provide relief in the form of repayment for the stakeholder every 10 days by depositing milk (taken from their cattle) to dairy factory owned by Cooperative.

Table 2 Biogas Reactor Payment

Reactor capacity [m^3]	Total payment (Rp)	Subvention (Rp)	Bill for stakeholder (Rp)	Payment per 10 days (Rp)	Time settlement (year)
4	4,720,000	2,000,000	2,720,000	30,000	2.5
6	5,290,000	2,000,000	3,290,000	34,500	2.6
8	5,920,000	2,000,000	3,920,000	42,000	2.6
10	6,890,000	2,000,000	4,890,000	53,000	2.6
12	7,400,000	2,000,000	5,400,000	60,000	2.5

Fig. 5 Slurry canal in apple garden Nongkojajar Indonesia [6]

The biogas results waste named slurry. Users can take an advantage of the slurry to fertilize their crops and fields (Fig. 5), and no need a chemical fertilizer that has a high price. Slurry can be used as the fertilizer to grass which is the main food for cattle, especially cows. It contents of many organic materials. That is not hazardous to the environment since it is easily decomposed by microbes in the soil. Many farmers also utilize biogas waste by selling it to other farmers and afford to increase the income if it is attempted for commercial business. Most of users, who own the farm, make a profit from their harvest by 20% every year.

5 Conclusion

The simple design-fixed dome of biogas reactor is convenient to the stakeholder, furthermore reduces the failure during the operation. The optimal condition of biogas reactor is reached if the contents of methane were 67.6% with 0.8 reactor effectivity that is capable to produce an average 3.6 m^3 of biogas a day. In terms of microeconomics aspect, the application of biogas technology can meet the energy requirements of the household with a relatively time exertion about 14 h a day. The concept of Cooperative became the main endorsement of biogas development in rural areas by applying the rules of installation repayments through the deposit of milk every 10 days.

Acknowledgments Many researches in the field of biogas are still being conducted in effort to co-develop the scientific and correct deficiencies that may occur during the deal of biogas production. One of them is the development of biogas purification tools to prevent rust on gas appliances. Carbon dioxide which occupies the second content in level under methane caused the deposition of gas stove or carbon crust on the gas pipeline which can lead damage of gas stoves and lamps. In addition, we need a device to keep the volume of excess biogas at the time of the reactor volume in maximum level, as well as converter-power storage unit in order to be implemented for electronic equipment. It is necessary for the support of all studies about biogas development from academia, technician, stakeholder, and all of parts of country for social welfare.

References

1. IEO. Indonesia Energy Outlook 2010. Pusat Data dan Informasi Energi Sumber Daya Mineral Kementerian Energi dan Sumber Daya Mineral. Jakarta; 2010.
2. Ghimire, Prakash. SNV Supported Domestic Biogas Program in Asia and Africa, Renewable Energy 2013; 49: 90–94.
3. CASINDO. Capacity Development and Strengthening for Energy Formulation and Implementation of Sustainable Projects in Indonesia, Technical Proposal; 2009. May 2009.
4. BIRU. (2010). User manual of biogas. Jakarta: Tim of Biogas.
5. Hivos. (2012). Indonesia Domestic Biogas Program. Indonesia.
6. JRP. (2013). Analyzing of Biogas Energy Potential and Energy Policy in KPSP Setia Kawan Nongkojajar Pasuruan Indonesia. Joint Research Program: 2013. ITS Surabaya-Carl von Universitaat Oldenburg.
7. Krishania, M. Analysis of different techniques used for improvement of biomethanation process: A review. Fuel 106, 2013: 1–9.
8. Muha, I. Mathematical modeling of process liquid flow and acetoclastic methanogenesis under mesophilic conditions in a two-phase biogas reactor. Bioresource Technology 106, 2012: 1–9.
9. Purwandari, FA. Pretreatment of oil palm empty fruit bunch (OPEFB) by N -methylmorpholine- N -oxide (NMMO) for biogas production: Structural changes and digestion improvement. Bioresource Technology 128, 2013: 461–466.
10. Bajgain, Sundar. (1994). Nepal Biogas Plant—Construction Manual: Construction Manual for GGC 2047 Model Biogas Plant. Biogas Support Program. Kathmandu, Nepal.
11. Widodo, T.W. Design and Development of Biogas Reactor for Farmer Group Scale, Indonesian Journal of Agriculture 2009; (2) 2: 121–128.

12. Hu, D. Mathematical modeling, design and optimization of conceptual configuration of soil-like substrate bioreactor based on system dynamics and digital simulation. Ecological Engineering 51, 2013: 45–58.
13. Lardon LA, (2011). iDynoMiCS: next-generation individual-based modelling of biofilms. Environmental Microbiology**13**: 2416–2434.

Business-Oriented Technological System Analysis (BOTSA) at Eindhoven University of Technology: An Innovative Learning Method to Foster Entrepreneurship

Mara Wijnker, Han van Kasteren and Henny Romijn

Abstract BOTSA is an innovative teaching method for students with technical background in the field of sustainable energy technologies and an interest in entrepreneurship. Two core features of this method, namely the connection between a technological analysis and a business case as well as the involvement of clients and business coaches, are explored in this paper. Case studies and results of evaluations among students who used this method to analyze their technological innovation are used to demonstrate the value of these two features and consequently the method as a whole.

Keywords Innovative learning method · Technological innovation · Entrepreneurial students

1 Introduction

The BOTSA method is an innovative teaching methodology for students who are studying sustainable energy technologies and combine this with an interest in entrepreneurship at the Eindhoven University of Technology in the Netherlands. It was developed between 2010 and 2013. This paper aims to briefly discuss the essential features of BOTSA and illustrate its value by means of some case studies.

M. Wijnker (✉)
Strategic Area Energy, Eindhoven University of Technology,
Den Dolech 2, 5612AZ Eindhoven, The Netherlands
e-mail: m.a.s.g.wijnker@tue.nl

H. van Kasteren
Innovation Lab, Eindhoven University of Technology,
Den Dolech 2, 5612AZ Eindhoven, The Netherlands

H. Romijn
School of Innovation Sciences, Eindhoven University of Technology,
Den Dolech 2, 5612AZ Eindhoven, The Netherlands

T. Taufik et al. (eds.), *ICoSI 2014*, DOI 10.1007/978-981-287-661-4_33

One of the programs at the university in which BOTSA became an important component is the minor "Entrepreneurship in Sustainable Energy." The case studies in this paper are drawn from this minor program. The core components of the minor program in which the BOTSA method has been applied are a technical project and a business case. In the minor program, these two core components are combined with 4 courses on entrepreneurship in combination with skills training.

Concretely, BOTSA provides a systematic way of guiding students through the ex-ante analysis of the technological aspects of a promising renewable energy innovation, after which the business aspects, its organization, and feasibility are studied and, in turn, followed by the analysis of various exogenous contextual factors. At each step, the scope of the analysis is broadened.

The starting point lies with the technological innovation itself, which focused on its hardware and technical functioning and its operation and maintenance parameters. This is followed by a wider investigation of what we can call the technological system and the market environment. This should lead to insights into the organizational embedding of the innovation, and the sector in which it will be implemented. At this stage, feasibility studies and sector studies are undertaken as this is the context in which the innovation needs to manifest itself and eventually hold its own in a market environment. A crucial aspect in this phase is the interactions between the technological part and the business part of the analysis. These have to be studied in tandem, as both need to support one another; both are developed in interaction with each other. Resolving an issue in one dimension can lead to questions in the other and vice versa. The third layer of assessment is the wider external context, sometimes called "landscape." This layer encompasses the national, regional, and global economic situation; socio-cultural influences (e.g., growing awareness about climate change threats); and the regulatory context. All these factors influence the potential of innovative energy technologies, but are themselves beyond the immediate influence of individual innovation projects.

2 Aim of the Paper

In a previously published paper [1], the development of the method in practice and its foundations in the evolutionary innovation theories and theories of (sustainable) entrepreneurship are discussed. The current paper explores whether BOTSA equips students with sufficient technological background for making business-related decisions and whether the students obtain a business-oriented perspective through the combination of the two parts of the analysis. The combination of the two different perspectives—technological and commercial—and the interaction with clients/business coaches during the investigations are the two essential features of the BOTSA method. Here, we analyze these elements with a view to assess the added value of these features compared to regular, theoretical, and business-related projects executed by the students.

The following two questions will guide our discussion:

Q1: Does the connected way in which the technological innovation and business case are analyzed in BOTSA lead to solutions that are closer to reality compared to only studying the business aspects?

Q2: Do project with interaction with clients and experienced entrepreneurs as coaches motivate the students more compared to project in which no clients and no coaches are involved?

3 BOTSA's Value in Practice: Evidence from Case Studies

Q1: Does the connected way in which the technological innovation and business case are analyzed in BOTSA lead to solutions that are closer to reality compared to only studying the business aspects?

BOTSA gives students the opportunity to understand the technological part of the innovation before considering its business aspects. Students are, however, aware of the pending questions of the clients which are mostly related to the business aspects. This leads students to switch between the two parts and make connections, thus allowing them to arrive at more realistic solutions. The importance of making this connection can be demonstrated by several examples.

One group of students focused on a thermal-driven cooling system named SolabCool.[1] This system consists of two thermochemical heat pumps, and an outdoor and indoor cooler. The cooling process takes place in two distinctive steps—a charging and discharging process. Both heat pumps follow these steps in opposing cycles. Each time one pump discharges, ensuring continuous cooling. The opposing cycles and using the thermal energy make the system very energy efficient.

During the technical project, the students assessed the production process by means of modeling each step in production through the 3M (man, machine, and material) model. Important conclusions were time-saving possibilities and necessary input to increase the production. The client was very interested in the model produced by the students and asked the students to intensify their work. Having finalized their technical project, the students, however, needed to continue with the business part of their assignment. During this part of the study, they were obliged to work according to a predetermined structure which did not allow them sufficient time to dive into the production process in more detail.

Because the students were sufficiently knowledgeable in terms of specifications of the product and the process, the students were actually better equipped to advise the client than if they would have only studied the business aspects of the case. Only looking into the production process without having the details of the technology itself could not have been given the same result: A more superficial advice would have been given which would have been less useful to the client. In this case, the client could not be completely satisfied as there was a need for a more detailed

[1]https://www.youtube.com/watch?v=zVTxTJN0cA0 and http://www.solabcool.com/?l=1.

study which could not be facilitated because of the educational objectives which needed to be reached combined with the limited possibilities to deviate from the predetermined structure.

Another example is the Sowise solar boiler.[2]

The Sowise solar boiler was invented 20 years ago and has technologically been optimized and tested throughout this long period but has not reached widespread introduction into the market. Students made a technical analysis in which they compared this solar boiler to other systems providing the same functions. Technologically, the system functioned well. The price of the system was high though, the students therefore suggested to reconsider the production process and search for optimization: possibilities to save time, money, materials etc. For the business part, students concluded that selling of the system would be difficult because it did not obtain good enough results in a standard performance test of solar boilers. The entrepreneur stated that this test was not suitable for his type of system, but there was no alternative. Students learned that this hampered a good market introduction and communicated this to the entrepreneur. The entrepreneur acknowledged this and applied for a follow-up project (also performed by university students) focused on how to position his system in the market. In this follow-up project, students applied the technological knowledge of the first group of students while focusing on the business aspects. In short, the advice of the students made the entrepreneur realise that also a proper market introduction and positioning of his system are important.

Another example is the business case of Cyclox.[3] Cyclox is a substance which can be added to fuels used in combustion engines and consequently reduces the soot and NO_x emissions. It can be made out of black liquor which is a waste stream of the paper industry. Because some components of Cyclox are above the present standard levels allowed for car fuels, use of Cyclox in conventional diesel engines is presently not allowed. For the short term, this means another market segment has to be looked for. Use of Cyclox in ship engines is possible because they have to obey less strict regulation concerning the concentration levels of certain compounds. In this case study, the students needed to focus on design of a Cyclox plant and several aspects related to bringing this product to the market.

The students familiarized themselves with the technological aspects of Cyclox. While working on their business case, it became clear that the process of making Cyclox out of black liquor as part of the paper industry was not yet feasible for larger quantities, mainly because there was not an optimized catalyst yet. Although the technological aspects were not solved, a customer with a large demand for ship fuels was already found by the entrepreneur, so the dimensions of a plant needed to serve at least this customer. The business case therefore focused currently on making Cyclox out of oil (conventional resources) while simultaneously developing

[2]https://www.youtube.com/watch?v=hkDIndrsn0Q and http://www.solarsowise.nl/.

[3]https://www.youtube.com/watch?v=B-rUEZnZu0I and http://www.greencarcongress.com/2010/04/cyclox-20100419.html.

a feasible method to extract black liquor in larger quantities from the waste of the paper industry. This meant that the projected carbon neutrality of the case and the system could not be realized immediately. However, Cyclox still contributes to reductions in emissions compared to using solely conventional fuels in combustions engines and therefore remains a greener alternative which is a comparative advantage that can still be used. The analysis of the students helped the entrepreneur, being not an expert in process technology but more in mechanical and combustion technology. Students thus showed that for the short term, the feasibility of a chemical plant converting black liquor out of waste of the paper industry into Cyclox was not realistic. So alternatives, of which one was presented in the business case of the students, had to be found. This offered a new challenge to the entrepreneur. This example demonstrates also that the students had a good understanding of the technological part and used these insights to reflect on their business case.

Students were asked whether the different parts of the minor program (the two projects as well as several business-related courses) were sufficiently related to each other. Most students responded positively to this question. The students perceived their courses as theoretical input for the "real case" which they dealt with during their projects. The students acknowledged that having sufficient technological background offers a more realistic solution to business-related issues.

Q2: Do project with interaction with clients and experienced entrepreneurs as coaches motivate the students more compared to project in which no clients and no coaches are involved?

From the evaluation, it became clear that students appreciate the support offered by coaches with experience in entrepreneurship. According to students, coaches maintain an overview and reflect critically on steps taken by the students. Furthermore, they direct students toward important business-related aspects, as taking the perspective of the customer into account and make students aware of professional behavior.

Interaction with clients is important because it motivates students to see that the advice originating from their research can contribute to a real product and/or decision-making process. To students, this adds an enormous challenge compared to working on theoretical cases and solutions within regular classes and projects. Direct interaction by discussion and reflection benefits both parties; students become more motivated; and clients ensure that the students continue to work on the challenge offered in the assignment (instead of losing themselves in for the client's less important details).

An example of the importance of working for a client is demonstrated here in the following cases. Two student groups worked on assignments of Eindhoven Airport.[4] One group analyzed possibilities of a transport system for people and luggage from Eindhoven-Acht, a future railway station, to Eindhoven Airport. The other group analyzed the possibilities of reducing the energy-related CO_2 emissions within the buildings of Eindhoven Airport to zero within 3 years. To both groups, the

[4]http://www.eindhovenairport.nl/nl/eindhoven-airport/corporate-responsibility/.

involvement of Eindhoven Airport, being well-known within the region, made this a special project for which the students became very motivated. The client was involved through provision of data and information on the airport, international policy and a first discussion. Although the students did not obtain much guidance afterward, throughout the process, they were very motivated and worked toward interesting solutions. After having finalized their work, both groups were requested to present their findings for an advisory board of the airport. To the students, this demonstrated the reality of their case and importance of their work. Furthermore, this offered them an interesting, new, and useful experience. To the airport, this work offered insight into the technical and economical possibilities of improving their airport.

4 Conclusions

Because the students analyze a technological innovation as well as a business case, they are able to arrive at conclusions more realistic than only studying a business case. This was demonstrated through several examples: modeling of the SolabCool cooling system, the business case of the Sowise solar boiler, and the adapted business model for a Cyclox production plant. Students reflect on the conclusions of their technical project and bring knowledge of the technology into the business case of their innovation. They gain knowledge not only through their client and coach but also through their research and combine this to objectively determine the most optimal business proposition or model.

Interaction with business coaches and clients has been demonstrated to be very important. It makes assignments more useful, practical, and interesting to students. The students feel their work has more impact and are motivated by their often ambitious clients. Business coaches are able to advise the students well in their process because of their large experience. This is a different experience than when students are confronted with hypothetical cases for which the outcome are of no importance to anyone. It thus strengthens their practical and business perspective and analytical skills. Evaluations of the minor program show that students want both projects to be smoothly connected and related. They see the importance of the connection between the business case and the technical project, thus making the BOTSA learning method a valuable experience.

Reference

1. Wijnker, M.A.S.G., Kasteren, J.M.N.v., Romijn, H.A. (2015), 'Fostering sustainable energy entrepreneurship among students: the Business Oriented Technological System Analysis (BOTSA) program at Eindhoven University of Technology', ISSN 2071-1050

Calcium Phosphate-Carbon Nanotube Composites for Load—Bearing Bone Implant Application

G. Gunawan and Iis Sopyan

Abstract The limited supply of bone graft is a major problem in human organ transplantation costing lives of many worldwide. Calcium phosphate (CaP) material has been considered the most favorable and best substitute in tissue engineering and orthopedics, thanks to its good biocompatibility and osteoconductivity. However, one of disadvantages of this material is its limited clinical applications for non-load-bearing bone implant due to low strength and brittleness. The objective of this study was to develop high mechanical strength calcium phosphate ceramics mechanically suitable for load-bearing bone implant application. Chemical and physical approaches have been undertaken to solve the mechanical weakness of calcium phosphate ceramics by doping with zinc (Zn) and compositing with carbon nanotube (CNT), respectively. The starting powders have been synthesized via sol-gel method. After uniaxial pressing, the compacted samples were sintered using conventional pressureless sintering. The samples were studied in terms of the phase stability, relative density, Vickers hardness, fracture toughness as well as in vitro test in SBF solution. The results showed that hydroxyapatite is the main phase and β-tricalcium phosphate is the secondary phase of obtained powders. The composites have relative density $93.34 \pm 0.50\%$ with a Vickers hardness as high as 5.00 ± 0.02 GPa and 2.78 ± 0.01 MPa.m$^{1/2}$ fracture toughness, which are in the range of human cortical bone's mechanical properties. Bioactivity test in a simulated body fluid for 1, 7, 14 and 30 days showed that the samples were covered by apatite cell since the first day, and the density of apatite increased by immersion time. Thus, from mechanical and bioactivity characteristic viewpoint, this CNT

G. Gunawan (✉) · I. Sopyan
Department of Manufacturing and Materials Engineering,
Faculty of Engineering, International Islamic University Malaysia,
P.O. Box 10, Kuala Lumpur 50728, Malaysia
e-mail: gunsyauqi@yahoo.co.id

I. Sopyan
e-mail: sopyan@iium.edu.my

G. Gunawan
Department of Mechanical Engineering, Faculty of Engineering,
Sriwijaya University, Indralaya 30662, Indonesia

T. Taufik et al. (eds.), *ICoSI 2014*, DOI 10.1007/978-981-287-661-4_34

reinforced Zn-doped calcium phosphate is mechanically and biologically suitable for load-bearing bone implant applications.

Keywords Biomaterials · Composite · Characterization · Zinc doping · Calcium phosphate · Mechanical · Porosity · Properties · SBF test

1 Introduction

Carbon nanotubes (CNTs) have emerged as outstanding reinforcement materials in nanocomposites production [1]. CNTs could improve mechanical integrity to the composite, without offsetting its bioactivity. Owing to excellent material properties in term of tensile strength, flexibility, and low density, CNTs make an ideal material to develop lightweight high-strength functional materials, such as bone. Along with the mechanical properties CNTs also have unique structural, electrical, thermal and chemical properties. CNTs in fact are biocompatible material [2] and can support neuron cell growth and differentiation of its own [3].

Calcium phosphate (CaP) bioceramics are attractive biomedical materials since they represent the inorganic part of human bones and teeth [4]. They are commonly used for dental and orthopedic implant applications due to their excellent biocompatibility and the nontoxicity of their chemical components. Among various phases of calcium phosphate, hydroxyapatite (HA, $Ca_{10}(PO_4)_6(OH)_2$) is the most common calcium phosphate ceramics applied for medical purposes [4, 5].

Despite these desirable properties, application of HA synthetic is limited due to its bioactivity and mechanical properties still need to be tailored for better performance [6]. Further, properties of HA, such as bioactivity, mechanical and adsorption properties can be improved by modifying composition via ionic substitutions [7].

Inclusion of trace metal elements has a significant role in improving physical and biological properties of bioactive ceramics. Zinc, an essential trace element in bone, has been used in the field of substituted apatites after considering the nature of all human bones. Zn has been demonstrated to have a variety of roles in human and other animals, associating with an activator in bone metabolism [8], promoting tissue growth in bone formation during in vitro and in vivo [9], and prevention of bone loss and osteoporosis [10].

Mechanical tests performed in HA, however, show very low values of fracture toughness of HA (0.8–1.2 MPa.$m^{1/2}$) [11] compared to 2–12 MPa.$m^{1/2}$ of human cortical bone [12]. Thus, this material is not suitable for heavy load-bearing applications. Therefore, compositing of HA is an appropriate strategy to improve its mechanical property. Numerous researchers have attempted to fabricate composites of HA reinforced with CNT with expectation of better mechanical properties implants. It has been a vast area of study and research lately and has showed outstanding results [13–15].

Currently, different approaches have been suggested in order to fabricate HA/CNT composites [16, 17]. This paper focuses on the development of a bulk Zn-doped HA product reinforced with CNT to be used in biomedical applications. We have reported the synthesis of Zn-doped HA powder elsewhere [18]. In this work, mechanical behavior and in vitro analysis of Zn-doped HA reinforced CNT composites is investigated.

2 Experimental

2.1 *Composite Preparation*

Zn-free HA and Zn-doped HA powders used to produce dense compacts were synthesized via a sol-gel technique [19]. Type of CNTs used in this study was multi-wall carbon nanotubes (MWCNT). The Zn-free HA–CNTs and Zn-doped HA–CNTs composites with various weight percentage of MWCNT were mixed in aqueous media. Each solution of HA and CNTs was first sonicated for approximately 30 min to avoid agglomerates. This procedure was followed by adding ammonium salt of polyacrylic acid (Duramax™ D-3005) as dispersant of carbon nanotubes. Finally, the solution was filtered and dried. After uniaxial pressing, the compacted samples were sintered via conventional pressureless sintering at temperature 1200 °C. The ramp-rate was fixed at 5 °C/min for both heating and cooling processes, soaking time of 2 h being adopted for each firing.

2.2 *Characterization*

Prior to characterization, all disk-sintered samples were then ground and polished manually using Mecapol (Presi, P260). The samples were ground using silicon carbide (SiC) paper (600–1200 grit) then polished using alumina slurry slightly until an optically reflective surface was obtained. The presence of the crystalline phase in the synthesized powder was analyzed by X-ray diffraction (XRD) (Shimadzu Diffractometer, XRD-6000). Thermogravimetric analysis was performed on a Hitachi apparatus (STA 7300 series). The powders were heated to 1200 °C at a constant heat rate of 5 °C/min. The bulk density of the sintered dense samples was measured by the Archimedes principle using a standard densimeter (Alfa Mirage, MD-300S). Subsequently, the average density was taken and the relative density was calculated by taking the theoretical density of HA as 3.156 g/cm^{-3}. For Vickers hardness (H_v) determination, the indentations were made using a pyramidal diamond indenter (Mitutoyo, MHV H-2) with an applied load of 200 g. The fracture toughness (K_{Ic}) of the sintered dense samples was determined using Vickers indentation method and the K_{Ic} value was calculated using the

Niihara equation [20]. The morphology of the simulated body fluid (SBF) samples was investigated by field emission scanning electron microscope (a JEOL JSM-6700F).

3 Results and Discussion

The XRD pattern of dense samples sintered at 1200 °C is shown in Fig 1. Sharp clear reflections proved the phase purity and crystallinity degree of the composite. It can be seen from XRD pattern that peaks of dense composite are attributed to HA (JCPDS 09-432). No peaks corresponding to phase β-TCP, $CaCO_3$ or CaO were found, suggesting that the conversion was complete when samples sintered at 1200 °C. Besides, the existence of graphite (CNTs) also can be seen from the XRD pattern. The peaks spotted at 2θ of 26.6° and 43.1° are assigned to graphite crystallographic (0 0 2) and (1 0 0) plane of CNTs, respectively [16].

Referring to Fig. 2, TG/DTA curves describe the weight loss along the investigated temperature range of 4 mol% Zn + CNTs composite. The weight loss during heating could be classified into three different stages. The first stage weight loss occurs between 30 °C and 300 °C due to evaporation of water. Then, it experienced about 1.5% weight loss for 4 mol% Zn + 1.5 wt% CNTs in the temperature range 600–700 °C, which is attributed to the crystallization of the powders

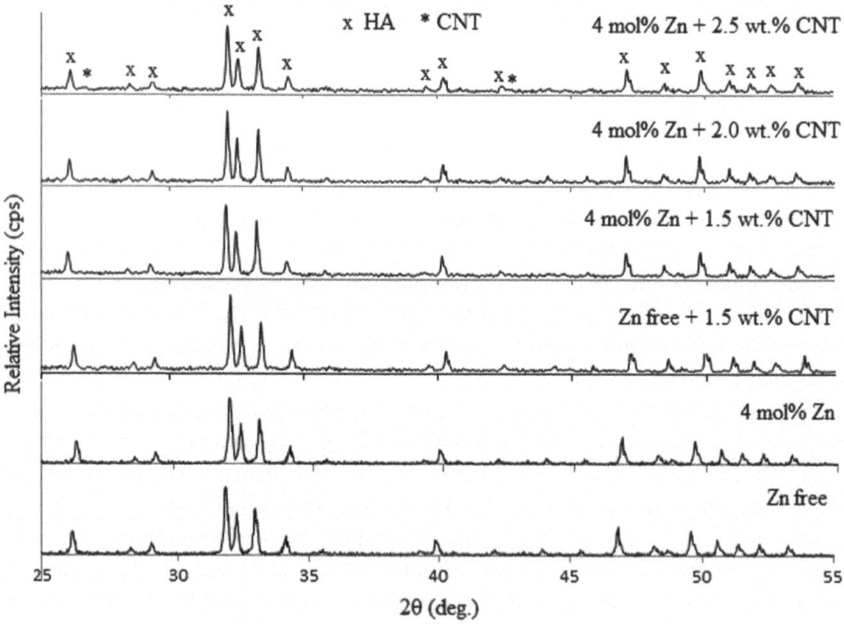

Fig. 1 XRD pattern of dense samples sintered at 1200 °C

from the amorphous phase. Meanwhile, 4 mol% Zn + 2.5 wt% CNTs experienced about 3.2% weight loss in the temperature range 400–500 °C. Up to temperature of 1000 °C, the 4 mol% Zn + 1.5 wt% CNTs increased a small amount of mass, whereas the 4 mol% Zn + 2.5 wt% CNTs lost a slight amount of mass. It is not clear, why the 4 mol% Zn + 1.5 wt% CNTs would have increased mass, as the air contained negligible water, however the 4 mol% Zn + 1.5 wt% CNTs may have lost weight due to some dehydroxylation. Finally, the dehydroxylation process stage final at temperature 1200 °C where the weight loss for 4% ZnHA + 1.5% CNTs is about 1.7% and 4 mol% Zn + 2.5 wt% CNTs is about 4.7% weight loss. Therefore, it can be concluded that the addition of CNTs concentration will make the final dehydroxylation rate increase. The vast majority of research involving HA + CNTs sintering has used either air or an inert gas (such as nitrogen or argon) bubbled through water as the sintering atmosphere. White et al. reported that HA + CNTs composite experience significant weight loss when sintered in the air compared to nitrogen atmosphere at temperature 1000 °C, about 8.5 and 7% respectively [14]. Meanwhile in this research, maximum weight loss is lower than that of nitrogen atmosphere reported. It is suggested that powder synthesis route affected hydroxylation rate, as the sol-gel method used to synthesize powder in this study experienced calcination at temperatures around 900 °C.

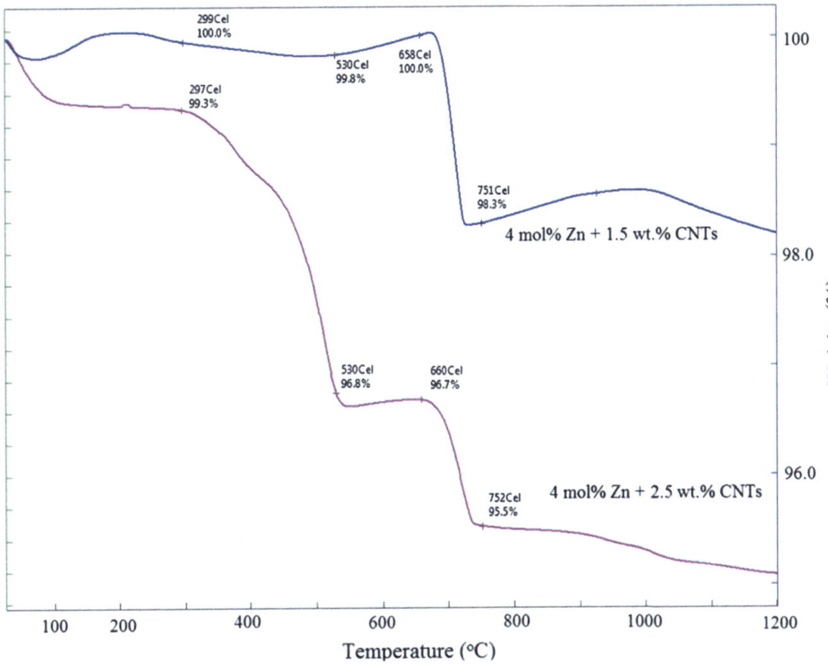

Fig. 2 TG/DTA result of the effect CNTs percentages on the weight loss

The density of the sintered samples can be affected by the difference in powder characteristics and CNTs content. Table 1 shows the relative density of 4 mol% Zn-doped HA–CNT with various CNTs concentrations sintered at 1200 °C. The addition of Zn in HA, particularly for the lower dopant concentration, was found to be beneficial in enhancing the relative density of the sintered body as showed by the higher value attained in these samples. The maximum relative density of 94.11 ± 0.01 was observed in the sample of 4 mol% Zn as compared to 87.29 ± 0.05 in Zn-free sample. The data also show a loose trend caused by CNTs loadings, with higher density values tending to correspond to low CNTs concentration and vice versa.

The Vickers hardness (H_v) and fracture toughness (K_{Ic}) of Zn-doped HA–CNT composite with various CNTs concentrations fired at 1200 °C is shown in Table 2. It can be seen that 4 mol% Zn composite shows higher Vickers hardness and fracture toughness as compared to Zn-free. Zn-free Vickers hardness and fracture toughness were 2.65 ± 0. 20 GPa and 0.99 ± 0.06 MPa.m$^{1/2}$, respectively, compared to 3.21 ± 0.10 of H_v and 1.43 ± 0.02 of K_{Ic} for 4 mol% Zn. Hence, it could be deduced that the incorporation of zinc in HA structure increases mechanical properties of HA. In general, the Vickers hardness and fracture toughness value of 4 mol% Zn composite increase with increase of CNTs loading. The influence of CNTs on the mechanical strength of composite was relatively noticeable when filled percentage of CNTs was 2.5%. The maximum Vickers hardness of about 4.30 GPa was observed in the sample when CNTs was introduced in 4 mol% Zn-doped CNTs composite. However, 4 mol% Zn + 3.0% CNTs could only attain a maximum hardness of 4.26 ± 0.30 GPa. It can be explained in detail via density results. It can be seen that the density of this sample has the lower value among the samples. Other possible factors that contributed to the low hardness are about matrix stabilities. Based on the data of mechanical strength versus amount of distributed phase in a composite, mechanical strength of composites increases until a peak point followed by a decrease of its value. Theoretically, mechanical strength of a composite decreases owing to deficiency in the matrix stability: The amount of reinforcement existing in the microstructure is interfering the matrix. The role of reinforcement as expected from a reinforced material may not be available anymore. At this point, the second phase may turn as a stress concentrator and as a consequence it results in descending of mechanical properties [17].

Table 1 Density of 4 mol% Zn-doped HA–CNT with various CNTs concentrations

Samples	Relative density (%)
Zn free	87.29 ± 0.05
4 mol% Zn	94.11 ± 0.01
Zn free + 1.5% CNTs	85.40 ± 0.20
4 mol% Zn + 1.5% CNTs	93.34 ± 0.50
4 mol% Zn + 2.0% CNTs	90.01 ± 0.20
4 mol% Zn + 2.5% CNTs	89.32 ± 0.30
4 mol% Zn + 3.0% CNTs	87.26 ± 0.20
4 mol% Zn + 3.5% CNTs	85.11 ± 0.40

Table 2 The Vickers hardness and fracture toughness of Zn-doped HA–CNT composite with various CNTs concentrations

Samples	Vickers hardness (GPa)	Fracture toughness (MPa.m$^{1/2}$)
Zn free	2.65 ± 0. 20	0.99 ± 0.06
4 mol% Zn	3.21 ± 0.10	1.43 ± 0.02
Zn free + 1.5% CNTs	2.74 ± 0.20	1.05 ± 0.04
4 mol% Zn + 1.5% CNTs	3.57 ± 0.02	2.51 ± 0.06
4 mol% Zn + 2.0% CNTs	4.17 ± 0.10	2.72 ± 0.07
4 mol% Zn + 2.5% CNTs	5.00 ± 0.02	2.78 ± 0.01
4 mol% Zn + 3.0% CNTs	4.26 ± 0.30	2.44 ± 0.08
4 mol% Zn + 3.5% CNTs	3.93 ± 0.05	2.34 ± 0.02

The results also show that the CNT were effective in improving the fracture toughness (K_{Ic}) of the synthesized HA. The 4 mol% Zn + 2.5% CNTs samples exhibited highest fracture toughness of 2.78 ± 0.01 MPa.m$^{1/2}$ as compared to 1.05 ± 0.04 MPa.m$^{1/2}$ measured for the Zn-free + 1.5% CNTs. It should be highlighted here that the K_{Ic} value obtained for the 4 mol% Zn + 2.5% CNTs is very promising, as most researchers have informed that the experimental K_{Ic} values

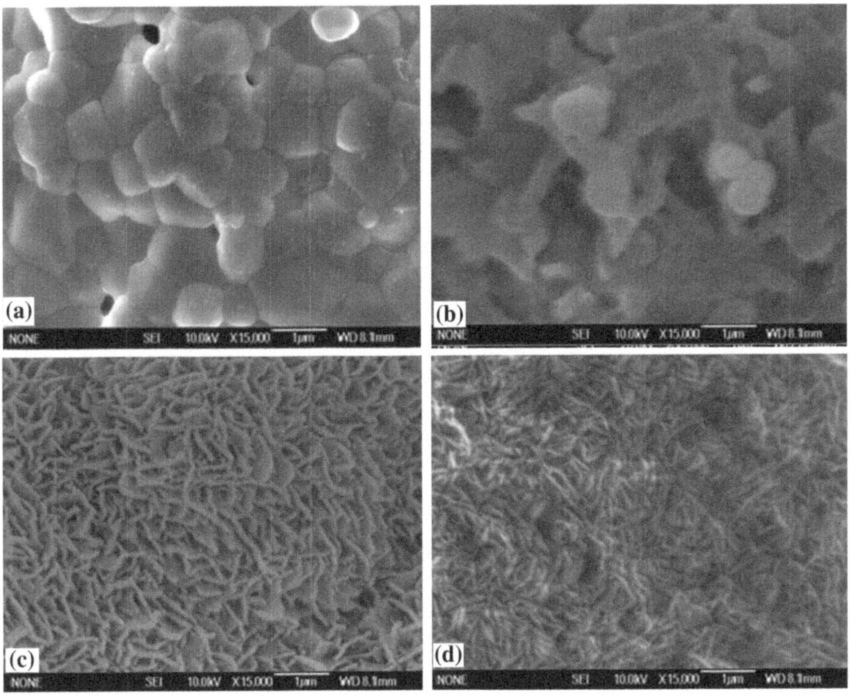

Fig. 3 Morphology of 4 mol% Zn + 2.5 wt% CNT composites (**a**) and samples after immersed in SBF solution for one day (**b**), 7 days (**c**) and 30 days (**d**)

for HA vary from 0.9 MPa.m$^{1/2}$ to about 1.2 MPa.m$^{1/2}$ [11, 21]. Moreover, the obtained value if K_{Ic} is in the range of human cortical bone (2–12 MPa.m$^{1/2}$).

Synthesis method and preparation of a Zn-doped HA–CNTs composite may affect the mechanical and bioactivity properties. In order to avoid the high cost of in vivo experiment, in vitro test has been performed to predict the bioactivity property of composite. The in vitro bioactivity analysis was evaluated using simulated body fluid (SBF) method [22]. It has been reported that the formation of a bone-like apatite layer on the surface of composite is essential for establishing bonding with natural bone [22]. Scanning electron microscope (SEM) of samples surface were taken at day 0, 1, 7, 14 and 30 days of the immersion in SBF to study the apatite formation rate. The morphology analysis of immersed sample in SBF solution is shown in Fig. 3. It was observed that apatite deposition occurred after first day of immersion in SBF as shown in Fig. 3b. In addition, after 7 days of immersing in SBF the surface of apatite layer changed through the growth of the apatite (Fig. 3c) and the density of apatite increased by immersion time (Fig. 3d).

4 Conclusions

In this study, Zn-free HA–CNTs and Zn-doped HA–CNTs composites with various weight percentage of MWCNT have been prepared. Overall, the phase of samples was hydroxyapatite and CNTs peaks also detected. From TG/DTA results shows that the addition of CNTs concentration will make the final dehydroxylation rate increase. The data also show a loose trend caused by CNTs loadings, with higher density values tending to correspond to low CNTs concentration and vice versa. Meanwhile, the maximum Vickers hardness and fracture toughness was attained 5.00 ± 0.02 GPa and 2.78 ± 0.01 MPa.m$^{1/2}$, respectively when sintered at 1200 °C. In vitro analysis shows that the samples have been covered by apatite cell since first day, and increased the density of apatite by immersion time. This study showed that CNTs reinforcing was beneficial in producing high toughness of Zn-doped HA.

Acknowledgements This work is supported by Ministry of Higher Education (MOHE) Malaysia through Fundamental Research Grant Scheme (Grant no FRGS11-025-0173) and we gratefully acknowledge the support.

References

1. Ku S.H.; Lee M.; Park C.B. Carbon-based nomaterials for tissue engineering, *Advanced healthcare materials* **2013**, *2* (2), 244–260.
2. Dubin R.A.; Callegari G.C.; Kohn J.; Neimark A.V. Carbon nanotube fibers are compatible with mammalian cells and neurons, *IEEE Transactions on Nanobiosciense* **2008**, *7* (1), 11–14.

3. Chao T.I.; Xiang S.; Chen C.S.; Chin W.C.; Nelson A.J.; Wang C.; Lu J. Carbon nanotubes promote neuron differentiation from human embryonic stem cells, *Biochemical Biophysical Reseach Communications* **2009**, *384* (4), 426–430.
4. Dorozhkin S.V. Biphasic, triphasic and multiphasic calcium orthophosphates, *Acta Biomaterialia* **2012**, *8* (3), 963–977.
5. Daculsi G.; Laboux O.; Malard O.; Weiss P. Current state of the art of biphasic calcium phosphate bioceramics, *Journal of Materials Science: Materials in Medicine* **2003**, *14* (3), 195–200.
6. Chen Y.; Miao X. Effect of fluorine addition on the corrosion resistance of hydroxyapatite ceramics, *Ceramics International* **2004**, *30* (7), 1961–1965.
7. Jiao M.J.; Wang X.X. Electrolytic deposition of magnesium-substituted hydroxyapatite crystals on titanium substrate, *Materials Letters* **2009**, *63* (27), 2286–2289.
8. Yamaguchi M.; Oishi H.; Suketa Y. Stimulatory effect of zinc on bone formation in tissue culture, *Biochemical Pharmacology* **1987**, *36* (22), 4007–4012.
9. Ito A.; Otsuka M.; Kawamura H.; Ikeuchi M.; Ohgushi H.; Sogo Y.; Ichinose N. Zinc-containing tricalcium phosphate and related materials for promoting bone formation, *Current Applied Physics* **2005**, *5* (5), 402–406.
10. Yamaguchi M. Nutritional factors and bone homeostasis: synergistic effect with zinc and genistein in osteogenesis, *Molecular and cellular biochemistry* **2012**, *366* (1–2), 201-221.
11. Ramesh S.; Tan C.Y.; Tolouei R.; Amiriyan M.; Purbolaksono J.; Sopyan I.; Teng W.D. Sintering behavior of hydroxyapatite prepared from different routes, *Materials & Design* **2012**, *34*, 148–154.
12. Wang X.; Nyman J.S.; Dong X.; Leng H.; Reyes M. Fundamental biomechanics in bone tissue engineering: Morgan & Claypool; 2010.
13. Shin U.S.; Yoon I.K.; Lee G.S.; Jang W.C.; Knowles J.C.; Kim H.W. Carbon nanotubes in nanocomposites and hybrids with hydroxyapatite for bone replacements, *Journal of Tissue Engineering* **2011**, *2011*, 1–10.
14. White A.A.; Kinloch I.A.; Windle A.H.; Best S.M. Optimization of the sintering atmosphere for high-density hydroxyapatite-carbon nanotube composites, *Journal of The Royal Society Interface* **2010**, *7 Suppl 5*, 1–11.
15. Zhao L.; Gao L. Novel in situ synthesis of MWNTs-hydroxyapatite composites, *Carbon* **2004**, *42* (2), 423–426.
16. Neelgund G.M.; Olurode K.; Oki A. A simple and rapid method to graft hydroxyapatite on carbon nanotubes, *Materials Science and Engineering C* **2011**, *31* (7), 1477–1481.
17. Osorio A.G.; dos Santos L.A.; Bergmann C.P. Evaluation of the Mechanical Properties and Microstructure of Hydroxyapatite Reinforced with Carbon Nanotubes, *Review on Advance Materials Science* **2011**, *27*, 58–63.
18. Gunawan; Sopyan I.; Suryanto; Naqshbandi A. Zinc-doped biphasic calcium phosphate nanopowders synthesized via sol-gel method, *Indian Journal of Chemistry Section a-Inorganic Bio-Inorganic Physical Theoretical & Analytical Chemistry* **2014**, *54A*, 152–158.
19. Bezzi G.; Celotti G.; Landi E.; La Torretta T.M.G.; Sopyan I.; Tampieri A. A novel sol-gel technique for hydroxyapatite preparation, *Materials Chemistry and Physics* **2003**, *78* (3), 816–824.
20. Niihara K. Indentation microfracture of ceramics—its application and problems, *Journal of the Ceramic Society of Japan* **1985**, *20*, 12–18.
21. Wang C.K.; Ju C.P.; Chem Lin J.H. Effect of doped bioactive glass on structure and properties of sintered hydroxyapatite, *Materials Chemistry and Physics* **1998**, *53*, 138–149.
22. Kokubo T.; Takadama H. How useful is SBF in predicting in vivo bone bioactivity?, *Biomaterials* **2006**, *27* (15), 2907–2915.

CFD Studies on the Flanged Diffuser Augmented Wind Turbine with Optimized Curvature Wall

M. Nurur Rochman, Aulia Nasution and Gunawan Nugroho

Abstract The Diffuser Augmented Wind Turbine (DAWT) offers potentials to cope with the wind availability situation like in Indonesia, i.e. with yearly average of 3–5 m/s. In this paper, computational CFD studies to get insight into the role of adding a flange (at the trailing edge of previously proposed diffuser with interior wall model [1]) to the wind velocity intensification inside the diffuser. Two models of flange are investigated, i.e. the flat-flange and airfoil-shaped flange. The role of angle-of-installment that provides the maximum velocity inside the diffuser is also reported. Results show that the additional flange to the trailing-edge of the diffuser will more step up the air velocity inside the diffuser. The installment of flat-flange will additionally increase the air velocity up to 29% higher. Meanwhile by optimizing the position of flange's angle, a more step up in velocity up to 3% can still be harvested (i.e. max at $\phi = 60°$). An even higher additional velocity (in compare to the installment of flat-flange) can be harvested by installment of flange with airfoil-shape, i.e. up to 31% (max at $\phi = 72°$).

Keywords Flanged DAWT · Optimized curvature wall · Flange's angle

1 Introduction

Energy security has become essential to support modern life, including for a developing countries like Indonesia. The fulfillment of energy needs for many countries worldwide are nowadays still dominated by fossil-based energy sources.

M. Nurur Rochman (✉) · A. Nasution · G. Nugroho
Department of Engineering Physics—FTI, Institut Teknologi Sepuluh Nopember (ITS) Surabaya, Kampus ITS Sukolilo, Surabaya 60111, Indonesia
e-mail: rochman12@mhs.ep.its.ac.id

A. Nasution
e-mail: anasution@ep.its.ac.id

G. Nugroho
e-mail: gunawan@ep.its.ac.id

T. Taufik et al. (eds.), *ICoSI 2014*, DOI 10.1007/978-981-287-661-4_35

The depletion of fossils-based energy deposits are clearly stated by many studies, which is also the case for Indonesia. It is estimated that within a period of 20 years, Indonesia's oil and gas supplies will run out. Therefore, it is necessary to revolutionize and boost the preparedness of the nation to fulfil its energy security from the abundantly available renewable resources that unfortunately have not fully explored yet. One of the potential resources to be further developed is the utilization of the wind energy.

Wind power conversion technology can be regarded as a cheap solution, since wind energy can be freely obtained, and the technology itself is not too complicated. The widely available wind energy conversion technology is known as the Horizontal Axis Wind Turbine (HAWT). However, this type can achieve its highest efficiency only for wind average speed of 10 m/s. Of course it is not suitable to be used in tropical regions like Indonesia, which has only wind potential with an average speed of 3–5 m/s [2]. Therefore, a more suitable wind energy extraction technology should be scrutinized and further explored, and the potency offered by the Diffuser Augmented Wind Turbine (DAWT) seems promising to be further investigated.

During the history of its development, the DAWT has been studied in their variative forms [3–8]. The concept of DAWT had been firstly developed in 1977 [3]. Foreman et al., had explained that the DAWT can reduce the air pressure behind the diffuser to be 0.6 times of the freestream's air pressure. This will allow the withdrawal of air that flows faster on the turbine.

Progress of the DAWT developments are now still going on, in order to achieve a more intensely increase in the power output. This technology offers an improvement in the wind energy's conversion efficiency to produce greater output power [6]. Increased of wind velocity at the diffuser can in principle increase the output power in a cube relationship, i.e. doubling the wind speed can yield an eight times higher of the output power, even for a relatively small-sized turbine [5]. So the increased power in the DAWT will be contributed due to the increased air velocity [1].

The developments in the DAWT design are directed to improve the performance of wind turbines, i.e. to create a stronger wind velocity inside the diffuser. One of the developmental designs that had been done was to study the flange addition, which can increase the speed of 1.6–2.4 times the freestream air velocity [8]. Flange of a certain size will intensify local vortices formation behind the diffuser, which will cause such a *suction effect* and leads the wind velocity that flows through the diffuser becomes larger [7, 9]. The length of the diffuser plays also a role in increasing the resulted output power. Ohya et al. mentioned that extending the DAWT length can increase the output power up to 4–5 times.

Previous studies were also done to get insight into the role of the curved interior wall of the diffuser. An optimized curvature profile of the diffuser's wall that will provide a maximum wind speed inside the diffuser was obtained, i.e. with geometrical profile that are similar to a 4-digit NACA airfoil geometry of NACA 5807. Based on the CFD simulation, it was obtained that this configuration can increase the wind speed inside the diffuser to be 65.5% higher in comparison to the diffuser with a flat wall as originally proposed by Ohya [7].

In this paper, further CFD studies to investigate the role of flange installment to further step-up the wind velocity flowing through the diffuser will be reported. Investigations are done in FLUENT platform. Two proposed form of flange will be used, i.e. a flat-flange (as originally used by Ohya [7]) and the airfoil-based flange (with a geometry of NACA 5807—that is similar to diffuser's wall). These two flanges are installed at the trailing-edge of the diffuser, and the effect of the variational angle-of-installment will also be reported.

2 Theories

(a) Diffuser Augmented Wind Turbine (DAWT)

The DAWT is actually a modification from the Horizontal Axis Wind Turbin (HAWT), i.e. by adding a hollow structure covering its rotor blades. This modification thus allows increasing the wind velocity that will strike the blades, causing a more wind power potential. As mentioned previously by Ohya [7], the increase in wind velocity is due to the formation of local pressure drop in locations of strong local vortices behind the diffuser, as mentioned by a schematic drawing in Fig. 1.

The flange addition at the rear (trailing-edge) of diffuser will create many local strong vortices formation in areas behind the diffuser. The strong vortices formation with a relatively small pressures will consequently creating such a *suction effect* to the wind that flows downstream across diffuser area [7].

(b) Airfoil NACA 4 Digit

NACA 4-digit airfoil's nomenclature system is coded by using a series of 4-digit numbers. Each digit of the numbers states certain information concerning the

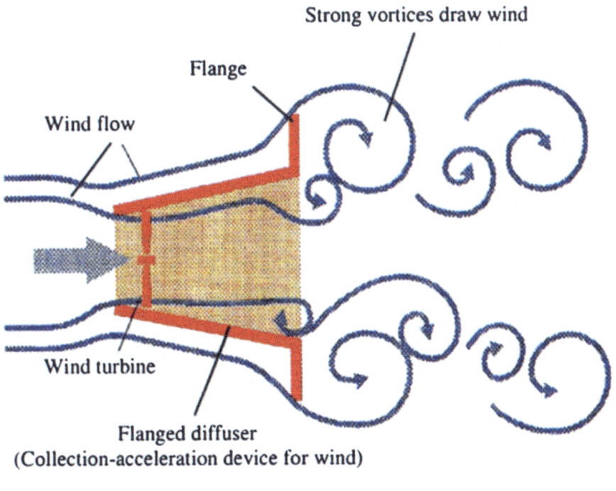

Fig. 1 A schematic view on the air velocity step-up mechanism inside the DAWT [7]

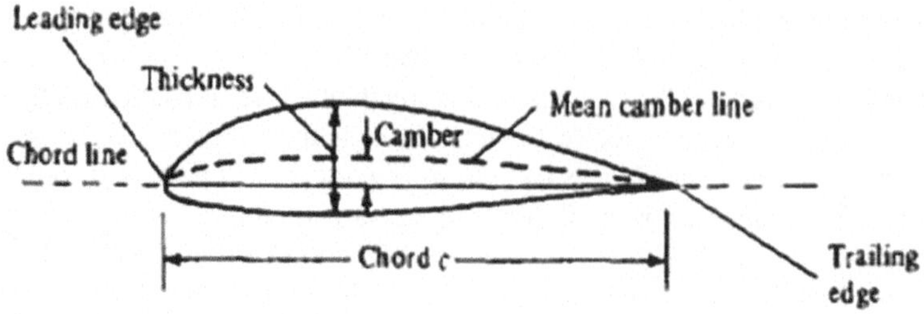

Fig. 2 Nomenclature of an airfoil NACA

geometry of the airfoil. The first digit represents the maximum value of the mean line (expressed in hundredths of the chord). The second number represents the position of the maximum mean line on the chord (expressed in tenths of the chord), while the last 2 digits represent the maximum thickness of the airfoil (expressed in hundredths of the chord). The airfoil 2412, for example, has a maximum chamber of 0.02c, located at 0.4c from the leading edge, and has a maximum thickness of 0.12c. The character c means a chord, i.e. the total length of the airfoil. Schematic drawing of NACA 4-digit airfoil's nomenclature system is depicted in Fig. 2.

(c) **Optimized Curve Interior for DAWT**

Previous efforts to enhance the wind velocity of the airflow through the diffuser were done by making a curved surface on the diffuser's wall. An optimizing algorithm was implemented to obtain the optimized airfoil's curvature geometry parameters that yield a maximum wind velocity. This optimization provides an optimized diffuser wall, i.e. as an airfoil surface with the following geometrical parameters:

- maximum camber: 0.05c
- Location of maximum chamber (from leading edge): 0.8c
- Maximum thickness: 0.07c, and
- Angle-of-attack (α) = 6°

Under these parameters, the value of $\frac{L}{D}$ is 15.507. A curvature geometry with these parameters is similar to the one of NACA 5807 airfoil.

3 Methods

(a) **Computational Conditions**

For the purpose of computational fluid dynamics (CFD) studies, a geometry of the diffuser with parameters that are similar to the one that previously proposed by Ohya was used. This geometry can be characterized in terms of its length (L), diameter (D), and the flange height (h), with a ratio of (L/D) = 1.5 and of

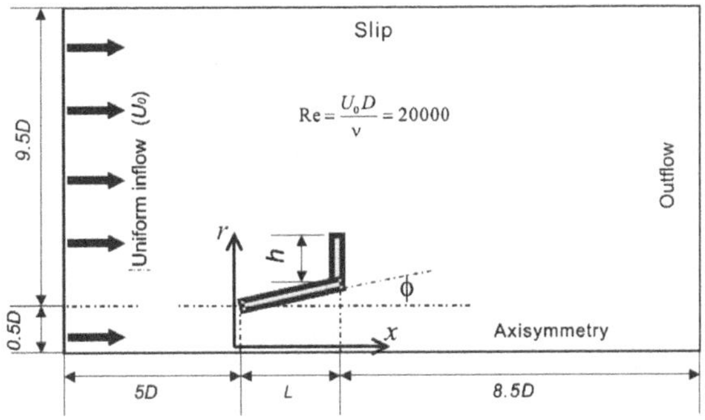

Fig. 3 Computational Conditions by Abe et al. [4]

(h/D) = 0.25 [7]. Meanwhile for the inside wall of the diffuser, we use a curved surface as obtained from previous study [1]. The schematic of computational conditions, as similar that are used by Abe et al. [4], can be seen in Fig. 3.

In the simulations, two form of flange geometry was installed at the trailing-edge of the diffuser with an optimized curvature wall, i.e. a flat-flange (similar to the one that was used by Ohya et al. [7]) and airfoil-form flange (with a shape similar to NACA 5807 airfoil). For both flanges, angle-of-installment will be varied to obtain the angle that will yield the highest wind velocity that flowing across the diffuser. Simulations were done by providing a uniform velocity airflow with an initial velocity U_o of 4 m/s. Schematic of geometrical description of the diffuser with the proposed flange installment used for the simulation studies is depicted in Fig. 4.

In a DAWT geometry, the angle φ is usually known as the *expansion angle*, i.e. an angle that represents the slope of changes in incremental areas of the diffuser's cross-section downstream with the flow direction. In terms diffuser with an inner wall of airfoil-shape, this angle also known as the *angle-of-attack*, and from our previous study, the most effective angle will be of 6° [1]. The variations in flange's angle-of-installment (θ) are set to start from 0° (i.e. angle that is parallel to the cord line of the airfoil: equal to angle-of-attack φ) up to $\theta = (90° - \varphi)$.

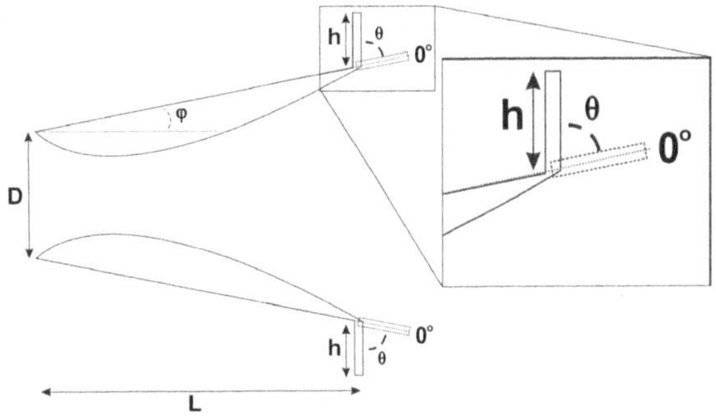

Fig. 4 Geometry DAWT with flange angle variation

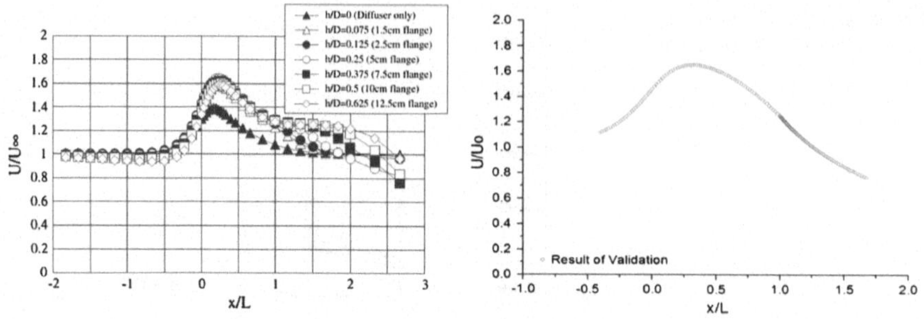

Fig. 5 Result of validation, (*left*) from Ohya et al. [7], *right* simulated with the developed codes

(b) **Validation**

In order to prior validate the developed CFD codes, a similar simulation configuration (as already published by Ohya et al. [7] i.e. a flat wall flanged diffuser with parameters of (L/D) = 1.5, and of (h/D) = 0.25) was used. Comparison of the result using the developed codes with the one that has been published in can be seen in Fig. 5. It can be shown that for the same diffuser's geometrical parameters, the maximum relative wind velocity of 1.65 can be obtained at the position of $x/L \approx 0.3$ from the inlet of diffuser. This computational results indicate and confirm that the developed codes are correct, and can be used for further simulating tasks with a high confidence level.

4 Results and Discussion

Results of the simulation to compare the effect of flange installment at the flat wall diffuser to the one with optimized curvature interior wall can be seen in Fig. 6.

Flanged diffuser with an optimized curvature wall will yield up to 7% velocity higher, in comparison to the one with a flat wall (i.e.$U/U_o = 1.72$ in comparison to $U/U_o = 1.65$). Meanwhile by varying the angle-of-installment, the obtained U/U_o inside the diffuser for both types of flange's geometry can be seen in Fig. 7 below.

The variation of the flange's angle-of-installment was ranging from 0° until 84°, with a 12° intervals. Results show that the highest wind velocity is obtained when the flange is positioned at angle of 60° (for the flat-flange), and at angle of 72° (for the case of airfoil-flange). A more insight into the influence of flange's angle-of-installment can be more clearly seen in Fig. 8.

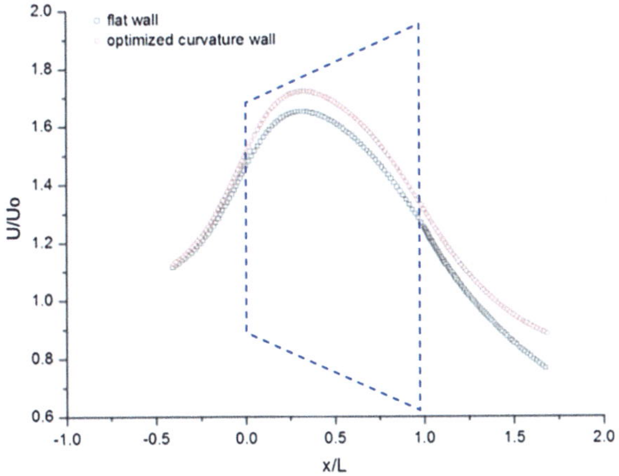

Fig. 6 Effect of flange installment at the diffuser with a flat and with optimized curvature interior wall

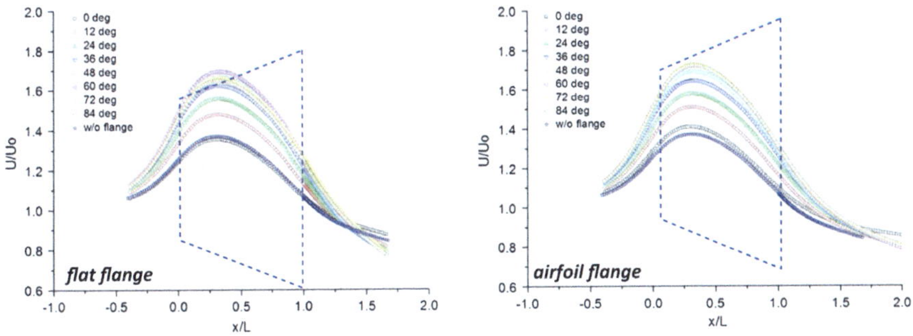

Fig. 7 Variation flange's angle-of-installment (*left* Flat-Flange, *right* Airfoil-Flange)

Tremendous increase in the wind velocity can be harvested by set the direction of the flange at a proper angle, i.e. between 60° and 84°. Installment of a flat-flange at the diffuser with an optimized curvature wall will step up the wind velocity from $U/U_o = 1.36$ up to maximum of $U/U_o = 1.65$ (up to 29% higher). Meanwhile using the airfoil-shaped flange will step up the wind velocity from $U/U_o = 1.41$ up to maximum of $U/U_o = 1.72$ (up to 31% higher).

The increase in wind velocity inside the diffuser is due to the creation of a higher pressure drop in the diffuser area. Simulation results confirm the phenomena of the pressure drop profiles (i.e. for different diffuser configuration), as shown in the following Fig. 9.

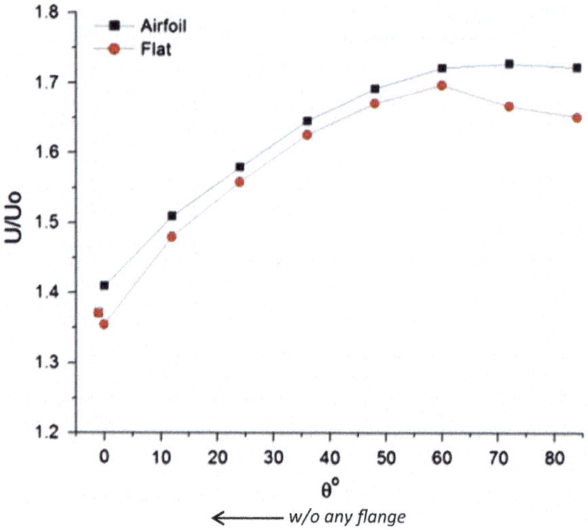

Fig. 8 Changing in the relative wind velocity in the diffuser as a function of flange's angle

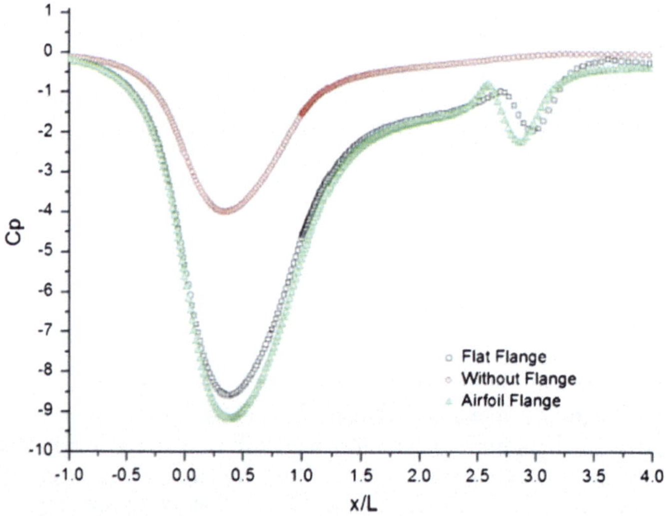

Fig. 9 Pressure Coefficient inside the diffuse for different diffuser's configuration

5 Conclusion

From the simulation results, the design of the curved diffuser with flat-flange obtained a greater improvement at the position $x/L \approx 0.3$, resulting in an increase of wind velocity up to 29% higher than the previous diffuser design without flange.

Meanwhile the installment of the airfoil-shaped flange to the same diffuser will increase the wind velocity up to 31%.

Future works will be devoted to perform experimental works to validate the simulation results, and to investigate the influence of flange installment at the leading-edge (diffuser's entrance), in addition to the one that installed at the trailing-edge.

References

1. Nasution, A., & Purwanto, D.W. (2011), Optimized Curvature Interior Profile for Diffuser Augmented Wind Turbine (DAWT) to Increase its Energy-Conversion Performance, IEEE First Conference on Clean Energy and Technology CET.pp. 315–320.
2. H. Suharta, (2007), Energy Data in Indonesia (In Indonesia), B2TE-BPPT.
3. Foreman, K. M., Gilbert, B., & Oman, R. A. (1978), Diffuser Augmentation of Wind Turbines, Solar Energy, pp. 305–311.
4. Abe, K.-i., & Ohya, Y. (2004), An Investigation of flow fields around flanged diffuser using CFD, Journal of Wind Engineering and Industrial Aerodynamic 92, pp. 315–330.
5. Toya H., T. Kusakabe, & T. Yuji (2007), Fluid Flow Analysis and Design of a Shroud for Wind Turbine Using Ansys, International Conference on Electrical Machines and Systems, pp. 298–301. Seoul.
6. Wang F., L. Bai, J. Flecher, J. Whiteford, & D. Cullen.(2008), Development Of Small Domestic Wind Turbine with Scope and Prediction of Its Annual Power Output, Renewable energy, pp. 1637–1651.
7. Ohya, Y., Karasudani, T., Sakurai, A., Abe, K.-i., & Inoue, M. (2008), Development of a Shrouded Wind Turbine with a Flanged Diffuser, Journal of Wind Engineering and Industrial Aerodynamics 96, pp. 524–539.
8. Ohya, Y., & Karasudani, T. (2010), A Shrouded Wind Turbine Generating High Output Power with Wind-lens Technology, Energies, pp. 634–649.
9. Kardous, M., R. Chaker, F. Aloui, & S. Ben Nasrallah (2013), On The Dependence of an Empty Flanged Diffusers Performance on Flange Height : Numerical Simulations and PIV Visualization, Renewable Energy, pp. 123–128.

Developing Academic Executive Information System Uses Kimball Methodology: Case Study in an Indonesia Higher Education System

Asroni, Noor Akhmad Setiawan and Sri Suning Kusumawardani

Abstract Muhammadiyah University of Magelang (UMM) has integrated academic information systems in terms of both data and applications. This system has been constructed 6 years ago and has been used by the service level to the policy level. However, until today, there is no data can be presented in an appropriate format for the decision-maker, since the data are functionally separated. In addition, reporting data is not easily used by decision-makers because it does not apply the concept of the data warehouse. So that the information can be used by decision-makers, the data warehouse must be constructed using appropriate methods. In this paper, Kimball methodology is chosen because the design starts from user needs. The methodology causes the system to be able to present complete and appropriate data presentation as well as the user wants. UMM Academic Executive Information System (EIS) has been successfully constructed using the data warehouse. As a result, the system is able to present the integrated data and is able to demonstrate factual data from a variety of dimensions as required by decision-makers. By using the EIS, the decision-maker can trace the problem source that happened in their education system.

Keywords Kimball methodology · Data warehouse · Academic executive information system · Problem source finding

---

Asroni (✉)
Information Technology Department, Universitas Muhammadiyah Yogyakarta,
DI Yogyakarta 55183, Indonesia
e-mail: asroni@gmail.com

N.A. Setiawan · S.S. Kusumawardani
Electrical Engineering and Information Technology Department,
Universitas Gadjah Mada, Jl Grafika 2, Sleman 55281,
DI Yogyakarta, Indonesia
e-mail: noorwewe@ugm.ac.id

S.S. Kusumawardani
e-mail: suning@ugm.ac.id

T. Taufik et al. (eds.), *ICoSI 2014*, DOI 10.1007/978-981-287-661-4_36

1 Introduction

Executive Information System (EIS) is a computer-based information system which is designed to provide ease of access to the data by executives. The presented data is a summary of the various sources of data that can be used for analysis, decision-making, and organization planning [1]. EIS grew from the need of information that required by the board, especially in strategic planning and decision-making. EIS includes aspects related to decision-making, among others: analysis and trend detection, measurement and indicator tracking, trace analysis, problem diagnosis, competitive analysis, and monitoring organizational performance [2].

The main problem in the development of UMM EIS is to determine the information that must be provided by the system due to the decision-makers have difficulty in interpreting their information needs. There are two strategies that can be used to identify needs, i.e. based on a business process or an activity. The data warehouse is a collection of data from various sources internal and/or external, arranged such that it can be processed into useful information for the running of the organization. EIS is one application that uses the principles of data warehouse. EIS can view the data in a special format that can be easily used by executives to analyze the problem appropriately [3, 4].

Decision-makers can only see the data in one dimension, but a decision would be better if based on the data showed from different dimensions. If the data can be viewed from a variety of dimensions, then a problem can also be viewed from various perspectives. UMM has a lot of academic data that has been collected over 6 years. It collected through the daily academic processes that supported by information system infrastructure. The data have been processed for annual report. The problems arise when the data must be processed for decision-making. The data presented in tabular form only without being able to see the information from different viewpoints [5]. It can be concluded that the format of the data required for decision-making is not sufficient. So the system is able to provide information to use a better structure, it would require a data warehouse system. Implementation of data warehouse requires proper design, because each institution has unique problems and needs including UMM.

There are six faculties in UMM, two of them have an average score of graduates were below of the standard. The decision-makers suspect this is caused by the low effort in student recruitment, lecturing, or student learning activities. The root cause should reveal first before decide the best action plan.

2 Theoretical Background

2.1 Design Approach

There are two important terms in the EIS development process, i.e. the data mart and data warehouse. Data mart has a smaller scale when compared with the data warehouse. Data mart is collected from a department while data warehouse is collected by the organization. A data mart is subject oriented (part of a business process) based on the needs of a particular department, while data warehouse is enterprise-oriented (mean of a business process).

In the practice of data warehousing, there are two commonly used methods of design, i.e. Inmon method or Kimball method. Both are very different in terms of their design pattern. Inmon method using top-down approach, meaning that the design process is done by conducting in-depth analysis based on needs before the data warehouse constructed; it can be said that the data mart is a unit drawn from the data warehouse. In this method, the technical aspect is more prominent because it leads to the process of data modeling. Kimball methods have different characteristics because using a bottom-up approach and oriented on business processes [3]. The data warehouse is built based on the needs of the organization so that the data mart is built first so that it can be used to compile the data warehouse. In the Kimball method, aspect needs are more emphasized. User information system need plays a greater role in providing input to the preparation of the data warehouse.

At UMM academic departments, data warehouse needs to be built because the decision-makers need appropriate data that can be used to find the problem source. Thus, Kimball method is more suitable than the Inmon method. Two other reasons Kimball method was chosen, are:

1. The time required to construct a specific data warehouse is shorter than to construct a comprehensive data warehouse.
2. To develop the data warehouse can be initiated from a specific requirement so that the development resources are less.

2.2 Kimball Methodology

Basically, according to the method of Kimball, a data warehouse is constructed from the information needs of the user needs in certain priority order. This method is divided into 9 steps [6]:

1. Choosing process as department request
2. Choosing the grain
3. Identifying and conforming the dimensions
4. Choosing the fact

5. Storing pre-calculated in the fact table
6. Rounding out the dimension table
7. Choosing the duration of the table
8. Tracking slowly changes of the dimension
9. Deciding the priority and mode of query.

The 9-step process results, in a new data structure, can be further processed into information for strategic decision-making process on a scale of the university.

2.3 *Development Method*

Refer to the problems and approaches, the implementation framework for constructing the UMM EIS is arranged as shown in Fig. 1.

3 System Development

Like other institutions, the UMM organization structure is essential to support the activities of a business process. UMM organization is divided into 3 levels, includes: operational, managerial, and decision-makers. EIS is designed for decision-makers only, i.e. rector, dean, and board as shown in Table 1.

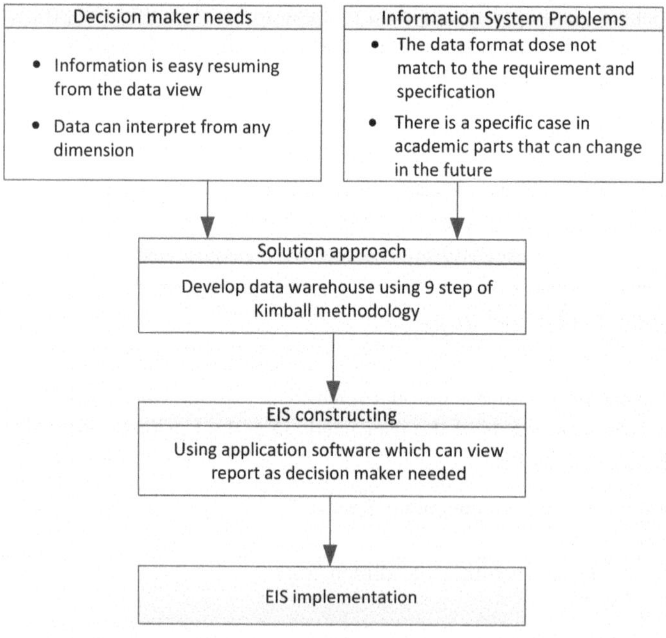

Fig. 1 Development method

Table 1 EIS User

User	Position	Data specification
Rector and board	*Decision-maker*	– Academic trend – Activity resumed
Dean	*Manager*	– Monthly activity report

3.1 Information Needs Analysis

Data determination and presentation in the data warehouse is a very important process because it involves the results of a report to be presented to decision-makers. The report must be accurate, easy to understand, and can be seen from several dimensions so that decisions will be taken to have a high accuracy.

Academic process is one of the core activities in a university. Due to the importance of this process, the UMM decision-maker needs information about student academic achievement in every year. The information can obtain from the interpretation of the data that has been filtered. Student academic achievement is influenced by: academic activities, human resources, and new student recruitment systems. There is highly prioritized information for the UMM academic decision-maker, namely

1. Why the GPA of graduates in a particular department is under the standards? Why the GPA is not increasing significantly in the last 5 years? Which process should be improved to increase the GPA?
2. What factors have dominant effects on GPA UMM graduates and what is the best way to improve the GPA so meet the standards?

3.2 Data Warehouse Architecture

The UMM data warehouse is developed with the goal of providing easy access to the specific data, accordance with the identified issues. Planning the development of the data warehouse architecture includes logical architecture and physical architecture. A logical architecture design stages of the data flow to data sources into a data warehouse, while the physical architecture is a description of the technical configuration of the data warehouse [7]. Figure 2 shows the logical architecture of data warehouse design. The operational data which use in this research are: academic data, human resource data, and prospective student data. The collected data is then filtered and loaded into a separated database (DATA_DUMP). The output data of this sorting process will be stored in the data warehouse.

The design of the physical architecture is shown in Fig. 3. Data dump and data warehouse was on the same machine and operating data are on different servers, so that the process of extraction, transformation and loading can be done separately.

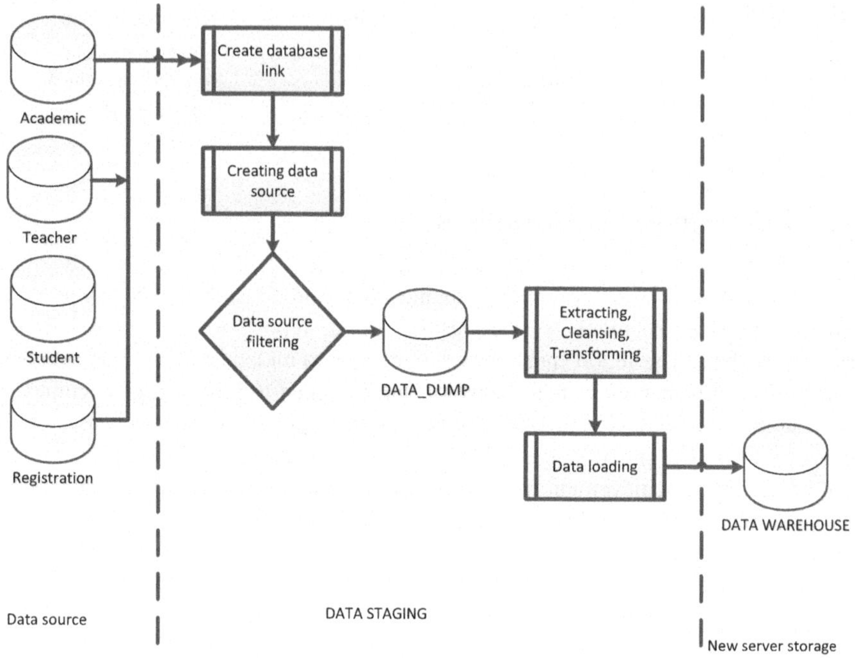

Fig. 2 Data warehouse architecture

This manner is to avoid a traffic jump issue. In other words, the users of the EIS were not directly related to the operational database [8].

4 Implementation and Result

The implementation phase of the data warehouse to support the EIS using Kimball method can be described as follows:

4.1 Choosing Process as Department Request

UMM business process run by departments as part of the organization known as a unit. There are 3 main units at UMM, namely internal management unit (financial, asset, human resources, student services), academic unit (lectures, research, services), and general affair (organization, cooperation). Academic information needed by decision-makers is the data from the most influential in the achievement of learning outcomes; they need data of tutorial activities, learning activities, and technical acceptance of new students.

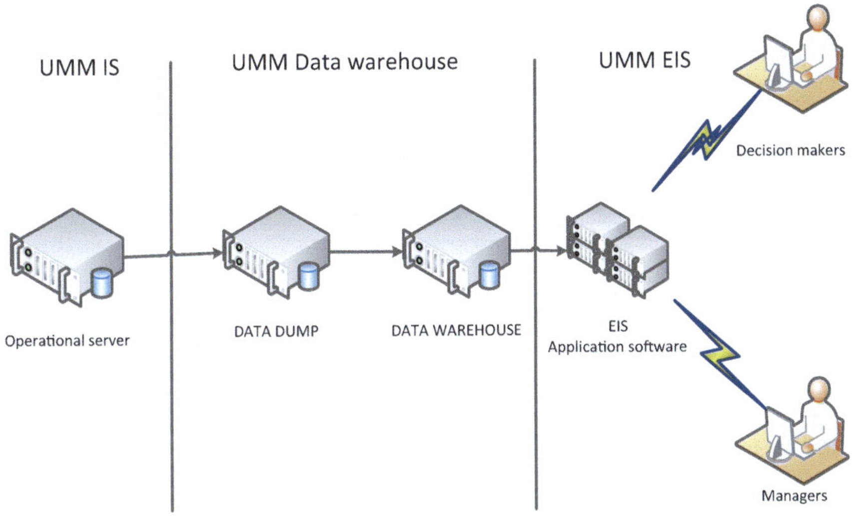

Fig. 3 EIS architecture

4.2 *Choosing the Grain*

This step is to select and specify the data that will be selected to enter the DATA_DUMB, including lecturing, student GPA, total credits, recruitment score, lecturer activities, and student hometown.

4.3 *Identifying and Conforming the Dimension*

In this step, the relationship between the dimensions of the source specified in the matrix as shown in Table 2.

4.4 *Choosing the Fact*

This step produced the facts that can be stored in the fact table. In general, the fact table contains entities which can be measured. The facts are set student achievement, student activities, lecturer performance, and prospectus student academic performance. The fact table also contains historical data that stored in the data warehouse.

Table 2 Sample of dimension table

Dimension	Grain					
	Learning plan	Student activities	Lecture activities	Lecture portfolio	Student information	Selection mode
Batch	√		√	√	√	√
High school					√	
Testing grades						√
Final grade					√	
Selection mode						
City of origin					√	
Total credit	√					
GPA	√	√				
Subject						
Attendance		√				
Subject grade	√	√				
Lecturer tenure						
Lecturer workload			√	√		

4.5 *Storing Pre-calculated in the Fact Table*

Data is chosen because may influence the achievement of learning. The selected data contain:

1. The fact of student achievement
2. Fact of lecturer workload
3. Registration system performance.

4.6 *Rounding Out the Dimension Table*

This step is to determine the relationship between the dimensions and sources in a matrix as shown in Table 3.

Table 3 Detail of dimension table

Dimension	Field	Description
Registration period	*daftar_id* *daftar_thn_ajar* *daftar_name_gel*	Registration is filtered based on registration period
Selection mode	*jseleksi_id* *jseleksi_nama* *jmasuk_id* *jmasuk_nama*	Selection data filtering based on prospect student and the selection mode
Origin of high school	*smuasal_id* *smuasal_nama* *daerahasal_nama*	Lecturer data filtering based on the student's origin of high school

4.7 Choosing the Duration of the Table

The time duration is determined by the needs of the user. The data has been successfully collected over a period of 6 years duration of the database can be selected so that a maximum of 6 years, but the decision-makers want the last 5 years of data.

4.8 Tracking Slowly Changes of the Dimension

Little changes may occur in a particular dimension because of business process. Data warehouses can be changed to follow the changes in the dimensions of the unit or because of changes in the decision-maker require new data. Changes in the dimensions of the data warehouse can be done in three ways: change the dimension of attributes, add new records, and add new columns.

Of the three basic types of such changes, the design of the data warehouse UMM uses changes in the dimensions to keep pace with the changing needs of the decision-makers.

4.9 Deciding the Priority and Mode of Query

Priority can be determined using the Extract-Transformation-Load (ETL). ETL process is designed to perform extracts from each unit to be used together. The results of the data presented in the format ETL according EIS application that will be used. Accurate ETL process design can make the process of establishing a data warehouse into an efficient, flexible, and can be treated easily. In general, the ETL process can be divided into four main modules [9–11] as follows:

1. Extraction: data were extracted from each unit and stored in a disk with a specific structure that is ready to be transformed into the data warehouse.
2. Cleansing: filter out unneeded data as incomplete data, missing data, incorrect data, and duplicate data.
3. Transform: the data that has been cleaned, then converted to fit the specified format.
4. Loading: data entered into the data warehouse that queried ready by the EIS application.

5 Discussion and Conclusions

Business Analysis is a visualization of mapping the data warehouse so that it has the same viewpoint. This process uses drill down technique to get the selected data. BI analysis is used to see the detailed data so that students can be summarized according the facts are determined. The BI Report provides two options appropriate reporting template provided by the software or specified by the user (Ad Hoc Report). If the ad hoc report selected, then the EIS will show the report in more pleasing appearance.

One of the main issues that occurred at UMM is the average graduate GPA under 3.0, while the standard is 3.0. The decision-makers decide to increase the current student GPA, but they have to look at the fact. The decision-maker should find out the root of the problem first. EIS can view graphs that show facts in many dimensions. The graphs can use by decision-maker to find the relationship between the student selection process, student activities, and lecturer activities and the average of student GPA. Figure 4 shows the comparison between the accepted

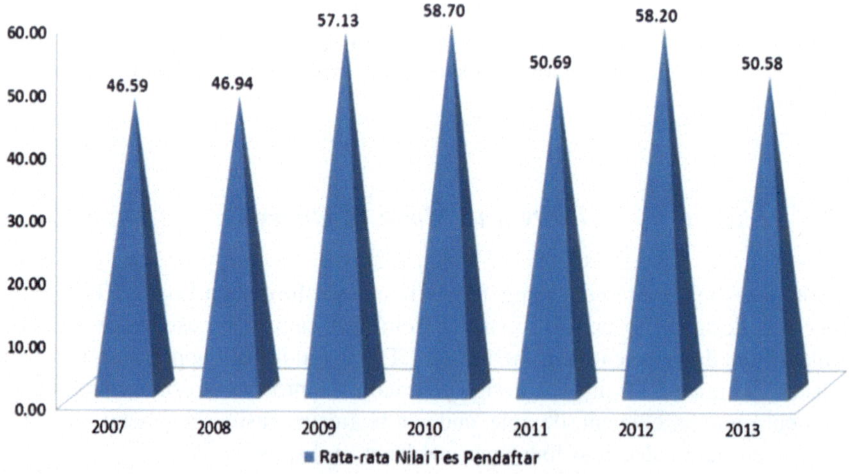

Fig. 4 Student selection score (Rata-rata nilai tes pendaftar) 2007–2013

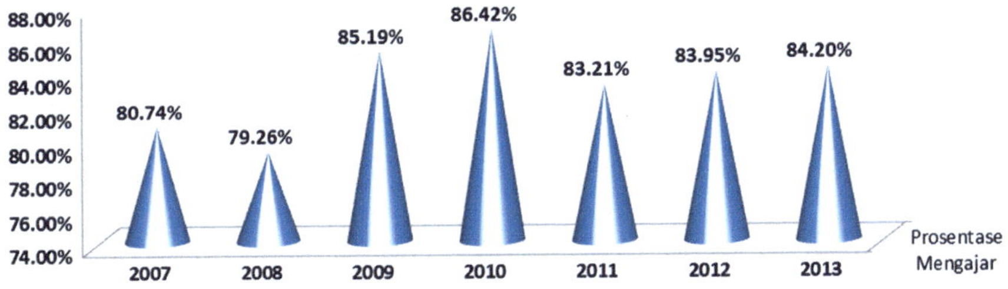

Fig. 5 Tutoring attendance (Prosentase mengajar) batch 2007–2013

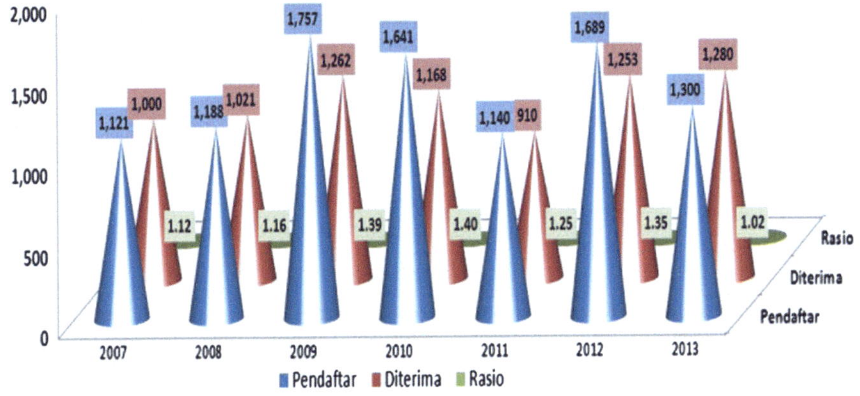

Fig. 6 Ratio (rasio) bertween total applicants (pendaftar) and successful applicants (diterima)

students to prospective student; the smallest ratio occurred in the year 2007. Figure 5 shows the attendance of 80% student than >75% total lecturing, accept students in batch 2007. From those data can conclude that at that time the student recruitment was ineffective.

Figure 6 shows the smallest average GPA also in the batch 2007. It also can be concluded that the low student selection, performance is influenced to the student GPA. There is a correlation between the results of student selection to the student motivation and achievement. The same case also happened in batch 2011. UMM decision-maker decided to enhance the student GPA need to increase the performance of student recruitment bureau first, before increase the lecturing performance.

According the implementation result can be summed up as follows:

1. The data warehouse is able to summarize, select, and obtain detailed data required by the EIS from the various data sources scattered in the UMM information system.
2. The data warehouse is created in the new format and a new structure that is suitable for EIS and matched to decision-maker request.

3. EIS, which is built using data warehouse concepts, has advantages in obtaining the required data. The data obtained are considered a summary of the various dimensions. Information obtained from the analysis of the data is in a holistic case so they can support better decisions.
4. EIS is easily used by decision-makers.

Acknowledgements Thank you very much for Prof. Ahmadi, rector of Universitas Muhammadiyah Magelang which has provided the opportunity to apply the EIS. Also, many thanks to the staff of IT department in their best efforts and cooperation as long doing this research.

Research Limitation

In this research, the 9-step method used for designing new EIS in the academic department only. The application of this method in other fields in different institutions and their impact on decision-making is still needed to see if this method suitable for design the EIS of educational institution in Indonesia.

References

1. Pervan, G.P. and Phua, R., 1996, January. Executive information systems in Australia: current status and some historical comparisons. In System Sciences, 1996., Proceedings of the Twenty-Ninth Hawaii International Conference on, (Vol. 2, pp. 110–119). IEEE.
2. Averweg, U.R. and Roldán, J.L., 2006. Executive information system implementation in organisations in South Africa and Spain: A comparative analysis. Computer Standards & Interfaces, 28(6), pp. 625–634.
3. Inmon, W.H., 2005. Building the data warehouse. John Wiley & Sons.
4. Turban, E., Rainer, R. K., and Potter, R. E. (2005). Introduction to information technology. New York, John Wiley & Sons.
5. Watson, H.J. and Frolick, M., 1992, September. Determining information requirements for an executive information system. In Challenges and strategies for research in systems development (pp. 323-336). John Wiley & Sons, Inc.
6. Connolly, T.M. and Begg, C.E., 2005. Database systems: a practical approach to design, implementation, and management. Pearson Education.
7. Kimball, R., 1998. The data warehouse lifecycle toolkit: expert methods for designing, developing, and deploying data warehouses. John Wiley & Sons.
8. McFadden, F.R., 1996, January. Data warehouse for EIS: some issues and impacts. In System Sciences, 1996., Proceedings of the Twenty-Ninth Hawaii International Conference on, (Vol. 2, pp. 120–129). IEEE.
9. Ponniah, P., 2004. Data warehousing fundamentals: a comprehensive guide for IT professionals. John Wiley & Sons.
10. Jun, T., Kai, C., Yu, F. and Gang, T., 2009, May. The research & application of ETL tool in business intelligence project. In Information Technology and Applications, 2009. IFITA'09. International Forum On (Vol. 2, pp. 620–623). IEEE.
11. Kimball, R. and Caserta, J., 2011. The Data Warehouse? ETL Toolkit: Practical Techniques for Extracting, Cleaning, Conforming, and Delivering Data. John Wiley & Sons.

Effect of Reservoir Sediment Flushing from Wlingi and Lodoyo Reservoirs on Downstream River Water Quality

Fahmi Hidayat, Surya Budi Lesmana and Ery Suhartanto

Abstract In order to cope with severe sedimentation problems in Wlingi and Lodoyo reservoirs in the Brantas River basin, East Java, Indonesia, sediment flushing has been conducted since August 1990, immediately after the eruption of Mt. Kelud in February 1990. The sediment flushing has been conducted mostly during the rainy season by coordinated gate operation to recover the reservoirs' capacity. The largest and most intractable downstream environmental impacts associated with sediment release from reservoirs are caused by reservoir emptying and flushing. When sediment flushing measure is performed, environmental effects should be considered comprehensively. To understand the adverse environmental effects of sediment flushing in Wlingi and Lodoyo reservoirs on the downstream reaches, an investigation was carried out to study water quality before, during, and after the sediment flushing operation. This study was conducted to investigate the adverse environmental effects on the downstream reaches' water quality of Wlingi and Lodoyo reservoirs based on the sediment flushing operation conducted on 17-19 March 2009.

Keywords Sediment · Flushing · Reservoir · River · Water · Quality

---

F. Hidayat (✉) · S. Budi Lesmana
Civil Engineering Department, Brawijaya University, Malang, Indonesia
e-mail: hidayat.f@jasatirta1.net

F. Hidayat
Jasa Tirta I Public Corporation, Malang, Indonesia

S. Budi Lesmana
Yogyakarta Muhammadiyah University (UMY), Yogyakarta, Indonesia
e-mail: surya_lesmana@umy.ac.id

E. Suhartanto
Water Resources Engineering Department, Brawijaya University, Malang, Indonesia
e-mail: erysuhartanto@ub.ac.id

T. Taufik et al. (eds.), *ICoSI 2014*, DOI 10.1007/978-981-287-661-4_37

1 Introduction

The reservoirs in the upper reach of Brantas River basin in Indonesia have lost their storage capacities rapidly, ranging from 35 to 80% of its original storage capacities due to severe watershed degradation, improper land management, and volcanic eruption. The total gross and effective storage capacities of the five reservoirs in the upper reach of Brantas River basin remain 50 and 63% of its original condition, respectively [1].

Wlingi Dam with a catchment area of about 2890 km^2 is located on the southern skirts of Mt. Kelud. The dam was constructed in 1977 for the purpose of electric power generation and irrigation water supply, and was also expected as storage of the erupted material from Mt. Kelud. After construction of Wlingi Dam, the sedimentation rate has been very rapid, and annual average sediment deposit in the reservoir up to 1988 was around 1.3 million m^3 [3].

In January 1990, just before the eruption of the Mt. Kelud, the gross storage capacity of Wlingi Reservoir was reduced to 19.2% of the designed gross storage capacity. After February 1990 eruption, Wlingi Reservoir was filled up completely by the sediment. As the result of the continuous removal works, the gross storage capacity in 2013 was recovered to 4.83 million m^3, which corresponds to 20.10% of the designed gross storage capacity of 24.0 million m^3.

Lodoyo Barrage is located at approximately 7.5-km downstream reach of Wlingi Dam. The barrage was constructed in 1983 for the purpose of afterbay of Wlingi Dam and power generation. The decreasing rate of storage capacity in Lodoyo Reservoir is also high as well as Wlingi Reservoir. The gross storage capacity of Lodoyo Reservoir in 2013 was reduced to 2.72 million m^3, which corresponds to 52.30% of the designed gross storage capacity of 5.20 million m^3. There was significant change in the decreasing rate of the storage capacity after the eruption of Mt. Kelud in 1990, though there was a trapping effect of Wlingi Reservoir for the erupted material from Mt. Kelud.

The coordinated sediment flushing has been conducted in Wlingi and Lodoyo reservoirs since August 1990 immediately after the 1990 eruption of Mt. Kelud. The flushing has been conducted mostly during rainy season by coordinated gate operation of the two dam reservoirs. The sediment flushing by gate operation has been, in terms of removed sediment volume, proved to be an effective measure to maintain reservoir effective storage capacity and to mitigate the riverbed degradation in the downstream reach due to dam trapping effect. Total of sediment volume flushed from Wlingi and Lodoyo reservoirs from 1990 to 2013 was 8.97 million m^3 and 2.15 million m^3, respectively.

There are two opposing groups of thought on the subject of reservoir emptying. One proposes that downstream environmental impacts should be minimized by keeping the trapped sediment within the pool. While this is feasible initially, sediment cannot be accumulated indefinitely, and at some point of time, sediment must be discharged or removed by dredging. The opposing school of thought is to allow sediments to escape during emptying, but to mitigate deleterious effects [2].

Water quality monitoring must be conducted before, during, and after flushing. Routine monitoring during flushing should focus on parameters such as temperature, pH, BOD, COD, TSS, turbidity, and total phosphate. Additional parameters may be required, depending on the site conditions. This paper presents the monitoring results before, during, and after flushing to study the adverse effect of reservoir sediment flushing on downstream river water quality in Wlingi and Lodoyo reservoirs in the Brantas River basin. By analyzing the monitoring results, we will understand the parameters fluctuated due to the sediment flushing, the spatial extent of impacts on water quality, and duration of impacts.

2 Research Method

To understand the adverse environmental effects of sediment flushing in Wlingi and Lodoyo reservoirs on the downstream reaches, an investigation was carried out for water quality assessment before, during, and after the sediment flushing operation on March 17–19, 2009, in 12 sampling sites, i.e., Wlingi Dam, Lodoyo Barrage, and Kademangan Bridge in Blitar; Ngujang Bridge in Tulungagung; Mrican Barrage in Kediri; Mekikis Bridge in Kertosono; Ploso Bridge in Jombang; Porong Bridge in Sidoarjo; Lengkong Barrage in Mojokerto; and Jebreng Bridge, Karang Pilang Water Treatment Plant (WTP), and Ngagel WTP in Surabaya. The time of sampling was based on the calculation of travel time of water flow in the Brantas River in order to have similar water samples in all locations. The location of sampling sites is shown in Fig. 1.

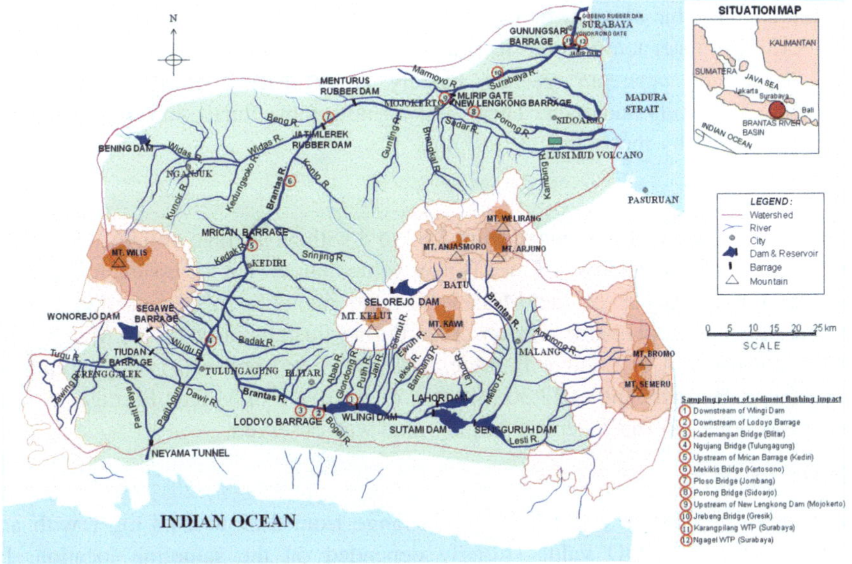

Fig. 1 Study area map and location of sampling sites

The water sampling was carried out following the Indonesian National Standard: SNI 06-2412-1991, using appropriate sampling apparatus to meet the stipulated requirements. The water quality test in laboratory was conducted following the Indonesian National Standard: SNI 06-2413-1991, using appropriate test equipments to meet the stipulated requirements. Interview survey with local inhabitants and market survey on freshwater fish were also carried out for further information.

3 Results and Discussion

The results of investigation are summarized as follows:

1. Noticeable environmental impact

During and after the sediment flushing operation, the following impacts were identified by the field observation:

- High turbidity and change of water color during the sediment flushing. This phenomenon was remarkable until Ploso Bridge (approx. 113 km from Lodoyo Barrage), but not observed at Porong Bridge (approx. 150 km from Lodoyo Barrage).
- Weakened or died fish was remarkable until Mrican Barrage (approx. 75 km from the Lodoyo Barrage).

2. Water quality of river and reservoirs

Temperature:

Measurement value of water temperature ranged from 26 to 33 °C with an average of 28.13 °C. Water temperature recorded relatively lower value (mostly less than 29.0 °C) at the upstream reaches, namely upper reaches of Mrican Barrage, including reservoirs and river. There was no noticeable fluctuation of water temperature caused by the sediment flushing.

pH:

Measurement value of pH ranged from 4.62 to 8 with an average of 6.96. The value of pH recorded was relatively lower (lower value than 7.1) at the upstream reaches, namely upper reaches of the Mrican Barrage, including reservoirs and rivers. A sharp decline in pH value was observed during flushing at 08:00 (of March 18), at the downstream of Lodoyo Barrage, at 18:00 (ditto) between Lodoyo Barrage and Mrican Barrage, and at 23:30 (ditto) at downstream of Mrican Barrage. As for other locations, no sharp decline in pH value was observed.

BOD:

Measurement value of BOD had a wide range from 2.27 to 28.4 mg/1 with an average of 6.01. BOD value strongly depended on the sampling location. It recorded higher value at the following locations: downstream of Lodoyo Barrage at

Kademangan Bridge, between Lodoyo Barrage and Mrican Barrage at Ngujang Bridge, and downstream of Mrican Barrage. There recognized a sharp increase in BOD value was observed during flushing at 15:00 (of March 18) at downstream of Lodoyo Barrage at Kademangan Bridge, at 18:00 (ditto) at Lodoyo Barrage and Mrican Barrage, and at 23:30 (ditto) at downstream of Mrican Barrage.

COD:

Measurement value of COD had the same tendency as that of BOD: a wide range from 4.71 to 382.82 mg/l, depending on the locations. It recorded higher value at the following locations: downstream of Lodoyo Barrage at Kademangan Bridge, between Lodoyo Barrage and Mrican Barrage at Ngujang Bridge, and downstream of Mrican Barrage. There recognized a sharp increase in COD value at the almost same timing as those of BOD during flushing.

TSS:

Measurement value of TSS had a wide range from 24.1 to 9990.8 mg/l. There recognized a noticeable feature: sampling points on the lower reaches, namely downstream of New Lengkong Barrage, Jrebeng Bridge, Karangpilang WTP, and Ngagel WTP, recorded less fluctuated values or stable value of TSS less than 100 mg/l. On the other hand, the sampling points on the upper reaches, including sampling points at Kademangan Bridge, Ngujang Bridge, Mrican Barrage, Mekikis Bridge, and Ploso Bridge, recorded extremely high value during sediment flushing but relatively low value before and after the flushing. There also recognized a sharp increase in TSS value at almost the same timing as those of BOD, COD, and pH, at sampling locations from Kademangan Bridge, Ngujang Bridge, Mrican Barrage, and Mekikis Bridge.

Turbidity:

Turbidity is one of the representative parameters which indicates the appearance of water. It is strongly related to TSS; hence, the value of turbidity had the same tendency as that of TSS. At sampling locations at Kademangan Bridge, Ngujang Bridge, Mrican Barrage, Mekikis Bridge, and Ploso Bridge, sharp increase in turbidity was recorded due to sediment flushing. But no remarkable increase was observed at other locations.

Total phosphate:

Measurement value of T-P ranged from 0.066 to 1.38 mg/1 with an average of 0.24 mg/1. Sampling points on the lower reaches, namely downstream of New Lengkong Barrage, Jrebeng Bridge, Karangpilang WTP, and Ngagel WTP, recorded stable measurement results at around 0.22 mg/l on average. On the other hand, sampling points on upper reaches including sampling points at Kademangan Bridge, Ngujang Bridge, Mrican Barrage, and Mekikis Bridge, recorded much variable figures. In terms of effect of sediment flushing, there was a remarkable

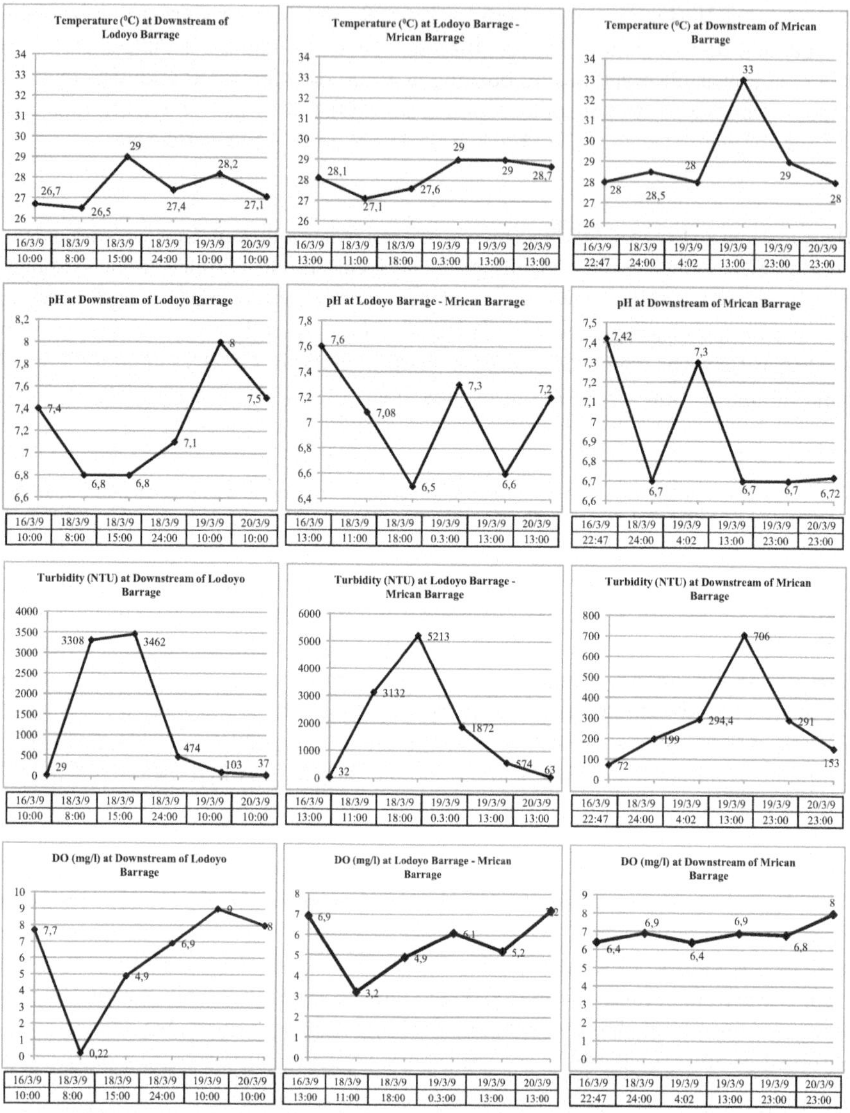

Fig. 2 Observed water temperature, pH, turbidity, and DO at downstream of Lodoyo Barrage, between Lodoyo Barrage and Mrican Barrage, and downstream of Mrican Barrage

increase in T-P at 23:00 (of March 20) at the location of Ploso Bridge. The graphics of some observed parameters, i.e., water temperature, pH, turbidity, and DO at downstream of Lodoyo Barrage, between Lodoyo Barrage and Mrican Barrage, and downstream of Mrican Barrage are shown in Fig. 2.

4 Discussion

1. Impacts on water quality

(a) Parameters fluctuated due to the sediment flushing

Most of the parameters analyzed remarkably and variably responded to the sediment flushing operation. Even if one parameter responded, the way of fluctuation varied depending on the sampling location. In addition, some parameters fluctuated sharply and the others did vaguely. All parameters, except for temperature, showed a fluctuation in response to sediment flushing operation at the locations of Kademangan Bridge, Ngujang Bridge, Mrican Barrage, and Mekikis Bridge. At sampling location of Ploso Bridge, only TSS and Turbidity values were fluctuated, but the fluctuations were not as sharp as the value observed at Kademangan Bridge, Ngujang Bridge, Mrican Barrage, and Mekikis Bridge sampling locations.

The reason or meaning of the fluctuation of these parameters is discussed below. The pH value of water in the Brantas River is slightly basic (alkali) because of the existence of limestone in the basin. During flushing, pH turned to acid slightly because the existing organic matter accumulated on the river and reservoir bottom was stirred up in the water and decomposed, which created CO_2 and resulted in causing acidity. However, after several hours, or some 10 h at most had passed, the pH value returned to normal since the created CO_2 was emitted to the air during the flow to the downstream reach.

BOD and COD:

BOD and COD parameters are used for pollution indicator by organic matter. During the flushing, the existing organic matter accumulated on the river and reservoir bottom was stirred up in the water as mentioned above and caused the increase in BOD and COD.

DO:

In response to the increase in BOD and COD, the O_2 in the water was consumed a lot. This is the reason why DO value decreased sharply. However, the O_2 concentration returned to the normal (before flushing) level in several hours in most of the cases (locations). It is presumed that the O_2 concentration returned to the normal level due to the aeration during flushing at dams and/or flowing downstream section in the river.

Total phosphate (T-P):

The same as that of pH, change of the measurement value due to stirring up of accumulated matter containing high nitrogen and phosphorous in the bed material of the reservoirs.

(b) Spatial extent of impacts on water quality

The remarkable fluctuation of water quality was recorded from the downstream of Wlingi Dam to the downstream of Mrican Barrage. As for other locations, only the fluctuation of TSS and turbidity was detected at Ploso Bridge, but this fluctuation was not sharp. According to the field observation before, during, and after the sediment flushing, the change of color and turbid water flow were observed even at New Lengkong Barrage as mentioned above. Taking these two facts into consideration, it is supposed that the water quality change at New Lengkong Barrage was too little to be detected as a change in concentration including TSS and turbidity. But it is important to recognize that the impact of sediment flushing on water quality reached down to New Lengkong Dam.

(c) Duration of impacts

It is difficult to say how long the impacts last because the water quality is affected a lot by natural variation. One measurement result is just an instantaneous value; therefore, it is necessary to accumulate measurement results for generalization. For most of the parameters in this case, however, the deterioration of water quality ceased within one week after the sediment flushing.

2. Impacts on aquatic organisms

Sediment fushing in Wlingi and Lodoyo reservoirs will have impacts on fisheries in the reservoirs and its downstream reach. Virtually, all the fish in a reservoir will be flushed downstream or killed during emptying, and fish nests will be desiccated. Fish downstream of the dams can be affected by: (1) physical blockage of gills by high sediment concentration, essentially causing suffocation, (2) reduced dissolved-oxygen levels, (3) washout of nests and fry if unseasonably large flows are released during flushing, (4) modification of the benthos due to sediment deposition, (5) modification of the aquatic population structure, (6) disruption of the aquatic food chain and elimination of food sources, especially on the benthos, and (7) incorporation of toxic substances into the food chain if reservoir sediments containing toxins are released during flushing (agricultural chemical, mine tailing, etc.) [2].

3. Evaluation of environmental impacts of sediment flushing

Environmental impacts caused by sediment flushing were evaluated and summarized below:

- Impact on water quality of river and reservoirs is considered to be significant, because it will cause extremely high concentration of SS, turbidity, sediment concentration, BOD, and COD. This deterioration of water quality, together with low DO, will cause secondary impact on aquatic organisms.
- Impact on aquatic organisms is considered to be significant, because many fish were killed or weakened due to respiration problem caused by extremely high TSS and low DO. Many fish were observed dead or weakened in the river, especially at around Mrican Barrage.

4. Mitigation measures

Sediment flushing and prevention of the impacts on aquatic organisms are, in a sense, in a relationship with antinomy. The primary purpose of the sediment flushing is to flush out the deposited sediment as much as possible. The main cause of the impacts on aquatic organisms is suspended solids generated by sediment flushing. Accordingly, there is no thorough solution to prevent from killing the aquatic organisms. In order to mitigate the impacts on aquatic organisms, there are two points to be considered: firstly, to avoid the breeding time of fish, especially important fish in terms of ecologically and economically, and secondly to minimize TSS concentration as much as possible by dilution effect.

Although the fish ecology of the Brantas River was not necessarily well clarified, the breeding time of fish is at the end of dry season. Some catfish in the river spawns when the flooding happens. Accordingly, sediment flushing should be executed avoiding the end of dry season and the beginning of rainy season, namely. As for the above second points, dilution effect can be expected in rainy season. Therefore, the sediment flushing should be executed during rainy season. Taking the two conditions into account, the sediment flushing should be executed in the late rainy season. This suggestion is only based on the environmental viewpoints. It is needless to say that the timing of sediment flushing should be determined comprehensively, considering all other conditions.

5. Recommendation for further study

This river environmental investigation did not include an inventory of dead or weakened fish caused by the sediment flushing. Therefore, the number and the species of dead/weakened fish were not made clear. The relationship between the fish species and tolerable TSS concentration is not clear at this stage. For the next sediment flushing, the fish inventory should be done to understand these points. It is also important to estimate the oxygen and suspended sediment concentration below a dam during flushing by modeling the system which includes parameters for transport rate, reaeration rate, and the reduction in sediment concentration due to deposition.

5 Conclusions

Because of a large amount of sediment inflow due to severe watershed degradation, improper land management, and volcanic disaster, the reservoirs in the upper reach of Brantas River basin have lost their storage capacities rapidly, ranging from 35 to 80% of its original storage capacities. The coordinated sediment flushing by gate operation in Wlingi and Lodoyo reservoirs has been proved to be an effective measure to maintain reservoir effective storage capacity. Impact of sediment flushing in Wlingi and Lodoyo reservoirs on downstream river water quality is considered to be significant because it will cause extremely high concentration of

SS, turbidity, sediment concentration, BOD, and COD. The fact that the impact is confined within several days is the reason for considering as not significant. The most significant impact by the sediment flushing on aquatic organisms was that it caused fish kills. In order to mitigate the impacts on aquatic organisms, it is considered to avoid the breeding time of fish.

Acknowledgements The authors would like to thank Jasa Tirta I Public Corporation, the Brantas and Bengawan Solo River Basins Management Agency, Indonesia, for providing the water quality data.

References

1. Harianto, Hidayat, F., Subagyo, E., Rianto, A., Taufiqurrachman, M., & Nugrahany, A. (2012). Management of Reservoir Sedimentation in the Brantas River Basin Indonesia. *International Seminar on Dam for a Changing World, ICOLD.* Kyoto, Japan.
2. Morris L.G. & Fan J. (1998). *Reservoir Sedimentation Handbook.* McGraw-Hill, New York.
3. Soekistijono, Aris Harnanto & Fahmi Hidayat, (2005). *Coordinated Sediment Flushing in Wlingi—Lodoyo Reservoirs and the Study of its Economics Benefits and Effect to Water Quality and Ecosystem in the Downstream Reaches,* Proceeding of International Symposium on Ecohydrology, Bali, 21–23 November 2005, Indonesian National Committee for IHP-UNESCO, Indonesian Institute of Science—LIPI and Ministry of Public Works GOI.

Entrepreneurship Creativity: An Implication for Architects' Competitiveness and Sustainability

Mudashir Gafar, Rozilah Kasim and David Martin

Abstract The success of an entrepreneur and architect often linked directly or indirectly to the philosophy of creativity. The entrepreneurs greatly add to the success of firms and organisations as architects contributed to the built environment worldwide. Both architect and entrepreneur oftentimes connected in the skills of creativity and design. The research methodological approach employed is a critical literature review regarding entrepreneurship, creativity and design as related to the need for an architectural enterprise development. In addition, the paper reviewed the process nature of entrepreneurship in conjunction with the creativity and problem solving skills inherent in both designers. Shapero and Sokol's entrepreneurial event model employed to provide a supportive explanation on the individual (architect and entrepreneur) behaviour process which may lead to an entrepreneurship creative process. Implications suggested as well as architect–entrepreneur professional practice in the current competitive economy recommended.

Keywords Entrepreneurship · Entrepreneur · Architect · Creativity · Professional practice

1 Introduction

This conceptual paper contributes to the understanding of creativity of an entrepreneur and architect, and its implication for professional development in architecture. The word creativity and entrepreneurial resourcefulness are catalysts for

M. Gafar (✉) · R. Kasim (✉) · D. Martin
Faculty of Technology Management and Business,
Universiti Tun Hussein Onn Malaysia, 86400 Parit Raja,
Batu Pahat Johor, Malaysia
e-mail: gafarmudashir@yahoo.com

R. Kasim
e-mail: rozilah@uthm.edu.my

D. Martin
e-mail: martin@uthm.edu.my

T. Taufik et al. (eds.), *ICoSI 2014*, DOI 10.1007/978-981-287-661-4_38

growth and development of most successful firms, organisations and even developed nations [1]. Creativity and entrepreneurship behaviour is long associated [2], and the literature acknowledged that creative individuals are more likely to engage in entrepreneurial behaviour [3]. Despite, formulating ways to raise creativity and innovative aptitude among the students of higher learning institution (HLIs) are of serious challenge to scholars in the academic setting worldwide [4].

Moreover, architect and entrepreneur often related in the skills of creativity, innovation and problem-solving skills recognised. Hence, several commentators have justified the benefit and necessity of creativity and innovation education in relation to entrepreneurship development, in the HLIs. Nevertheless, operational framework for the process nature of entrepreneurship in conjunction with the creativity and problem-solving skills inherent in both designers are still uncommon. In fact, what is the lesson from the development of creativity and design to impart entrepreneurship and vice versa? Is there any common ground within the vast literature of entrepreneurship that may provide the basis for comprehending the entrepreneurial process in a logical way that is useful for both academicians and practitioners?

However, entrepreneur and architects create significant value, directly and indirectly for the benefit of the society, and value created by both creators are frequently related to the concept of creativity. On this account, this conceptual paper presented critical literature review on entrepreneurship creativity in relation to design process of an architect and recommend as a possible strength for architectural entrepreneurship outlook in the current competitive economy.

2 Research Methodology

A critical literature review regarding entrepreneurship, creativity and design in relation to the need for architectural enterprise development is discussed. Shapero and Sokol's [5] entrepreneurial event model employed to provide a supportive explanation on the individual (architect or entrepreneur) behaviour process which may lead to an entrepreneurship creative process. Subsequently, the perspectives of scholars on the importance of meeting the reality of a competitive economy for the sustainability of the architectural profession are presented. Presented are some recommendations for the training of the new generation of architects to be architect–entrepreneur creative thinkers.

3 Literature Review on Entrepreneurship Process

Scholars are formulating more ways to learn more about the entrepreneurial process to critically establish the character and driving forces of entrepreneurs [6]. Three major points of information provided the keywords related to the entrepreneurial process and perspective. On one side, the vital source is research-based publications and other scholastic documents of academic journals, textbook on entrepreneurship, autobiographies of entrepreneur, conferences proceedings and government publications. On the other side, observation of an entrepreneur in practice is the second source. The analysis of these experiences provided insights on the personalities of entrepreneurs and will substantiate the discovery of commonalities that helps to explain the view.

An entrepreneurial process is a willingness to start a venture, or the unambiguous business idea that was being pursued [7]. However, such business ideas may be outwardly motivated resolution, an aspiration to start a business, or a self-motivated exploration for business opportunities. Shane and Venkataranam [8] emphasised on two stages of the entrepreneurship process: "discovery and exploitation," while [9] further stressed that discovery is itself a process. He explained that possible developmental stages to "discovery" as opportunity identification, idea generation through opportunity development and fine-tuning. In fact, business idea is not always love at first sight, instead is formed through a sequential, changeable process.

Matthews [10] reaffirmed that an entrepreneurship process of idea discovery involves value creation idea, value appropriation idea and entrepreneur commitment to business start-up. He submitted that planning, projections and collection and analysing relevant information are activities that are unavoidable in the business formation process. The implementation of the idea occurs at the exploitation stage of the entrepreneurship process. Notable components or activities in the exploitation stage are actualization effort in term of sourcing for capital and human resources, coordination and management of resources, and establishing market share. Despite that discovery and exploitation stages outlined separately, their operationalization may be simultaneously. Berglund and Wennberg [11] noted that the issue is connecting entrepreneurial process to the idea, environment and character of the creator.

Therefore, Davidsson [9] further explained factors that influence entrepreneurship thought or creativity which are able to appraise business ideas development, planning and possible societal barriers during implementation stage. The basis of the re-evaluation on idea discovery and exploitation process is to ascertain whether the designed process can deliver the desired venture creation result. Entrepreneurs' visionary capacity to assess the process depends practically on their past and related know-how and flexibility to adjust against the environmental barriers.

From the literature perspective, entrepreneurship process is a game of planning [12] and scholars even emphasised the inseparable of planning from the process of business start-up [13–15]. To complement this assertion, some scholars also agreed that creative process of business creation has a strong connection with individual's personality traits [16, 17]. By implication, personal-execution possibility pointer best determine the success of any venture creation and this called critical thinking capacity of an individual. Moreover, Zhao [16] acknowledged that "in the popular imagination, entrepreneurs are regarded as heroes who pursue their creative vision" even in the face of environmental constraints. As a matter of fact, "entrepreneurs' creativity are used to solve day to day problems" of the entrepreneurial process.

From research-based perspectives and in general says, both indicated that every individual is gifted and has the potential to become an entrepreneur as postulated in Shapero and Sokol's entrepreneurial event model [5]. In fact, research proved that individuals confronted with an opportunity that matches their proactive personality skills and background, and close association with people of same interest have strong capacity to trigger them to pursue a career as entrepreneurs [9, 18].

However, Shapero and Sokol's entrepreneurial event model provides a supportive explanation on the individual behaviour process which may lead to an entrepreneurship creative process whether as a result of negative (push) or positive (pull) factors towards new business creation [19]. For the purpose of this paper, the assumption on the "negative displacement event" such as job loss that could trigger desirability to be self-employed as a last employment option is not the focus of this paper (see Fig. 1).

Instead, the interest is on the positive pull, perception of desirability and feasibility to become an entrepreneur as a result of similarity of purpose in the problem-solving skills and high level of creative thinking capacity inherent in both architect and entrepreneur. In this context, the philosophy focuses on the graduating students of architecture, who need to undertake the potential of entrepreneurship education programmes because they oftentimes have no clear idea of what they want to do after graduation. The challenge that most of them faced is high competitive job market. The perception of desirability is that the level of value ultimately impacted on individual's understanding of what is attractive or desirable and what is not. In fact, Shapero and Sokol [5] identified friends, background, tutor, disposition of society, particularly, perceived internal loci control inherent in the personality traits of an individual as strong predictors of behaviour. In addition, perceived desirability is directly linked to "subjective norms" as presented in the theory of planned behaviour and empirical studies showed that experiences strongly impact what is desirable or vice visa [20]. In fact, it determines attraction to creative enterprise outlook.

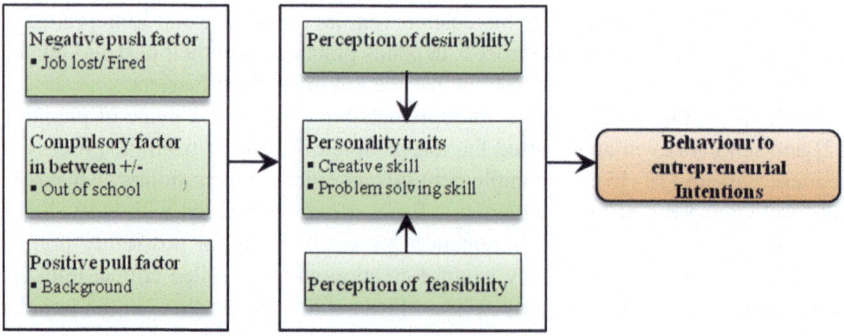

Fig. 1 Entrepreneurial Event Model [5]

Conceptually, the process nature of entrepreneurship creativity in terms of perceived feasibility and desirability could generate mutual understanding of the entrepreneurial process in a systematic and comprehensive way that is useful for both the architects and entrepreneurs (scholars and practitioners). As suggested by Shapero and Sokol [5] that internal locus of control is a predictor of propensity to act, however, commonality of creativity and problem-solving capacity are also strengths to act for the two creators; even though, propensity to act is an issue.

In sum, the processes of opportunity recognition to idea formulation and exploitation of the identified opportunity are all the required knowledge for the creativity process of entrepreneurship. Though, building such creative process is the problem of contemporary entrepreneurship education's research.

3.1 The Philosophy of Creativity Process

Notable past research on creativity process identified creative thinking process and creative problem-solving as the two key components. Wallis (1949) cited in [10] classified creative problem-solving as preparation, incubation, clarification and confirmation. In the same document, Matthews [10] critique focus on the failure of Wallis' classification, noted limitations in the area of psychological processes such as understanding of problems, ability to generate different idea alternatives. Other shortcomings are limited capacity to change one's mental set; ability to reorganise and handle intricacy and; ability to evaluate the ideas generated with provision of possible alternatives. Despite the disagreement, creativity process is illuminated as divergent thinking in academic circles.

However, creativity training in the higher learning institutions' programs usually is of critical research interest lately as a result of the need for entrepreneurship and innovation in the competitive economy worldwide. In this regard, some training in techniques that promote divergent thinking is of paramount focus in the educational landscape.

According to Garavan and O'Cinneide [21] learning process of a beginner in an entrepreneurship program is similar to channel of problem discovery, idea development, solution development and action planning (idea to business start-up) for an entrepreneur in practice. They explained that creative problem-solving training for young entrepreneur need to combine various training techniques (lecture, case studies, team play, entrepreneurs' site visitation and learning by doing) in order to advance to reality of venture creation. In order to depict the creative problem-solving learning process preference of an entrepreneur, Table 1, presented Hynes' [22] entrepreneurship process model which focused on input, content, teaching approach and desire output.

The entrepreneurship creativity training HLIs depend on the four components of entrepreneurship learning process, and to create students' entrepreneurial mindset is far beyond idea generation only [23]. The process of idea selection, evaluation of idea and implementation process are crucial to personal and organisational success

Table 1 Entrepreneurship learning process [22]

Inputs students	Content focus	Teaching focus	Outputs
1. Prior knowledge base	Entrepreneurship defined	Didactic (read/lecture)	Personal confidence, effective communication
2. Motivation	Role of entrepreneur	Skill building through (Case, study group discussion, presentation, problem solving, simulations, teamwork)	Dynamic knowledge (enterprise, Initiative, self-employment, business, consciousness, management and marketable skills, analytical problem solving skills, decision making, risk taking)
3. Personality	Innovation		
4. Needs/interests	Product development		
5. Independent	Idea generation		
6 Attitudes	Market research		
7. Parental influence	Finance and management	Discovery through (Brainstorming, personal goal setting, career planning, consultancy)	Career development (broader knowledge on career options)
8. Self esteem	Teamwork-networking		
9. Values	Business marketing		
10. Work experience	People management		

and in studies of entrepreneurship and innovation in real-life experience. Other distinctions include idea blending, idea summation, selection of the relevant idea and transformation of the idea to reality. At the same time, motivation, creative skills in enterprise and domain-relevant skills are the other components that cannot be ignored in the entrepreneurship creativity development in HLIs. In fact, to impact entrepreneurship creative skills on an individual is much easier to build on the skills and abilities of that individual, even though, inherent motivation in exploiting such competence to a situation is another critical factor.

3.2 Design as a Creative Process

Design in the context of architecture or entrepreneurship only means a creative process tailored toward particular aspiration to create solutions and the essence is to add values. In fact, nobody has a monopoly of knowledge about the field of design. To compliment this, Lawson [24] reaffirmed "we all can and do design" and even "we can learn to design better." Obviously, creative process of design could be endless since it contains subjective value opinion.

However, review of design and design thinking process did not focus on architectural design processes, instead as an approach to the new world of possibilities in entrepreneurship outlook for scholars and practitioners. Hence, the focus is to highlight the inherent strength in the graduates and practicing architects' design creative capacity for entrepreneurship success. Majority of architects are yet to exploit the potential of their design skills, time and resources management to discover new possibilities [25] despite that they have some similarities with entrepreneurs in creative design process [26].

Matthews [10] and Sarasvathy [26] claimed that business individuals need to become designers and we argued that architects' design thinking need to be transformed and incorporated an entrepreneurship creative component to their problem-solving ability. Hence, designer (architect) needs to be an entrepreneur as much as an entrepreneur needs to be a designer. In fact, this is a confluence of creativity. Subsequently, diversity of the architect's design skills and problem-solving capacity's skills may provide entrepreneurial opportunities for the reality of competitive economy, if explored. In a matter of fact, design thinking application in redesign of products, processes, structures and forms is particularly useful in corporate entrepreneurial situations. The conclusion is that the design process is a way of embryonic idea (dream) to action (reality). It is a viewpoint which is open to the issue of developing new venture creation by experimenting with a different approach in the professional practicing.

4 Convergence of Creativity, Architecture and Entrepreneurship

It is a long practice of relating entrepreneurship and new creation of business behaviour as an act of creativity [27–29]. Though several scholars described entrepreneurship as a multi-directional word and homogeneity in meaning, as a process through which opportunities to create new business ideas and the reality of reaching positive proceeds in the market are discovered, evaluated and explored.

Novelty in business is the concerned of entrepreneurship as creativity is also concerned with the conception of novelty and value. Entrepreneurship and creativity's research has followed a similar trend in terms of the focus on the methods used [30]. Some of the general attributes of creativity and entrepreneurship are found in the creative industries or agency of the individual (architecture, art, design, designer fashion, music and performing arts) or group that produces novelty and value. Henry [31] recognised creative people with nonconformist ideas lower entry barrier to new firm creation and most importantly play a role in "a firm's innovation process and also in corporate venturing processes".

However, architecture falls within the creative industries or enterprises. The personality in this profession is identified with its background in individual's creativity, skills and competence for job and wealth creation by the way of formulation or exploitation of intellectual property [32]. The creation of new idea practically involves opportunities' finding in the existing fields that could sometimes lead to the creation of new venture or business opportunities. More so, Hearn [33] and Hamidi [4] stated that early stages of generating new ideas are usually characterised by divergent thinking in creativity, architecture and now entrepreneurship.

By implication, creativity is dynamic processes that entail changing patterns of activities in both architecture and entrepreneurship as fields of study.

Concurrently, creativity stressed as a vital to professional success. On the assertion of the supremacy of creativity and entrepreneurship in the professional sustenance, we argued that architectural creativity may have a connection to the entrepreneurial creativity. Subsequently, there is a high possibility of the graduating architect's creativity to boost their entrepreneurial business creation intention. At this postmodernism-economic driving society, it is intellectually necessary for examining entrepreneurship movement of the global scholastic interest on the training of the students of architecture discipline. In fact, creativity is central to professional success in the creative economy [33, 34]. It is equally necessary to explore it is possible contribution and correlation to architectural entrepreneurship development in the higher institutions of learning.

The existing knowledge does not limit entrepreneurship, creativity and architecture; all could benefit from the critical analysis of the confluence of knowledge, challenges and extend previous understanding in developing new ideas, processes and application. Therefore, every individual in the three fields need skills of persuasion or motivation to influence others within and outside the firm/organisation to support or invest in new ideas development.

5 Implication for Architectural Profession

The training need of the new generation of architect is better built on Carey and Matlay [35] assertions "that creative disciplines education is characterised by experiential, project-based learning environments and regular peer reviews". In addition, "it is frequently led, by educators and practitioners in their respective fields" [36]. Objectively, this is the strength for the architecture educators to build on their creative assessment approach and blends entrepreneurship initiatives into their curriculum structure.

Notable scholars described an architect-art-entrepreneur as someone who holds tactics of knowledge in competence, commitment and creativity-based mindsets [37–39]. The assertion is philosophical and if explored, it will not only be self-rewarding, but add value to architects' competitiveness and sustainability in the society and economic development of nations.

Hence, scholars have asserted the benefits of entrepreneurial activities as self-challenge and self-discovery, thus, serving as a pathfinder for self-sufficiency and self-satisfaction [26, 40–42]. In this regard, freedom, financial benefit and economic uplifting of society are the valuable rewards for entrepreneurial success [43]. In line with Robinson [44] and Brown [45] ideology, we suggested enterprise education for the future graduates as a pathway for the academic and professional success in the modern age.

6 Conclusions

Based on the authors' understanding and literature, reviews highlight a number of pertinent suggestions on the need to blend creativity teaching in architecture and entrepreneurship as to promote entrepreneurship orientation

- At the national level, the government's education policymakers should review the university's curriculum to embrace entrepreneurship orientation at all level of education.
- At the university level, the university's curriculum design and program content developers should review and emphasis enterprise course content into both studio and non-studio courses.
- Consideration should be on how students explore multiple enterprise ideas blending idea summation, and selection of the relevant idea and transformation of an idea to reality within their professional training context.
- Students' ability to present and justify ideas to peers and lecturers in the environment, which potentially represents a good rehearsal for their future professional careers.
- Educators should develop strong industry–university linkages to advance professional and real-life experiences of students.

The guarantee of success in this new age is for the new generation of architects to be architect–entrepreneur creative thinkers. In sum, as architects who have the capacity to create cutting-edge enterprise by transfusing their domains of the profession (architecture) and inherent creative skills to become the title holder of this competitive era.

Acknowledgements The authors would like to recognise the research unit of the Universiti Tun Hussein Onn Malaysia for supporting this research under the Postgraduates Incentive Grant.

References

1. Hitt, M. A., Ireland, R. D., Sirmon, D. G. & Trahms, C. A. (2011). Strategic entrepreneurship: creating value for individuals, organizations, and society. *The Academy of Management Perspectives*, *25*(2), 57–75.
2. Amabile, T. M. (1996). Creativity in Context, Westview, Boulder, Colorado.
3. Zhou, J. (2008). New Look at Creativity in the Entrepreneurial Process. Strategic Entrepreneurship Journal, 2, 1–5.
4. Hamidi, D. Y., Wennberg, K. & Berglund, H. (2008). Creativity in entrepreneurship education, *Journal of Small Business and Enterprise Development*, *15*(2), 304–320.
5. Shapero, A. & Sokol, L. (1982), *The social dimensions of entrepreneurship*, in Kent, C.A., Sexton, D.L. and Vesper, K.H. (Eds), Encyclopaedia of Entrepreneurship, Prentice-Hall, Englewood Cliffs, NJ, pp. 72–90.
6. Gartner, W. B. (2001). Is there an elephant in entrepreneurship? Blind assumptions in theory development. Entrepreneurship Theory and Practice, 25(4), 27–39.

7. Timmons, J. A. & Spinelli, S. (2006). New Venture Creation: Entrepreneurship for the 21st Century. McGraw-Hill.
8. Shane, S. & Venkataranam, S. (2000). The promise of entrepreneurship as a field of research. Academy of Management Review, 21, 217–226.
9. Davidsson, P. (2006). *The Types and Contextual Fit of Entrepreneurial Processes*, In Burke, A.E. (Ed.) Modern Perspectives on Entrepreneurship, Senate Hall Academic Publishing.
10. Matthews, J. H. (2009). *What are the lessons for entrepreneurship from creativity and design?* In Solomon, G. (Ed.) Proceedings of the 2009 Academy of Management Annual Meeting: Green Management Matters, Academy of Management, Chicago, Illinois.
11. Berglund, H. & Wennberg, K. (2006). Creativity among entrepreneurship students: comparing engineering and business education. International Journal of Continuing Engineering Education, Vol. 16 No. 5, pp. 366–79.
12. Achtenhagen, L. & Johannisson, B. (2013). Games in entrepreneurship education to support the crafting of an entrepreneurial mindset. *New Pedagogical Approaches in Game Enhanced Learning: Curriculum Integration, 20.*
13. Nga, J. K. H., & Shamuganathan, G. (2010). The influence of personality traits and demographic factors on social entrepreneurship start up intentions. *Journal of Business Ethics, 95*(2), 259–282.
14. Majid, I. A., Kamaludin, M. H. & Saad, M. M. (2012). Sustainability-Driven Entrepreneurship: The Mediating Effect of Opportunity-Based Management Structure on the Relationship between Entrepreneurial Orientation and Environmental Sustainability Management of SMEs: A Conceptual Framework. *European Journal of Business and Management, 4*(13), 148–155.
15. Au, K., & Kwan, H. K. (2009). Start-Up Capital and Chinese Entrepreneurs: The Role of Family. *Entrepreneurship Theory and Practice, 33*(4), 889–908.
16. [Zhao, H., Seibert, S. E. & Lumpkin, G. T. (2010). The relationship of personality to entrepreneurial intentions and performance: A meta-analytic review. *Journal of Management, 36*(2), 381–404.
17. Lee, S. Y., Florida, R. & Acs, Z. (2004). Creativity and entrepreneurship: a regional analysis of new firm formation. *Regional studies, 38*(8), 879–891.
18. Summers, D. F. (2013). *Forming Entrepreneurial Intentions: An Empirical Investigation of Personal and Situational Factors*. Routledge.
19. Kolvereid, L. (1996a). Organizational Employment versus Self-Employment: Reasons for Career Choice Intentions. *Entrepreneurship Theory & Practice*, 20(3), pp. 23–31.
20. Krueger, N.F., Reilly, M.D. & Carsrud, A.L. (2000), Competing models of entrepreneurial intentions, *Journal of Business Venturing*, Vol. 15, pp. 411–32.
21. Garavan, T. N. & O'Cinneide, B. (1994). Entrepreneurship Education and Training Programmes: A Review and Evaluation–Part 1. *Journal of European industrial training, 18*(8), 3–12.
22. Hynes, B. (1996). Entrepreneurship education and training: introducing entrepreneurship into non-business disciplines. *Journal of European Industrial Training*, Vol. 20 No. 8, pp. 10–17.
23. Hynes, B. and Richardson, I. (2007). Creating an entrepreneurial mindset: getting the process right for information and communication technology students, in Lowry, G. (Ed.), Information Systems and Technology Education: From the University to the Workplace, IGI Global, Hershey, PA, Chapter VI.
24. Lawson, B. (1997). *How Designers Think: The Design Process Demystified*, (3rd Ed). Architectural Press, Reed Elsevier, Oxford.
25. Quirk, V., Daily, A. & Jett, M. (2012). After The Meltdown: Where Does Architecture go From Here? Retrieved from: http://www.archdaily.com/226248/after-the-meltdown-where-does-architecture-go-from-here/.
26. Sarasvathy, S. (2004). *Making it Happen: Beyond theories of the firm to theories of design.* Entrepreneurship Theory and Practice, Winter 519–531.

27. Gafar, M., Rozilah, K. & David, M. (2014). *Beyond Architecture: Entrepreneurship diversity crosswise multidisciplinary nature of the profession*. Proc. of the Intl. Conf. on Advances in Social Science, Economics & Human Behaviour, pp. 33–39.
28. Ward, T.B. (2004). Cognition, Creativity and Entrepreneurship. *Journal of Business Venturing*, Vol. 19, pp. 173–88.
29. Moroz, P. W. & Hindle, K. (2012). Entrepreneurship as a Process: Toward Harmonizing Multiple Perspectives. *Entrepreneurship Theory and Practice*, *36*(4), 781–818.
30. Florida, R. (2012). The connection between creativity and entrepreneurship. The Atlantic Cities Place Matters, Retrieved from: www.theatlanticcities.com/jobs-and-economy/2012/08/creativity.
31. Henry, C. (2008). *Entrepreneurship in the Creative Industries: An International Perspective*, Edward Elgar Publishing, Cheltenham. Howkins.
32. DCMS (2006b). *Developing Entrepreneurship for the Creative Industries: The Role of Higher and Further Education*. Department of Culture, Media and Sport, Creative Industries Division, London.
33. Hearn, G. & Bridgstock, R. (2010). Education for the Creative Economy: Innovation. *Education in the creative economy: Knowledge and learning in the age of innovation*.
34. Belz, F. M. & Binder, J. (2013). Sustainability Entrepreneurship: A Process Model. *Available at*: https://papers.ssrn.com.
35. Carey, C. & Matlay, H. (2007). Entrepreneurs as educators: The case of the creative industries in the UK. *Industry & Higher Education*, *21*(6), 435–443.
36. Carey, C. & Naudin, A. (2006b). Enterprise curriculum for creative industries students: an exploration of current attitudes and issues. *Education + Training*, Vol. 48 No. 7, pp. 518–31.
37. Carey, C. & Matlay, H. (2010). Creative disciplines education: a model for assessing ideas in entrepreneurship education? *Education + Training*, *52*(8/9), 694–709.
38. Parkman, I. D., Holloway, S. S., & Sebastiao, H. (2012). Creative Industries: Aligning Entrepreneurial Orientation and Innovation Capacity. *Journal of Research in Marketing and Entrepreneurship*, *14*(1), 95–114.
39. Aggestam, M. (2008). *Art-entrepreneurship in the Scandinavian music industry*, in Henry, C. (Ed.), Entrepreneurship in the Creative Industries: An International Perspective, Edward Elgar Publishing, Cheltenham.
40. Gibb, A. (2005). Towards the Entrepreneurial University, Entrepreneurship Education as a lever for change, National Council for Graduate Entrepreneurship. *Policy paper*, *3*, 1–46.
41. Gafar, M., Rozilah, K., & David, M. (2012). *Toward a more Entrepreneurship Architectural Education in Malaysia*. Proceedings International Conference of Technology Management, Business and Entrepreneurship, pp. 755–770.
42. Zhao, H. & Seibert, S. E. (2006). The big five personality dimensions and entrepreneurial status: A Meta-Analytical Review. *Journal of Applied Psychology*, *91*(2), 259.
43. Jones, C., Matlay, H. & Maritz, A. (2012). Enterprise education : for all, or just some? *Education + Training*, Vol. 54 No. 8, pp. 813–824.
44. Robinson, D., Jamieson, C., Worthington, J. & Colander, C. (2010). The Future for architects? Building Future, RIBA. Retrieved from: www.buildingfuture.org. UK pp. 1–45.
45. Brown, R. (2007). *Promoting entrepreneurship in arts education. Entrepreneurship in the Creative industries: An international perspective*. Google book, Edward Elgar Publishing.

Green Energy Approach for Batik Industry in Order to Increase Productivity and Maintain a Healthy Environment

Ramadoni Syahputra and Indah Soesanti

Abstract Batik is a piece of cloth applied by means of a dye-resist technique using "batik-wax" as the resisting medium. Indonesian batik was designated by UNESCO as a Masterpiece of Oral and Intangible Heritage of Humanity. As part of the acknowledgment, UNESCO insisted that Indonesia has preserved their heritage. This fact should be grateful and responded with efforts to develop and preserve Indonesian batik. One effort to develop and preserve Indonesian batik is to adopt green energy technologies in the production process. As generally in Indonesia, batik industry is currently still a class of small and medium enterprises (SMEs). One of the quite popular in Indonesia is Bantul batik. Of the many batik SMEs in Bantul, there are two SMEs that have the potential to develop, namely Ida Lestari Batik and Arjo Munir Batik. The issue of the SMEs is that the production process is still using kerosene stoves, while the price of kerosene is very expensive and difficult to obtain. Therefore, in this paper the application of environmentally friendly energy sources for the production process of batik is described. The energy source is the installation of solar home system. The system is used to distribute electrical power to the batik electric stove. The use of solar home system and batik electric stoves is more practical and economical than kerosene stove. Another advantage is to reduce dependence on fossil fuels and help preserve the environment with the use of green technology.

Keywords Green energy · Solar home systems · Smes batik · Batik electric stove · Community development

---

R. Syahputra (✉)
Department of Electrical Engineering, Faculty of Engineering,
Universitas Muhammadiyah Yogyakarta, Yogyakarta 55183, Indonesia
e-mail: ramadoni@umy.ac.id

I. Soesanti
Department of Electrical Engineering and Information Technology,
Faculty of Engineering, Universitas Gadjah Mada, Yogyakarta 55281, Indonesia
e-mail: indah@mti.ugm.ac.id

T. Taufik et al. (eds.), *ICoSI 2014*, DOI 10.1007/978-981-287-661-4_39

1 Introduction

Use of renewable energy and energy efficiency, which is a part of the green energy technologies, is very important to reduce global warming and protect the ecosystem. Green technology will be able to reduce carbon emissions. Green energy technologies by utilizing alternative energy sources such as wind, solar, hydroelectric, and others reduce CO_2 emissions by improving energy efficiency. Climate change is now occurring due to uncontrolled carbon emissions that affect the world economy. Because of temperature increases, agricultural products will fall, damage from floods and storms will increase, tropical diseases will become more common and access to water will become more of a problem for more and more people. Costs incurred for our environment is greater and the loss could not be changed. Earth's flora and fauna will suffer both directly from the higher temperatures and indirectly through the destruction of their habitat. Even a small increase in temperature will lead to coral bleaching and threatening some amphibians. Temperature rises of 3 or 4 °C and more will lead to major extinctions around the globe [1].

A popular one in green energy technologies is solar energy. Solar energy has shifted impressive technology. Early solar technology consists of small-scale photovoltaic cells. The latest technology of large-scale PV systems has been fed into the power grid. Technology costs have fallen substantially over the last 30 years [2]. The rapid expansion of the solar energy market is the result of government policies that support instruments, increased price volatility and environmental externalities of fossil fuels, especially greenhouse gas emissions.

Basically the solar energy resource potential that far exceeds the entire global energy demand [3]. Although the technical potential is very large and the recent growth of the market, the contribution of solar energy for the global energy supply mix is still negligible. Numerous studies, including de Vries et al. [4], have been discussing various issues related to solar energy. This study provides an overview of the synthesis of the existing literature and presents an economic analysis to look at the competitiveness of solar energy compared with fossil energy. Although the presence of the clean development mechanism of the Kyoto Protocol has been contributed to assist the implementation of several projects of solar energy, its role in promoting solar energy is very small compared to that of other renewable energy technologies as cost competitiveness. An existing study has shown that the share of solar energy in the world energy supply mix could exceed 10% in 2050. This share would still be a small part of the total world energy supply. This share is also a small fraction of renewable supply.

2 Batik Industries and Their Problems

Batik is a fabric sheet made by drawing designs on fabric using dots and lines of hot wax (malam). This wax is used to reject dye by soaking a cloth in one color. Once the color is applied, the wax is removed using boiling water and repeat if some of the desired color. Batik has been known to the world community that comes from Java. Custom pattern often has a symbolic meaning that is used in certain ceremonies, while the coastal pattern draws inspiration from various cultures. Batik has been used as everyday clothing since ancient times, and is still used and very loved by many Indonesian people today in events ranging from formal to casual. In October 2009, UNESCO has set Indonesian batik as a Masterpiece of the Oral and Intangible Heritage of Man. As part of the recognition, UNESCO asserted that Indonesia preserve their cultural heritage [5].

In this study, to test the application of solar home systems are burdened with an electric stove in two SMEs batik industries, namely Ida Lestari Batik SMEs and Arjo Munir Batik SMEs. Both SMEs located in the village of Wijirejo, Pandak subdistrict, Bantul districts, Daerah Istimewa Yogyakarta province, Indonesia. The existing conditions of the two SMEs are described as follows:

1. Ida Lestari batik SMEs capable of producing batik ranging from 200 to 400 pieces of batik cloth per month. Low productivity in the SMEs is not on the ability of the human resources and the difficulty of obtaining raw materials and wax batik cloth, but rather:
   (a) A number of constraints in producing batik is because they only rely on kerosene stoves and firewood and even use coconut fiber burnt. Since the government of Indonesia has implemented a program for kerosene to gas in 2008, then one of the sectors that are directly affected by the program is batik industry. In Bantul district, kerosene prices trending up from time to time (i.e., current prices in April 2014 reached Rp 11,000 per liter), and often there is a shortage of kerosene.
   (b) Ida Lestari batik SMEs actually must have a single electric stove for batik, but not reliable. In addition to the amount that is only one piece, another issue is the continuity of the electricity. Electric stove used in the SMEs is not equipped with a battery, so it really just relies on commercial power. If commercial power is lost, the stove cannot operate. The flow of electricity in the location of these SMEs often experience blackouts, both in weather conditions of rain and sunny weather, because of the low level of reliability.
2. Arjo Munir Batik SMEs capable of producing batik ranging from 150 to 260 pieces of batik cloth per month. Relatively low productivity is caused by:
   (a) So far, it only relies on kerosene stoves and even using firewood and coconut fibers are burned, the high price of kerosene and kerosene shortages often occur.
   (b) Actually the SMEs still rely kerosene stove in the production of batik and batik yet have an electric stove.
3. Both SMEs uses electrical energy derived from the PLN. Ida Lestari batik SMEs subscription with a power capacity of 450 VA and Batik Arjo Munir subscription with the same power capacity. However, during the frequent power

outages that at times cannot be estimated, in the rain and sunny weather at the time either day or night. This situation will be very difficult if they want to use an electric stove in the process of batik production.

3 Methods

In order to increase production capacity, market expansion, and capacity building activities in Ida Lestari Batik SMEs and Arjo Munir Batik SMEs, then the programs carried out the following activities:

1. Batik production capacity of both SMEs can be improved:
   (a) Provide an electric stove specifically for batik SMEs.
   (b) Provide and install solar cell home systems to ensure continuity of the flow of electrical power to distribute electrical power to the stove and also home lighting batik production. Provision the solar cell home system is also useful to overcome the problems in both SMEs because it has a low power capacity installed and frequent power outages, while each SMEs will use electrical power continuously for 2 pieces of 125 W electric stoves.
2. Improving human resource capacity in the respective SMEs.

The problem-solving step in the batik industry in this research is shown in Fig. 1.

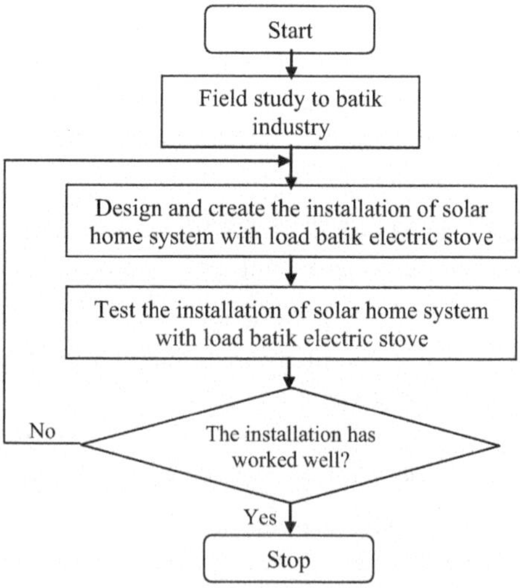

Fig. 1 The problem-solving steps in the batik industry in this research

4 Test Results

The geographical condition of Indonesia which is located in the tropical region is traversed equator particular gift for the people of Indonesia if it is associated with potential sources of electrical energy derived from sunlight. On the bright midday sun radiation is able to reach 1000 W/m^2. If a semiconductor device with area of 1 m^2 has an efficiency of 10% then the solar cell module is capable of delivering 100 W of power [6–8]. Currently commercial module efficiency solar cells ranged from 5 to 15% depending on the constituent materials [9, 10]. On a national scale, the government of Indonesia has a serious program in renewable energy, one of which is the solar energy [11]. The government of Indonesia has targeted that power generator from renewable energy resources will be on-grid for up to 5% by the year 2025.

In this work, the create-design of solar home system with loading of batik electric stove is done. The scheme of solar home system in this work is shown in Fig. 2. Solar cell home systems installation for each SME is to ensure the

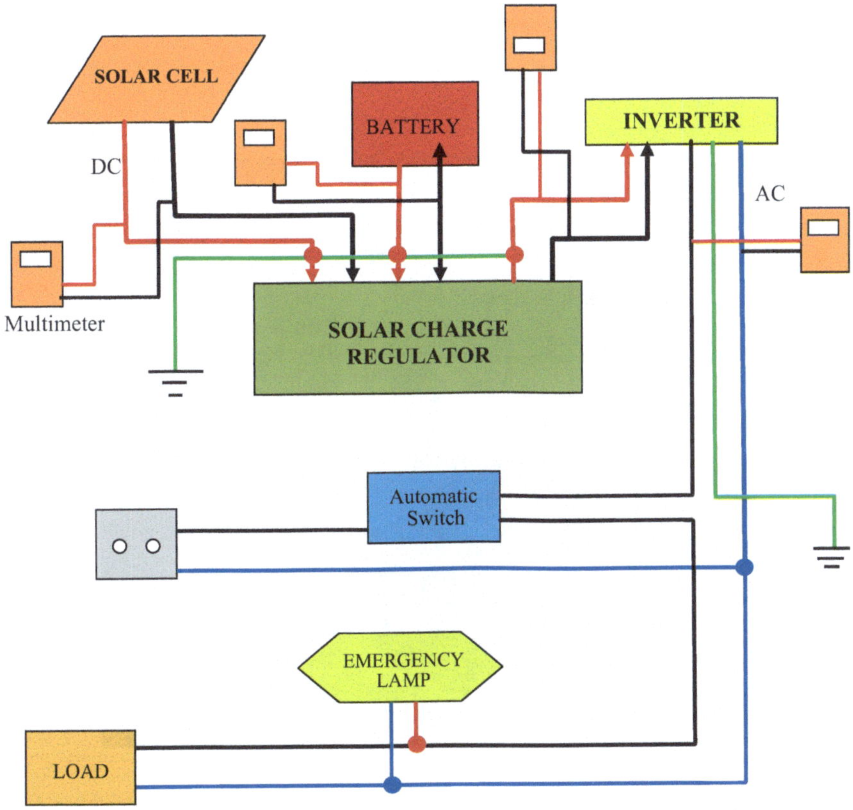

Fig. 2 The scheme of solar home system in our work

availability of electric current for batik electric stove and lighting the batik production houses. The power capacity of solar cell in the installation is 100 Wp. The technical specifications of 100 Wp solar cell panel for batik industry has described in Table 1. The main burden of solar home systems is batik electric stove of 125 W, 220 V, as shown in Fig. 3.

The main function of solar home systems are the electrical power supply to the load on the batik electric stove with a power capacity of 125 W each burner. Production of electric energy solar home systems is highly dependent on sunlight. In one day, the most effective sunlight to generate electrical energy in just over 5 h, as shown in Fig. 4. Solar insulation (Fig. 4) can be determined as follows:

Table 1 Technical specifications of 100 Wp solar cell panel for batik industry

Parameters	Values	Units
Maximum power	100	W
Open circuit voltage	21.6	V
Short circuit current	5.70	A
Maximum system voltage	1000	V
Dimension	835 × 540 × 28	mm
Test conditions	AM1.5 1000 W/m225 °C	–

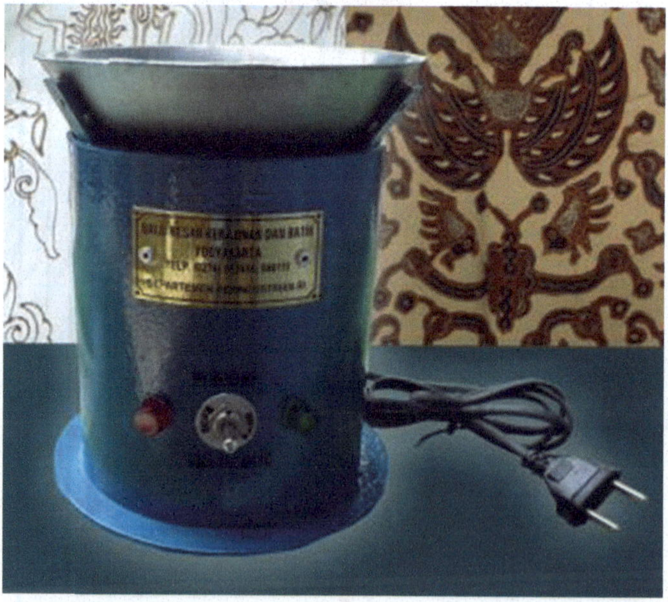

Fig. 3 Batik electric stove of 125 W, 220 V

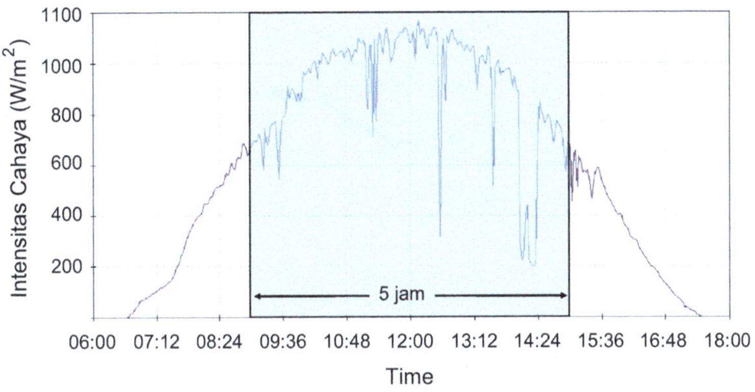

Fig. 4 Daily solar insulation characteristics [10]

Table 2 Technical specifications of solar home system load

Parameters	Values	Units
Maximum power	125	W
Nominal voltage	220	V
Nominal current	0.60	A
Frequency	50	Hz

Load type: Electric stove for batik
Place of Testing: Ida Lestari batik SMEs

Table 3 Test results of solar home system installation

Parameters	Values	Units
Maximum power	100	W
Nominal voltage	220	V
Maximum drying time	5	h
Frequency	50	Hz
Energy produced	100 × 5 = 500	Wh
Long duration usage load	500/125 = 4	h

1. Insulation sunlight can be estimated from that of predicted weather conditions.
2. Weather forecast data in the form of long sun emits rays can be obtained from meteorological and geophysics agency.
3. The length of time the sun radiates its light is converted into the peak sun hours

Testing of solar home systems with a single load electric stove for batik described as shown in Tables 2 and 3.

After the sunlit solar cells for 5 h, then the solar cells can be used with a load power of 125 W for 2 h, which is described in Table 3.

Figure 5a, b show the output voltages and the output currents of solar cell under study, respectively. Testing of the solar cell with loading of 125 W batik electric

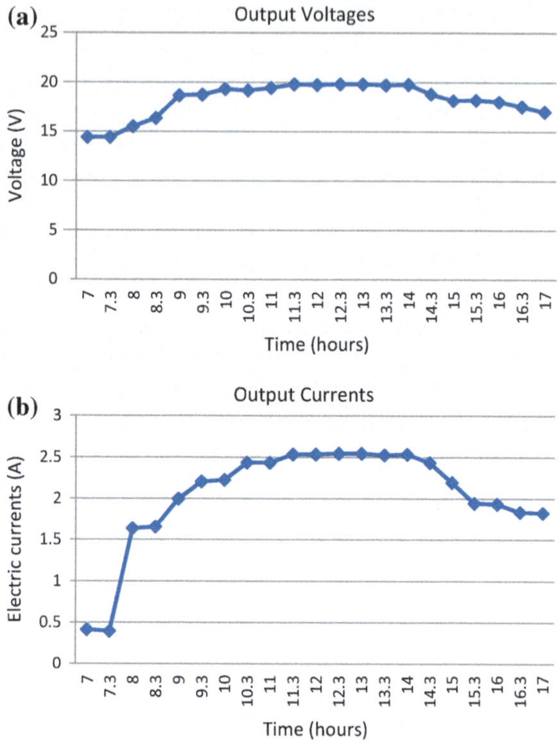

Fig. 5 **a** Solar cell output voltage, **b** solar cell output currents

stove starts from 7.00 AM to 17.00 PM. As shown in Fig. 5a that output voltage of solar cell has varied from 14.45 to 19.88 V. Magnitude of the voltage dependent on sunlight, where the maximum voltage occurs at 12.00 AM and the minimum voltage occurs at 7.00 PM.

Thus, application of solar home systems in Ida Lestari Batik SMEs and Arjo Munir Batik SMEs help continuity of batik production. This is in accordance with the needs of both SMEs who have often experienced a power outage that disrupts batik production process, especially if using an electric stove for batik.

Results of IBM's service activities have felt the impact and benefits for both Ida Lestari Batik SMEs and Arjo Munir Batik SMEs. The impact and benefits have been felt through service activities are:

1. Batik production capacity of both Ida Lestari Batik SMEs and Arjo Munir Batik SMEs has increased with the following details:

   (a) Ida Lestari Batik SMEs, if prior to the work only capable of producing combination stamp and write batik and pure batik with a total capacity of approximately 200 pieces of cloth batik per month, then after our activity is able to increase its production up to 250 pieces of cloth batik per month.

The increase in productivity has been accompanied by an increase in gross profit of an average SME Rp 5 million per month to an average of Rp 6 million per month.

(b) Arjo Munir Batik SMEs, if prior to the work only capable producing combination stamp and write batik and pure batik with a total capacity of approximately 400 pieces of cloth batik per month, then after our activity is able to increase production to 450 pieces of cloth batik per month. The increase in productivity has been accompanied by an increase in gross profit of an average SME Rp 9 million per month to an average of Rp 10.5 million per month.

2. Expansion of each SME market has increased. Both SME market has dared not only to supply the batik cloth batik shops around Jalan Malioboro and Beringharjo mart, Yogyakarta, but also through exhibitions in several cities such as Jakarta and Bandung are often sponsored by the Department of Industry and Local Government of Bantul.
3. Human resources capacity in each SME has increased in quality, mastery of which operating and maintenance of solar home systems installation and electric stoves for batik.

5 Conclusion

This work has proved to be very beneficial for both Ida Lestari Batik SMEs and Arjo Munir Batik SMEs. Solar home systems installation and procurement of electric stoves for batik has managed to increase both production capacity and gross profit of the SMEs. In a period of ten months in 2013, Ida Lestari batik SMEs productivity increased from an average of 200 pieces of batik cloth to an average of 250 pieces per month. Similarly, Arjo Munir Batik SMEs, where productivity increased from an average of 400 pieces of batik cloth to an average of 450 pieces per month. The increase in productivity has been accompanied by an increase in gross profit of the SMEs.

References

1. R. Syahputra, I. Soesanti, M. Ashari. (2016). Performance Enhancement of Distribution Network with DG Integration Using Modified PSO Algorithm. *Journal of Electrical Systems (JES)*, 12(1), pp. 1−19.
2. Uddin, S. K. and Taplin, R. (2008). Toward Sustainable Energy Development in Bangladesh. *The Journal of Energy and Development*, 17 (3), 292–315.
3. R. Syahputra, I. Soesanti. (2016). Application of Green Energy for Batik Production Process. *Journal of Theoretical and Applied Information Technology (JATIT)*, 91(2), pp. 249−256.

4. I. Soesanti, R. Syahputra. (2016). Batik Production Process Optimization Using Particle Swarm Optimization Method. *Journal of Theoretical and Applied Information Technology (JATIT)*, 86(2), pp. 272–278.
5. R. Syahputra, I. Soesanti. (2016). Design of Automatic Electric Batik Stove for Batik Industry. *Journal of Theoretical and Applied Information Technology (JATIT)*, 87(1), pp. 167–175.
6. A. Keyhani, M.N. Marwali, dan M. Dai, (2010). Integration of Green and Renewable Energy on Electric Power Systems, *John Wiley & Sons*, New Jersey.
7. R. Syahputra, I. Robandi, M. Ashari. (2015). Performance Improvement of Radial Distribution Network with Distributed Generation Integration Using Extended Particle Swarm Optimization Algorithm. International Review of Electrical Engineering (IREE), 10 (2). pp. 293–304.
8. R. Syahputra, I. Robandi, M. Ashari. (2014). Optimization of Distribution Network Configuration with Integration of Distributed Energy Resources Using Extended Fuzzy Multi-objective Method. International Review of Electrical Engineering (IREE), 9(3), pp. 629–639.
9. J. Twidell and T. Weir. (2006). Renewable Energy Resources, Second Edition, Taylor & Francis, New York.
10. R. Syahputra, (2012). Distributed Generation, State of the Arts in Electrical Energy Services, LP3 M UMY, Yogyakarta.
11. R. Syahputra, I. Robandi, M. Ashari. (2015). Reconfiguration of Distribution Network with DER Integration Using PSO Algorithm. TELKOMNIKA. 13(3). pp. 759–766.

Optimization Potential Value Added of Sunda Strait Bridge Construction in Accessibility Function Model

W. Wangsadinata, D. Ma'soem and D.M. Lestari

Abstract In the implementation of Sunda Strait Bridge (SSB) development as a public transportation facility, there is a consistent purpose with the acceleration vision and the expansion of Indonesian Economic Development that is "Creating an Independent, advanced society, equitable and a prosperous Indonesia Community", as measured qualitatively in the document of MP3EI (Master Plan for the Acceleration and Expansion of Indonesian Economic Development). In order to optimize the *connectivity program* in the national territory, it is necessary to measure the quantitative value of the MP3EI *document substance* in the accessibility functions model to accelerate the country development in achieving its prosperity. The main indicator of the accessibility measurement is subjective and qualitative, therefore it is necessary to carry out quantitative assessment to identify the accessibility, thus the difference between those with and without the construction of Sunda Strait Bridge can be measured. The current study is a case study of the accessibility function modeling of the geographic transportation covering five model countries, namely Indonesia (Suramadu Bridge), Japan (Akashi Kaikyo Bridge), Korea (Incheon Bridge), China (Hangzhou Bridge), and Malaysia (Penang Bridge). The purpose of this study is in the context of the determination of value-added gap measurement tool, and as a control function in an effort to minimize the regional connectivity failure due to the bridge infrastructure construction. Value-added gap is the accessibility value which is constantly growing and becoming a priority, so it is necessary to optimize the activity management control on the variables and the identified function accessibility. The main research focuses on the indicator formulation of value added in the functional accessibility that takes into account various aspects of both qualitative and quantitative. The validation results were obtained from the questionnaires survey and interviews. The data were

W. Wangsadinata (✉) · D. Ma'soem · D.M. Lestari
University Tarumanagara, Jakarta, Indonesia
e-mail: wiratman@wiratman.co.id

D. Ma'soem
e-mail: cpi_abidin@yahoo.com

D.M. Lestari
e-mail: diniharimulyalestari@yahoo.com

T. Taufik et al. (eds.), *ICoSI 2014*, DOI 10.1007/978-981-287-661-4_40

obtained from two (two) sources, the qualitative methods which derived from the questionnaire surveys and interviews result combined with the quantitative methods which the data obtained from BPS (Statistics Indonesia), coupled with references from various sources, and the entire data were then holistically tested using dynamic modeling system with Powersim Studio 7 software. The results of research analysis are of the form of quantitative formulation policy especially in transportation management, where the construction of Sunda Strait Bridge can increase the accessibility value in line with the increase in the value of GRDP, and if it is not with it, the accessibility value is low. Therefore to support the strengthening national territory connectivity to achieve the maximum result, it is absolutely necessary to construct the Sunda Strait Bridge.

Keywords Accessibility function · The value-added gap · Sunda strait bridge · Dynamic system models

1 Background

The implementation of Sunda Strait Bridge infrastructure development can create a value-added problem, namely accessibility function that continuously increases and becomes a priority. It is necessary to have an activity control management over the identified problems and variables through the increasing accessibility value of a region. Based upon this, the study is structured as an attempt to formulate an accessibility value model, so that there will not be any unsuccessful regional connectivity as the impact of Sunda Strait Bridge infrastructure, as on Suramadu Bridge.

The relationship between the population and GRDP was included in the economic indicators, whereas mobility was the transportation indicator [1]. The basic indicators are subjective and qualitative accessibility, therefore the quantitative measure is necessary to represent the accessibility so that the value-added gap of Sunda Strait Bridge construction can be measured as it is shown in Fig. 1.

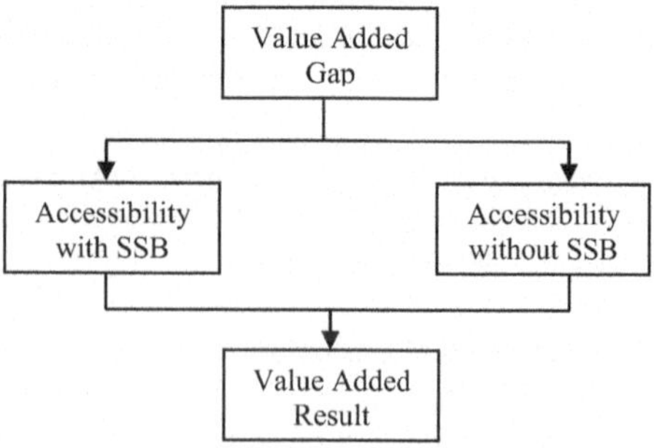

Fig. 1 Value-added gap

2 Research Objectives

The purpose of this study is to produce a policy formulation related to the optimization of value-added potential in GRDP function in generating the accessibility function value model formulation in line with the implementation of regional autonomy conditions, by setting the measuring instruments for regional development (value-added gap), and as a control tool (forward and backward linkage) as an effort to minimize the connectivity failure between two regions, so that Sunda Strait Bridge construction will generate a great contribution nationally. It is expected to obtain a more applicable technical results in supporting MP3EI documents quantitatively and suitable to be applied in Indonesia within the scope of the development management of regional transportation infrastructure.

3 Assumptions

The accessibility indicators used are subjective and qualitative, so it is necessary to have their quantitative measure to find out the accessibility (gap). It is assumed that the five case study models of the bridge from various countries (China—Hangzhou Bridge, Malaysia—Penang Bridge, Korea—Incheon Bridge, Japan—Akashi Kaikyo Bridge and Indonesia—Suramadu/with and Sunda Strait Bridge/without) being utilized in this study possess very similar characteristics, then the bridges will be formed into a *Physical Gap* Identification Model in the form of accessibility value associated with some influenced accessibility function $F(A)$ indicators such as, broad area (km^2), GRDP (Trillion), population growth rate (people), inflation/HDI (%/year), distance (min), bridge length (km), and the speed (km/h). From the accessibility, data contributes to the bridge in each of these areas, there will appear the *Physical Gap* to the accessibility and *Physical Gap* between the accessibility function and GDRP Function for Indonesia, which needs to be optimized to achieve the required level of development with or without Sunda Strait Bridge in the future.

4 Literature Review

The policy of bridge infrastructure construction can give a direct impact on the regional accessibility which further will increase the regional development [2] as shown in Fig. 2.

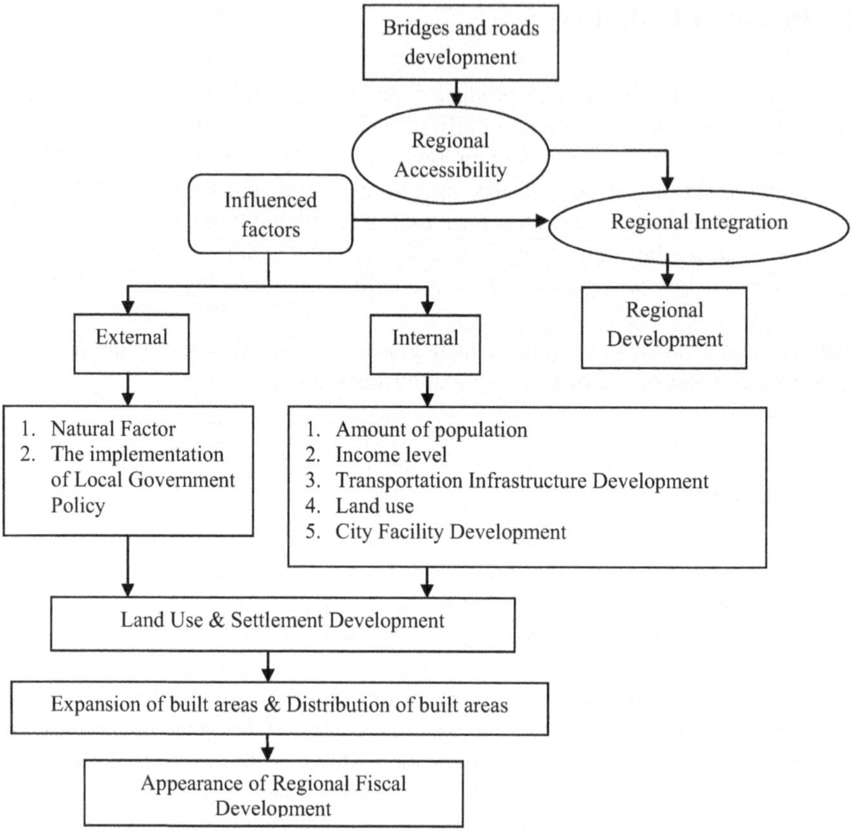

Fig. 2 Conceptual framework for bridges as regional connectivity

5 Research Methodology

5.1 Research Settlement Steps

1. Identification of Physical Gap
   In this study, the initial physical gap is ΔVA which is in the form of analysis modeling on the five bridges so that the *X* factor (ΔVA) and the *Y* factor (accessibility function) in each country can be analyzed along with their respected final physical gap in the form of impact/cause analysis to obtain a hypothesis. The details input of mental map formulation in increasing the physical gap control like value-added gap modeling to the bridges, as described in Fig. 3.
2. Expert Validation Questionnaire and Questionnaire Respondents
   The questionnaire is validated by a number of experts in order to get the identification of the determinant variable (*X*) and the target variable (*Y*) as a

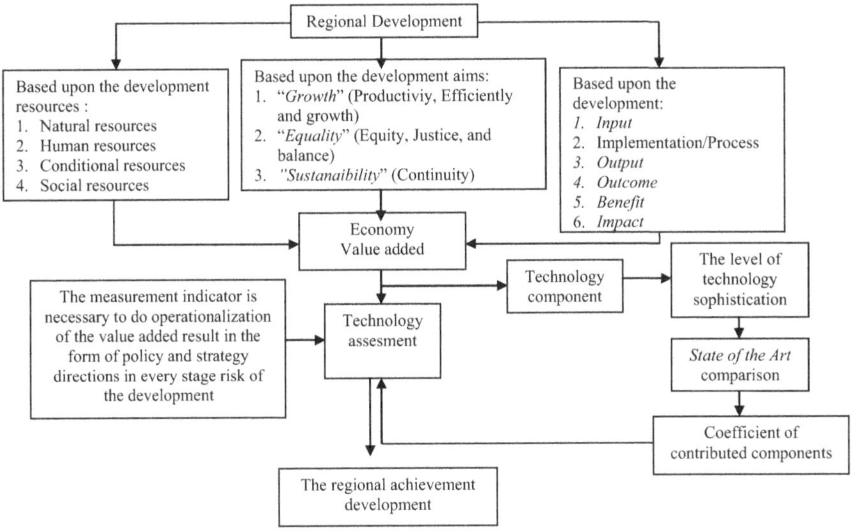

Fig. 3 Formulation of mental map input to improve physical gap control

source for preparing the questionnaire that would be presented to the respondents of various agencies and stakeholders.

3. Data processing

   (a) Identification of variables (the relationship between *Y* and *X*)
   The dependent variable *X* (the width of an area). The independent variable *Y* (the population, the income level, the length of roads, the number of ground vehicles, the extensive use of land built, the city facilities growth).

   (b) Sub Model Integration Dynamic Systems
   The performance models *Yi*, *Y*2, *Y*3, *Y*4, *Y*5, *Y*6, *Y*7 referred to as the sub-system models, next these sub-systems are integrated in a holistic model of the dynamic system.

   (c) Identifying Mental Map
   Essentially, in this study there are seven independent variables and one dependent variable, the independent variable are the socioeconomic, the environmental, the social, the political, the demographic, the investment and technology, the transportation, and the nature. The dependent variable is the accessibility with or without the presence of the bridge and including value added. The models will be divided as one major model and seven sub-models. The main model is a mental map and seven sub-models, i.e., *Y*1, *Y*2, *Y*3, *Y*4, *Y*5, *Y*6, *Y*7 are the output of the sub-sub-models that will be studied further (Fig. 4). Stages of the research process described more detail in Fig. 5.

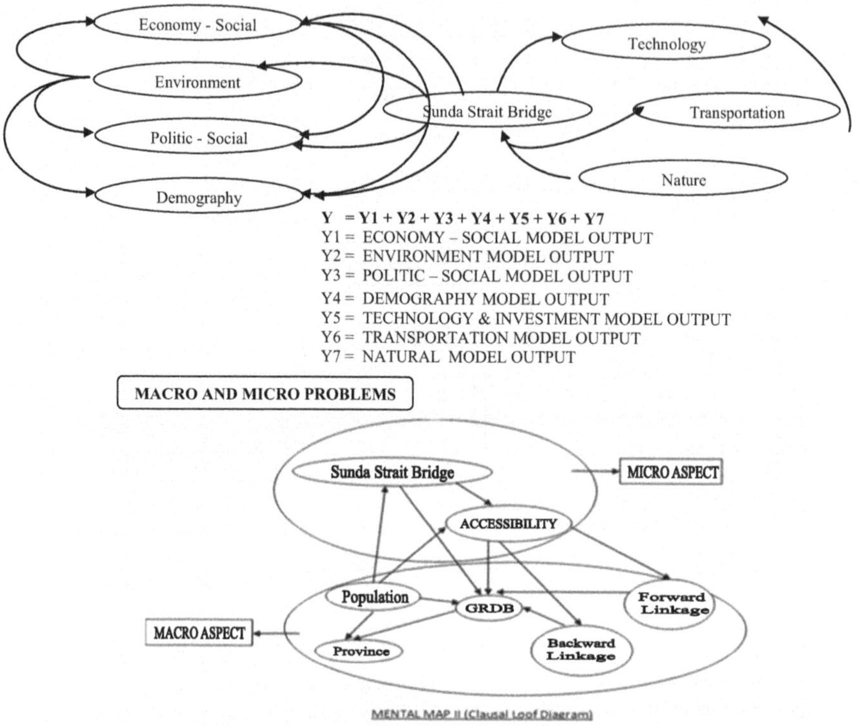

Fig. 4 Mental Map I and II (clausal loop diagram)

6 Results and Discussion

6.1 Physical Identification Gap

The Initial Physical gap is ΔVAA (value-added gap) in the form of analysis modeling on the five bridges, so the *X* factor (value-added gap) and the *Y* factor (accessibility function) in each country can be analyzed along with the physical final gap in the form of the impact/cause analysis to obtain a hypothesis. In the modeling of the five bridges, the variables are identified in the form of Physical Gap illustrations which influence the accessibility function so that the value-added gap can be measured. The assumption is based on the research gap results; this stage needs to be performed before the first phase of questionnaires distribution. The Physical Gap Illustration is *F*(*A*) = (GRDP, natural resources, human resources, technology, investment and demographics), where the accessibility function is influenced by certain factors, such as the technology, natural resources, human resources, demography, technology, investment, and GRDP (presented in Fig. 5 and Table 1).

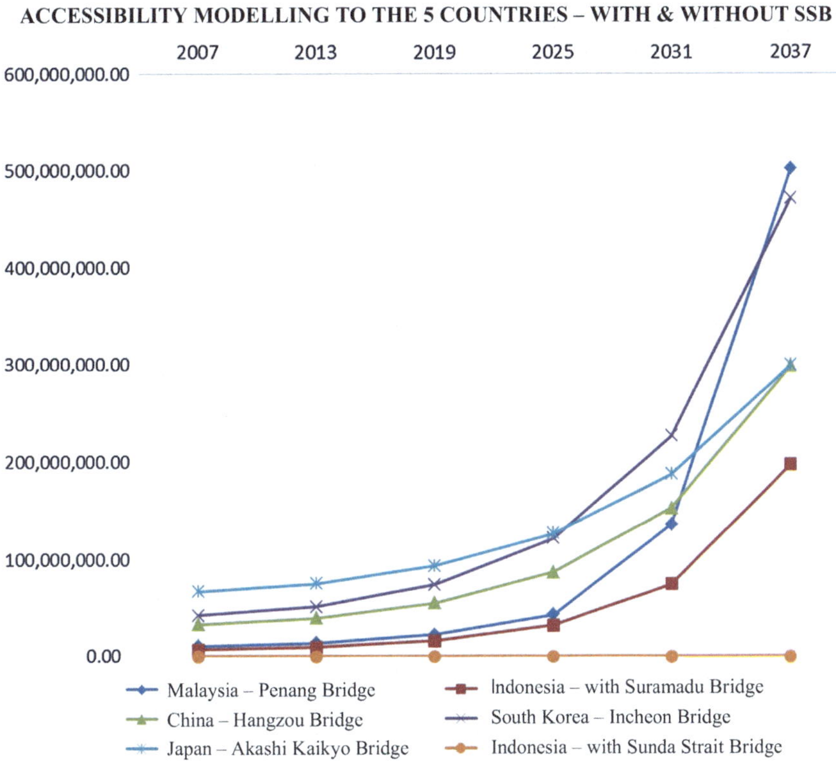

Fig. 5 Physical modeling chart gap

6.2 *Expert Validation Questionnaire and Questionnaire Respondents*

Following the initial questionnaire validation carried out by the expert, then the determinant variable (X) and the target variable (Y) were identified to obtain the conformity between X and Y variables as well as the suitability between indicator of each X variable. The next stage, the questionnaire for the respondents were distributed to the construction doers to get the results in the form of diagrams/pictures (Fig. 6).

6.3 *Analysis of I–O Table (Input–Output Table)*

The strengthening of regional connectivity indicators measured in the area of accessibility function model which in this study represent each of the zones, covering 16 provinces in Sumatra and Java Island. The data from BPS includes 11 data

Table 1 Preliminary data modeling related with the accessibility functions based research gap

Year	$f(A)$	Indonesia with SSB	Malaysia	Cina	Korea	Jepang	Indonesia without SSB
		Suramadu Bridge	Penang Bridge	Hangzou Bridge	Incheon Bridge	Akashi Kaikyo Bridge	Sunda Strait Bridge
2007	Area (km^2)	2654	902	686	560	572	2226
2013	Population (people)	2,055,087	116,917	605,370	731,546	85,120	980,039
2019	Density population	741.73	1267.11	962.34	1901.85	1531.73	707.56
2025	GRDP (trillion)	566.82	708.52	808.81	2,632.35	1040.00	160.31
2031	Inflation (%/year)	5.50	9.20	3.20	3.60	2.00	5.50
2037	Distance (min)	6.53	10.13	5.25	13.79	2.99	22.125

Source Dinihari (2013)

which will be presented into 11 tables and each table consists of 16 matrices, those are distance data between the zone (min), population, GDRP, agriculture, mining, industry, utilities, construction, trade, finance, and services. The priority data which needs to be prepared include:

1. The data by province are needed (for the parameters outside the zone distance, population, and GRDP) that describe transactions between provincial zone.
2. The industrial data which have a relation between raw-based materials and processing-based businesses on each zone are needed to describe the forward and backward linkage.
3. The time series data is needed containing the same year for all parameters.

The interactional zone graph of industrial potential data based, which have a relation between raw-based materials and processing-based businesses to describe the forward and backward linkage. The zone interaction graph of potential data is based on province (the distance between the zones for the parameters in km) which describes the transactions between provincial zones.

7 Conclusions and Recommendations

1. VAA is a measure of the bridge infrastructure success in increasing the value added to increase the connectivity in Indonesian territory. Value added of accessibility (VAA) is a measure of the performance based accessibility value that describes the accessibility functions of several related variables to a

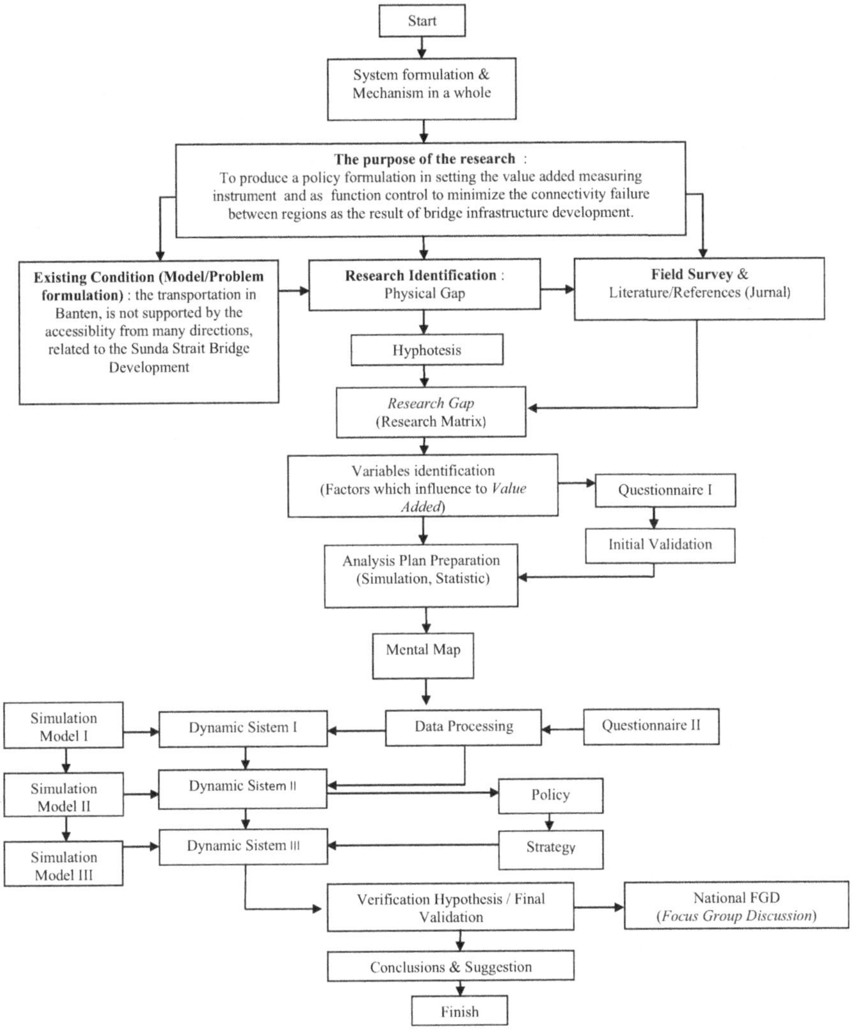

Fig. 6 Research process flowchart

particular period. Because the accessibility model is implemented on the bridge infrastructure, therefore the period used must be in long term, and the period which will be implemented in Sunda Strait Bridge is the period between the year 2007–2037.

2. Sunda Strait Bridge brings positive impacts such as to unite the nation in a variety of socioeconomic activities, regional and as the front and rear national territory or hinterland including the isolated territory. It is also useful to equalize the employment opportunities and decrease the poverty to the area which broadly still has low accessibility.

3. Some factors that may lead to the economic and social activities increase of a region which is stimulated by transportation infrastructure, namely, Sunda Strait Bridge is due to the mobility and accessibility increase of the region. The potential traffic rising in the presence of Sunda Strait Bridge is expressed from the increase of GDRP in the region of both ends of the bridge. Therefore, Sunda Strait Bridge is worthy built.

References

1. Rashidi, L.H., & Samimi A. (2012). The Relationship between Economic Indicators and Transportation and Freight Infrastructure Productivity Growth. Journal of Urban Planning and Development, ASCE.
2. Benung, R. M. (2004). Pengaruh Jembatan Kahayan terhadap Perkembangan Wilayah Perkotaan Palangkaraya. Yogyakarta: Tesis, MPKD, UGM. Dardak, H. (2012, May 9). Strategic Infrastructure and Regional Development Sunda Strait. Writing presented at the National Conference, Wisma Makara UI, UI Campus, Depok.

Rainwater Harvesting System Implementation for Domestic Water Use: The Environment and Financial Benefits

Imroatul C. Juliana, M. Syahril Badri Kusuma, M. Cahyono, Hadi Kardhana and Widjaja Martokusumo

Abstract Rainwater harvesting (RWH) system is a technology that focuses on sustainability and supports the sustainable environment development. The implementation of RWH systems provides many environment and financial benefits. Some of the environment benefits of RWH system are as follows: reduce the surface runoff, reduce the burden of soil aquifer, and provide the availability of clean water. This study analyzed the RWH system implementation benefits in both environment and financial sides. The financial benefits of RWH system implementation are calculated based on rainwater that can be used to replace the need for clean water. The environment benefits are defined by the reduction in main water tap use and the reduction in generated roof runoff volume. This study used a simple RWH system that uses the roof as a catchment area, the pipeline as a distribution system, and the tank as a storage system. The water use is for domestic potable and nonpotable for a household with up to four occupants in Bandung. The catchment area is taken up to 70 m^2. A water balance model for various scenarios was developed to calculate the algorithm of the system. The costs taken in RWH system include the construction, installation, maintenance, and operational costs. The analysis shows that the implementation of RWH systems provides advantages over the use of conventional systems. It can save clean water usage up to 54.92% and provides runoff reduction up to 71.53%. RWH system requires additional costs approximately 0.66% from the value of the house. It was found that it is possible to achieve payback in RWH system implementation under several scenarios.

Keywords Rainwater harvesting · Benefits · Environment · Financial · Domestic use

I.C. Juliana (✉) · M. Syahril Badri Kusuma · M. Cahyono · H. Kardhana
Faculty of Civil and Environmental Engineering, Bandung Institute of Technology, Bandung, Indonesia
e-mail: icjuliana76@gmail.com

W. Martokusumo
School of Architecture, Planning, and Development Policy, Bandung Institute of Technology, Bandung, Indonesia

T. Taufik et al. (eds.), *ICoSI 2014*, DOI 10.1007/978-981-287-661-4_41

1 Introduction

Water is a vital element in human life. A person cannot survive without water because water is one of the life supports for humans. Although water covers the earth, more than 97% tastes salty, 2% of freshwater is locked in snow and ice, so that only 1% can be used for human needs [1]. The availability of water resources is waning. Projected by 2025, there will be a water crisis in several countries, including Indonesia even though Indonesia is a country with a lot of water [1]. Increased human population growth pushes the usage of natural resources without regard to the availability for future generations. In addition, problems of water resources become more serious because of the lack of good management in water resources.

Rainwater is one natural resource that is often neglected. Rainwater is a source of high-quality and quantity water which is available every rainy season and has the potential to reduce the usage of freshwater sources. Rain is often regarded as a nuisance that should be promptly disposed of. Therefore, potential rainfall often redundant into a runoff. These paradigms need to be changed with regard to rainwater as an asset that needs to be collected as much as possible. Rainwater can be harvested, stored, and then reused for further use. The concept of rainwater harvesting is a concept that emphasizes sustainability. Previous studies have shown that the utilization of rainwater provides economic and environmental benefits. By the concept of rainwater harvesting, rainwater can be used for nonpotable water usage, which will reduce the usage of groundwater and provide the financial benefits [2, 3]. The concept of rainwater harvesting gives a good impact on the surrounding environment [4] and is considered as one of the adaptive strategies for climatic change related to water shortage.

In Indonesia, the reuse of rainwater is rarely performed. Many factors and obstacle arise. One is the assumption that the installation of RWH system is expensive and not significantly provides benefits from the financial side. Stakeholders are still reluctant to apply the concept, and utilizing groundwater or tap water is still the best preference. Due to the absence of a comprehensive study on the use of rainwater and the lack of proof, many people still hesitate to implement the concept. People do not realize the benefits that can be obtained with the use of rainwater. This study aims to analyze the benefits of implementing RWH systems for household or domestic water use, especially in terms of environment and financial overview, so that the benefits of RWH systems implementation can be more evident.

2 Rainwater Harvesting (RWH) System Concept

The term water harvesting is used to describe the process of collecting and storing water from an area that has been treated to increase precipitation runoff [5]. RWH system has several major components: the catchment area to catch rainfall, the water delivery system from the catchment area to storage, storage tanks for storing water, and tools to flow water to storage [6]. RWH concept has traditionally been long applied in arid areas. As an example, in Taiwan, the practice of rainwater harvesting has traditionally applied in areas that have limited surface water or groundwater [7]. In India, RWH system has been carried out more than a hundred years ago in a very arid area known as the Golden Desert. In Gunung Kidul Indonesia, due to the nature of the area that causing water shortage, one of the way that made by the community to meet the water demand is collecting rainwater.

Several years ago, RWH is only used as an alternative way to compliance water in areas with water shortage. Along with the advances in technology and the increasing environmental awareness, the concept of RWH started to be used as an alternative source to meet the needs of the primary and supplementary source of water. RWH concept can be applied into large- and small-scale catchment areas. Rainwater harvested from roofs can still be used for drinking, cooking, and bathing because it has a quality that is not much different from the water coming from the taps [8]. The main usefulness of RWH is divisible as a major source of clean water, additional water sources for the use of clean water, and additional water sources for use instead of clean water [9].

3 Methodology

This study used a simple RWH system (Fig. 1). The storage tank capacity is an important component because it determines the overall system performance and cost. The tank's performance is influenced by the characteristics of the catchment area, rainfall, and water demand required [3].

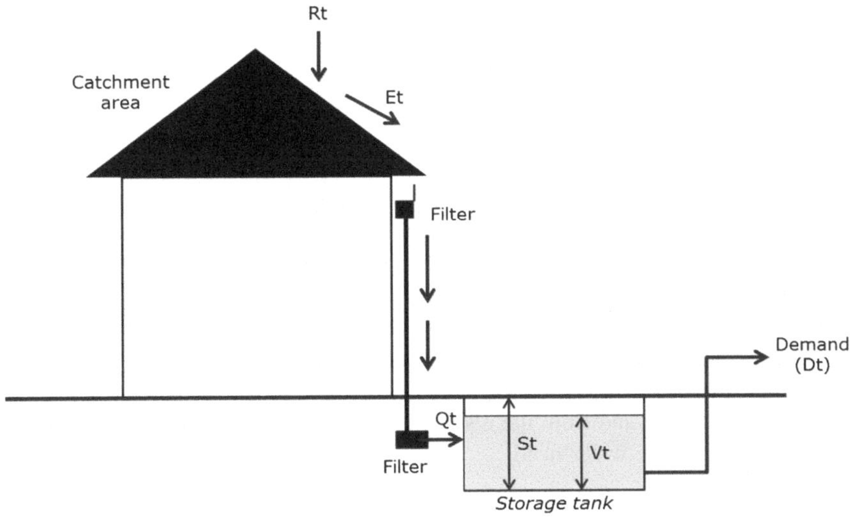

Fig. 1 Simple RWH system. Adapted: [3]

3.1 System Performance Model

Effective rainfall volume is obtained by multiplying the rainfall with the wide catchment area, and the ratio between the volume of water that and can be captured by the catchment area (roof) and the water that flowing out as:

$$\mathrm{ER} = R.A.C \tag{1}$$

where ER is the effective runoff (m^3), R is the rainfall (mm), A is the area of the catchment area (m^2), and C is the ratio between the volume of water that is captured by the catchment.

RWH system performance calculated based on the concept of water balance. Water balance simulation algorithms are used to explain the operation of RWH systems performance. The changes of the storage tank volume can be expressed as follows:

$$V_t = V_{t-1} + Q_t - D_t - E_t - L_t \tag{2}$$

In this study, the storage tanks are below ground level so that the effect of evaporation and losses can be eliminated as follows:

$$V_t = V_{t-1} + Q_t - D_t \tag{3}$$

where $0 \leq S \leq V_t$ and V_t is the storage volume at time t (m^3), V_{t-1} is the storage volume at time $t-1$ (m^3), Q_t is the volume that enters the storage in a time interval t (m^3), D_t is a water requirement release in lapse of time t (m^3), E_t is evaporation (m^3), L_t is the losses (m^3), and S is the storage capacity (m^3).

The volume of water will be collected in a time equal to the volume of water in the past ($t-1$) plus the incoming water flow and reduced with the outflow at a certain time. A model of behavioral analysis simulates an algorithm of the system operation volume in storage tank based on the concept of mass balance by a certain time interval.

RWH system performance indicators can be expressed with the reliability and effectiveness of the tank. The storage tank reliability can be expressed in units of time or volume [10]. Time indicator reliability can be used as a basis for proper system performance for domestic RWH system, which, in this case, does not use too much water [7]. In this study, the system performance is in the unit of time which is expressed by the number of days where the RWH system met with the water needs. This system performance indicated with the storage tank performance which expressed the behavior of the tanks. The system behavior shows the storage tank volume per day, and from the results, it can draw a number of days which the RWH system met the water need.

3.2 Financial Analysis

The RWH system implementation will provide economic benefits. The more use of rainwater will increase the financial benefits [11]. Financial analysis is performed to calculate the economic feasibility of the system. There are various economic methods that can be used to calculate the investment feasibility. The benefit–cost ratio method can be used to calculate the financial benefits of RWH system implementation. Benefit–cost ratio (BCR) is a method that often used in the early stages of investment plan evaluation or as an additional analysis with the aim to validate the results of the evaluation that has done previously [12]. BCR method gives an emphasis to the aspect ratio between the value of benefits that would be obtained with the aspect of cost (and losses) incurred by these investments. BCR values can be calculated from the year that the system would return on investment.

The cost component is the capital cost, maintenance, and operational costs. Annual cost is all cost spent along the age of the system. Benefits were calculated based on the number of rainwater volume that can be used to replace the water use.

4 Results and Analysis

4.1 Study Site Overview

The study area is the city of Bandung which has two seasons: dry season between April and October and rainy season in October through April. Bandung is the capital city of West Java Province with the surface area 16,730 ha and comprises of 30 districts and 152 subdistricts. The total population is 2,455,517 people, with average population density 14,676 people/km^2. Clean water supplies for the community generally provided by main water taps and by utilizing groundwater. The average annual rainfall ranges from 1000 mm in the central and southeast regions to 3500 mm in the north of the city. The case study is in Ujung Berung, a small district in East Bandung with the area of 18,500 m^2. The area had been selected as a pilot project of a rainwater harvesting system.

4.2 Rainfall Data

Daily rainfall data in Ujung Berung station in 2000–2012 were analyzed. From the analysis, the year of 2012 has taken because it was considered as a dry year that is expected to reflect a critical condition of the RWH system. Maximum daily rainfall in 2012 was equal to 80 mm with an average 6.8 mm.

4.3 Domestic Water Use

The average daily domestic water consumption per each person in Indonesia is 144 L/person/day. Further research suggests that the consumption can reach 147.74 L/person/day [13]. Based on the results, in this study, the approach of required amount of total water use is taken 150 L/person/day (0.15 m^3/person/day) and the need for nonpotable taken by 60% [3] of the total water requirement.

4.4 RWH Systems Performance

The calculation is divided into a number of scenarios based on tank capacity and the number of occupants of the house and water use. The water use is divided into total and nonpotable water uses. The need for total water use is 0.15 m^3/person/day and for nonpotable is 0.09 m^3/person/day. The catchment area is 70 m^2 with the ratio of rainfall captured by the roof is 0.95 [5]. The number of occupants is divided into 1 and 4 persons per house. Based on the storage tank performance analysis for several capacities, the size of 2 m^3 capacity was taken. The size is considered feasible for a household to build. The 2 m^3 size is taken with the consideration that for each water demand, the average performance of the RWH system with 2 m^3 storage tank has reached 65.2% (Fig. 2). Rainwater use, runoff volume, system performance, and financial analysis will also be analyzed for each scenario.

Initial analysis is conducted by analyzing the system behavior for each scenario. System performance is calculated based on the time, which is the number of days that water needs to be met by the RWH system. This result shows with a small number of deficit days (Fig. 3). From Fig. 2, it is shown that the lesser the rainwater used, the greater the system performance will be. With 2 m^3 tank capacity, the best system performance is 80% for 1 person nonpotable water use (NP1) and followed by 1 person total water use (P1), 4 persons nonpotable water use (NP4), and 4 persons total water use (P4), respectively, for 70, 64, and 43% system performance

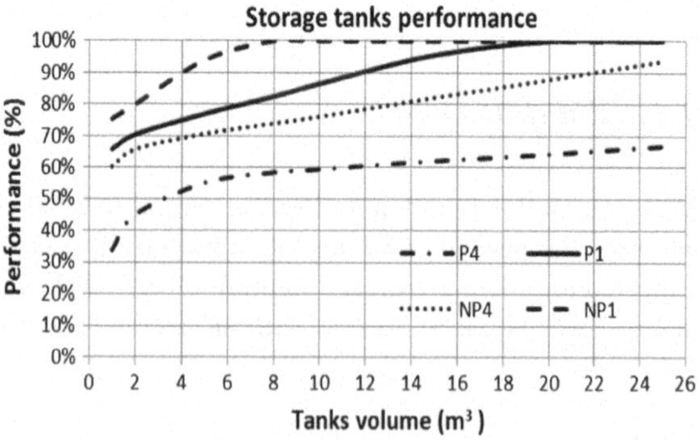

Fig. 2 Storage tanks performance for each water use

(Fig. 2). If the reliability value of the system is 60%, then the RWH systems with 2 m^3 storage tank capacity could meet the water needs up to 0.36 m^3/day/house (P1, NP4 and NP1).

The implementation of RWH system can reduce the use of main water taps or groundwater. RWH system can replace 99 m^3 water use for 4 occupants with total water need of 219.6 m^3, or it can save 54.92% of water use. RWH system can save water need from 28.96% (NP1) to 54.92% (P4). The implementation of RWH system can also reduce the runoff volume. The more use of rainwater, the more reduction in runoff generated. With the conventional system without RWH, the volume of roof runoff is 157.39 m^3/year from total potential rainfall 174.88 m^3/year. With RWH system implementation for 4 persons potable water use, the generated roof runoff is 44.81 m^3/year, or in another way, there is a 71.53% reduction in runoff volume. Conversely, when rainwater is used only for 1 person nonpotable water use, the runoff reduction is only 15.97%. The detail of saved clean water use and reduced roof runoff is given in Table 1 and Fig. 4 for P4, P1, NP4, and NP1 respectively.

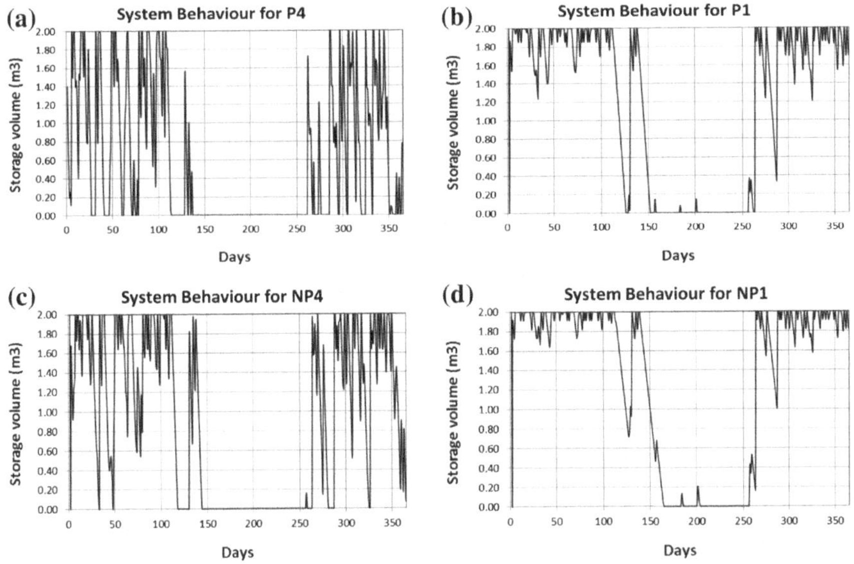

Fig. 3 System behavior for **a** P4, **b** P1, **c** NP4, and **d** NP1. Annotation: NP1 = 1 person nonpotable water use. P1 = 1 person total water use. NP4 = 4 persons' nonpotable water use. P4 = 4 persons' total water use (P4)

Table 1 System performance, saved clean water use, and reduced run

Water use		System performance (Y) to met water need	Rainwater use	Reduced runoff with RWH	Reduced clean water use
Person	Volume (m^3/day)	%	Volume (m^3/year)	%	%
P4	0.60	47	99.00	71.53	54.92
P1	0.15	71	36.15	25.11	34.15
NP4	0.36	61	75.96	52.25	42.35
NP1	0.09	76	23.40	15.97	28.96

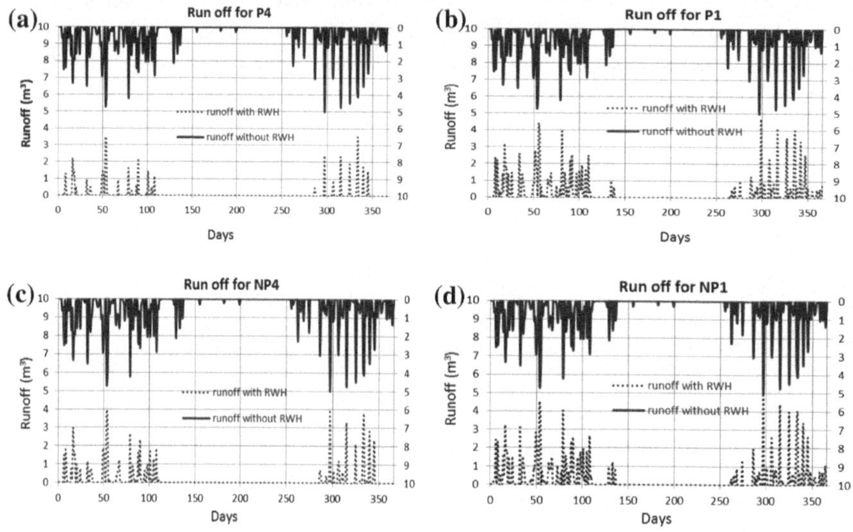

Fig. 4 Generated roof runoff volume for **a** P4, **b** P1, **c** NP4, and **d** NP1

4.5 *Financial Analysis*

Financial analysis conducted on the costs and benefits. Cost consists of the initial capital cost, maintenance cost, operational cost, and rainwater treatment price. Maintenance is primarily important for performance and safety reasons. Annual maintenance cost is the cost per year at 4% [14], and operational cost is the cost of rainwater treatment with the water treatment plant taken Rp. 2000/m^3 [15]. The interest rate is 12%, and the depreciation rate is 4.5%. Benefits are taken from the number of main water taps volume that can be replaced by rainwater multiplied by the price of water Rp. 6000, [15].

From the simulation results, it can be seen that with the higher use of rainwater, the system will become more profitable as more benefits are generated so that the

Table 2 Cost items

Items	Capital cost				Maintenance cost	Operational cost
	Unit price	Volume	Unit	Total	Total/year	Total/year
Storage tank (2 m^3)	Rp. 2,500,000	1	pc	Rp. 2,500,000	Rp. 100,000	
Pipes	Rp. 25,000	20	m	Rp. 500,000	Rp. 20,000	
Elbow	Rp. 10,000	3	pc	Rp. 30,000	Rp. 1200	
Pump and accessories	Rp. 600,000	1	pc	Rp. 600,000	Rp. 24,000	
Gutter	Rp. 45,000	5	pc	Rp. 225,000	Rp. 9000	
Gutter filter	Rp. 30,000	1	pc	Rp. 30,000	Rp. 1200	
Soil excavation	Rp. 75,000	4	m^3	Rp. 300,000		
Man labor	Rp. 100,000	10	man/day	Rp. 1000,000		
Rainwater use						
Potable 4	Rp. 2000	99.00	m^3			Rp. 198,000
Potable 1	Rp. 2000	36.15	m^3			Rp. 72,300
Nonpotable 4	Rp. 2000	75.96	m^3			Rp. 151,920
Nonpotable 1	Rp. 2000	23.40	m^3			Rp. 46,800
Total				Rp. 5,185,000	Rp. 155,400	

Table 3 Benefits of the RWH system

Water use		Rainwater use	RWH benefit	Payback year	BCR
Person	Volume (m^3/day)	Volume (m^3/year)	Rp/year	Years	
P4	0.60	99.00	Rp. 594,000	9	1.09
P1	0.15	36.15	Rp. 216,900	17	1.02
NP4	0.36	75.96	Rp. 455,760	11	1.09
NP1	0.09	23.40	Rp. 140,400	22	1.05

breakeven point value more quickly obtained. For the value of BCR > 1, the most rapid payback years are for the use of P4 which is within 9 years, and the system will have more benefit than the cost to be incurred. Successively, the fastest payback value for NP4, P1, and NP1 was 11 years, 17 years, and 22 years (Table 3).

If the capital cost of RWH system is Rp. 5,185,000, and the sales price of type 70 house in Bandung for an average are Rp. 707,000,000 [16], the initial costs to be incurred to construct RWH system are less than 1% or only 0,73% of the sales price. This analysis shows that the housing developers should build RWH systems because the cost is quite cheap and affordable. Detail analysis of the cost and benefits of RWH system is shown in Tables 2 and 3.

5 Conclusion

At the case study, the RWH systems with 2 m^3 storage tank capacity could meet up to 0.36 m^3/day/house water needs. From the previous analysis, a conclusion could be drawn that the implementation of RWH systems provides an environment and financial benefits. The more the rainwater used, the more the benefits. On the environment side, RWH system can save clean water use up to 54.92% and reduce the volume of runoff comes from the roof up to 71.53% per year when compared to conventional systems. The implementation of RWH systems also provide financial benefits where the initial capital cost is only 0.73% of the house sales price, and at a certain stage, the amount of benefits generated will be bigger than the costs. The intangible benefit of the system can be bigger if viewed for further environmental sustainability.

References

1. U.S. Environmental Protection Agency (USEPA), *Air: dunia yang dahaga, 2010,* National Geographic Indonesia, Jakarta.
2. Roebuck, R.M., dan Ashley, 2006, *Predicting The Hydraulic and Life Cycle Cost Performance of Rainwater Harvesting System Using a Computer Based Modelling Tool,* Proceeding 7th International Conference of Urban Drainage Modelling, Australia.

3. Roebuck R.M., 2010, *A Whole Life Costing Approach for Rainwater Harvesting Systems*, Dissertation, Bradford University, Australia.
4. Angrill, Sara, Farreny, Ramon, Dan Gasol, 2011, *Environmental Analysis of Rainwater Infrastructures in Diffuse and Compact Urban Models in Mediterranean Climate*, International Journal of Life Cycle Assess, Germany.
5. Frasier, G.W., Myers, L., 1983, *Handbook of Water Harvesting*, US Department of Agriculture, USA.
6. Gould, J., dan Nissen-Peterson, E, 1999, *Rainwater catchment systems for domestic supply: design, construction and implementation*, Intermediate Technology Publications, London.
7. Liaw, C.H., Y.L. Tsai. 2004. Optimum Storage Volume of Rooftop Rainwater Harvesting Systems for Domestic Use. Journal of the American Water Resources Association, 40(4): 901–912.
8. Meera V and Ahammed, M.M.. (2006) *Water Quality of Rooftop Rainwater Harvesting Systems: A review*, Journal of Water Supply Research and Technology—Aqua, 55, 257–493.
9. Fewkes, A., 2006, *Modelling the performance of rainwater collection systems: towards a generalised approach*, Urban Water Journal Vol 4 (323–333), UK.
10. Fewkes, A., Butler, D., 1999, *The Sizing of Rainwater Stores Using Behavioural Models*, Proceedings of the 9th International Rainwater Catchment Systems Conference, Brazil.
11. Rahman, A., Dbais, J., and Imteaz, M., 2010, *Sustainability of Rainwater Harvesting Sistems in Multistorey Residential Buildings, American Journal of Engineering and Applied Sciences 3* (73–82), United States.
12. Giantman, 2006, *Ekonomi Teknik*, P.T. Raja Grasindo Persada, Jakarta.
13. Mangiring, Dian, 2007, *Kajian Pola Konsumsi Air Bersih Rumah Tangga di Kelurahan Setiamanah Kota Cimahi sebagai Masukan Bagi Upaya Konservasi*, Master Thesis, ITB, Bandung.
14. U.S. Environmental Protection Agency (USEPA), *Laporan Kebijakan Moneter Triwulan IV Bank Indonesia, 2013*, BI, Jakarta.
15. PDAM Tirtawening, 2013. *Kapasitas Produksi PDAM Tirtawening Kota Bandung*.
16. *Daftar Harga Rumah di Bandung*, 2014, www.rumah.trovit.co.id.

The (In-)Visible Hand: A Governance Perspective on Low-Carbon Transitions in the PRC

Linda Katrin Westman

Abstract The multi-level perspective on sociotechnical transitions (MLP) is a framework that has been widely applied in analysing the dynamics of innovation and change in large technical systems. The model has proved useful for explaining complex mechanisms of interaction between sectors and levels but has been criticized for a lack of focus on political and governance aspects likely to form part of these dynamics (Smith et al. in Res Policy 34(10):1491–1510, 2005 [1]; Bulkeley et al. in Cities and Low Carbon Transitions. Routledge, London, 20011 [2]). This paper discusses transition dynamics from a governance perspective by analysing an empirical example of low-carbon development. Using the case study of the city of Rizhao, the paper provides a mapping of institutional governance arrangements involved in low-carbon development processes, focusing in particular on the urban planning and the shift towards energy conservation and recycling technologies in the city's industrial sector. A couple of points are suggested in this paper. Firstly, planning arrangements, although not explicitly part of transition dynamics, influences low-carbon dynamics through broader forms of impact on development pathways. Secondly, policy and planning instruments appear to play an important role in shaping low-carbon pathways on an urban level in China, and different forms of policy tools and enforcement strategies in turn have different effects on transitions mechanisms. Third, in spite of the dominant top-down nature of China's political system, network interactions contribute considerably to spreading low-carbon solutions on an urban level. A key importance of these links is to provide channels for information transfers that enable the local multiplication of low-carbon practices. Finally, the paper includes a theoretical discussion on whether, and in that case how, a governance perspective may complement an MLP analysis of urban low-carbon development processes.

L.K. Westman (✉)
Development Planning Unit, University College London, London, UK
e-mail: linda.westman.11@ucl.ac.uk

T. Taufik et al. (eds.), *ICoSI 2014*, DOI 10.1007/978-981-287-661-4_42

Keywords Low carbon · MLP · Sustainable transitions · Governance · Urban development · China

1 Introduction

In the wake of the fifth IPCC report, international media and academic research are once again increasingly focusing on climate change issues. In the midst of an apparent political status quo of global emissions negotiations and an increasingly allusive two-degree target, many nations and regions are embracing local low-carbon agendas. China is adopting low-carbon targets based on various concerns. With a rapidly expanding, industrializing, fossil fuel-based economy, the country has been on a high-carbon growth trajectory for the past decade. As the world's largest energy consumer and, by far, the world's largest carbon dioxide emitter, China is also facing the prospect of dwindling energy domestic resources and rampant pollution issues [3–6]. In order to confront these issues, China's central government has issued a number of low-carbon policies, plans and targets. Within the country's top-down system, these are delegated by the central government to provincial and municipal levels, where local leaders are in charge of realizing these agendas. This study uses an empirical example to explore how low-carbon issues are linked with urban governance arrangements by mapping out the actors and interests involved in decision-making processes and the interaction and cooperation between societal sectors required in enforcing low-carbon policy.

2 Case Selection and Data Collection

Rizhao is chosen as a case study because of its reputation as a forerunner in terms of environmental and energy performance, which suggests that it is likely to display several mechanisms involved in urban low-carbon development in China. Rizhao is also a medium-sized, medium-income city, which has followed development trajectories typical of cities in the region (in terms of population, economic and urban growth, and economic restructuring since the end of the 1980s), and therefore potentially displays development trends representative of the urban regions along China's East coast.

The data used in this paper were collected during field research in Rizhao between January 2013 and June 2013. The main method used for data collection was semi-structured interviews with key informants, such as urban planners, academics and representatives of local authorities and enterprises. These data were complemented with secondary data from Rizhao Statistical Yearbooks (1986–2011), which was mapped out to show various development trends in the city. These data were further complemented through a comprehensive review of policy documents and newspaper articles related to the issues of the study.

3 Governance and a Multi-level Perspective (MLP) on Transitions

Governance is a concept used frequently in academia and policy documents, often in reference to "good governance" or as a collective description for government structures or activities. The most common way of understanding the concept is by the method of categorization, the most widely employed one dividing governance systems into top-down/traditional models (describing linear, state-centred control) and bottom-up models (describing administrations' influence on government agendas). A third, interaction-based "network" model understands steering to involve the activities of civil society, markets, and multiple societal groups in cooperation around policy processes. Theories about environmental governance have generally evolved from the views of top-down management towards network-oriented models. For example, governance of "sustainability" issues tends to be marked by uncertainty, ambivalence and complexity, thus deviating from linear approaches and often being based on complex interaction [7].

Low-carbon development is a form of sustainability development likely to involve considerable amount of change within multiple sectors in society: infrastructural systems, technologies, engineering practices and user behaviours [8]. A theoretical framework used to explain how system-level change occurs is the multi-level perspective on sociotechnical transitions (MLP). The MLP model derives out of a synthesis of insights from sociology and history of technology and evolutionary economics [9–11]. It describes how innovation and technological change develops across three analytical levels, referred to as sociotechnical landscapes, sociotechnical regimes and innovative niches [12]. Sociotechnical regimes represent physical aspects of technological systems and their integration into society through reinforcing elements such as engineering practices, product standards, policy, and legislation and user habits [8, 12]. Sociotechnical landscapes are made up of broad cultural, sociopolitical, economic and physical contexts that stabilize regimes. The framework explains the process through which landscape factors place regimes under external pressure for change and, simultaneously, how new innovations develop in protected niches in the regime. New practices are upscaled through co-evolutionary processes of interaction between levels and sectors that involve learning processes and the formation of supportive actor networks [8, 13]. Eventually, the new technological innovation realigns with other elements into a new sociotechnical regime.

A branch of sociotechnical transition research engaged specifically in governance issues is the transition management (TM) literature. TM theory explains how governments can manage technological change by generating transition "visions" and protecting niche innovations by stimulating learning processes and efforts to create networks that protect new innovations [14, 15]. In critique of the TM literature, it has been suggested that the formulation of goals by "visionary forerunners" creates a technocratic and depoliticized view of transition dynamics. Changes in the organization of large sociotechnical systems are likely to involve highly contested processes associated with and strong resistance to change [2, 16, 17]. There remains a void of explanations relating to the sources and changes of interests and power in transition processes, as understood, for example, in the literatures on policy networks and power and conflict studies [1, 17, 18]. In order to highlight these dimensions, this paper considers which actors are included in decision-making processes related to low-carbon development processes by using a policy network perspective. The policy network literature highlights particular aspects of state–stakeholder interactions by describing the nature of borderlines and patterns of connections between the state and the private sector [19]. This paper focuses on a couple of key dimensions of decision-making processes, including which actors/institutions are included, dominant forms of interaction and conflicts of interests [20].

In terms of policy realization, the MLP framework suggests that the upscaling of niche activities is created through coordination and co-evolvement of heterogeneous elements in society. However, from a governance perspective, it is not apparent how this form of coordination is achieved, where co-evolvement occurs and who is included in it, especially in the context of a top-down political system. This paper aims to explore the connection between "co-evolution" and cooperation-based management by analysing how "alignment" between societal sectors may look in a centralized, comparatively authoritarian political system. The second section of the case study considers policy enforcement from an instrumental governance approach—that is, by the categorization of enforcement strategies associated with urban low-carbon plans.

4 Rizhao: Governance Arrangements

4.1 Decision-Making

In China, an increasing level of authority and responsibility has been delegated to local governments as part of a decentralization process that has taken place over the past decades [21]. However, when forming municipal plans, the strongest influence remains the top-down transfer of guidelines from higher levels of government, which occurs through a couple of mechanisms. One is the direct adoption of targets delegated to municipal departments by their provincial counterparts. These are integrated into policy documents, such as the municipal Five-Year Plans (FYPs), which contain guidelines on resource conservation, ecological protection, pollution, and emission and energy-saving targets. Another mechanism is the involvement of higher level planning institutes. For example, Rizhao's two most recent urban Master Plans (1992, 2006) were drafted by the central institute China Academy of Urban Planning (CAUPD). External planners represent the interests of the central government, or in their own view, "objective" opinions, which may differ from that of local decision-makers. A planner states that their rationale for preserving the environment in Rizhao was for the sake of long-term planning and sustainability considerations [22]. For example, the conservation of Rizhao's coastline was suggested by CAUPD planners when drafting the 1993 Master Plan and has contributed to the long-time preservation of the area and enabled resistance to development demands of local stakeholders [22].

Standard planning procedures include meetings that are used to resolve the issues of jurisdictional overlaps between departments and incorporate demands of local stakeholders. For example, in drafting Rizhao's eco-city construction plan, the external planners "worked mostly with the Environmental Planning Bureau (EPB), but we visited many other units, the agricultural and urban planning departments, industrial representatives and the fisheries… After the plan came out, we asked many times for their comments and tried to work out amendments according to their ideas" [22]. In spite of efforts to minimize contradictions, outcomes may not always be ideal. Crucially, the negotiation of viewpoints in informal meetings allows for dominant interests to compete on uneven terms. The interviews of this study suggest that the main power over development directions reside with the Mayor's office. A planner confirms that "planning decisions should be based on the opinion of the majority of stakeholders involved, but the local government office has more influence over final decisions" [22]. For a number of reasons, an inclination of the top leadership is a primary preoccupation with promotion of economic growth [23, 24]. One is that economic growth is emphasized over other achievements in China's cadre evaluation system. Another is the need to raise municipal revenues [5]. The easiest way for municipalities to generate funds is through land concessions, which creates a disproportionate willingness of urban governments to invest into land development and to support major enterprises that expand the local tax base [23].

Evidence from interviews in this study suggest that the larger the share of GDP is, tax income and contribution to local employment from the industrial sector play a more important role than environmental considerations in decision-making processes. As stated by a local academic, "the decision-making is a process of 'in the day-time we research, in the night-time we vote', because in the end, even if there is pollution, the government needs to consider GDP, economic growth, factories… they still need to keep them there" [22]. A similar line of conflict exists in relation with the priority to develop land or improve ecological protection. In determining the spatial layout of the area, the government will consider ecological aspects, but the planning bureau is under very high pressure from investors applying to develop the region. Regarding the influence of citizen interests on the decision-making process, a local policy-maker expresses that "the strong demand of the citizens was an important factor that pushed the government to pay attention to these issues" [22]. However, citizen opinions will only be considered to a limited extent based on a small number of participants and mainly in a reactive fashion. Oppositional voices are especially unlikely to be heard, as stated by a planner, "there are people who do not want the factories, like the ex-villagers who have to live right next to them…. The government will have a meeting to allow for opposing opinions, but often these voices will not be considered" [22].

This discussion suggests a couple of points. Although the main line of influence in planning processes runs in a top-down direction, higher government guidelines are combined with considerable local influence. For example, although the FYPs contain targets for ecological protection and industrial expansion, these are negotiated between interests when translated into specific projects. This form of decision-making allows for dominating state–industry relations to skew urban planning in a high-carbon growth direction. Crucially, the top leadership's emphasis on the creation of economic expansion contributes to this tendency. External planners may sometimes exercise restraints on these processes and act in favour of green development alternatives. Also, the growing influence of the public may also act in favour of environmental protection, but this impact remains weaker than that of powerful local stakeholders.

4.2 Policy Enforcement

Municipal departments are in charge of enforcing low-carbon policy that falls within their jurisdiction, where the main activities consist of monitoring and promotion. The actual realization of environmental policy depends on the activities of the private sector as the units meant to reduce pollution, cut energy use, adopt recycling technologies and construct buildings that are mainly companies. Monitoring of the industrial sector is essentially associated with traditional command-control instruments. An example is the supervision of energy efficiency standards in industrial units. Energy-saving quotas are allocated to the municipality's key energy using enterprises and failure to meet levels will result in leaders of state-owned units immediately being blocked from awards and promotions [24]. According to the recently adopted energy conservation law, industrial energy efficiency standards have also become a legal requirement, meaning that units that fail levels can be punished with fines or blacklistings. These legal tools appear to have had an impact on industrial energy use in recent years, particularly for key energy-consuming units, which have reduced their energy use intensity after 2009 (Fig. 1).

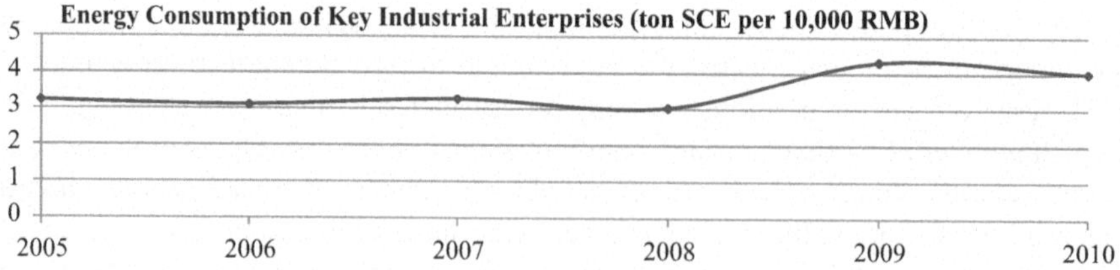

Fig. 1 Rizhao energy consumption of key industrial enterprises. *Source* Rizhao Municipal Yearbooks

There are, however, a number of problems associated with command-control tools, including limited municipal funding that cause difficulties in comprehensive monitoring (e.g. lack of staff, lack of technical capacity or lack of reliable statistics) and companies failing environmental standards because of lacking funds [22]. A promotional strategy to guide companies towards low-carbon solutions is to provide financial incentives, which takes the form of central-level subsidies and provincial awards for energy conservation retrofits. Economic incentives are also created by providing financial benefits for the high-technology sector, based on preferential taxes and land policy for green-tech enterprises. An effect of this support is the facilitation of spread of low-carbon technologies through the establishment of companies that sell energy-saving technologies or promote energy-saving solutions to high energy-consuming industrial units. For example, the energy efficiency retrofits of Rizhao's steel factory were facilitated through cooperation agreements signed with three domestic enterprises that specialize in thermal energy and metallurgical engineering technologies [25].

Another effect of these preferential financial policies is the creation of channels for technology diffusion. Rizhao has seen a rapid increase in FDI since 2007, almost doubling the total amount in between 2007 and 2010, although exports and imports have grown even faster [26]. The city's exports and imports are largely made up of energy-intensive, high-carbon products such as steel, auto-related products, chemical products and coal. Increased trade in these sectors contributes negatively to the city's energy profile; however, by attracting high-technology enterprises, the city also receives investments from companies with a high environmental performance. An example is the adoption of the most advanced and largest scale industrial recycling technologies in the city, carried out by Shandong Yatai Senbo Pulp Paper Co. The pulp factory is part of an international umbrella organization, which operates according to the environmental guidelines above those enforced in the cities in China. The firms' recycling equipment, imported from leading European firms, allows the factory to produce its own electricity. This has ended its reliance on coal and instead shifted to selling biowaste-generated power to the local grid. Technology may also be channelled from other regions in China. For example, Jiejing Ocean Biotechnology Company, another of Rizhao's leading firms in terms of environmental performance, is part of a larger business association operating on a provincial level, which can transfer knowledge, experience and higher environmental standards to its municipal branch [27]. Other promotional strategies used by the municipal authorities are training and joint research. For example, local authorities engage companies in training sessions focused at identifying inefficient energy use and introducing new energy management systems. There are also examples of cooperation between the academic and private sectors, such as staff training in adoption of PV technology, promoted by Rizhao Technical College in cooperation with the private enterprise Shandong Lanjing Yitan Co [28]. The deployment of methane-to-energy technology has been encouraged by state–private

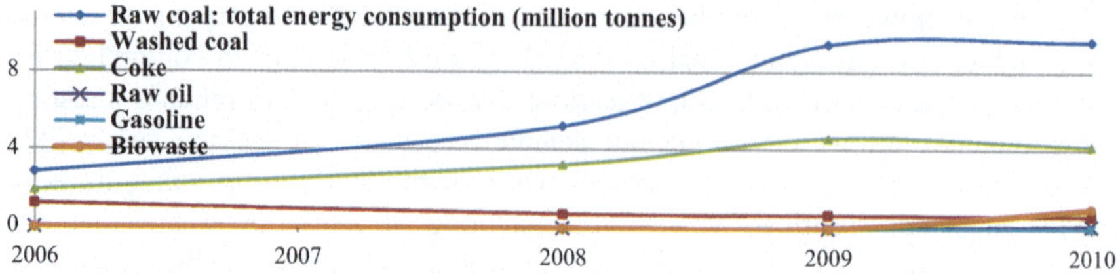

Fig. 2 Rizhao industrial energy use by source. *Source* Rizhao Municipal Yearbooks

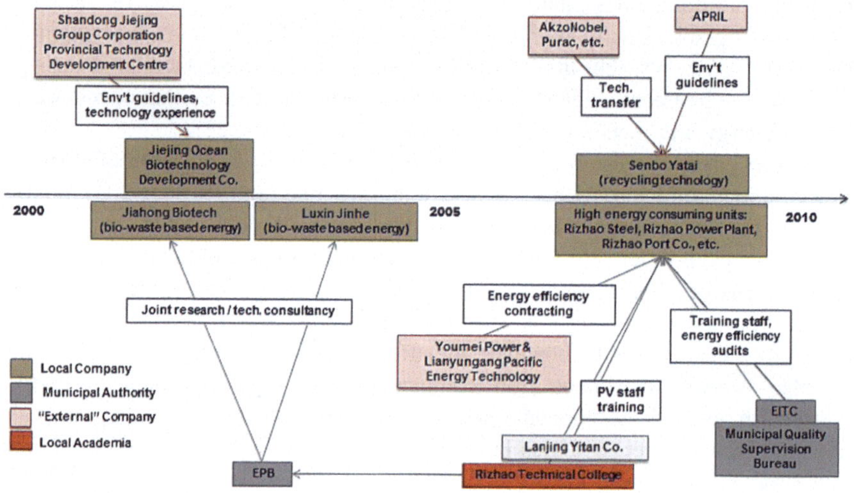

Fig. 3 Rizhao low-carbon related information transfers in the industrial sector. *Source* Rizhao Municipal Yearbooks

cooperation in R&D. Two of the city's chemical industrial units, Jiahong Biotechnology and Luxin Jinhe, have developed biowaste to energy technologies in cooperation with the municipal EPB. These forms of promotion have contributed to a marked increase in the use of biowaste as an energy source in the industrial sector since 2009 (Fig. 2).

Rizhao's financial policies, training programs and joint research have indirectly contributed to the transfers of information that facilitate learning processes related to the increased adoption of low-carbon technologies. The examples are outlined in the figure below. Although the figure contains only a small number of examples, they suggest that the adoption of low-carbon technologies involves connections between authorities, firms and academia, which function as channels for information transfers (Fig. 3).

5 MLP Mechanisms from a Governance Perspective

The MLP framework is primarily used to understand processes in which new technologies become mainstreamed and eventually stabilize into new dominant systems; policy-making processes are not necessarily an integral part of these dynamics. However, an explicit focus on policy processes is essential for understanding how underlying political and economic structures shape the development directions of an urban system. In Rizhao, the negotiation of continued expansion of heavy, fossil fuel-dependent enterprises, backed up by powerful state-industrial interests, conflicts with the creation of low-carbon development trajectories. In terms of sociotechnical regimes, the fossil fuel-based industry can be considered a dominant system where policy processes occupy an integral and reinforcing function. Power relations may be seen as playing a key role in explaining the obduracy of incumbent regimes. It should also be noted that it makes a crucial difference which actors are included in the "transition arena". As development interests differ between sectors and actors, the creation of visions differs depending on who is influencing them. It should be emphasized, however, that this study focuses on the processes taking place at an urban level, where decision-making is limited by the administrative capacities of the city government. If processes are studied from a national perspective, these interactions would most likely constitute a more central part of transition dynamics, as decision-making would be directly related to the construction of low-carbon visions (e.g. support of different forms of technologies or practices). In the case of this urban case study, many low-carbon policy choices are already given, and actors need to negotiate their interests based on these guidelines.

An overlap between the governance and the transitions literatures can be found by considering how different governance strategies contribute to different MLP mechanisms. According to evidence from this study, control-demand instruments may act as direct pressures for change, which may be effective in a highly controlled top-down system. Economic support instruments may contribute to the protection for niche technologies and also contribute to the establishment of channels for information transfers, which, according to the MLP, are required in the processes of sociotechnical change. Training campaigns and state–private cooperation strategies can also be used to develop niche technologies and facilitate learning processes. Returning to the debate on the "manageability" and "co-evolution" of transitions, this study suggests that low-carbon processes are strongly influenced by policy strategies. Smith has noted that energy policy-makers may alter regulations, but the responses of actors determine policy outcomes [17]. From this perspective, it can be argued that although the actions of firms are not directly controlled by policy-makers, they are conditioned by policy contexts. The examples above illustrate firms' responses to highly deliberate attempts at shifting their behaviour in a low-carbon direction: responses to command-control instruments, economic incentives and the integration of new information into business strategies. Also, establishing connections between sectors may further transition dynamics in a top-down context. Although transition dynamics involve the interaction of a large amount of actors within different sectors, yet such linkages and learning processes can have a direct connection with the development strategies. It should be noted, however, that the manageability of transitions depends on the sector in question. For industrial development, especially in a city dominated by a number of large state-owned enterprises, control over development is less fragmented. In a transition involving changes in consumer lifestyles (e.g. habits related to transport, resource consumption), the realization of change is much more likely to be characterized by unpredictability and lack of control. However, the messy reality of low-carbon development in China does seem to be guided by not altogether invisible hands of change.

References

1. Smith, A. & Stirling, A. & Berkhout, F. (2005), "*The governance of sustainable socio-technical transitions*", *Research Policy*, Vol. 34, Issue 10, Pp. 1491–1510
2. Bulkeley, H., Castan Broto, V., Hodson, M. & Marvin, S. (2011), *Cities and Low Carbon Transitions*, Routledge, London
3. International Energy Agency (IEA), (2007), "WORLD ENERGY OUTLOOK 2007: China and India Insights", IEA, Paris
4. Rosen, D.H. & Houser, T. (2007), "China energy: a guide for the perplexed", Centre for Strategic and International Studies and Peterson Institute for International Economics, Washington DC
5. World Bank (2012), *Sustainable Low-Carbon City Development in China,* Edited by Baeumler, A. & Ijjasz-Vasquez, E. & Mehndiratta, S., The World Bank, Washington
6. International Energy Agency (2013), Redrawing the Energy-Climate Map, World Energy Outlook Special Report 2013, IEA, Paris Accessed 2013.11 at: http://www.iea.org/publications/freepublic
7. Newig, J., et al. (2007), "Editorial: Governance for Sustainable Development in the Face of Ambivalence, Uncertainty and Distributed Power: an Introduction", *Journal of Environmental Policy & Planning,* Vol. 9, Issue 3–4, Pp. 185–192
8. Geels, F. (2004), "From sectoral systems of innovation to socio-technical systems: insights about dynamics and change from sociology and institutional theory", *Research Policy*, Vol. 33 (6–7), Pp. 897–920
9. Hughes, T.P. (1983), *Networks of Power: Electrification in Western Society, 1880–1930*, Johns Hopkins University Press, Baltimore
10. Hughes, T.P (1987), "The Evolution of Large Technological Systems", In: Bijker, W.E., Hughes, T.P., Pinch, T., *The Social Construction of Technological Systems: New Directions in the Sociology and History of Technology*, MIT Press, Cambridge
11. Nelson, R. & Winter, S. (1977), "In search of useful theory of innovation", Research Policy, Vol. 6, Issue 1, Pp. 36–76
12. Rip, A. &Kemp, R. (1998), "Technological change", in Human Choice and Climate Change, Vol. 2, Resources and technology, (Eds) Rayner, S & Malone, E., Battelle Press, Washington
13. Elzen, B. & Geels, F.W. & Green, K. (2004), *System Innovation and the Transition to Sustainability,* Edward Elgar, Cheltenham
14. Kemp, R. & Rotmans, J. (2004), "Managing the transition to sustainable mobility", In Elzen, B. & Geels, F.W. & Green, K., *System Innovation and the Transition to Sustainability,* Edward Elgar, Cheltenham

15. Loorbach, D. and Rotmans, J. (2010), "The practice of transition management; examples and lessons from four distinct cases", *Futures*, Vol. 42, Issue 3, Pp. 237–246
16. Smith, A. & Stirling, A. (2008), *Social-ecological resilience and sociotechnical transitions: critical issues for sustainability governance,* STEPS Working Paper 8
17. Smith, A. & Voss, J. & Grin, J. (2010), "Innovation studies and sustainability transitions: The allure of the multi-level perspective and its challenges", *Research Policy*, Vol. 39, Issue: 4, Pp. 435–448
18. Kuzemko, C. (2013), "Understanding the Politics of Low Carbon Transition: Context, Paradigms and Power", EPG Working Paper: 1301, University of Exeter
19. Dowding, K. (1995), "Model or Metaphor? A Critical Review of the Policy Network Approach", *Political Studies*, Vol. 43, Issue Supplement 1, Pp. 136–158
20. Van Waarden, F. (1992), "Dimensions and types of policy networks", *European Journal of Political Research*, Vol. 21, Issue 1–2
21. Tsang, S. & Kolk, A. (2010), "The evolution of Chinese policies and governance structures on environment, energy and climate", *Environmental Policy and Governance*, Vol. 20, Issue 3, Pp.180–196
22. Researcher's interviews with policy-makers, local officials, companies and urban planners (2013)
23. Qi, Y. & Ma, L. & Zhang, H. & Li, H. (2008), "Translating a global issue into a local priority: China's local government response to climate change", *Environmental Development*, Vol. 17, Pp. 379–400
24. SPG (2008) Shandong People's Government: 首页 > 执法监管 > 环境 > "山东省公布2007 年度 17市节能目标责任考核情况", Accessed 2013.10 at: http://www.gov.cn/zfjg/content_1027149.htm
25. Rizhao Steel (2011): 首页 > 社会责任 > 节能减排 > "日照钢铁加大节能减排力度 建设全流程清洁生产企业纪实", Accessed 2013.10 at: http://www.rizhaosteel.com/News_info.asp?SclassId=34&ID=35
26. Rizhao Statistical Yearbooks (2007–2006), China Statistics Press, Rizhao
27. SJG (2012) Shandong Jiejing Group > About > Rizhao Jiejing Ocean Biotechnology Development Co., Ltd., Accessed 2013.10 at: http://www.china-jiejing.com/enu/related/jjhyswjs.html
28. RMPG (2009), Rizhao Municipal People's Government: 首页 > 新闻动态 > 港城资讯 > "我市新能源开发利用进入新阶段", Accessed 2013.10 at: http://www.rizhao.gov.cn/ContShow.php?category_id=183&aiticle_id=3904

The Effective Mixture of Anatase–Rutile Nanoparticles as Dye-Sensitized Solar Cell (DSSC) Using Natural Dye of *Garcinia mangostana* and *Rhoeo spathacea* Extract

Bibit Lestari, Irana Eka Putri, Ruri Agung Wahyuono, Dyah Sawitri and Doty Dewi Risanti

Abstract A synergistic effect between anatase and rutile TiO_2 is known to be able to improve light harvesting and the overall solar conversion efficiency. Dye-sensitized solar cell (DSSC) was fabricated using the mixture of anatase–rutile phase of TiO_2 nanoparticles prepared through co-precipitation method with $TiCl_3$ precursors. The particle's sizes were 30 nm for anatase and 44.9 nm for rutile as characterized by using XRD (X-ray diffraction). The natural dyes employed were *Garcinia mangostana* pericarp extract and *Rhoeo spathacea* extract. The two selected dyes were used because of their abundant availability and good performance for being used as photosensitizer as indicated by their absorbance wavelength of 392.5 nm for *G. mangostana* extract and 432 and 658 nm for *R. spathacea*. Both dyes contain anthocyanin pigment, which is a known visible-light trapper that undergoes charge excitation, and upon complexation with a metal oxide semiconductor like TiO_2, a charge-transfer process occurs. Our results show that the best performance of DSSC having anatase/rutile ratio of 90:10 (η = 0.076%) and (η = 0.063%) was obtained for samples sensitized with *G. mangostana* extract and *R. spathacea* extract, respectively. This ratio has the largest surface area and smallest particle size. The synergistic effect of the two phases is due to the ease of electron migration from rutile to anatase conduction band and larger surface area allowing more dye being able to be absorbed.

Keywords DSSC · Titanium dioxide · *Garcinia mangostana* · *Rhoeo spathacea* · Efficiency

B. Lestari (✉) · I.E. Putri · R.A. Wahyuono ·
D. Sawitri · D.D. Risanti
Faculty of Industrial Technology, Department of Engineering Physics, Institut Teknologi Sepuluh Nopember, Kampus ITS Keputih Sukolilo, Surabaya 60111, Indonesia
e-mail: bibit10@mhs.ep.its.ac.id

T. Taufik et al. (eds.), *ICoSI 2014*, DOI 10.1007/978-981-287-661-4_43

1 Introduction

DSSC (dye-sensitized solar cell) was introduced by Grätzel dan O'Regan, and its progressive development increasing its efficiency. Basic principle of DSSC likes photosynthesis process, and it absorbs spectrum of sunlight to produce high current and voltage [1]. Solar radiation can excite electron at highest occupied molecular orbital (HOMO) to the lowest unoccupied molecular orbital (LUMO). Meanwhile, dye absorbs photon energy of visible light to inject electron [2]. TiO_2 (titanium dioxide/titania) is chosen because of its electron reactivity when it is reacted with anthocyanin pigment. The highest efficiency of DSSC was reported to be 2.6–11% by using Ru-complex as its sensitizer [3].

TiO_2 is semiconductor material that has bandgap of 3.0–3.2 eV. It has three main phases, namely anatase, rutile, and brookite. The thermodynamic stability for anatase and rutile phases depends on the crystalic size, i.e., 11 and 35 nm, respectively [4]. TiO_2 is easy to be synthesized, electron reactive to light, corrosion resistant, and high stability [5].

Although natural dyes in DSSCs have shown overall efficiencies below 1%, still they are considered for being alternative rather than using the expensive Ru-complex dye. Among other natural dyes, extract of *Garcinia mangostana* pericarp and *Rhoeo spathacea* can achieve the efficiency of 1.17 and 1.49%, respectively [6]. With respect to the availability of production, *G. mangostana* takes a whole year with average production of 79.07 kilotons/year [7]. *R. spathacea* can be easily grown in the Indonesia and available as decorative plants.

DSSC performance and its efficiency can be developed by many aspects by mixing nanoporous metal oxide semiconductor to obtain larger surface area and enlarge absorbance wavelength dye. In this paper, DSSC performance can be observed by using variation of TiO_2 anatase–rutile volume fraction and anthocyanin pigment of *G. mangostana* pericarp and *R. spathacea* extract.

2 Experiment

TiO_2 was prepared by mixing 10 ml $TiCl_3$ precursor, 0.3 ml hydrochloride, and 4.7 ml aquades at 45 °C until they become homogen. 50 ml hydrochloride and 50 ml NH_4OH solution were added to the solution. When the solution's color was changed to white and solid phase appears, the mixing process was stopped. TiO_2 solution was washed by using aquades for several times until the pungent odor has gone. Then, the annealing process was given at 400 °C/4 h and 1000 °C/7 h, for anatase phase and rutile phase, respectively. The anatase/rutile ratio studied was listed in Table 1.

G. mangostana pericarp powder was extracted using Soxhlet extractor with aquades. The ratio between *G. mangostana* and aquades was 1:5. This process was done at temperature 200 °C for 5 h. Similar process also applies to *R. spathacea*.

Table 1 Variation of anatase/rutile ratios studied

Name	Dye	Name	Dye	Anatase/Rutile ratio
G1	*Garcinia mangostana*	R1	*Rhoeo spathacea*	95%:5%
G2		R2		90%:10%
G3		R3		85%:15%
G4		R4		80%:20%
G5		R5		70%:30%
G6		R6		65%:35%

The obtained TiO_2 powder was characterized by using X-ray diffractometry (Philips X'pert MPD using Cu-Kα with wavelength of 1.54 nm) and Brunauer-Emmet-Teller (NOVA Quantachrome Instrument Version 10.01). XRD is used to determine particle size and TiO_2 phase identification. Particle size can be calculated by using Scherrer's equation.

$$D = \frac{0.89\lambda}{\beta\cos\theta} \tag{1}$$

Where D is the particle size, λ is wavelength of Cu-Kα (1.54 nm), β is full width at half maximum, and θ is the angle of corresponding peak in rad.

Then, BET characterization is used to determine pore size, pore volume, and surface area of TiO_2 nanoparticles by using N_2 gas adsorption. Surface area can be calculated by using:

$$S = \frac{n}{\rho D_{\mathrm{AV}}} \tag{2}$$

Where S is surface area, n is particle's factor (n = 6), ρ is the density of TiO_2 (4.23 g/cm^3), and D_{AV} is the particle size average [8]. Dye-sensitized solution was characterized by using UV–Vis (UV1100) spectrophotometer to measure sunlight absorbance wavelength.

IPCE (incident photon to current conversion efficiency) is used for determining photoconversion efficiency. The IPCE measurement employs halogen lamp (GR-150), monochromator (CT-10T, JASCO), and multimeter (Sanwa PC500a) to obtain current density measurements. IPCE can be calculated by using

$$\%\ \mathrm{IPCE} = \frac{1240\ [\mathrm{eV\,nm}] \times J_{sc}\ [\mu\mathrm{A\,cm}^{-2}]}{\lambda\ [\mathrm{nm}] \times P_{\mathrm{light}}\ [\mu\mathrm{W\,cm}^{-2}]} \tag{3}$$

where J_{sc} and P_{light} are photocurrent density and incident power density of light, respectively, at the corresponding wavelength (nanometer) [9].

3 Results and Discussion

Figure 1 shows the XRD patterns of TiO_2 powder. It is seen that the obtained TiO_2 has anatase (A) and rutile (R) phases for the given annealing process. According to Eq. (1), TiO_2 particle size is 30 nm and it is 44.9 nm for anatase (101) and rutile (110), respectively.

Figure 2a shows the absorption spectra of *G. mangostana* and *R. spathacea*. The absorption peak of *G. mangostana* is 392.5 nm, while that of *R. spathacea* is 432 and 658 nm. The difference in the absorption peak is because *G. mangostana* and *R. spathacea* contain different types of anthocyanin pigment. Anthocyanin pigment has five type, i.e., pelargonidin (orange), cyanidin (orange-red), delphinidin (blue-red), petunidin (blue-red), and malvidin (blue-red) [10]. *G. mangostana* containing cyanidin type of anthocyanin has absorbance peak in the range of 400–500 nm [11], while anthocyanin in *R. spathacea* has absorbance peak in the range of 410–420 nm and 660–670 nm [12].

Figure 2b shows the FTIR spectrum patterns of dye extracts. *G. mangostana* and *R. spathacea* have transmission wave number of 3334.52 and 3343.29 cm^{-1}, respectively, indicating that the dye extracts contain anthocyanin [13].

The photovoltaic performance of DSSCs using natural dye as sensitizer was performed by measuring current–voltage (I–V) under sunlight. Figure 4a, b shows the I–V curves of DSSC using *G. mangostana* (G) and *R. spathacea* (R), respectively. In a DSSC, the overall solar conversion efficiency (η) depends on the short-circuit current density (I_{sc}), open-circuit photovoltage (V_{oc}), fill factor (FF), and the total light incident on the cell (P_{light}), according to Fig. 3:

$$\eta = \frac{J_{SC} \times V_{OC} \times FF}{P_{Light}} \quad (4)$$

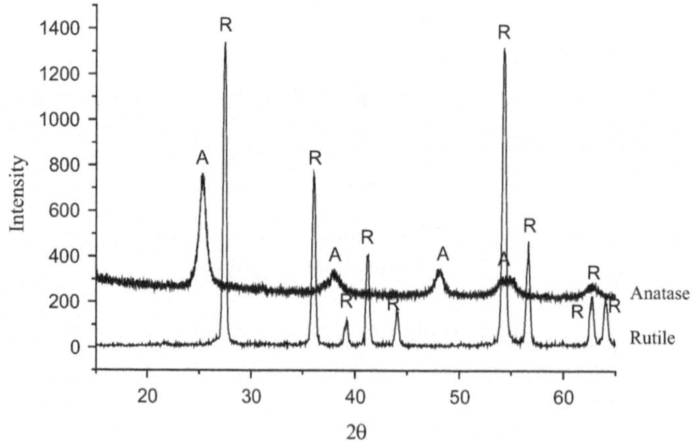

Fig. 1 XRD patterns of anatase and rutile TiO_2

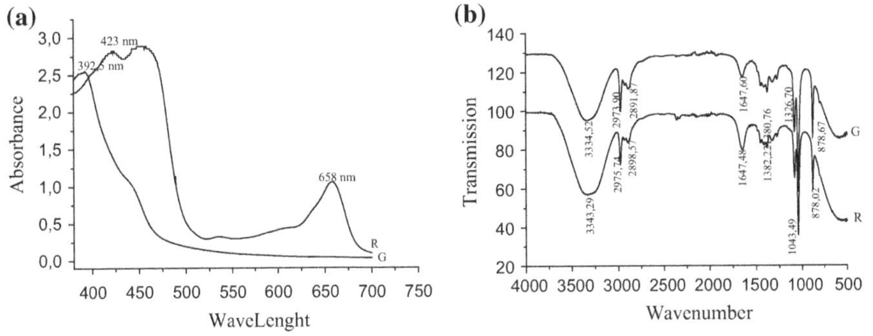

Fig. 2 **a** UV–Vis and **b** FTIR of *G. mangostana* (G) and *R. spathacea* (R) of dye extracts [14]

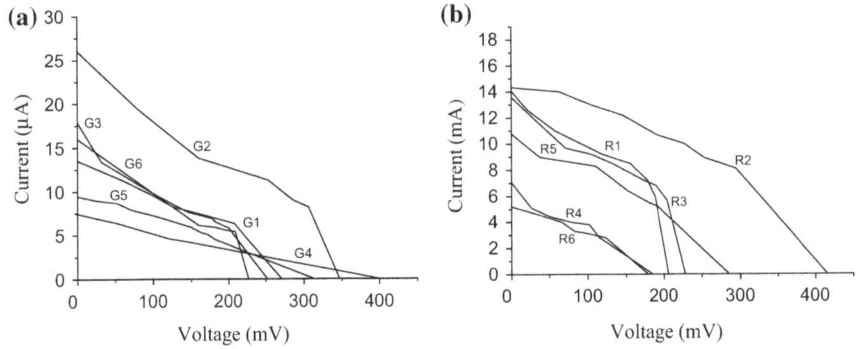

Fig. 3 Current–Voltage (I–V) curves using **a** *G. Mangostana* and **b** *R. spathacea*

Table 2 shows the I–V parameters of DSSCs prepared with different anatase/rutile ratio and both natural dye extracts. From this table, the highest efficiencies ($\eta = 0.0076\%$) and ($\eta = 0.0063\%$) were obtained from G2 and R2, respectively. G2 and R2 cell have similarity; it has rutile contents of 10%. These results agree well with the earlier findings [15] indicating that for rutile percentage between 10 and 15%, the electrons easily migrated from the efficient light harvester rutile, to the anatase conduction band; in this case, the electrons are more mobile.

The DSSCs using 100% anatase TiO_2 or rutile TiO_2 has lower efficiency than DSSCs using mixed-phase TiO_2 nanocomposites (Dalton 2009). Figure 4a shows the efficiency of DSSC with various rutile contents. From these figures, the best performance of DSSC was obtained at rutile content of 10%. This was caused by DSSC using TiO_2 containing 10% rutile has greater surface area (155.65 m^2/g) and has smaller particle size (9.11 nm) (Table 3 and Fig. 4b). Samples with larger surface area will absorb much dye leading to more photon being converted to electrical energy and eventually higher efficiency.

Table 2 Performance of DSSC with various anatase/rutile ratios using the studied dyes

Name	FF (%)	J_{sc} (mA/cm^2)	V_{oc} (mV)	% IPCE	η (%)
G1	36	0.054	271	0.00155	0.035
G2	32	0.104	347	0.00186	0.076
G3	32	0.06	252	0.00207	0.033
G4	22	0.03	400	0.0017	0.017
G5	30	0.038	314	0.00134	0.024
G6	31	0.064	227	0.00198	0.03
R1	30	0.056	206	0.00229	0.035
R2	40	0.057	415	0.00285	0.063
R3	33	0.054	228	0.00177	0.035
R4	30	0.028	185	0.0018	0.01
R5	32	0.043	286	0.00323	0.027
R6	37	0.021	179	0.00062	0.009

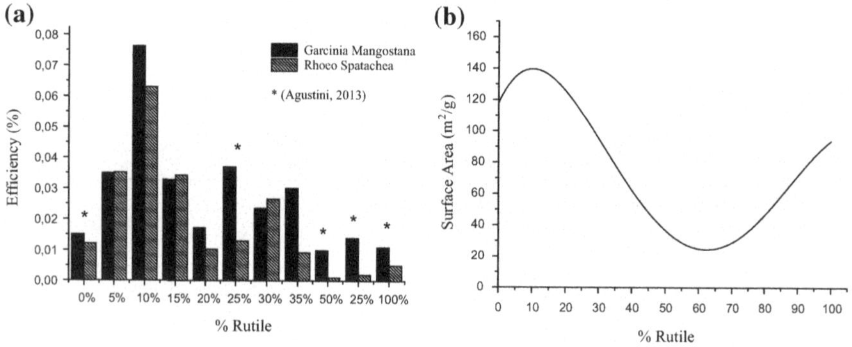

Fig. 4 **a** Efficiency DSSCs and **b** surface area of variation of anatase/rutile ratio

Table 3 D_{Average} as function of rutile percentage

% Rutile	0	10	25	30	50	75	100
D average	12.55	9.11	20.05	11.2	39.02	40.09	15.09

4 Conclusion

The best performance of DSSC was obtained at rutile content of 10% with the cell efficiency (η = 0.0076%) and (η = 0.0063%) for photosensitizer *G. mangostana* and *R. spathacea* dye, respectively. The DSSCs with rutile content of 10% has greater surface area (155.65 m^2/g) and smaller particle size (9.11).

Acknowledgements The authors wish to thank for the funding of the Directorate of Higher Education Ministry of Education and Culture of Republic of Indonesia on National Strategic Grants 2013 with contract number 013516/IT2.7/PN.08.01/2013.

References

1. O'Regan, B., Grätzel, M., (1991). A Low-Cost, High-Efficiency Solar Cell Based on Dye Sensitized Colloidal TiO_2 Films. *Nature*. Vol. 353, pp. 737–740.
2. Grätzel, M. (2003). Review Dye-Sensitized Solar Cells. *Photochemistry and Photobiology C,* vol. 4, pp. 145–153.
3. Halme, Janne. (2002). Dye-sensitized Nanostructured and Organic Photovoltaic Cells: Technical review and Preliminary Tests. *Master Thesis,* Helsinki University of Technology.
4. Zhang, H. Banfield, J F. (1998). Thermodynamics Analysis of Phase Stability of Nanocrystalline Titania. *J. Mater. Chem.* Volume 8, pp. 2073–2076.
5. Dholam, Rupali S. (2010). Development of Solar Sensitive Thin Film for Water Splitting and Water Heating Using Solar Concentrator. *Doctoral Dissertation of University of Trento,* Italy.
6. Narayan Rita Monishka. (2001). Dye Sensitized Solar Cells Based on Natural Photosensitizers. Makara, Technology vol. 11. No.2, page. 78–84.
7. Direktorat Jenderal Pengolahan Dan Pemasaran Hasil Pertanian, (2006). Road Map Pengembangan Agroindustri Manggis.
8. Tanaka, Y., Suganuma, M. (2001). Effect of Heat Treatment on Photocatalytic Properties of Sol-Gel Derived Polycrystalline. *J. Sol-gel Sci and Tech.* Vol. 22, pp. 83–89.
9. Varghese, K.O and Grimes, A.C. (2007). Appropriate Strategies for Determining The Photoconversion Efficiency of Water Photoelectrolysis Cells: A Review With Examples Using Titania Nanotube Array Photoanodes. Solar Energy Materials & Solar Cells 92 (2008) 374–384.
10. J.M.R.C. Fernando., Sendeera, G.K.R. "Natural Anthocyanins as Photosensitizers for Dye-Sensitized Solar Devices". Res. Comm. Current. Sci., 95, 663–666, (2008).
11. Chang, H and Lo, Y.J., (2010). Pomegranate Leaves and Mulberry Fruit as Natural Sensitizernfor Dy-Sensitized Solar Cell. Journal of Solar Energy vol. 84, page 1833–1847.
12. Lai, H.W. (2007). Commercial and Natural Dyes as Photosensitizers for a Water-Based Dye-Sensitized Solar Cell Loaded with Gold Nanoparticles. Journal of Photochemistry and Photobiology A: Chemistry 195 (2008) 307–313.
13. Agustini, S., Wahyuono, R.A., Sawitri, D., Risanti, D.D. (2013). Effect of Nano Anatase-Rutile Tio_2 Volume Fraction With Natural Dye Containing Anthocyanin on The Dye Sensitized Solar Cell. Padjadjaran International Physics Symposium 2013 (PIPS.2013). AIP Conf Proc 1554 Lpp 66–69 2013.
14. Ali R.A.M and Nayan.(2009).Fabrication and Analysis of Dye-Sensitized Solar Cell Using Natural Dye Extracted from Dragon Fruit. International journal of integrated engineering (issue on electrical and electronic engineering) page. 55–62.
15. Li, G., Richter, C.P., Milot,R.L., Cai, L., Schmuttenmaer, C.A., Crabtree, R.H., Brudvig, G. W., Batista, V.S.. (2009). Synergistic Effect Between Anatase and Rutile Tio2 Nanoparticles in Dye-Sensitized Solar Cells. *Dalton Trans.*, 2009, 10078–10085.

The Prospective Analysis of Coastal Town Development Based on Waterfront City (Case Study: Bantan Sub-District, Bengkalis–Riau Province)

Puji Astuti, Mardianto Manan, Febby Asteriani, Cihe Aprilia Bintang and Dharfrimadil Akhyar

Abstract Bantan Sub-district is one of area in Bengkalis District, Riau Province. Located on the coast called the Selat Baru Beach, ecological island separated from its parent, in this case is the Island of Sumatra, remote and insular nature. Having a large number of endemic species and the diversity of high value has a relatively small catchment area so that most of the surface water and sediment flow into the sea. In terms of social, economic and cultural, small islands is unique compared to its parent island. Bantan Sub-district, the potential as a small island is extremely limited and vulnerable to environmental changes, so must be optimized utilization and carefully constructed. The advantage of this area is located adjacent to Malaysia and there is an international ferry port to Malaysia. By positioning the region directly opposite the Selat Baru Beach, the right concept for supporting any limitation in this area is Waterfront City, followed by realizing sustainable conservation of environment. But, before doing some studies to support the concept, it is very important to measure prospectively of Bantan Sub-district based on coastal town in creating a waterfront city. The method used in this study is deductive method with the intention of seeing prospectively coastal city development activities based on waterfront city. The results of analysis, coastal town development based on waterfront city in Bantan Sub-district are very prospective business to develop. The business is well positioned as a mature and profitable strategic advantage profile. This analysis is a combination between matrix ETOP and SAP position. The

P. Astuti (✉) · M. Manan · F. Asteriani · C.A. Bintang · D. Akhyar
Department Urban & Regional Planning, Riau Islamic University, Pekanbaru, Indonesia
e-mail: pujiastutiafrinal@yahoo.co.id

M. Manan
e-mail: manan_mardianto@yahoo.co.id

F. Asteriani
e-mail: febby411@yahoo.com

C.A. Bintang
e-mail: aprilia_cihe@yahoo.co.id

D. Akhyar
e-mail: dharfrimadilakhyar@yahoo.com

T. Taufik et al. (eds.), *ICoSI 2014*, DOI 10.1007/978-981-287-661-4_44

integration of the supporting elements can create a sustainable coastal town. As the leading sectors and capital develop, the region favored the tourism sector, especially beach tourism. With the object of coastal tourism can be a tourist attraction in the Bantan Sub-district, then this can be used as a strategic area of tourism development that will be able to bring significant profits to the advancement and improvement of the economic potential of tourism locals.

Keywords Prospectively · Coastal city · Waterfront city · Straits · Tourism

1 Introduction

Bantan as a small island is extremely limited and vulnerable toward environmental changes, so it must be an optimized utilization and carefully constructed. Based on geographical territory conditions, the right concept to optimize the utilization of the tourism region is a waterfront city. Another advantage is Bantan area directly facing the Malacca Strait as an international shipping route. So the development of the region is associated with a larger area and a consideration of its own in its development. The fact in Bantan Sub-district is associated with the development of the region:

1. Favorable geographical location.
   According to Law No. 27 of 2007 on the Management of Coastal Areas and Small Islands, Bantan is a group of small islands with an area of 424.4 km^2 and its unity ecosystem. Located on the Bengkalis coastal, dealing directly with international waters with good tourism potential. This layout gives difficulties in management but it gives a big advantage if managed in a proper way.
2. The lack of supporting sector to develop.
   Supporting sectors in Selat Baru Beach to develop coastal tourism still remained at a very low stage, because the factors that affect the utilization of the tourism region are not identified yet. Thus, the efforts should be made by the government, private sectors and communities. Bantan can improve and promotes the area by integrating of various elements.
3. It has not been optimal utilization of the tourism region.
   This is not a part of the issues that emerged relevant in terms of the management of the tourism sector by the relevant agencies, the lack of cross-sectorial coordination, tourism development policies that have not been fully implemented. Thus, integration among the supporting elements, the potential of tourism in the Bantan can create the mainstay tourism sector and serve as capital to develop the region and be able to improve and develop all potential tourist attraction there.

2 Research Methods

Based on the problems, the question is about how the future prospects Bantan in tourism development based on the concept of waterfront city? Has the development of the area and the use of the concept of waterfront city already provided prospective opportunities for the region? The aims of the study aim to measure prospectively tourism development on the base on waterfront city in the Selat Baru Beach.

2.1 The Prospectivity Analyzed Using SWOT ETOP Analysis

The stage of the analysis was done to identify the potential analysis, opportunities, threats, and weaknesses of the study area conditions. To identify the problem, the analysis used table analysis that describes each object or potential locations and describes each object that contains qualitative considerations. The SWOT ETOP analysis advantages are more measurable results using measuring devices that have been defined (Table 1).

(a) **External Variables Analysis (ETOP Matrix)**
The ETOP matrix (Environmental Threats Opportunity Profile) can be seen the position of a business development. Determining ETOP matrix is as follows: (1) Identify the elements that are the opportunities and threats. (2) Assigning weights to each element. (3) Multiplying the weight by value, to obtain the weighted value. (4) Adding up the value of opportunities weighted (Fig. 1).

Table 1 SWOT analysis

External factor	Internal factor	
	Strength (S)	Weakness (W)
Opportunity (O)	SO	WO
Treath (T)	ST	WT

Source Oktora [12]

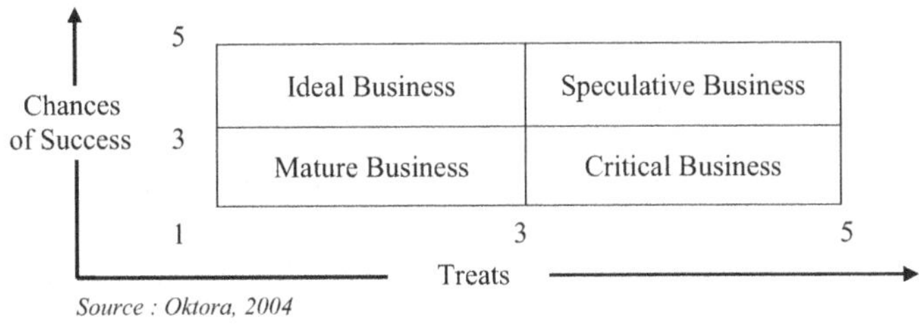

Fig. 1 ETOP matrix. *Source* Oktora [12]

(b) **Internal Variable Analysis (SAP Matrix)**

The steps of internal variables analysis can be done with the same procedure with the determination ETOP matrix, but using the elements of strength and weakness based on internal variables then can be a strategic advantage profile or Strategic Advantage Profile (SAP). To determine SAP, we need to hold the classification of internal variables as follows:

1. Determination Scale
   I = (maximum scale−the scale at least)/number of classification
   I = (5−1)/6
   I = 0.67
2. Grouping competitive position, can be seen in the following Tables 2 and 3.

(c) **SWOT Matrix Analysis**

SWOT matrix analysis is an analytical matrix that is used to determine whether or not prospective effect of the coastal town development based on waterfront city. In the analysis, the variable use strengths, weakness, opportunities, and threats. The SWOT analysis is shown in Table 4.

Table 2 Grouping position SAP

Value	Position
1–1.66	Avoid
1.67–2.33	Weak
2.34–3	Tenable
3.01–3.67	Favorable
3.68–4.34	Strong
4.35–5.01	Dominant

Source Oktora [12]

Table 3 Explanation SAP

No.	Position	Explanation
1.	Avoid	Business development effort that has many risks. Strategy development used to be really guaranteed
2.	Weak	Development of a business that is in a weak position has quite a lot of hurdles to make a profit. Moreover, in this position, the development requires substantial investment sources
3.	Tenable	To develop a business that is in this position, and then there must be aggressive measures to keep this development effort
4.	Favorable	Development in the advantages position means that the developments is at a very low uncertainty and have a little extra effort
5.	Strong	Development of this position must be followed by the flexibility designed anticipation with environmental controls and a careful evaluation of the routine so that it can respond to the needs and opportunities that further strengthen the position

Source Oktora [12]

Table 4 SWOT matrix

SAP	ETOP			
	Ideal	Mature	Speculative	Critical
Avoid	Prospective (P)	P	P	P
Weak	P	P	P	P
Tenable	P	P	P	U
Favorable	P	P	U	U
Strong	P	U	U	U
Dominant	Unprospective (U)	U	U	U

Source Oktora [12]

Table 5 Explanation of ETOP matrix

Ideal business	Speculative business
Development goals can be enjoyed by all the stakeholders and the expected benefits to be achieved	Implementation of good development, helped rescue a poor strategy formulation, or prevent failure
Mature business	Critical business
Implementation is lacking in development will hinder a good strategy	Poor development strategy is characterized by an inability to execute/ implement good

Source Oktora [12]

The Value Score of SWOT ETOP is using some elements like Strength, Weakness, Opportunity, and Threat. This value is measured by the amount of the interest rate and the effect of the elements that have been defined. On that score use two values using the following assessment (Table 5):

1. The value based on interest rate of elements:

a. Not important (1)
b. Less Important (2)
c. Neutral (3)
d. Important (4)
e. Very Important (5)

2. Value based on effect of elements:

a. Not important (1)
b. Less Important (2)
c. Neutral (3)
d. Important (4)
e. Very Important (5)

2.2 *Analysis Dan Interpretation*

2.2.1 The Analysis of Internal Variables

The analysis is used to identify and analyzed the potential, opportunities, threats, and weaknesses of the study area conditions. To identify the problems is used table analysis that describes each object or location of potential as well as to determine whether the tourism development in Bantan at unprospective or prospective for the future (Tables 6, 7, 8, 9, and 10).

Table 6 Internal elements of coastal tourism development in Bantan sub district

No.	Internal elements	
	Strengths	Weaknesses
1.	The tourism activities in Bantan can develop coastal region	Lack of processing of tourism region in the Bantan Sub District
2.	Selat Baru Beach existence of coastal tourism can increase region economic	Lack of infrastructure and facilities support in the area of tourism led to the region rarely visited tourism
3.	Potential of cultural and natural resources to create attraction in tourism region	Tourism area development plan is still not implemented
4.	The existence of tourism area in Bantan allow the emergence of new business opportunities	Lack of support activities that would turn the region
5.	The tourism area in Bantan organized by the government	Lack of cooperation between private and the government, resulting in under-developed tourism region

Source Analysis 2013

Table 7 Calculation of strength elements

No.	Strength elements	Weight (W)	Value (V)	(W × V)
1.	The tourism activities in Bantan can develop coastal region	0.11	3.90	0.43
2.	Selat Baru Beach existence of coastal tourism can increase region economic	0.10	3.50	0.35
3.	Potential of cultural and natural resources to create attraction in tourism region	0.10	3.80	0.38
4.	The existence of tourism area in Bantan allow the emergence of new business opportunities	0.09	3.90	0.35
5.	The tourism area in Bantan organized by the government	0.09	3.80	0.34
Totally				1.85

Source Analysis 2013

Table 8 Description of strength elements

No.	Strength elements	Description
1.	The tourism activities in Bantan can develop coastal region	Selat Baru Beach Tourism Regions in Bantan can improve rapid growth in Bengkalis District
2	Selat Baru Beach existence of coastal tourism can increase region economic	Selat Baru coastal resorts that exist in Bantan to boost the region economics
3.	Potential of cultural and natural resources to create attraction in tourism region	Selat Baru Beach tourist area in Bantan has good characteristic and potential of natural resources and culture that could be an attraction so visitors always wanted to visit there
4.	The existence of tourism area in Bantan allow the emergence of new business opportunities	The tourism region existence can make an effort along the Selat Baru Beach for the communities
5.	The tourism area in Bantan organized by the government	Selat Baru coastal region in Bantan has been arranged by Spatial Planning, Detail Spatial Plan so that land-use changes could occur then the environmental aspects of the most important aspects to be considered

Source Analysis 2013

Table 9 Calculation of weakness elements

No.	Weakness elements	Weight (W)	Value (V)	(W × V)
1.	Lack of processing of tourism region in the Bantan Sub District	0.11	3.60	0.40
2.	Lack of infrastructure and facilities support in the area of tourism led to the region rarely visited tourism	0.10	3.80	0.38
3.	Tourism area development plan is still not implemented	0.10	3.30	0.33
4.	Lack of support activities that would turn the region	0.10	3.30	0.33
5.	Lack of cooperation between private and the government, resulting in under-developed tourism region	0.10	3.20	0.32
Totally				1.76

Source Analysis 2013

From the analysis, the obtained weighted value (score) of the elements of strength is 1.85, while the scores for the element weakness is 1.76. So that the total score of the analysis of these two elements is 3.61. Based on these scores, the position of the coastal city of tourism development in Bantan with the concept of a waterfront city is in the interval 3.01–3.67, it means that the development in Selat Baru Beach have an advantageous position. The development is located in a favorable position means that the development is at a very low uncertainty

Table 10 Description of weakness elements

No.	Weakness elements	Description
1.	Lack of processing of tourism region in the Bantan Sub District	Management is very important in the development, less optimal processing in management will cause the area is not well developed
2.	Lack of infrastructure and facilities support in the area of tourism led to the region rarely visited tourism	Lack of facilities and supporting infrastructure can lead to less than optimal function of the area of tourism in Bantan
3.	Tourism area development plan is still not implemented	The development plan has not been implemented due to lack of funds and limited management
4.	Lack of support activities that would turn the region	The existence of tourism support activities is one factor that can turn the tourism area, with efforts to revive tourism region can provide activities
5.	Lack of cooperation between private and the government, resulting in under-developed tourism region	Government and private should be able to work together and agree, if the government does not cooperate with the private sector is not the tourism regions of the developing and developed

Source Analysis 2013

condition and has little resistance. The strengths and weaknesses, the development will benefit if the existence is of a good management. The development will also increase investment and have the opportunity to create an attractive environment for the community.

2.2.2 The Analysis External Variable

Through ETOP matrix can be seen the position of a business development. The following are ways of determining ETOP matrix, namely: (a) Identification of the opportunities and threats elements, (b) Assigning weights to each element, (c) multiplying the weight by value, to obtain the weighted value, (d) Summing values of opportunities (Tables 11, 12, 13, 14, and 15).

Based on analysis, the obtained weighted value (score) of the element of opportunities is 1.93, while the scores for the element threat is 1.70. So that the total score of the analysis of these two elements is 3.63. It means that the position of tourism development in Selat Baru Beach in Bantan Sub-district is mature businesses that development must be a companied by a good implementation (Bonoma in Salusu 1984).

Table 11 External elements of coastal tourism development in Bantan sub-district

No.	External elements	
	Opportunities	Treats
1.	The Ferry international will be able to facilitate access to Bengkalis	If there is no plan or area development of tourism, the region does not develop
2.	The preparation and direction of development based on Bengkalis Spatial Planning as Tourism Zone in Bantan Sub District	Lack of transparency between the government and the private sector because the private sector to buy public land at low prices
3.	Bengkalis is at a strategic location adjacent to the neighboring countries	Still not drafting a tourism master plan to make tourism development policy direction that has not been planned carefully and clearly
4.	Based on the Spatial Detail Plan Bantan Sub District the policies and programs that make the Bantan as tourism region	The behavior of people who lack awareness of the environment would thus undermine the sustainability of coastal

Source Analysis 2013

Table 12 Calculation of opportunities elements

No.	Opportunities elements	Weight (W)	Value (N)	(W × N)
1.	The Ferry international will be able to facilitate access to Bengkalis	0.14	4.1	0.57
2.	The preparation and direction of development based on Bengkalis Spatial Planning as Tourism Zone in Bantan Sub District	0.13	3.8	0.49
3.	Bengkalis is at a strategic location adjacent to the neighboring countries	0.12	3.50	0.42
4.	Based on the Spatial Detail Plan Bantan Sub District the policies and programs that make the Bantan as tourism region	0.12	3.7	0.44
Totally				1.93

Source Analysis 2013

2.3 The Prospective Analysis of Coastal Town Development Based on Waterfront City

This analysis is used to determine whether the business development of coastal town in Selat Baru Beach Bantan is a prospective or unprospective business. This analysis is a combination between matrix ETOP and SAP position. Based on the analysis that had been done previously, the development of tourism in Bantan was in a position as a mature business profile and strategic excellence profitable, it was found that the development of tourism in Bantan is a prospective business. Based on analysis of the obtained SAP, tourism development position Selat Baru Beach is

Table 13 Description of opportunities elements

No.	Opportunities elements	Description
1.	The Ferry international will be able to facilitate access to Bengkalis	The existence of the International Ferry port to facilitate travel or accessibility to Bengkalis and Malacca Strait
2.	The preparation and direction of development based on Bengkalis Spatial Planning as Tourism Zone in Bantan Sub District	Formulation development direction should be based on Bengkalis Spatial Planning, so if there is no that Tourism Regions cannot be developed
3.	Bengkalis is at a strategic location adjacent to the neighboring countries	Bengkalis strategic location adjacent to Malaysia so as to give effect to the promotion of tourism development
4.	Based on the Bantan Spatial Detail Plan Sub District the policies and programs that make the Bantan as tourism region	Bantan Spatial Detail Plan can make Bantan as regional scale tourism area, with development of infrastructures and facilities comes developing areas in Bantan

Source Analysis 2013

Table 14 Calculation of treats elements

No.	Treats elements	Weight (W)	Value N)	(W × N)
1.	If there is no plan or area development of tourism, the region does not develop	0.13	3.20	0.42
2.	Lack of transparency between the government and the private sector because the private sector to buy public land at low prices	0.12	3.40	0.41
3.	Still not drafting a tourism master plan to make tourism development policy direction that has not been planned carefully and clearly	0.12	3.50	0.42
4.	The behavior of people who lack awareness of the environment would thus undermine the sustainability of coastal	0.13	3.50	0.46
Totally				1.70

Source Analysis 2013

located in the strategic position. The position obtained from the ETOP analysis is mature businesses. It means, concept of waterfront city suitable to apply in Bantan Sub District.

Based on the research it can be concluded as follows:

1. Bantan Sub-district is a small island located in the coastal region that has Selat Baru Beach, located in Desa Selat Baru and not far from the capital city of Bengkalis and dealing directly with international water.
2. Determining of tourism area in concept waterfront city based on the location of the coastal areas measured by prospective analysis placing Selat Baru Beach in prospective position for development. And SWOT ETOP analysis and SAP, the development of this region are in the business profitable and mature.

Table 15 Description of treats elements

No.	Treats elements	Description
1.	If there is no plan or area development of tourism, the region does not develop	There is no tourism regional development plans in Bantan, the region will not develop and it also can not increase the economy in the district and Sub District
2.	Lack of transparency between the government and the private sector because the private sector to buy public land at low prices	Authorities fear that the private sector buy land near the area of tourism with a very cheap price because the government does not want to open up to the private sector
3.	Still not drafting a tourism master plan to make tourism development policy direction that has not been planned carefully and clearly	Master Plan of Tourism Region development plan is a guide to the development of tourism in the region, the Coastal Tourism development in Selat Baru Beach has not planned carefully and clearly, if the plan does not mature and obviously it could not have happened tourism development of the area
4.	The behavior of people who lack awareness of the environment would thus undermine the sustainability of coastal	Human behavior is less awareness of the environment is a factor that is quite influential in affecting the quality of the environment, if this attitude is maintained, it will cause environmental degradation in the area of tourism

Source Analysis 2013

3. The development of this coastal city is oriented to the development of public welfare in Selat Baru. The efforts to protect the water front area (waterfront city) of damage are necessary to have the cooperation and unity of vision from stakeholders, namely the public, government and private sector to realize the character of the waterfront area in the Selat Baru Beach.

References

1. Adisasmita, Rahardjo. (2006). *Pembangunan Kelautan dan Kewilayahan*. Yogyakarta; Graha Ilmu.
2. Adisasmita, Rahardjo. (2008). *Pengembangan Wilayah*. Yogyakarta; Graha Ilmu.
3. Adjisasmito, Sakti. (2010). *Perencanaan Infrastruktur Transportasi Wilayah*. Yogyakarta; Graha Ilmu.
4. Akhyar, Dharfrimadil et al. (2013). Final Report Urban Planning Studio*: Pengembangan Pariwisata Pantai Selat Baru Kabupaten Bengkalis*. Urban and Regional Planning Department, Riau Islamic University; Pekanbaru.
5. Documents Of Regional Spatial Plan Riau Province in 2012.
6. Documents Of Regional Spatial Planning Bengkalis District in 2012.
7. Document Bengkalis in Figure, 2012.
8. Law No.53 of 1999.

9. Law No.12 of 2009.
10. Law No.26 of 2007.
11. Mulyanto. (2006). *Prinsip-Prinsip Pengembangan Wilayah*. Yogyakarta; Graha Ilmu.
12. Oktora, Halimah. (2004). *Identifikasi Faktor-Faktor Prioritas Pengembangan Taman Ronggowarsito sebagai ruang Terbuka Publik Tepian Sungai Bengawan Solo*. Theses. Planologi Diponegoro University; Semarang.
13. Riyadi, Masykur. (2000). *Pembangunan Daerah Melalui Pengembangan Wilayah*. Yogyakarta.
14. Suprapto dan Pasaribu. (2005). *Penataan Ruang Secara Global*. Yogyakarta; Bumi Aksara.
15. Tarigan, Robinson. (2005). *Perencanaan Pembangunan Wilayah*. Jakarta; Bumi Aksara.
16. Warpani, Suwardjoko. (1980). *Analisa Kota dan Daerah*. Bandung; Penerbit ITB.

User Interface Evaluation of Indonesian Online News Portals: Case Study of Vivanews and Detikcom

Muhammad Yazid, Rizky Arya Pratama and Slamet Riyadi

Abstract Following the advance of information technology, nowadays online news portal is very popular. By accessing news portal, people can easily read news from their smartphone, smart tab, and other mobile devices. In terms of human–computer interaction aspect, news portal should serve an effective interaction between the website and the readers. News portal should be designed based on standard interface design rules, which is commonly called Eight Golden Rules. The objective of this research is to evaluate the interface of two major Indonesian online news portals, Viva News and Detikcom. Both news portals were evaluated using heuristic method based on eight criteria; consistency, shortcut availability informative feedback, design dialog, error handling, easy reversal action, sense of user control, and reduce short-term memory load. The result shows the evaluation of each news portal and the comparison between both news portals. The evaluation result is then useful to improve the usability and functionality of the portals.

Keywords User interface design · Eight golden rules · Online news portal · Usability and functionality

1 Introduction

Penetration of Internet users in Indonesia is increasing significantly. Indonesian Association for Internet User (APJII) in collaboration with Indonesian Statistics Center (BPS) reported that in 2013, user penetration reached 28% of 248 million

M. Yazid (✉) · R.A. Pratama · S. Riyadi
Department of Information Technology, Faculty of Engineering, Universitas Muhammadiyah Yogyakarta, Yogyakarta, Indonesia
e-mail: muhammadyazid26@gmail.com

R.A. Pratama
e-mail: rizkyarya66@gmail.com

S. Riyadi
e-mail: riyadi@umy.ac.id

T. Taufik et al. (eds.), *ICoSI 2014*, DOI 10.1007/978-981-287-661-4_45

citizens of Indonesia [1]. This percentage includes users who access news from online portals. In line with the increment of users, the number of news portals also increases significantly. In the beginning of 2014, there are more than 30 news portals which are managed professionally. Others are news portals which are developed and managed by communities or personal.

In order to assure the usability and functionality of the portal, news portals should serve effective interactions between readers and portals. However, in fact there are many portals that do not take this matter into account. Therefore, this article intends to evaluate the user interface of two major online news portals, namely Vivanews and Detikcom [3]. The evaluation results can be used as references by developers to improve the usability of their portals developed.

2 Methodology

This research was conducted by evaluating the interface of news portals using heuristic evaluation method, which is known as Eight Golden Rules of user interface design [2]. The rules are as follows:

(a) *Strive for consistency*
Interface should be designed consistently in the following aspects: fonts, color, capitalization, layouts, and terminology. Sequence of actions should also be made consistently to perform similar task.

(b) *Enable frequent users to use shortcuts*
Frequent users will intensively interact with the system. To enable users in performing their tasks quickly, the system should provide abbreviations and function keys.

(c) *Offer informative feedback*
The system should provide feedback as response for the user's actions. The feedback ensures interaction between users and interface is in two ways.

(d) *Design dialog to yield closure*
When the system performs sequence of actions, it is important to give to users an informative feedback that the action was completed. This feedback satisfies users and makes certainty of task completion.

(e) *Offer simple error handling*
The interface should not allow a user to make a serious error. For example, the interface should not allow alphabetic characters to be entered in numeric entry fields. However, if the user makes an error, the system should be able to detect the error and offer some mechanism for recovery.

(f) *Easy reversal of actions*
The interface should allow reversal of an action. This feature helps relieve anxiety. The users know that if an error occurs on performing an action, the action can be undone. This encourages exploration of unfamiliar options.

(g) *Sense of user control*
Experienced users like to be in control of the system. Therefore, you should design the interface to make the users the initiators of actions rather than responders.

(h) *Reduce short-term memory load*
You can reduce short-term memory load by.

- Designing simple screens.
- Consolidating multiple page displays.
- Reducing window-motion frequency.
- Providing sufficient training time for codes, mnemonics, and sequences of actions.
- Providing online access to command syntaxes, codes, abbreviation, and other information.

3 Results and Discussion

This section provides detailed evaluation result for the Vivanews and Detikcom portals according to eight rules which were presented previously. In addition, a comparison between both portals is also discussed at the end of this section.

(A) *Consistency*
The result for this rule is divided into three aspects which are as follows:

1. Judging from the font, color, capitalization, and consistent placement of menu. This website has consistent fonts since each page uses the same font. However, color, capitalization, and placement of menu are not consistent as indicated in Figs. 1 and 2. For example, when entering into the main portal of viva website, when the mouse is moved into the title of the top menus (news) submenu appears as shown.
But when choosing one of the news categories in the main portal, display of menu on the top of web page is different.
It means viva website is not so consistent; because there are differences at menu options on the main portal and the news portal. Not just menu option being different but also the difference in font, color, capitalization, and placement of news items.
2. From similarity of terminology that used on menu of the website interface. It is very affect to user because when a website has a different terminology in every page, it will be difficult for users to find the information that we looked for. In addition, the admin will likely difficult to set up a website that was previously developed by other developer. For viva website, the menu on the top of main portal displays several information categories (news, football, life, blogs, forums, buying and

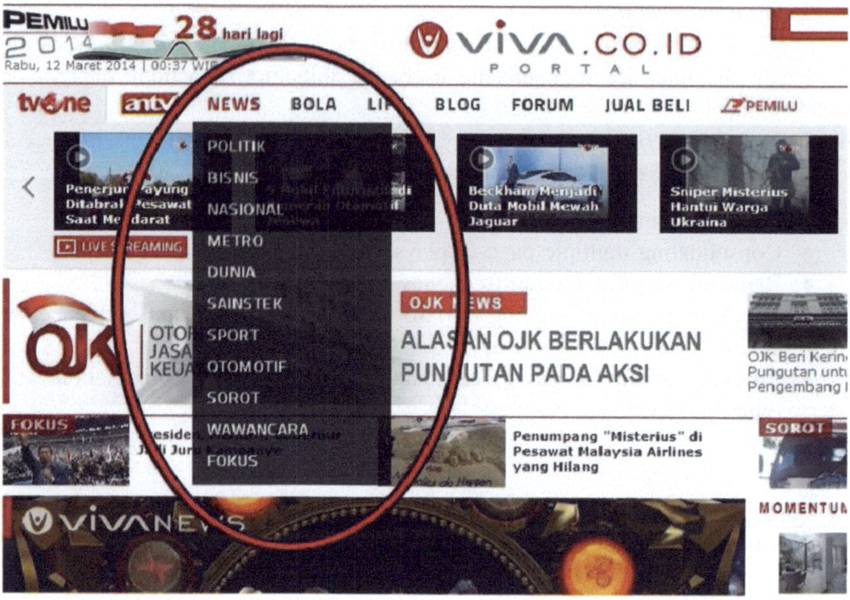

Fig. 1 Interface submenu of the menu News

Fig. 2 Interface menu on one of the news portal

selling), then each of those categories, there are subcategories of information will appear instantly when mouse moved into the category. But subcategories only appear on the main portal, it does not appear in the portal of subcategories information. It also includes inconsistent in similarity of terminology (Figs. 3 and 4).

3. The last aspect of website consistency is instruction in each page or portal. In viva news website, there are parts that are not consistent with the instructions, when we want to access the other portals from menu on the top of the main portal website. In the main portal, when the mouse is moved into the title menu submenu appears. But when in other portals (vivanews, vivabola, vivalife) just display the title only and instantly go to information from the selected categories. For examples Figs. 1 and 2.

(B) *Shortcut Availability*

When the user accesses a website, the user wants to fasten their interaction with the system. Shortcut here means in the form of a button. Shortcut availability is helpful when the user wants it. For viva websites those facilities already exist. However, there is a shortage at *home shortcut* button on the vivanews portal looks very small. So, it can make the user does not know it is a shortcut or not (Fig. 5).

Fig. 3 Categories menu in the main portal

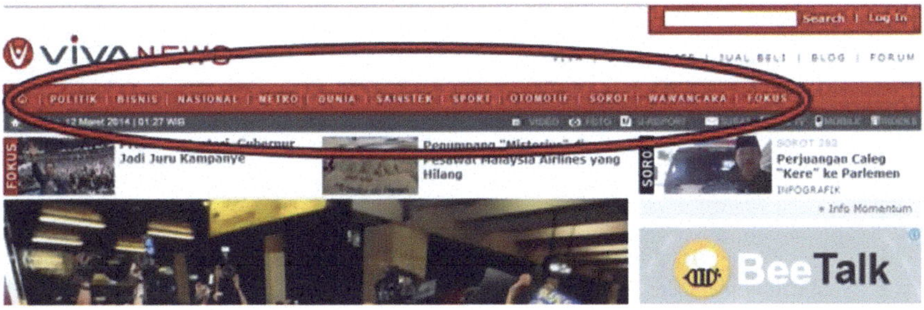

Fig. 4 The category menu on the portal of the category News

(C) *Informative Feedback*

When a website wants to improve the quality, it should be there is a reciprocal of the user that will improve the website quality. The viva website there are no a page suggestions and criticisms, this website only provides comments for a news post only. Things that cause the developer does not know that the website is developed can be easily understood by the user and many suggestions to improve the website quality (Fig. 6).

Fig. 5 Button home button is very small

Fig. 6 Notification members have been successful registration

Fig. 7 Notice that the username is already in use. It will prevent the user from a mistake that can make an error in the website

D. *Design Dialog*

When the user creates a website, the website should display a notification when the account is successfully created. It was practiced in member registration of the viva website.

E. *Error Handling*

An interface should not allow the user to make a fatal mistake. For example, in member's registration of viva website, when username already exists and cannot use the same username even though password is different. This will cause a fatal error on the website. Viva website has error handling for this case (Fig. 7).

F. *Easy Reversal of Actions*

A website should allow reversal of an action easily. When a user wants to read or comment, then the user wants to return to the instruction before they do not want them to come back easily. But it was not found in this website that is one of the weaknesses of this website. Although that action can be done through the browser, if that action is in the website, it will be different through the browser.

G. *Sense of User Control*

Users may feel that they are controlling the website. But it is opposite to this website. When the user opens a portal (vivanews) then an information appears, but that will change automatically in a few minutes later. That weakness can avoids the user from having a sense of control in this website. In addition, ease to operate the website also affects the sense of user control. Ease of use can be judged from the interface design on every page of the website. Because the interface design of website greatly affects the user, whether the design is easy to understand or not can

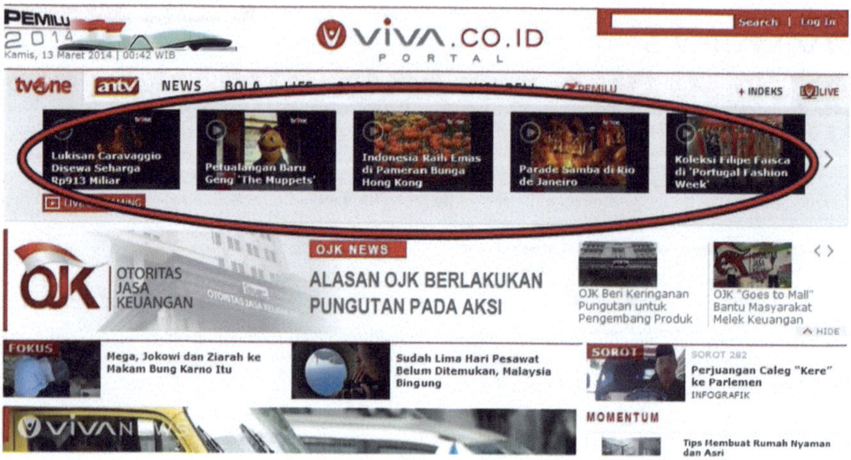

Fig. 8 Choice live streaming news

be judged from the placement of each menu. In the viva website portal, some live streaming videos appear that annoys the user because the video is not the headline of information. The video will be useless, because in the middle section of website display the news headlines for live streaming. If the video is a headline news, it should be visible when the website is accessed by the user. But the video should be considered, because the Internet connection in Indonesia is generally slow. This will increase the load time and the memory can be burdened (Fig. 8).

To view the latest news and the most popular news, on right side of the website should appear too much news list. It will be useless because any user looks for the headline information and not for the popular news (Fig. 9).

Overall the placement is good enough, but when the main portal shows too much headline, it is not good for a news website that is already well known. In addition, it will increase the load time, also it will make an appearance too long for a main portal, because every news seen by the user is the latest news. If there is the latest news on the child domain of viva website, the headline should not be displayed in the respective category of main portal, but it is displayed on the large space of main portal. For example, on bottom of the portal page viva.co.id website as shown in Fig. 10.

(H) *Reduce Short-term Memory Load*

Research shows that humans can hold not more than ten piece of information in short-term memory. To solve this problem website should be made simple and when in the main portal it is only displayed the most recent news. Because if it is too much news from each portal displayed in the main portal, it will reduce the speed of user access.

For viva website, too many indexes that are not supposed to be headlines. The most recent news from the other portals will appear until there is more new news with a large size. So, that it makes access to the website to be longer. With the video

Fig. 9 Indexes on major portals

on the main portal, most likely when the user wants to get the news but it is difficult to access the website will reduce the intention of the user to get the news from viva.

Here are the results and discussion of web analytics www.detik.com based on Eight Golden Rules of User Interface Design.

(I) *Consistency*

Consistency in website can be divided into three aspects as below.

1. Judging from the font, color, capitalization, and consistent placement of the menu. The detikcom website has been consistent in font, capitalization, and placement of the menu, where each page has same font, capitalization, and placement of the menu. Inconsistent only in the category of news on the main page of the detikcom website. Difference only in color on each category, as to differentiate and provide the right atmosphere in the depiction of each category. But the placement of the menu, a selection of news, and news categories are very good and consistent. Can be said for the level of consistency of this website is good. It will be easier for the user to use this website.
2. From the similarity of the terminology used in the menu on the website interface. On the detikcom website, a lot of resemblance or similarity of terminology, for example when you want to search for news, the search field located at the same place of each category of the website.
3. From the instruction on every page, the detikcom website has instructions that are relatively same on every page. For example, the placement of news search field similar to other pages, on the column news category also has the same instruction that is one click it will show the desired news category. And the important thing in a news website, the instructions are not too much or too long.

Fig. 10 Pieces display the bottom of the website

Overall the interface of detikcom news website has been consistent if it is seen from the three categories above. It will be easier for the user to gain experience reading or watching the news, it will make the website visitors will increase because of the simplicity. Here is an example of detikcom website display that looks consistent. Examples are drawn from the main page of the category detikNews, detiki-net, detikSport, and detikHealth (Figs. 11, 12, 13, 14, 15 and 16).

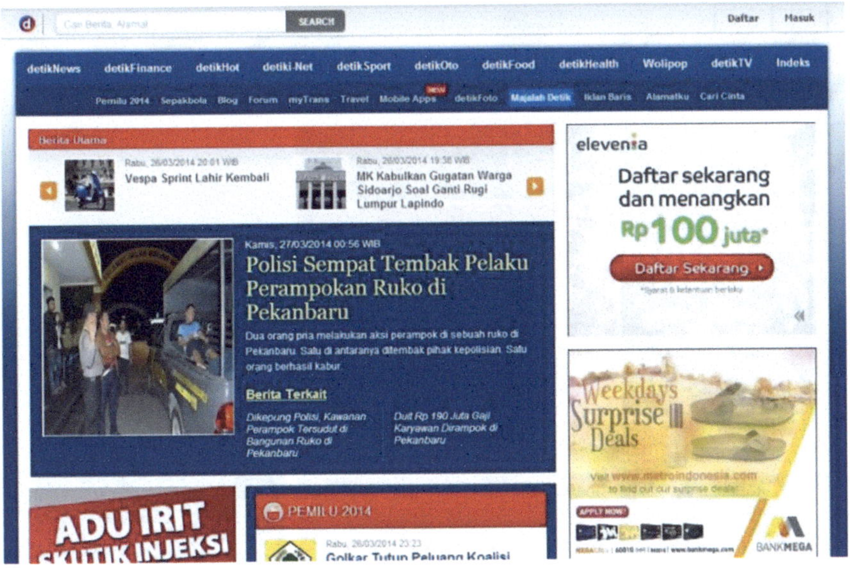

Fig. 11 The main interface detikcom news website

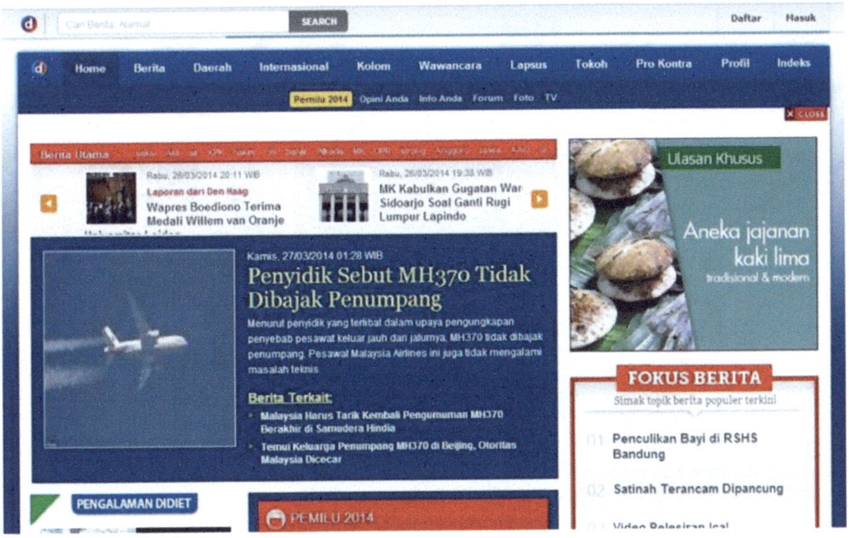

Fig. 12 Interface detikNews news category

(II) *Shortcut Availability*

For the detikcom website, there are some shortcut buttons with the logo. In the page header on the website there is a shortcut in the form of detikcom's logo, the shortcut

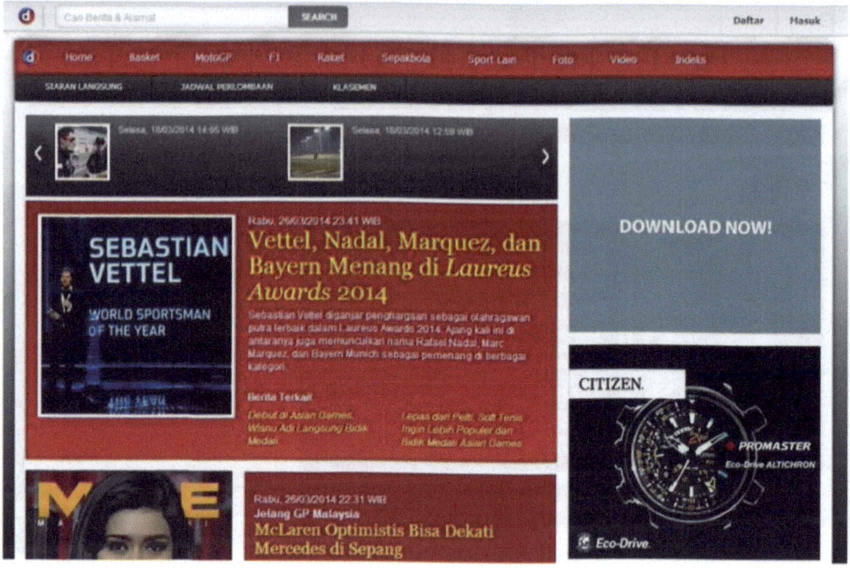

Fig. 13 Interface detikSport news category

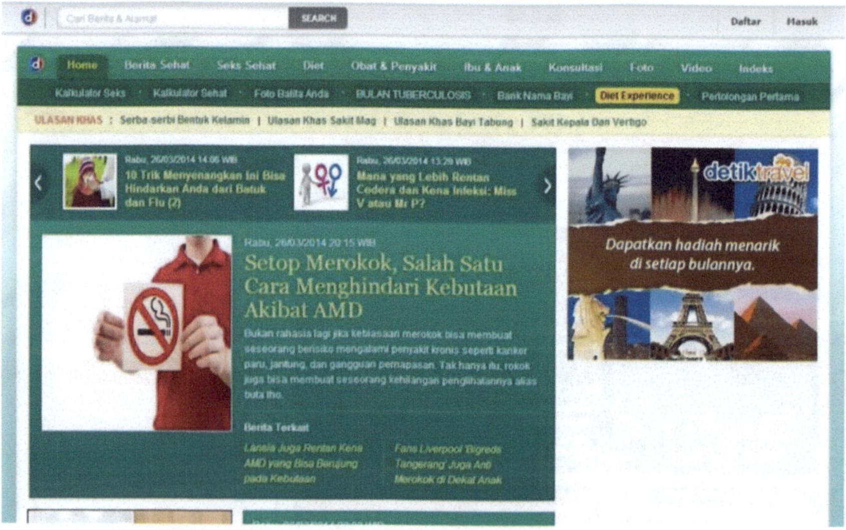

Fig. 14 Interface detikHealth news category

function to go to the main portal of detikcom. When scrolling the page, the header is not moving. So, it will shorten to return to the main portal using that shortcut. Although there are several other shortcuts that look small, but it is quite noticeable

Fig. 15 Interface detiki-Net news category

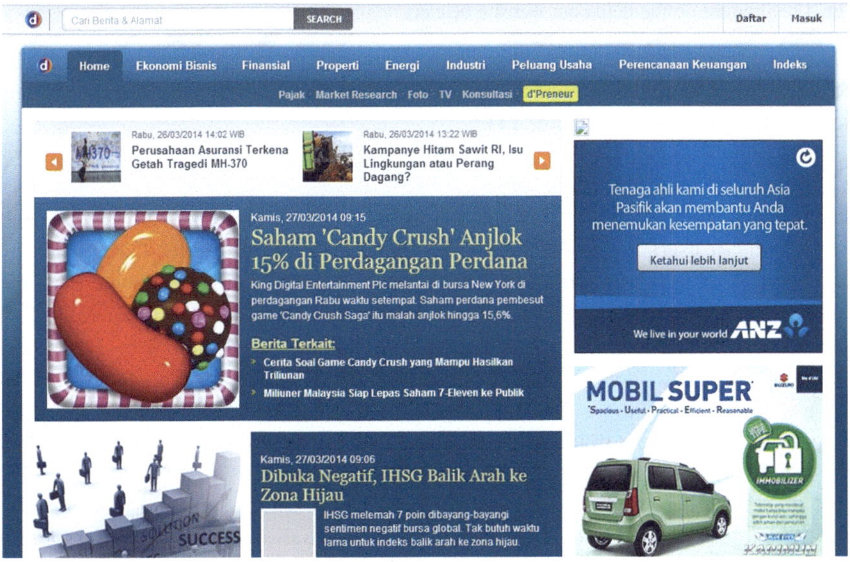

Fig. 16 Interface detikFinance news category

Fig. 17 Shortcut on the detikcom website header

because the color and shape of a unique logo that makes it easy to look and know that the logo is a shortcut (Fig. 17).

(III) *Informative Feedback*

Detikcom website does not provide a specific page for feedback comments and suggestions of users and does not display information about the feedback from the users who responded by detikcom. If on a news website, there is a special page for it, so to improve the usability and functionality will be easier. It is also called the method of observation.

(IV) *Design Dialog*

Design dialog can be a notification or instructions of usability. For example, when the process of creating an account on a website succeeds or fails to make the account, in detikcom's website, it already contained with a sentence that is easily understood by users. Here is a picture of a registration form on the detikcom account, which comes with a neat dialog on the form and instructions of detikcom's accounts (Figs. 18 and 19).

(V) *Error Handling*

Similar to viva website, detikcom website also has good error handling. On the detikcom website, the testing of error handling is on search. When searching for a

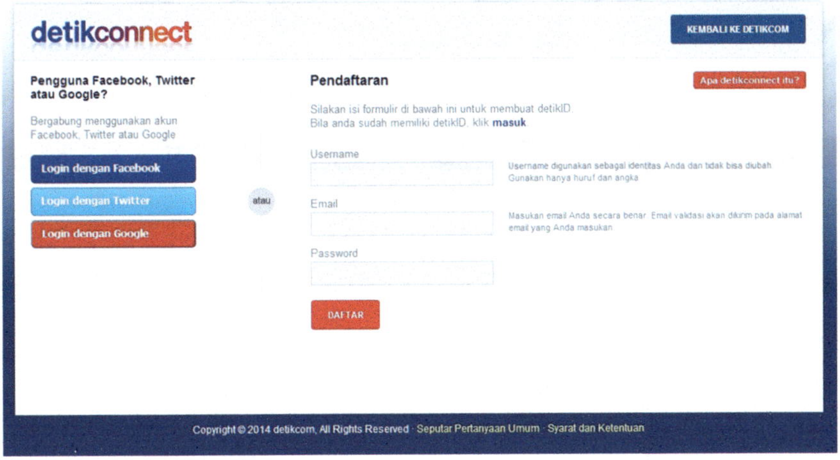

Fig. 18 Account registration form detikcom

Fig. 19 Instructions on detikcom account

keyword that does not exist on the website, detikcom gives notice that the search results are not found as shown in Fig. 20.

(VI) *Easy Reversal of Actions*

On the search page, detik.com did not find instructions to return to previous search results. Because when they want it, the user can do it through the browser interface, not through the website interface. It would be nice on the interface of the website

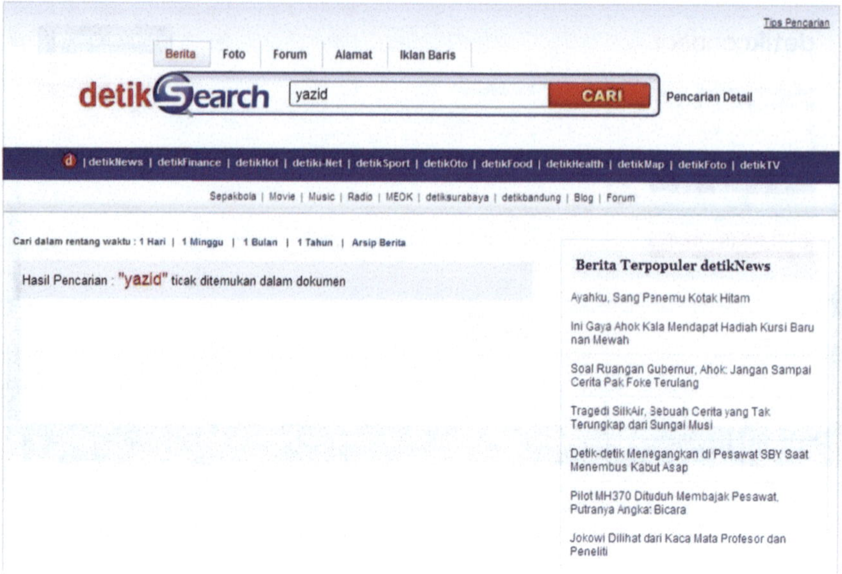

Fig. 20 Searches were not found

that also provides instructions for the return of order; this will improve the quality, usability, and functionality of the interface or the website system.

(VII) *Sense of User Control*

Each interface of the news websites is required to make the user feel like a main controller of the website. Because the news websites are used by various users, for the detikcom website, it makes the user feel uncomfortable in the placement of ads in a block and there are some ads that are dispersed. That iss what makes the website interface not look tidy enough. It will reduce the sense of control and comfort to the users who use the website (Figs. 21 and 22).

(H) *Reduce Short-term Memory Load*

For the memory load on the website interface, it can be seen from the fast or the slow process of accessing the website. Access to the Detikcom news website is not so slow, because the interface on the main portal and the category does not load too large images or videos. But there are several things that can slow down the access to the main portal interface in Detikcom, one of them is the main portal interface display that is too long. Overall, the Detikcom websites has faster access than viva website for mobile modem network connection.

From the evaluation of these, two websites are based on the Eight Golden Rules of User Interface Design, and Detikcom website is better than viva website. The general results of the evaluation on both websites can be seen in the comparison table.

Fig. 21 Ads are different but one block

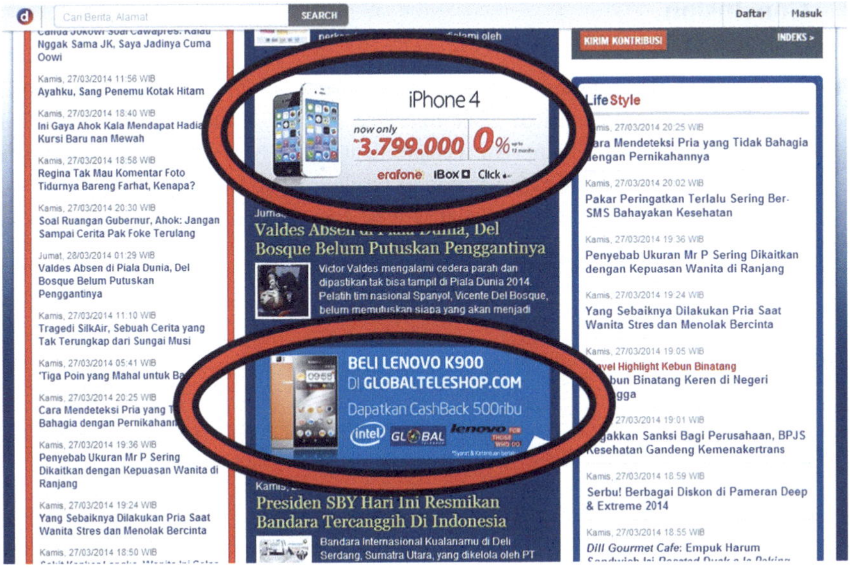

Fig. 22 Separate ads irregular

Item	Golden rules of UI design	Website			
		VIVA	DETIK	Grade	Information
1.	Consistency	2	4	1	Very bad
2.	Shortcut availability	4	5	2	Bad
3.	Informative feedback	1	1	3	Normal
4.	Design dialog	5	5	4	Good
5.	Error handling	5	5	5	Very good
6.	Easy reversal of actions	1	1		
7.	Sense of user control	3	4		
8.	Reduce short-term memory load	2	4		

Proven through statistics of top sites in Indonesia from alexa.com that the detikcom news website is ranked 9 and viva is ranked 20 overall taken of all the websites that are in Indonesia [4]. Whereas in 2013, from the same source for news in the Indonesia website category, detikcom is ranked first followed viva at rank 2.

Through the above statistics, it is clear that many news websites in Indonesia are competing and deliver benefits. Also, showing the development of technology in Indonesia is growing every year.

4 Conclusion

From the analysis and comparison of the two websites which use heuristic methods based on Eight Golden Rules of User Interface Design, it can be concluded that detikcom website is better than viva website. But not only viva website that needs to be fixed, detikcom website also needs to be developed in order to increase the usability and functionality of web. So, it will be easier for all users to use and seek news from the web.

References

1. Desy Saputra, 2014, APJII: Pengguna internet Indonesia terus meningkat, http://www.antaranews.com/berita/414167/apjii-penguna-internet-di-indonesia-terus-meningkat, 20 April 2014.
2. NIIT, Eight Golden Rules of User Interface Design page 2.4, Book Human Computer Interface Design NIIT 2011.
3. http://goodispost.blogspot.com/2013/08/10-website-portal-berita- terbaik-di.html - 10 Website Portal Berita Terbaik di Indoneisa 2013.
4. http://www.alexa.com/topsites/countries/ID - Top Sites in Indonesia.

Waste Prevention Effectiveness of Batik Production in Yogyakarta, Indonesia

Mohammad Rianda Al Rasyid and Retno Widowati Purnama Asri

Abstract As a form of art rooted deep in Indonesian culture, Batik has evolved to its modern form and become internationally recognized. Their production consequently has experienced an increase as global demand continues. Such an increase in popularity comes with the price of environmental waste produced when creating Batiks. This paper explores the effectiveness of green batik initiative implemented by the Indonesian government in managing the waste coming from Batik industry. In particular, the study reported in this paper covers Batik waste management in Yogyakarta. Results of the study demonstrate that although the initiative has helped in managing the waste; however, some actions by the government will still be needed to suppress the impact further to environment.

Keywords Batik · Waste Management · Environmental Impact

1 Background

Batik has been both an art and craft for centuries, and it is part of an ancient tradition. The word batik is derived from the Javanese word "amba," means "to write," and the suffix "titik" means little dot or to make dots. For the Javanese people, batik is not only a popular decorative textile, but it also has a deeper philosophy that has a deeper meaning for them. We could found in history that Javanese used batik in some important events such as marriage, birthday, and also during the funeral.

M.R. Al Rasyid
Magister Management, Universitas Muhammadiyah Yogyakarta,
Yogyakarta, Indonesia
e-mail: andhamohammad@gmail.com

R.W.P. Asri (✉)
Department of Management, Universitas Muhammadiyah Yogyakarta,
Yogyakarta, Indonesia
e-mail: wati630421@yahoo.com

T. Taufik et al. (eds.), *ICoSI 2014*, DOI 10.1007/978-981-287-661-4_46

The advantages of batik had been transformed into a bonanza since The United Nations Educational, Scientific and Cultural Organization (UNESCO) acknowledged on October 02, 2009, that batik is purely originated from Indonesia. Batik contributes a large amount of foreign revenue to Indonesia. Batik contributed about 10% of Indonesia textile export of US$12 billion in 2012.

However, producing batik means creating waste as well, and producing large number of batik means creating large amount of waste. How to manage a balance condition between production and its impact toward the environment is being a headline focus of government, society, and also batik entrepreneur. This paper deals with the discussion on how the green batik initiative has been implemented effectively and how the government takes responsibility in order to save the environment from the batik waste. The focus of this paper is to study the Batik waste management efforts in Yogyakarta Special District as a major spot where Batik production takes place.

2 Set of Problems

As a cultural heritage and economic commodity batik is one of the important assets of Indonesia, it is urgent to keep batik industries growing positively in both production and environment aspects. This paper puts forward the following topics:

1. How to preserve batik as one of the cultural heritages of Indonesia as well as world intangible cultural heritage?
2. How batik could contribute to the Indonesian economic growth?

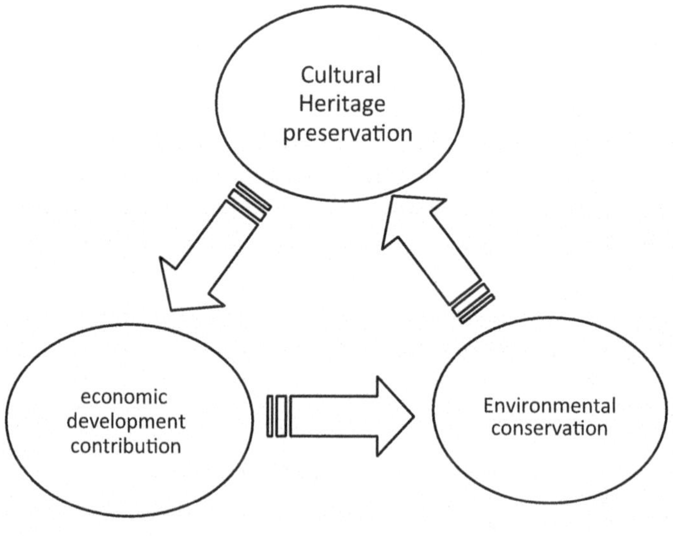

Fig. 1 Batik sustainability scheme

3. How could batik create a bad impact to environment, and how to prevent it?

From these three points the batik sustainability scheme, as shown in Fig. 1).

2.1 Cultural Heritage Preservation

According to the UNESCO, batik is one of the intangible cultural heritages for humanity, because batik had been classified as a cultural heritage that has a deep meaning inside its creation. An intangible cultural heritage for humanity is one of the classifications from the UNESCO to protect all of cultural heritages all over the world, because they are not only tangible but also intangible, such as human tradition and expression from many groups of people around the world that inherited verbally to their generations.

Many efforts have been conducted since the acknowledgement from the UNESCO to conserve batik in Indonesia. One of the "quick wins" was creating the blue print on how to preserve and develop Indonesian batik, by the Ministry of Trade on October 2011. According to the blue print, there are some points to preserve batik as a cultural heritage:

1. Taking the action of appreciation to wear batik as one of "a must worn uniform" at least once a week.
2. Developing the unit of batik education, not only in Pekalongan but also in the entire region of Indonesia.
3. Mapping the batik pattern and make the inventory of batik books or references.
4. Claiming traditional batik motifs to be patented, in order to keep away from the bad intention of some private business unit to get an advantage.

Every element of the society has the responsibility to preserve batik, so that the government as the policy maker and everybody have the same position to preserve batik as an Indonesian cultural heritage.

2.2 Contributing to Economic Development

According to the Official Press of the Ministry of Trade, batik has a high economic value potential. Based on the 2010 data, the number of workforce in batik industries was 916.783 people with their production value of Rp. 3.9 billion. Between 2006 and 2010, the batik demand had been increased by about 56%. Batik has been the national economic mobilization by contributing about US$69 million export value with the main state partners the United States of America (USA), Belgium, and Japan.

Noting that 99.39% out of 55,912 units of batik industries is coming from small- and micro-scale industry, the government has considered that Indonesian batik will be the crucial economic mobilization in 2025. The government wishes that batik not only as the living tradition of Indonesian society, but also as the mobilization of societal economy. Besides that, batik also has another advantage especially for empowering women prosperity, because based on the direct survey, most of the batik makers are women, both the home-based and industry-based production.

2.3 Environmental Conservation

Batik coloring material is derived from a dangerous composition such as *remasol*, *naptol*, and *indigosol* which could naturally threaten the green environment, for example, the extinction of water biota. Because the waste of batik industries will commonly be disposed to the water (river, sea), the dangerous waste could threat the environmental safety.

Batik production process takes some stages to produce a noble decorated fabric, such as waxing, dyeing, and de-waxing (removing the wax), as well as preparing the cloth, tracing the designs, stretching the cloth on a frame, waxing the area of the cloth that does not need dyeing, preparing the dye, dipping the cloth in dye, boiling the cloth to remove the wax, and washing the cloth. Finally, at the end of the stages, the wastes will be disposed to the river or environment. The final stage conditions will determine whether or not the wastes endanger the environment. Thus, before disposing the waste to the environment, there should be some treatment on processing the waste of batik to be environmentally friendly such that it could freely be thrown to the river.

There are different processes from both natural and synthetic coloring of batik along with their respective advantages and disadvantages. Traditionally, the natural coloring materials being used to dye the Javanese batik consisted primarily of beige, blue, brown, and black which are made from indigenous plants. The blue, which is the oldest color being used to make traditional batik, is made from the leaves of the Indigo plant. The leaves are mixed with molasses, sugar, and lime and left to be fermented overnight. The second coloring material that is applied when making traditional batik is soga, a brown color which can range from shades of yellow light to a dark shade of brown.

The synthetic coloring is taken from a selected chemical processes in which the substances of the materials will not harm the waxing process on batik dyeing [1]. The chemical substances that are used by the batik maker will be elaborated in the literature review.

3 Purposes

Practically, this paper expected to enrich the knowledge and provide insight into the waste prevention of batik production.

4 Literature Review

4.1 The Concept of Effectiveness

Effectiveness is the requisite of reaching the goals which has been determined by the organization. Effective is the measure of which goals are reached or come true [2].

4.2 Batik

Batik is "writing using wax" that decorates a fabric by covering some parts of it with a coat of wax and then dyeing the fabric. The technique intended to cover parts of fabric so that those parts will not receive color, and the waxed areas keep the original color and then, when the wax is removed, the contrast between the dyed and undyed areas forms the pattern [3].

The word batik is thought to be derived from the word "*ambatik*" which means "a cloth with little dots," and the suffix "*tik*" means little dot, drop, point, or to make dots. Batik may also originates from the Javanese word "*tritik*" that describes a resist process for dyeing where the patterns are reserved on the textiles by tying and sewing areas prior to dyeing, similar to tie dye technique [4].

4.3 Batik Waste

Batik production uses chemical coloring substances which are not safe to the environment. The synthetic coloring materials of batik which are used by the batik craftsmen include remasol, indigosol, and naptol. They use the batik synthetic coloring materials because of their ease of use and simplicity of application in comparison with natural coloring materials.

There are some classifications of synthetic coloring substances:

1. Naptol
   The component of naptol is diazonium salt or naptol salt, which could not be dissolved by the water, and to dissolve the substance, another subsidiary substance like caustic soda is needed. In batik processes, naptol is used to gain a sharp color.
2. Indigosol
   Indigosol is one of the coloring materials used in batik coloring processes and to obtain the expected color, it needs to be mixed with another chemical substance that naturally has the dangerous effect on both the body of craftsman and environment, i.e., HCI or H_2SO_4 (acid solvent) and $NaNO_2$ (sodium nitrite). The risks of mixing these substances are irritation to the skin and fatal poisoning of the body of the craftsman.

3. Caustic soda
   Sodium hydroxide is available in the form of flakes (100% concentrate) or in the form of liquid with the various concentration. Such substance may cause irritation to nose, pneumonitis, and intercellular edema (for human body) and have deadly impact on water organism.
4. Hydrochloric acid (HCI)
   Hydrochloric acid is a yellowish liquid that has a very corrosive nature; corrosive means very reactive and could cause fatal breakage to human body and environment. In batik coloring process, this substance is used as the saponification element.

5 Discussion

5.1 Batik, Waste and Its Contamination

Batik industry is one of the producers of liquid waste which resulted from the color dyeing processes, besides containing high and harmful coloring substances; the batik waste also contains synthetic elements which are very difficult to be dissolved in the water. In general, the batik pollutant classified as the hard metal, the turned-suspension of solid material, or organic substance. After the finishing of the dyeing process, the waste of the dyeing water turned into a turbid thing, and if this final turbid water disposed directly to environment without other neutralizing processes, it will damage the biological environment.

Due to the damaging risk of the wastewater, it is urgent to do some effort to keep the environment safe. The obligation of environmental preservation conforms to the government policy through the points of legislation.

The legal bases of the environmental preservation are as follows:

1. The Legislation no. 5 in 1990 about conservation of natural resources and biological environment and its ecosystem.
2. The legislation of Republic of Indonesia no. 32 in 2009 about the preservation and managing biological environment.
3. Government regulation no. 82 in 2001 about the management of water quality and controlling of water contamination.
4. The regulation of State Ministry of Health of Republic of Indonesia no. 415/MENKES/PER/IX/1990 about the requirements and quality of ground water.

Based on the law and regulation of Republic of Indonesia, it is urgent to know how far the regulation and legislation being implemented in the society especially in DIY.

5.2 The Late Response and Effort of Batik Waste Prevention by the Government

Government regulations regarding the waste prevention in Yogyakarta and Indonesia have been drafted since 1990 (about conservation of natural resources, biological environment, and its ecosystem), and since 2001 (about the management of water quality and the controlling of water contamination) that mean, long before the UNESCO recognition the government of Indonesia has set the waste legislation including batik waste prevention. But, why the environment is still polluted by the waste of batik? In Yogyakarta, some batik home industries disposing their wastewater directly to the river near their place. One of the examples is batik home industry located in Ndiro, Bantul Yogyakarta; its wastewater is directly disposed to a nearby river using a long drainage piping system.

Due to the late response, the result can be observed now in Yogyakarta. The environment prevention surrounding batik industries is still in process to be applied around those cities. These are the facts that the writer found when doing a direct verbal interview to batik Sri Kuncoro in Imogiri Yogyakarta on April 2014. The batik wastewater treatment facilities were built in 2009, and the wastewater treatment facilities are the simplest one. Mrs. Imaroh, the owner, said that the batik wastewater treatment facility was granted by an environmental nonprofit organization in DIY that responded quicker than the government did. The traditional wastewater treatment facility uses filtration technique which also called small IPAL (installation of wastewater treatment) or the household scale of IPAL. The process in this facility composes of some steps: equalization, coagulation, sedimentation, aeration, sedimentation, filtration, which are presented in Fig. 2 [5].

Another effort is generating the awareness to batik entrepreneurs in DIY to produce batik more efficiently and minimize the impact of batik wastewater. The last is improving people awareness of advantages of using natural coloring materials. The environmental awareness campaign sometimes did not match with the problems the batik producer faced. In fact, the supply of natural coloring materials from indigofera was very limited and more expensive than those of synthetic coloring materials. On the other hand, there were no other alternative solutions from the government on how they could get the cheap natural coloring materials.

The communal scale of IPAL (huge IPAL system) in Gulurejo village, Lendah district, Kulon Progo Regency, DIY has been completed its Detail Engineering Design (DED) and in the year of 2014 Kulon Progo will start the construction. The goal is that all the construction will be implemented in 2014. The structure will cost approximately Rp. 600 million, where the program is supported by the Java Eco-region Management Centre (Pusat Pengelolaan Ekoregion Jawa/PPEJ).

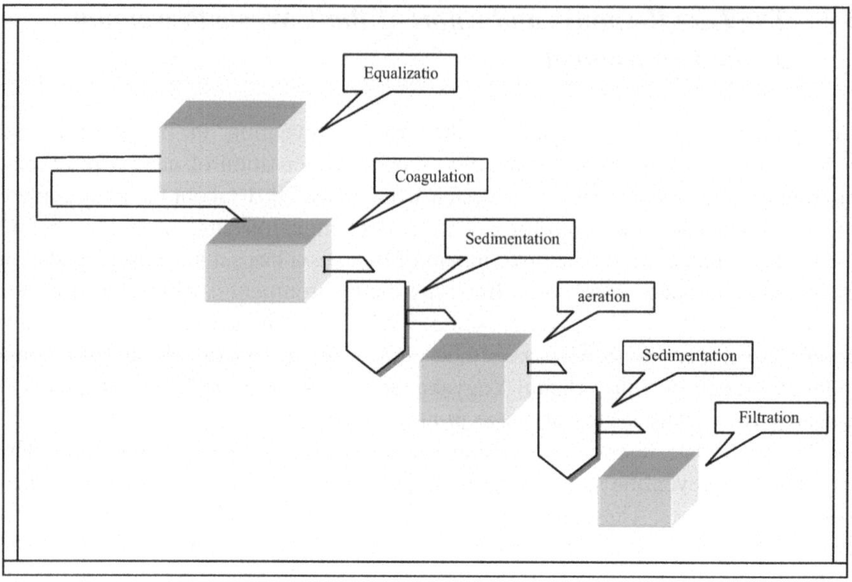

Fig. 2 Household scale of IPAL

5.3 Some Constraints Faced by the Government During the Implementation

Efforts both by the government and by batik maker, such as how to install a batik wastewater treatment facility, how to use it, how to communicate with the whole people of batik community, and many more, have so far always met the deadlock.

Here are some obstacles:

1. The lack of human insight and knowledge. Some of the batik craftsmen are coming from the impoverished family, and they have limited education that make difficult to communicate.
2. High-cost water treatment facility makes difficult to implement the program. To design and propose a program may be easy and not time consuming, but to spend a large amount of money in implementing it will have some consideration and take a longer time.
3. Natural coloring materials have a complex processing route and even they need larger amount of precursors to gain the expected colors.
4. There was not enough law enforcement from authority regarding the law violations committed by the batik industry including disposing the wastewater into the river.

We realize that the process of preventing the waste of batik was not easy; many programs have been planned, but some of the plans could not effectively applied.

The government need to response quickly in preventing the batik unfinished problem in DIY. The sluggish response will not cope with the problems and even it showing to the society that government has low awareness to react and give a positive feedback to the society.

5.4 Conclusion and Suggestions

There are some actions that the government needs to do, which are as follows:

1. The government has to create a mutual relationship with nonprofit organizations that concerns on the batik wastewater purification.
2. Build batik wastewater treatment facilities or giant IPAL, in every center of batik village, and granting small IPAL to every private home industry in DIY.
3. Apply firm and decisive actions in law enforcement against law violations.
4. Cultivating natural coloring material crops such as indigofera, soga, mengkuda, secang, mango leaves, and many other natural coloring material crops.
5. Present appreciations annually or regularly to create a higher environmental awareness among batik craftsmen and present them a tribute or recognition.
6. Conduct researches and deep observations in accordance with batik wastewater management.

We believe that as long as the government keeps a good communication and control to the society and conducting the programs that has been planned, the dream of creating the sustainability among cultural preservation, economic development, and environmental conservation will come true.

References

1. www.batikdirectory.com/batik-indonesia/pewarnaan-batik.html
2. (Soewarno Hayadiningrat S: 1994)
3. Ministry of Trade of Republic of Indonesia: "Indonesian Batik, the Cultural Beauty of Indonesia", 2008, booklet.
4. www.expat.or.id/info/batik.html
5. Laporan Akhir: "*Penyusunan Detail Engineering Design (DED) Instalasi Pengelolaan Air Limbah Industri Batik di Dusun Pengkol, Desa Gulurejo, Kec Lendah Kabupaten Kulon Progo*", Pusat Pengelolaan Ekoregion Jawa, 2013.

Watershed Erosion Analysis (Case Study of Citarum Upstream Watershed)

Ana Nurganah Chaidar, Indratmo Soekarno, Agung Wiyono and Joko Nugroho

Abstract This paper presents the results of the study in partial yearly erosion rate changes, and a simulation of land use according to the pattern space. Watershed ecosystem changes are represented by the modification of land cover and adapted to the spatial conditions. The area studied is the Citarum Upstream Watershed. This study is part of a dissertation which aims to get the value of the condition of soil erosion in a few years and landscaping scenarios in the upstream subCitarum most optimal, so that the reservoir can function properly in accordance with the design life of the reservoir. Analysis was conducted using soil erosion modeled with USLE using GIS models. This analysis provides the form of a map of soil erosion in the study area spatially. In order to use this model for estimating the value of USLE and the actual distribution of soil erosion in a river basin, the approach used is to use a GIS system, where it needs information related to spatial distribution of each parameter USLE. Various spatial characteristics of each USLE parameters obtained from maps available digitally inserted in the form of thematic maps and overlays in the GIS system. Grid cell system is available on further functions significant in GIS, where for each grid cell on the map overlay can be obtained by the value of the rate of erosion per unit area of grid cell is then converted to the rate of erosion per square kilometer. Depiction of the results of the erosion rate map using the USLE method-based Arc-GIS of the obtained values shows that tert erosion increases with land-use changes that occur. The increased is extreme enough that the 4 year

A.N. Chaidar (✉) · I. Soekarno · A. Wiyono · J. Nugroho
Civil Engineering Department, ITB Bandung, Bandung, Indonesia
e-mail: ananurganahchaidar@yahoo.com

I. Soekarno
e-mail: indratmosoekarno@yahoo.com

A. Wiyono
e-mail: agungwhs@si.itb.ac.id

J. Nugroho
e-mail: joko@si.itb.ac.id

T. Taufik et al. (eds.), *ICoSI 2014*, DOI 10.1007/978-981-287-661-4_47

increased almost 50% from 6.4 million ton/year in 2006 to 31 million ton/year in 2010. The simulations performed the pattern space Citarum River Flows Regional Plan and obtained sediment 4.36 million ton/year.

Keywords Erosion · USLE · GIS (Geographic information System)

1 Introduction

Erosion is the process of eroding the topsoil by rainfall which is influenced by the soil erodibility, rainfall erosivity, land slope, vegetation types and plan management. While, the sedimentation processes are the amount of sediment entering the river due to the land erosion.

The upstream Citarum watershed is at 107° 15′ 46.27″ to 107° 57′ 1.99″ E and 6° 43′ 8.65″ to 7° 14′ 32.09″ S and an area of ± 230.802 ha. Citarum Watershed Management Plan issued Citarum–Ciliwung watershed Management Agency in 2009 by using the USLE method (4.1.5), it was found that the entire region of Citarum watershed Upper section has a very bad erosion index class that is 491 tons/ha/year with a total erosion amount of 112 million tons/year. Based on the classification of the index value limit erosion class, almost all of the subwatershed approach the threshold value is very poor (>480 tonnes/ha/year). The high erosion values are obtained from many types of land cover in the form of vegetables cultivated and open ground. Erosion index based on class, region Citarum watershed Upper section has a total area of 43.7% in very good condition, 8.6% in good condition, the condition being 4.1, 12.7% in poor condition and 30.9% in very bad condition as shown in Fig. 1.

The conditions of the upstream Citarum watershed characteristics are as follows:

1. Administratively, the upstream Citarum basin entered the Bandung District. West Bandung, Bandung, the City of Cimahi, District of Sumedang. A small portion enters the Garut District.
2. Daily Average Temperature is 22–23 °C, while relative humidity ranges from 25–83%. Monthly wind speeds ranges between 23 and 106 km/day with a top wind speed of 213 km/day and the lowest is 7 km/day. Average monthly solar radiation ranged from 1.289 to 1.687 J/cm^2/day. Spatial distribution of rainfall in the upstream Citarum upstream watershed. It often happens with high intensity rainfall events in a place, whereas in other parts the rain did not happen. The annual rainfall varies between 1.966 and 2.600 mm.
3. In general, the upstream Citarum watershed climate are
    - Classified as type C according to the classification based on climate Schmith —Fergusson or type Am based on classification of Koppen.

Fig. 1 Location of the upstream Citarum watershed within the scope of the administrative area

- According to Oldeman, climate classification refers to the number of months of wet precipitation (>200 mm) and dry months (<100 mm),
- The upstream Citarum watershed classified into agro-climatic zones B2 with 3–2 wet months and 2–5 months dry. (double with number 4)

4. In general, the upstream Citarum watershed climate is classified as type C according to the classification of climate Schmith–Fergusson or type according to the classification of Koppen Am. Meanwhile, according to Oldeman climate classification based on the number of months of wet precipitation (>200 mm) and dry months (<100 mm), the upstream Citarum watershed classified into agro-climatic zones B2 with 3–2 months wet and 2–5 months dry.

5. Morphology Citarum watershed upstream section can be classified into two main units, namely the volcanic units and structural
6. Land cover in the upper reaches of Citarum watershed is dominated by rice fields (28.75%), either rainfed or irrigated rice. For dry land farming such as field is just 16:36%, and gardens/plantations around 16.88%.
7. Geology of the region is characterized by volcanic activity and ocean sediment Miocene to Holocene and Recent.
8. The upstream Citarum watershed aquifer dominated by the flow through the gaps and spaces between grains (50.4% or 114.683 ha) next to the aquifer flow through the space between the grains (28.6% or 64.940 ha).
9. Based on USDA soil map ,1:250,000 scale Citarum upstream watershed section is composed of land dystropepts, eutrandepts, eutropepts, hydraquents, paleudults, tropaquepts and tropudults but is dominated by soil type dystropepts (42.73%), hydraquents (21.74%) and soil tropaquepts (20.8%). Land hydraquents and tropaquepts is Gleysol, while dystropepts soil is soil association andosol.

In this study, we only discuss land cover and erosion change.

2 Methodology

2.1 Analysis of Land Erosion Rate

USLE (Universal Soil Loss Equation) model is an empirical erosion model estimation that was developed at the Center for Surface Flow and Erosion Data National Research Department of Agriculture, U.S. Department of Agriculture (USDA) in cooperation with Purdue University in 1954. The model was developed based on the research results of the erosion on small plots (Wischmeier plots) in a long-term study were collected from 49 locations. Based on the data, the erosion is estimated using rainfall data, soil, topographic, and land management. Descriptively, the USLE model is formulated as follows:

$$E_a = R \times K \times \mathrm{LS} \times C \times P \tag{1}$$

where

E_a amount of soil eroded per unit area per unit time (ton/ha/year)
R rainfall index and runoff erosivity
K erodibility index (sensitivity) of land
LS length–slope factor
C land crop cover and crop management factor
P specific soil conservation measures factors

USLE erosion model estimators have been widely used in Indonesia. Besides used as a model for probe erosion area, the model is also used as the basis of policy decision maker of soil and water conservation, although the accuracy of the use of these models in predicting watershed erosion is doubtful.

2.2 *Mapping Land Erosion*

Delineation process information on a flat rate of erosion can be done using GIS (Geographic Information System). GIS is an information system based on spatial data or geographically referenced data. Every data point on the earth's surface can be referred as spatial data, for example, the data density of population of a region, a city road network, the distribution of the sampling locations.

In order to use the model to estimate the value of USLE and the actual distribution of soil erosion in a river basin, the approaching use a GIS (Institutional Strengthening For Integrated Water Resources Management (IWRM) In the 6 Ci's River Basin Territory) where it needs information about the spatial distribution of each parameter USLE. Various spatial characteristics of each USLE parameters are obtained from available digital maps.

GIS with cell grid system is available on the further significant functions, where for each cell grid on the map overlay can be obtained by the value of the rate of erosion per unit area. The grid cell is converted to the rate of erosion per square kilometer.

Here is a map layer that will be used in the erosion rate:

- Map of erosivity R-Thiesen polygon and rainfall analysis;
- Soil Type Map Citarum–Ciliwung watershed Management Agency;
- Digital Elevation Map from SRTM (resolution 90 m); and
- Map of land use from Citarum–Ciliwung Watershed Management Agency.

2.3 *Erosion Rate Analysis USLE Methods by GIS*

Analysis is conducted using land erosion which is modeled with USLE using GIS models. This analysis will provide results in the form of a map of soil erosion in the study area spatially.

Modeling of erosion mapping consists of several stages as follows:

1. **Raster Fill Process**

Digital elevation map available often still has imperfections in spatial data. Therefore, these imperfections are used to complement the function of raster fill in the GIS system. Raster fill is used to identify the depth and surface repairing. The input for raster fill is SRTM digital elevation map of 2000th.

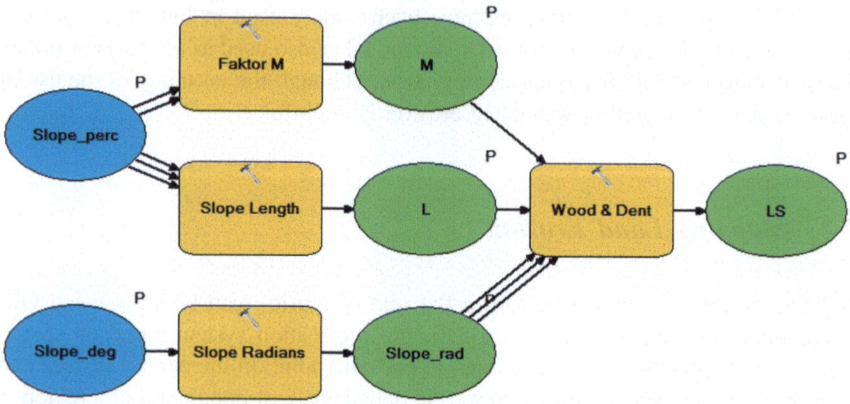

Fig. 2 LS calculation scheme

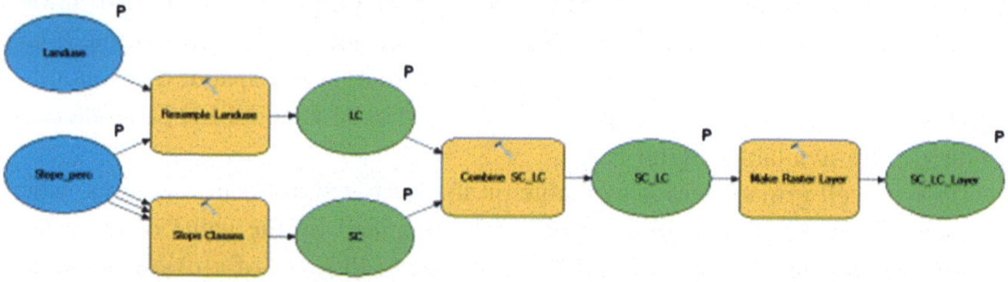

Fig. 3 Combination of land-use scheme and slope

2. **Feature to Raster Process**

Feature to raster process is used to convert the available data (shape files) into raster data. Input data for the feature to raster is the iso-erodent map (*R*) and soil map (*K*).

3. **LS Calculation Process**

LS slope characteristic factor is calculated using the formula Wood and Dent [1]. LS raster calculation scheme as shown in Fig. 2.

4. **Combination Process Grid Cell**

The data to calculate the CP factor is the combination of grid cell between land use and slope. Therefore, both of these factors need to be combined as shown in Fig. 3.

5. **CP Classification Process**

The value of the CP land management factors refer to reference Java Erosion Models—USLE 30 m: Application of the Universal Soil Loss Equation—2012. The factor of CP management is a function of land use and slope. The values are classified based on the level of land management. CP raster classification scheme as shown in Fig. 4.

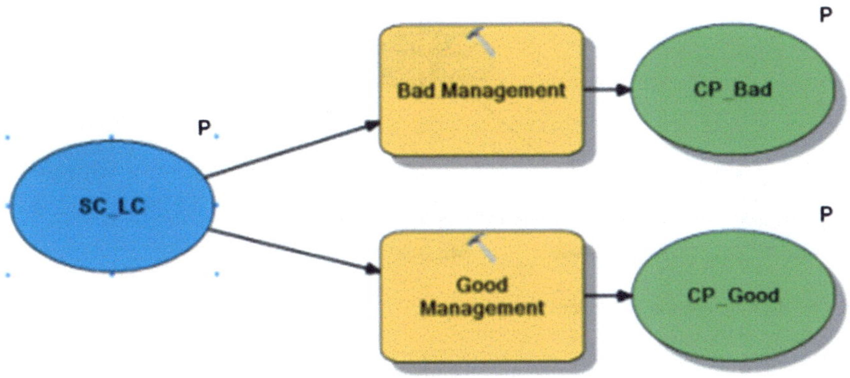

Fig. 4 Classification of CP process

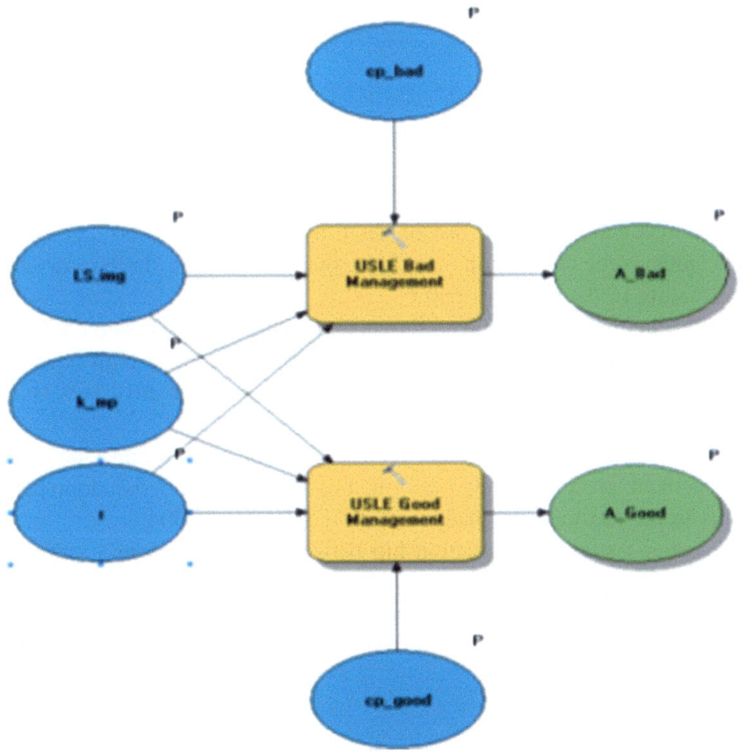

Fig. 5 Erosion rate calculation process

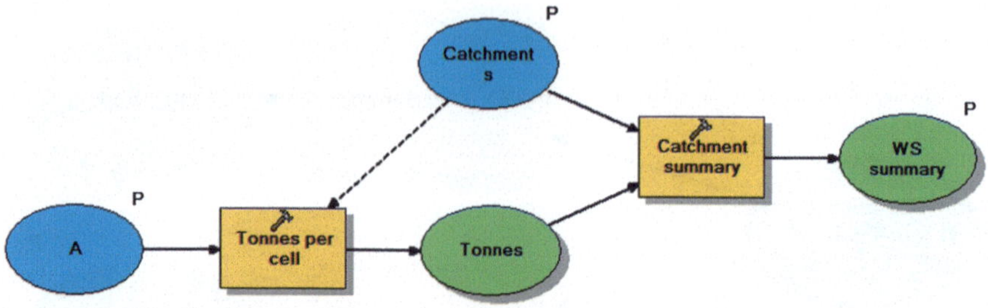

Fig. 6 Resume erosion rate calculation

6. **Erosion rate calculation process**

The value of the actual distribution of erosion and soil erosion can be seen by clicking on the fourth overlay raster layer that has been obtained, i.e., raster R, K raster, LS raster and CP. Multiplication of these values produce an erosion value as shown in Figs. 5 and 6.

3 Result

Figures 7 and 8 show the land use of Citarum upstream in 2006 and 2010. The extreme change is the conversion of forest to open land (bush, shrub or dry land). It occurred mainly in the southern part.

From the condition of the land as shown above, the rate of erosion could be seen in Figs. 9 and 10 where there is an additional land that has a high attrition rate. Erosion rate in 2006 is 6.4 million tons/year, while the year 2010 is 9.31 million tons/year.

Discussion of the relationship between land cover change and changes in the Citarum upstream erosion already performed by Bechar and Poerbandono [2] which the results in the erosion rate changes in 2001 relative to 2004.

As an initial analysis, the relationship between soil erosion and sediment that will occur in the river up to the dam will be determined. The results of this study obtained the trend of the rate of erosion as shown in Fig. 11. Figure 12 shows the erosion rates according to the existing land use throughout 6 years and $R = 0.9495$.

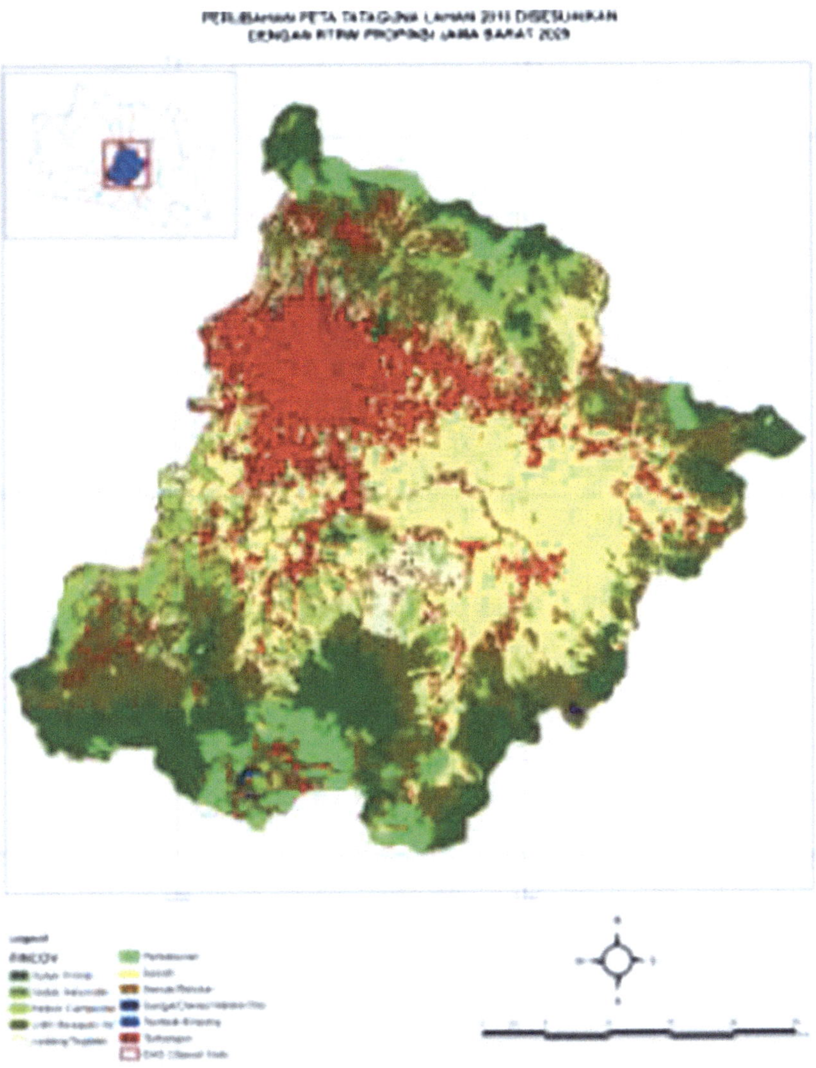

Fig. 7 P land-use map of 2006

To see the adjusted conditions, the Spatial Pattern of Space 2009 and the condition of the land in a manner simulating land use in 2010 are adapted to a space on the Spatial Pattern of West Java in 2029. The results from these simulations are shownin Fig. 13, where the rate of erosion is 4.36 million tons/year.

The simulation results on land use in 2010 adapted with the Spatial 2029 are shown in Table 1 where the wide catchment area changed from 19 to 37%.

Table 1 Area closure Lahan on land management 2010 and adjusted with the spatial

No.	Nama Tutupan Lahan	Luas (Ha)		Selisih (ha)	% terhadap Lu-2010	KET	% terhadap Lu-2010	
		Land-uses Land-uses 2010 merujuk RTRW	Landuse 2010					
1	Hutan Primer	26,279	26,279	–	–		14.37	14.37
2	Hutan Sekundar	443	443	–	–		0.24	0.24
3	Kebun Campuran	18,473	18,471	–	–		10.10	10.10
4	Ladang/Tegalan	7416	28,512	(21,096.02)	(73.99)	Pengurangan - Sesual RTRW	4.06	15.59
5	Perkebunan	8494	8494	–	–		4.64	4.64
6	Sawah	48,827	51,797	(297,026)	(5373)	Pengurangan - Sesual RTRW	26.70	28.32
7	Semak/Belukar	2060	12,191	(10,13,165)	(83.11)	Pengurangan - Sesual RTRW	1.13	6.67
8	Sungai/Danau/Waduk/Situ	553	553	–	–		0.30	0.30
9	Tambak/Empang	18	18	–	–		0.01	0.01
10	Terbangun	36,113	36,113	–	–		19.75	19.75
11	LNH Resapan Air	34,198	0	34,19,793	–	Pengurangan - Sesual RTRW	I8.70	–
	Jumlah	182,872	182,872	–			100	100

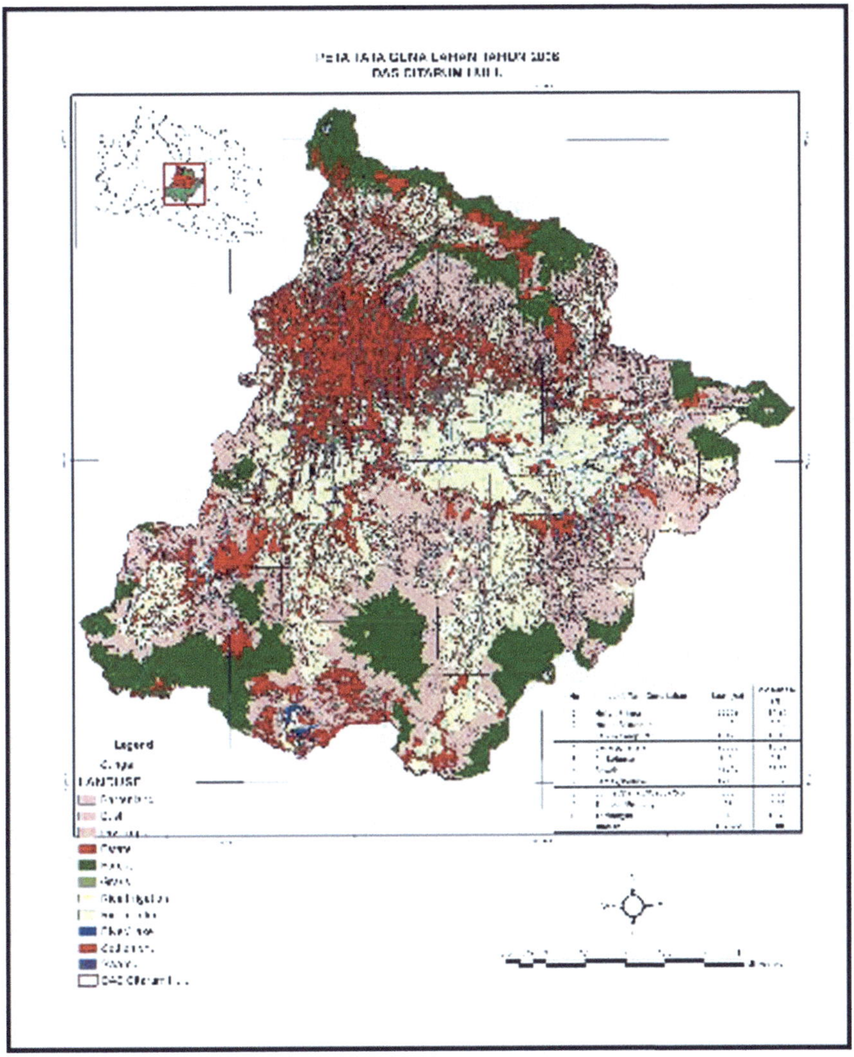

Fig. 8 Land-use map of 2010

4 Conclusion

In this paper, we have shown erosion study behavioral changes in Upper Citarum Watershed in accordance with the different land-use change layout. From the analysis, the conclusions are as follows:

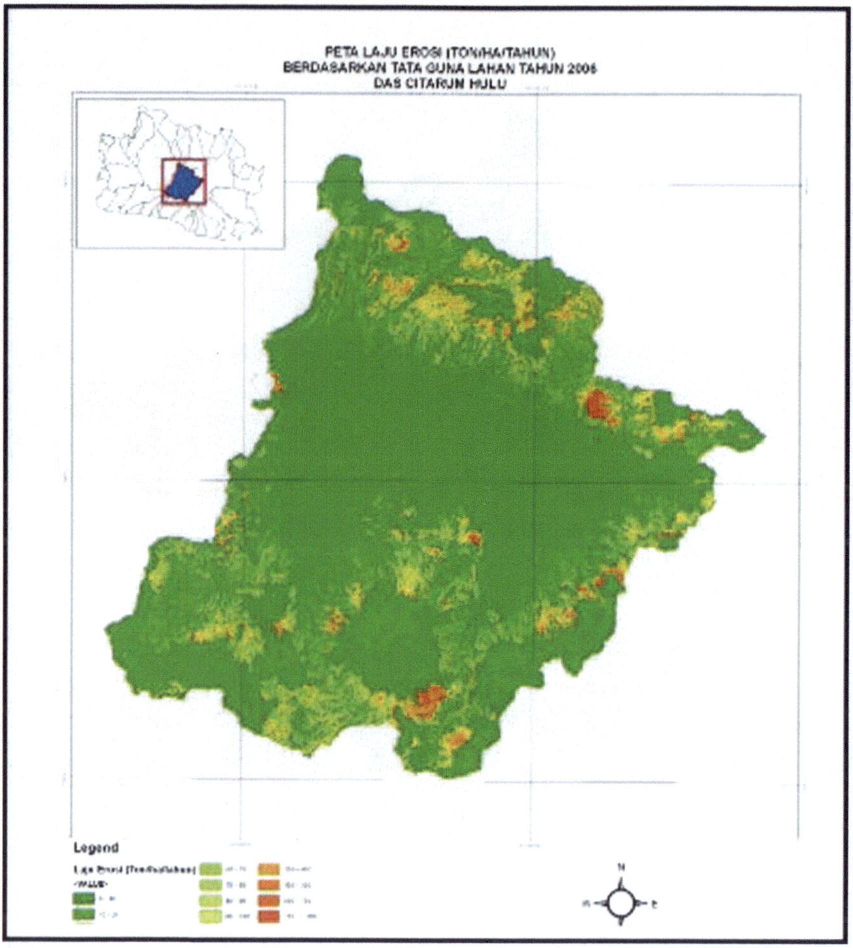

Fig. 9 Rate of erosion-based land classification map of land in 2006

1. The increase rate of erosion on any change in land use indicates worsen land conditions.
2. From the calculation of corresponding existing land use, erosion rate is very high, i.e., 50% every 4 years.
3. The results performed on the pattern space planned by the Revenue District of West Java in 2029, the decline in erosion rate in the same year, i.e., 2010 from 9.31 million tons/year to 4.36 million tons/year.
4. The results on land use in 2010 indicates that it was adjusted to 2029 spatial plan soil catchment conditions. It further shows that the changes from 19 to 37% in accordance with the plan of the West Java government that has green open space of at least 30%.

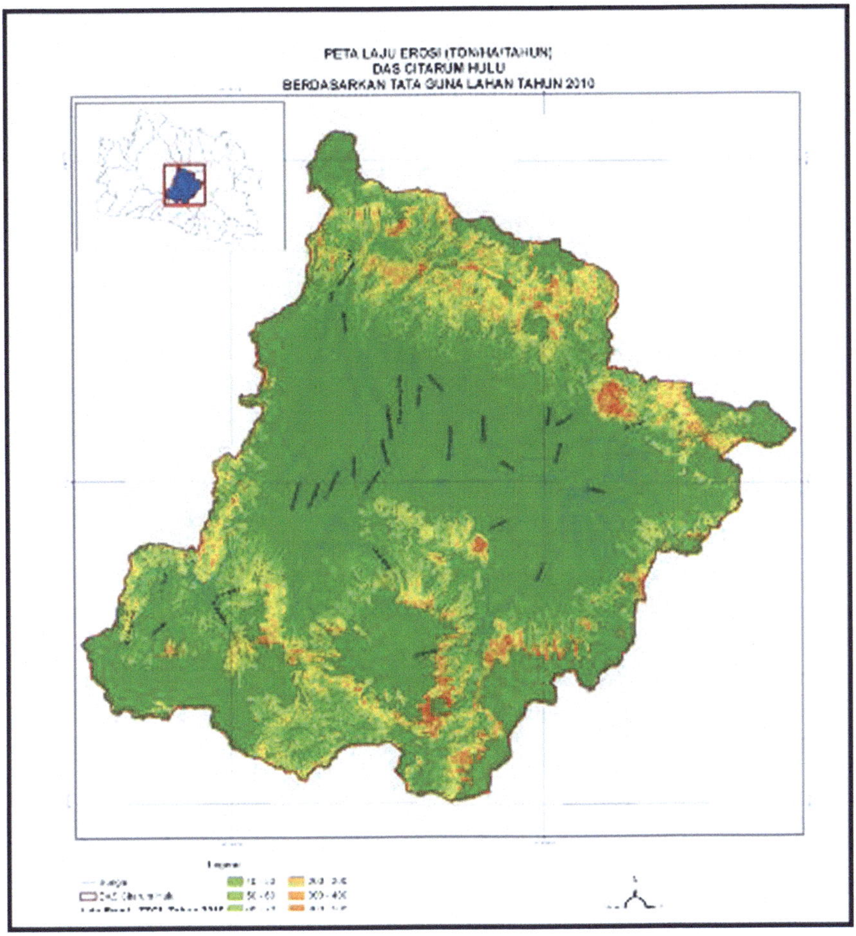

Fig. 10 Rate of erosion-based land classification map of land in 2010

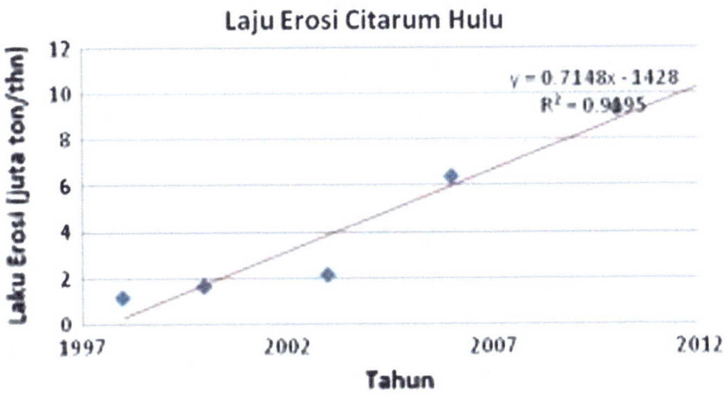

Fig. 11 Land-use map of 2010 adjusted with spatial

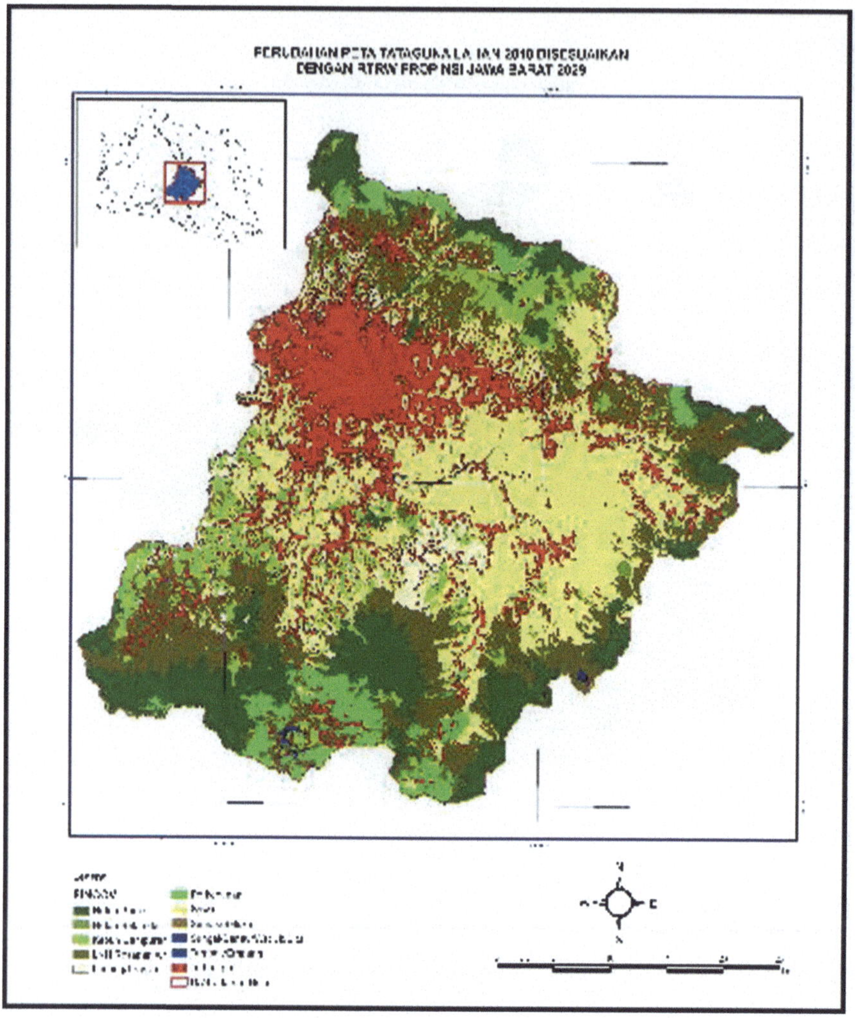

Fig. 12 Erosion rate conditions at Citarum Upstream

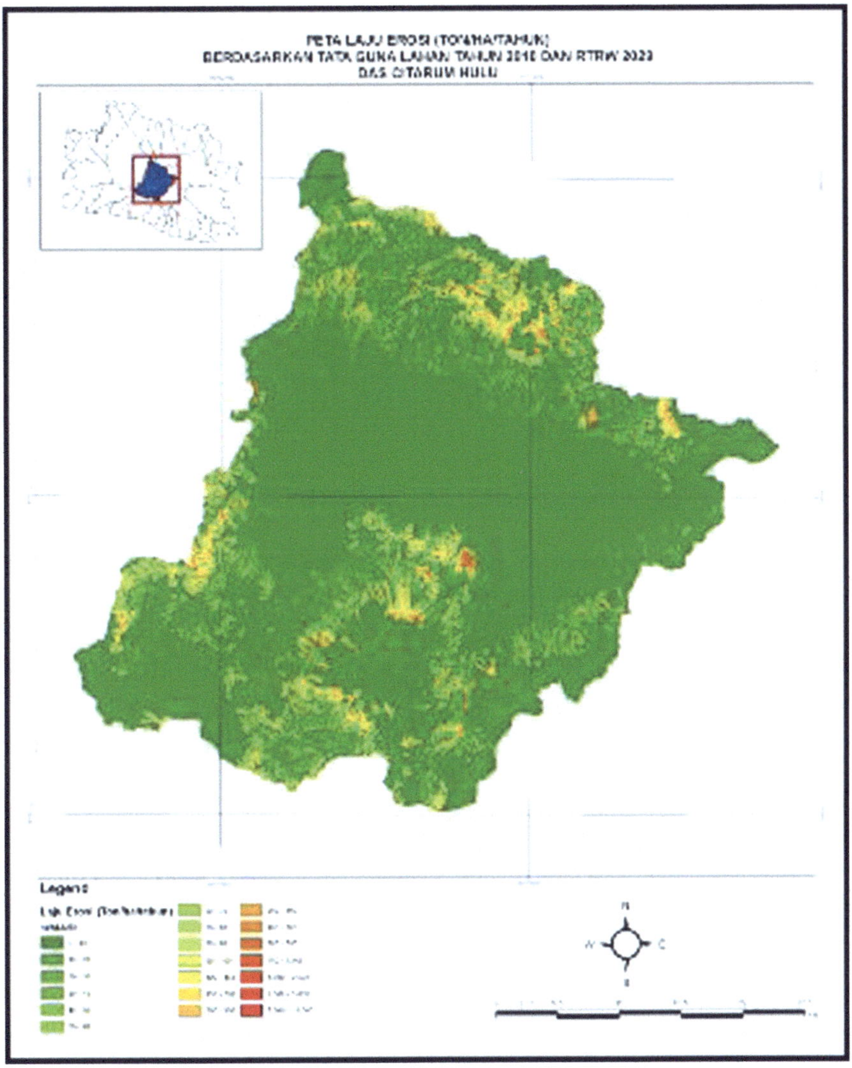

Fig. 13 Rate of erosion-based land classification map of land in 2010 adapted with the spatial

References

1. Wood, S.R, & Dent, F.J, 1983, *LECS:A Land Evaluation Computer System Methodology*, Soil Research Institute, Bogor, Indonesia.
2. Poerbandono, P, Basyar, A, Harto, A.B, 2006, Spatial modelling of sediment transport over the upper citarum catchment. *Journal of Engineering and Technological Sciences,* vol. 38, no.1, pp.11–28.

The manufacturer's authorised representative in the EU is Springer Nature Customer Service Centre GmbH, Europaplatz 3, 69115 Heidelberg, Germany. If you have any concerns regarding our products, please contact ProductSafety@springernature.com

Printed and bound by CPI Group (UK) Ltd, Croydon, CR0 4YY
30/09/2026
02225448-0015